Biological Barriers to
Protein Delivery

Pharmaceutical Biotechnology

Series Editor: Ronald T. Borchardt
The University of Kansas
Lawrence, Kansas

Biological Barriers to Protein Delivery

Edited by

Kenneth L. Audus

University of Kansas
Lawrence, Kansas

and

Thomas J. Raub

The Upjohn Company
Kalamazoo, Michigan

Plenum Press • **New York and London**

Library of Congress Cataloging-in-Publication Data

Biological barriers to protein delivery / edited by Kenneth L. Audus
and Thomas J. Raub.
        p.   cm. -- (Pharmaceutical biotechnology ; v. 4)
    Includes bibliographical references and index.
    ISBN 0-306-44368-6
    1. Protein drugs--Physiological transport. 2. Drug delivery
systems.   I. Audus, Kenneth L.   II. Raub, Thomas J.   III. Series.
RS431.P75B56   1993
615'.3--dc20
                                                    92-38532
                                                        CIP

ISBN 0-306-44368-6

©1993 Plenum Press, New York
A Division of Plenum Publishing Corporation
233 Spring Street, New York, N.Y. 10013

Printed in the United States of America

Contributors

María Susana Balda • Department of Internal Medicine, School of Medicine, Yale University, New Haven, Connecticut 06520

Ajay K. Banga • Department of Pharmacal Sciences, School of Pharmacy, Auburn University, Auburn, Alabama 36849

Laurence T. Baxter • Department of Radiation Oncology, Harvard Medical School, Steele Laboratory, Massachusetts General Hospital, Boston, Massachusetts 02114

Richard C. Boucher • Division of Pulmonary Diseases, Department of Medicine, The University of North Carolina at Chapel Hill, Chapel Hill, North Carolina 27599

Richard D. Broadwell • Division of Neurological Surgery, Department of Surgery, and Department of Pathology, University of Maryland School of Medicine, Baltimore, Maryland 21201

Sandra A. Brockman • Department of Biological Sciences, Carnegie Mellon University, Pittsburgh, Pennsylvania 15213

Marcelino Cereijido • Department of Physiology and Biophysics, Center for Research and Advanced Studies, Mexico City, Mexico

Yie W. Chien • Controlled Drug-Delivery Research Center, College of Pharmacy, Rutgers—The State University of New Jersey, Piscataway, New Jersey 08855

Rubén Gerardo Contreras • Department of Physiology and Biophysics, Center for Research and Advanced Studies, Mexico City, Mexico

David C. Dahl • Department of Medicine, University of Minnesota, and Regional Kidney Disease Program, Hennepin County Medical Center, Minneapolis, Minnesota 55415

M. R. García-Villegas • Department of Physiology and Biophysics, Center for Research and Advanced Studies, Mexico City, Mexico

Lorenza González-Mariscal • Department of Physiology and Biophysics, Center for Research and Advanced Studies, Mexico City, Mexico

Steen H. Hansen • Structural Cell Biology Unit, Department of Anatomy, The Panum Institute, University of Copenhagen, DK-2200 Copenhagen N, Denmark

Nigel M. Hooper • Department of Biochemistry and Molecular Biology, University of Leeds, Leeds LS2 9JT, United Kingdom

Rakesh K. Jain • Department of Radiation Oncology, Harvard Medical School, Steele Laboratory, Massachusetts General Hospital, Boston, Massachusetts 02114

Conrad E. Johanson • Cerebrospinal Fluid Research Laboratory, Department of Clinical Neurosciences, Program in Neurosurgery, Brown University/Rhode Island Hospital, Providence, Rhode Island 02902

Larry G. Johnson • Division of Pulmonary Diseases, Department of Medicine, The University of North Carolina at Chapel Hill, Chapel Hill, North Carolina 27599

Michael W. Konrad • The deVlaminck Institute, Lafayette, California 94549

Jean-Pierre Kraehenbuhl • Swiss Institute for Cancer Research and Institute of Biochemistry, University of Lausanne, CH-1066 Epalinges, Switzerland

Asrar B. Malik • Department of Physiology and Cell Biology, The Albany Medical College of Union University, Albany, New York 12208

Dirk K. F. Meijer • Department of Pharmacology and Therapeutics, University Centre for Pharmacy, Groningen, The Netherlands

Hans P. Merkle • Department of Pharmacy, Swiss Federal Institute of Technology, CH-8092 Zurich, Switzerland

Shozo Muranishi • Department of Biopharmaceutics, Kyoto Pharmaceutical University, Misasagi, Yamashina-ku, Kyoto, 607, Japan

Robert F. Murphy • Department of Biological Sciences, Carnegie Mellon University, Pittsburgh, Pennsylvania 15213

Marian R. Neutra • Gastrointestinal Cell Biology Laboratory, Children's Hospital and Department of Pediatrics, Harvard Medical School, Boston, Massachusetts 02115

Hiroaki Okada • DDS Research Laboratories, Research Pharmaceutical Division, Takeda Chemical Industries, Ltd., Osaka, 532, Japan

Sjur Olsnes • Institute for Cancer Research, Norwegian Radium Hospital, Montebello, 0310 Oslo 3, Norway

Ralph Rabkin • Department of Medicine, Nephrology Section, Stanford University School of Medicine, Stanford, and Palo Alto Department of Veterans Affairs Medical Center, Palo Alto, California 94304

Octavio Ruiz • Department of Physiology and Biophysics, Center for Research and Advanced Studies, Mexico City, Mexico

Kirsten Sandvig • Institute for Cancer Research, Norwegian Radium Hospital, Montebello, 0310 Oslo 3, Norway

Alma Siflinger-Birnboim • Department of Physiology and Cell Biology, The Albany Medical College of Union University, Albany, New York 12208

Bo van Deurs • Structural Cell Biology Unit, Department of Anatomy, The Panum Institute, University of Copenhagen, DK-2200 Copenhagen N, Denmark

Gregor J. M. Wolany • Department of Pharmacy, Swiss Federal Institute of Technology, CH-8092 Zurich, Switzerland

Akira Yamamoto • Department of Biopharmaceutics, Kyoto Pharmaceutical University, Misasagi, Yamashina-ku, Kyoto, 607, Japan

Kornelia Ziegler • Department of Pharmacology and Toxicology, Justus-Liebig University, Giessen, Germany

Preface to the Series

A major challenge confronting pharmaceutical scientists in the future will be to design successful dosage forms for the next generation of drugs. Many of these drugs will be complex polymers of amino acids (e.g., peptides, proteins), nucleosides (e.g., antisense molecules), carbohydrates (e.g., polysaccharides), or complex lipids.

Through rational drug design, synthetic medicinal chemists are preparing very potent and very specific peptides and antisense drug candidates. These molecules are being developed with molecular characteristics that permit optimal interaction with the specific macromolecules (e.g., receptors, enzymes, RNA, DNA) that mediate their therapeutic effects. Rational drug design does not necessarily mean rational drug delivery, however, which strives to incorporate into a molecule the molecular properties necessary for optimal transfer between the point of administration and the pharmacological target site in the body.

Like rational drug design, molecular biology is having a significant impact on the pharmaceutical industry. For the first time, it is possible to produce large quantities of highly pure proteins, polysaccharides, and lipids for possible pharmaceutical applications. Like peptides and antisense molecules, the design of successful dosage forms for these complex biotechnology products represents a major challenge to pharmaceutical scientists.

Development of an acceptable drug dosage form is a complex process requiring strong interactions between scientists from many different divisions in a pharmaceutical company, including discovery, development, and manufacturing. The series editor, the editors of the individual volumes, and the publisher hope that this new series will be particularly helpful to scientists in the development areas of a pharmaceutical company (e.g., drug metabolism, toxicology, pharmacokinetics and pharmacodynamics, drug delivery, preformulation, formulation, and physical and analytical chemistry). In addition, we hope this series will help to build bridges between the development scientists and scientists in discovery (e.g., medicinal chemistry, pharmacology, immunology, cell biology, molecular biology) and in manufacturing (e.g., process chemistry, engineering). The design of successful dosage forms for the next generation of drugs will require not only a high level of expertise by individual scientists, but also a high degree of interaction between scientists in these different divisions of a pharmaceutical company.

Finally, everyone involved with this series hopes that these volumes will also be useful to the educators who are training the next generation of pharmaceutical scientists. In addition to having a high level of expertise in their respective disciplines, these young scientists will need to have the scientific skills necessary to communicate with their peers in other scientific disciplines.

RONALD T. BORCHARDT
Series Editor

Preface

The application of native proteins or polypeptides as therapeutically useful drugs, e.g., insulin, factor VIII, has been in practice for several decades; however, with the recent advent of recombinant DNA technology, the number of protein and peptide drugs and drug candidates has increased exponentially. Currently, eight recombinant proteins are commercially available: human growth hormone, insulin, interferon-α, tissue plasminogen activator, erythropoietin, granulocyte colony stimulating factor, factor VIII, and hepatitis B vaccine. In 1988, it was estimated that over 250 companies worldwide were developing nearly 100 recombinant proteins[1] as human and veterinary products. Today, there are 132 biotechnology-based drugs and vaccines in development[2] with 21 of these awaiting FDA approval[2]; this number is much larger with many near FDA approval. In addition, nonrecombinant proteins, such as antibodies, and peptides are being used and developed for therapeutic purposes. These peptide and protein products of biotechnology present a unique spectrum of problems for the pharmaceutical scientist as well as the pharmacologist. Besides the fact that these molecules are often involved in many biological responses and activities which are not fully understood, the size and complex physiochemical nature of these molecules result in significant delivery limitations enforced by physiological phenomena. These biological barriers are a composite of cellular linings, e.g., epithelia and endothelia, metabolism, immunology, clearance, and the physical laws that govern solute diffusion. Consequently, the traditional approaches to drug delivery are not likely to be successful. Rational approaches to development of delivery strategies for proteins evolve from an understanding of the nature of these biological barriers at a fundamental level.

The scope of this volume is a current and critical review of information regarding peptide and protein transport and metabolism as it relates to delivery of endogenous (physiological ligands) and recombinant proteins to mammalian organs, tissues, and cells. Although not always possible due to the paucity of information available, each chapter emphasizes mechanisms of transport including quantitative evidence and structure/function (cause/effect) relationships. This volume is intended

[1]Copsey, D. N., and Delnatte, S. Y. J., Genetically Engineered Human Therapeutic Drugs, Stockton, New York, 1988.
[2]Pharm. Manufact. Assn., Products in the Pipeline, *Bio/Technology* **9**:947–949, 1991.

to be a comprehensive, state-of-the-art treatise for use by academic and industrial scientists who either are just beginning or are entrenched in the field of protein drug delivery. The objective was neither to promote particular model systems nor to attempt to develop *in vivo–in vitro* correlations. Rather, appropriate examples of both *in vivo* and *in vitro* systems are used to convey our present understanding of the various barriers from a structural and functional viewpoint. Our hope is that this volume will further entice collaborative interactions between pharmaceutical scientists and cell biologists, biochemists, physiologists, and immunologists.

The book begins with the role of tight junctions in limiting and controlling the diffusion of peptides/proteins across cellular barriers. Albeit an insignificant pathway for most proteins due to their size, the tight junction is an important physical barrier which may potentially be manipulated pharmacologically. Several chapters in the section on cellular characteristics focus on the vesicular trafficking of proteins into the cell via endocytosis and the various environmental attributes that the protein encounters in this pathway. Besides the implications for targeting intracellular sites, the process of membrane trafficking is paramount to later chapters detailing the transcellular movement of proteins in epithelium and endothelium. The latter route is generally regarded as useful only for delivery of potent agents due to the limited capacity of this process. Within this section is included general metabolic, mostly proteolytic, activities associated with cell surfaces and intracellular compartments. All of these chapters not only point out attributes of the barrier that require avoidance but also identify those attributes that can be exploited for the purpose of targeted delivery.

Subsequent chapters key on specific cell types that comprise the major cellular linings of skin, oral cavity, intestine, rectum, vagina, nasal passages, trachea, lung, and vasculature. Although other epithelial cell barriers may be the subject of delivery strategies, e.g., ophthalmic, we generally decided to restrict the scope to barriers where the blood compartment is the primary target. The vasculature-related chapters are concerned with peripheral endothelium and blood–brain fluid barrier which encompasses both endothelial and epithelial cell barriers. Each of these chapters emphasizes different aspects of endothelial cell physiology with regard to protein transport by highlighting different experimental approaches. Chapter 10 also describes the cellular mechanisms that control tight junction function.

The fact that most protein therapeutics are administered parenterally by intravenous or subcutaneous injection or infusion raises the issue of the fate of the molecule once it has reached the vascular compartment and what governs its movement into and out of the interstitial compartment. A combination of cellular (transport phenomena) and metabolic barriers contribute to the clearance of proteins from the blood by liver and kidney, each of which are detailed in separate chapters. Related to clearance is the multifaceted barrier represented by the immune system and this too is discussed in this book. The movement of proteins within tissues has been separated into two chapters. One chapter reviews those variables controlling diffusion of protein within most tissues, including those with lymphatic involvement, using tumors as a model. The other chapter focuses on the movement of proteins within the unique brain parenchymal tissues and should prove timely in light of the interest in treatment of neurological disorders with neurotropic proteins.

Last, we thank of all of the authors for their valuable and timely contributions. We hope the information will provide the reader with an essential appreciation for the obstacles, and the potentially useful attributes, that characterize these various barriers and that face today's protein drug delivery scientist.

KENNETH L. AUDUS
THOMAS J. RAUB

Contents

Chapter 3

Endosomal and Lysosomal Hydrolases

Sandra A. Brockman and Robert F. Murphy

Chapter 4

Protein Uptake and Cytoplasmic Access in Animal Cells
Bo van Deurs, Steen H. Hansen, Sjur Olsnes, and Kirsten Sandvig

Part II. **Epithelial Barriers**

Chapter 5

Transepithelial Transport of Proteins by Intestinal Epithelial Cells
Marian R. Neutra and Jean-Pierre Kraehenbuhl

Chapter 6

Intraoral Peptide Absorption
Hans P. Merkle and Gregor J. M. Wolany

Chapter 7

Macromolecular Transport across Nasal and Respiratory Epithelia
Larry G. Johnson and Richard C. Boucher

Chapter 8

Dermal Absorption of Peptides and Proteins
Ajay K. Banga and Yie W. Chien

Chapter 9

Rectal and Vaginal Absorption of Peptides and Proteins
Shozo Muranishi, Akira Yamamoto, and Hiroaki Okada

Part III. Vascular Barriers

Chapter 10

Vascular Endothelial Barrier and Its Regulation
Asrar B. Malik and Alma Siflinger-Birnboim

Chapter 11

Transcytosis of Macromolecules through the Blood–Brain Fluid Barriers *in Vivo*
Richard D. Broadwell

Part IV. Elimination Barriers

Chapter 12

Renal Uptake and Disposal of Proteins and Peptides
Ralph Rabkin and David C. Dahl

Chapter 13

Mechanisms for the Hepatic Clearance of Oligopeptides and Proteins: Implications for Rate of Elimination, Bioavailability, and Cell-Specific Drug Delivery to the Liver
Dirk K. F. Meijer and Kornelia Ziegler

Chapter 14

The Immune System as a Barrier to Delivery of Protein Therapeutics
Michael W. Konrad

Part V. **Tissue Barriers**

Chapter 15

Extravasation and Interstitial Transport in Tumors
Rakesh K. Jain and Laurence T. Baxter

Chapter 16

**Tissue Barriers: Diffusion, Bulk Flow, and Volume Transmission of
Proteins and Peptides within the Brain**
Conrad E. Johanson

I

Cellular Barriers

Chapter 1

The Paracellular Pathway

A Small Version of the Kidney Nephron

Marcelino Cereijido, Octavio Ruiz, Lorenza González-Mariscal, Rubén Gerardo Contreras, María Susana Balda, and M. R. García-Villegas

1. INTRODUCTION

The paracellular pathway is an aqueous, extracellular route across endothelia and epithelia that is followed by substances according to their size and charge. Although under certain circumstances it may be traversed by objects as large as leukocytes, it is generally used by water and small solutes. Under normal circumstances, proteins can only permeate through endothelia and a few types of epithelia, such as the glomerular and choroidal ones (Renkin and Gilmore, 1973; Kluge *et al.*, 1986). It is limited by a tight junction (TJ) placed at the outermost end of the intercellular space (Fig. 1). Since this structure was first—and for many years—studied in epithelia like the frog skin and the urinary bladder, where it almost completely obstructs the passage of all substances, it is usually called a "tight" or "occluding" junction. In epithelia that separate two compartments with different compositions (e.g., the frog skin, the colon mucosa, the kidney collecting tubule), the TJ is in fact very restrictive and the amount

Marcelino Cereijido, Octavio Ruiz, Lorenza González-Mariscal, Rubén Gerardo Contreras, and M. R. García-Villegas • Department of Physiology and Biophysics, Center for Research and Advanced Studies, Mexico City, Mexico.   *María Susana Balda* • Department of Internal Medicine, School of Medicine, Yale University, New Haven, Connecticut 06520.

Biological Barriers to Protein Delivery, edited by Kenneth L. Audus and Thomas J. Raub. Plenum Press, New York, 1993.

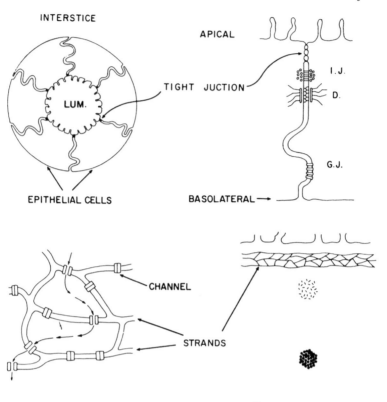

Figure 1. (Clockwise) Schematic representations of an epithelium. Six epithelial cells form tight junctions that severely restrict the escape of substances from the lumen (LUM.). Two adjacent epithelial cells form tight junctions, adherent junctions (I.J.), desmosomes (D), and gap junctions (G.J.). Lateralside of an epithelial cell: while the filaments of the tight junctions form a continuous belt that completely surrounds the cells at the apical/basolateral border, desmosomes and gap junctions only occur at a discrete spot. Strands are represented at a higher magnification to illustrate hypothetical ion channels, which can be in an *open* or a *closed* state. Current may only flow through the channels that are open at a given moment.

of substances flowing through the paracellular pathway is almost negligible. On the contrary, in epithelia separating two compartments with similar ionic composition, and that have to translocate a relatively large volume of fluid in a short time (e.g., the kidney proximal tubule, the gallbladder mucosa), the TJ is quite leaky and the paracellular pathway may account for up to 90% of the transepithelial flux. When the TJ is so leaky, the rate of permeation through the paracellular pathway may also be restricted by the narrowness and tortuosity of the intercellular space, as well as by the composition of the extracellular matrix (Cereijido, 1991).

TJs are also observed in endothelia, in the myocardium of the fetal heart, and even between two domains of the same cell, such as in Sertoli cells of the adult human testis. This review focuses on TJs established between neighboring epithelial cells.

2. THE STRUCTURE OF THE TIGHT JUNCTION

The tightness of the TJ, evaluated through the value of the transepithelial electrical resistance (TER) (see Reuss, 1991) or through the transepithelial flux of tracers (see Contreras et al., 1991), may be as low as 8 $\Omega \cdot cm^2$ (e.g., the proximal tubule of the kidney; Boulpaep and Seely, 1971) or as high as several thousand ohms per square centimeter (e.g., the colonic mucosa, the epithelium of the urinary bladder; Civan and Frazier, 1968). Thirty years ago, the low TER and the high permeability of epithelia like the mucosae of the small intestine and the gallbladder were attributed to cell damage due to experimental handling. However, the observation that in spite of their low TER and high permeability these epithelia do transport an appreciable amount of water, sugars, amino acids, and so forth, led to the concept that they possess a route of high permeability that is not artifactual. Eventually, several lines of evidence coincided in demonstrating that this route of high permeability proceeds through the paracellular space and not through the cytoplasm of the epithelial cell: (1) Epithelia like the intestinal mucosa have pores with radii of some 3–4 nm (Lindemann and Solomon, 1962), yet these large pores are not found in the plasma membrane, whose pores have a radius of only 0.4 nm (Paganelli and Solomon, 1957; Sidel and Solomon, 1957). (2) The electrical resistance offered by the transcellular route (the sum of the resistances of the apical and the basolateral plasma membranes arranged in series) is much higher than the electrical resistance measured through the whole epithelium (Stefani and Cereijido, 1983), indicating that the current circumvents the cells. (3) When the current flowing through the epithelium is scanned with a glass microelectrode moved close to the apical surface of the cells, it is observed to cross through intercellular spaces and not through cell bodies (Frömter and Diamond, 1972; Cereijido et al., 1980a).

In transmission electron microscopy (TEM), the TJ appears as a series of punctuate contacts between neighboring epithelial cells at the limit between the apical and the basolateral side (Fig. 1). TEM also shows that diffusion of extracellular markers such as hemoglobin, ruthenium red, and peroxidase is stopped at the level of the TJ (Miller, 1960; Kaye and Pappas, 1962; Kaye et al., 1962). Permeation of these markers is only possible when the TJ is opened by removal of Ca^{2+}, modification of the osmolarity of the bathing media, pharmacological agents, etc.

In the P face of freeze-fracture replicas, the TJ appears as a flat meshwork of anastomosing strands that are grouped in a narrow belt that runs at the limit between the apical and the basolateral side, close to the free surface of the epithelium, and that surrounds the whole cell. In the E face of freeze-fracture replicas, the TJ appears as a complementary meshwork of anastomosing grooves. Transitions between P and E faces show that strands and grooves are in register, justifying the assumption that strands occupy the grooves in the fresh tissue. The strands observed in freeze-fracture replicas correspond to the punctuate contacts observed by TEM. Their chemical nature is not known. However, the observation that inhibitors of protein synthesis block junction formation (Cereijido *et al.*, 1978, 1981) and that in some epithelia junctional development depends on the presence of proteases (Polak-Charcon, 1991), suggest that strands are constituted by polypeptides. Although there are three peptides known to be closely and specifically associated with the TJ, they are located too far from its lips to constitute the strands themselves (Stevenson *et al.*, 1989b; Citi *et al.*, 1988; Chapman and Eddy, 1989). Kachar and Reese (1982) and Pinto da Silva and Kachar (1982) have proposed instead that each strand is a long cylindrical micelle in which the polar heads of lipid molecules are oriented toward the axis.

The number of strands in the junctional belt of different tissues varies from one to ten or even more, and the TER value is observed to increase accordingly (Claude and Goodenough, 1973). Yet this increase is not linear, as would be expected from the addition of resistors in series, but exponential. This led Claude (1978) to suggest that strands may be traversed by aqueous channels that fluctuate between an open and a closed state. Figure 2A illustrates four segments of TJ with one channel each; obviously, the TJ is conductive only in the fourth situation, i.e., when both channels are simultaneously open. Therefore, a junctional belt with two strands would not have twice the electrical resistance of a belt with one strand, but four times as much. However, this explanation also holds for TJs with *a single* channel in each strand. A TJ several dozens of microns long (Fig. 2B) is expected to have instead a large population of channels, and current may cross successive strands by any of the channels that may be open at a given moment. For this reason, we have suggested (Cereijido *et al.*, 1989a): (1) that strands may be constituted by segments that are electrically compartmentalized, so that the situation described in Fig. 2B may be avoided; and (2) that this compartmentation may be afforded by anastomosing strands that cross from a given row to the next (Fig. 2C). This role of resistive segments that are not primarily interposed in the route of the flowing current, would seem at first paradoxical; yet Figs. 3 and 4 illustrate that, according to this model, the resistance of the paracellular pathway is highly sensitive to the electrical compartmentation of the strands, which is afforded by anastomosing segments. In this way, while the number of strands varies from one to ten, the TER value in different epithelia may range over several orders of magnitude, as is the case with transporting epithelia.

The permeability of the paracellular pathway also varies with the amount of intercellular cleft (Fig. 5). Other aspects being equal, an epithelium whose cells are smaller and with borders more interdigitated, would have a longer intercellular cleft per square centimeter and, accordingly, a higher paracellular permeability. A dramatic example of these differences is given by the mammalian ileum and the toad urinary bladder, two epithelia that exhibit roughly the same average number of

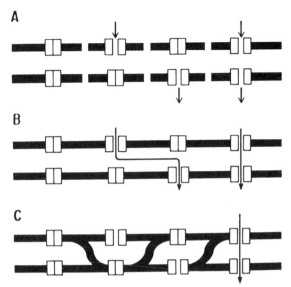

Figure 2. (A) Segment of a TJ with two strands, at four different moments. It is only conductive when the two channels coincide in the open state. (B) Since each strand is supposed to contain a large number of channels, current (arrows) may flow through any channel that is open at a given time. Assuming that the channel has an open probablity of 0.5, the conductance of a TJ with two strands is one-half that of a TJ with one strand. (C) Because of the compartmentation provided by frequent anastomoses between strands, current flowing through a given segment of strand can only utilize a restricted number of channels to cross the next strand. If the open probability of the channel is 0.5, a TJ with two strands will have one-quarter the conductance of a TJ with only one strand.

strands in their TJs, but whose TERs are around 200 and 10,000 Ω/cm² respectively (see Madara, 1991).

A given epithelium is usually constituted by more than one cell type, and the structure of the TJs that they establish may show considerable variations in the number and arrangement of their strands. Marcial *et al.* (1984) developed a morphological/electrical analysis that predicts the TER value of a given epithelium by adding the partial contribution of each segment according to its length and number of strands. This analysis allowed Madara and Dharmsathaphorn (1985) to correctly predict increments in TER of monolayers of T_{84} cells (human colonic) in which the pattern of the TJs varied as a function of time after plating.

The number and arrangement of the strands, as well as the length of the intercellular cleft, may not be the only factors affecting the permeability of the paracellular pathway. Thus, two subclones of the same cell type [Madin–Darby canine kidney (MDCK)], plated at equal density, have TJs that are almost identical, yet the TER of one is 300% higher than the other (González-Mariscal *et al.*, 1989;

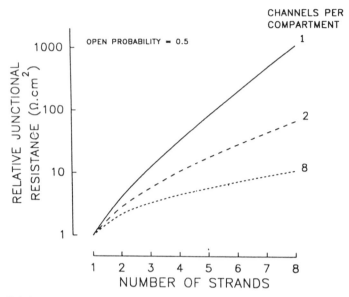

Figure 3. Relative junctional resistance offered by a TJ, as a function of the number of strands. Curves have one, two, and eight channels per compartment, respectively, with an open probability of 0.5.

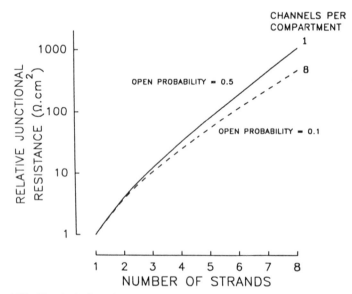

Figure 4. A TJ with a single channel per compartment (——)which spends 50% of the time in the open state ($P_{open} = 0.5$), offers approximately the same resistance as a TJ with eight channels (-----) which spend only 10% of the time in the open state ($P_{open} = 0.1$).

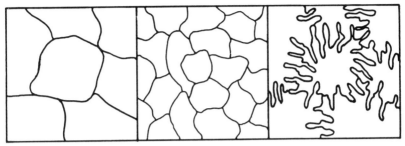

Figure 5.   An epithelium may have a much higher paracellular permeability than another epithelium, with identical TJs, if its cells are smaller and more interdigitated, by virtue of having a larger amount of paracellular pathway per square centimeter.

Stevenson *et al.*, 1989a). Furthermore, the TER value in a given monolayer of MDCK cells may be reversibly increased by 305%, without a noticeable modification of the freeze-fracture pattern, by changing the incubation temperature from 37°C to 4°C and back to 37°C (González-Mariscal *et al.*, 1984). It is interpreted that these variations in TER, which cannot be accounted for by alterations of the anatomical structure of the TJ, may be due to differences in the behavior of the flickering channels depicted in Fig. 2.

3. JUNCTIONS AND CELL-ATTACHING MOLECULES THAT COMPLEMENT THE ROLE OF THE TIGHT JUNCTION

The observation that the diffusion of extracellular markers such as hemoglobin, peroxidase, and ruthenium red is stopped at the level of the TJ, together with the fact that this structure is the only continuous formation that completely encircles the cell, indicate that it is indeed a diffusion barrier. Whether it is also a cell–cell attachment element is not known. Cell–cell attachment seems to be due instead to structures such as adherens junctions (ZA) and desmosomes (D). There seems to be a high degree of coordination between these structures and the TJ, as well as with gap (communicating) junctions. Thus, raising the Ca^{2+} concentration in the incubating medium from the micromolar to the millimolar range provokes drastic changes in the distribution of vinculin and actin and results in the sequential formation of ZA, D, and TJs (O'Keefe *et al.*, 1987; González-Mariscal *et al.*, 1985; Pasdar and Nelson, 1988a,b).

Cell–cell adhesion also depends on molecules of uvomorulin (or L-CAM), localized mainly on the extracellular side of the ZA (Boller *et al.*, 1985), though it might also be found in other positions along the lateral surface of the cells (Wang *et al.*, 1990). The use of antibodies against uvomorulin blocks the assembly of all elements of the junctional complex, emphasizing the high degree of interdependence.

In contrast, the TER of epithelial monolayers is not appreciably affected by treatment with antibodies to uvomorulin, suggesting that while this molecule is essential for *de novo* assembly of TJs, it might not be required to maintain already established junctions (Gumbiner *et al.*, 1988).

4. THE PHYSIOLOGY OF THE PARACELLULAR PATHWAY

The paracellular pathway predominates over the transcellular route in epithelia that separate two compartments with analogous chemical composition, and that translocate large amounts of fluid. Because of the similarity of the two media separated by these epithelia, gradients are very small, and large fluxes are due to the relatively high permeability of the TJ. Translocation through the paracellular pathway is primarily passive, i.e., fluxes are driven by electrochemical potential gradients (the sum of potentials arising from differences in concentration, electrical potential, and hydrostatic pressure between the two sides of the epithelium). The degree of discrimination between chemical species flowing through this route is very low. Thus, while the $Na^+ : K^+$ discrimination of a carrier or a pump may be higher than 100:1, in the paracellular pathway it is only 2:1. Under some circumstances, the flux of a substance may be propelled by the force arising from the electrochemical potential gradient of a second substance. This coupling between fluxes and forces gives rise to phenomena such as electroosmosis, streaming potentials, solvent drag, codiffusion, and so forth.

Of course, this does not mean that when a neutrophile crosses the paracellular route its movement is independent of metabolic energy, nor that fluxes might not be *secondarily* driven by active processes. Thus, NA^+,K^+ pumps located on the lateral sides of the cells, may pour large quantities of Na^+ into the interspace; since this compartment has a small volume, its concentration of Na^+ plus the accompanying co-ion (usually Cl^-) becomes high, and this condition has two consequences: (1) the high osmolarity draws water from the outer bathing solution across the TJ; and (2) it develops a hydrostatic pressure that drives the content of the interspace, in bulk, toward the serosal side (Curran and McIntosh, 1962; Diamond, 1962). Reuss (1991) has made a detailed description of these processes as well as the methods used to study them.

The structure of the TJ and the permeability of the paracellular space are known to change drastically to accommodate physiological needs. For example, the TJ between chloride cells of aquatic organisms is relatively simple and shallow when fish are adapted to sea water, but changes to a wider and elaborated pattern when they are transferred to estuaries or rivers and the osmotic gradient between their plasma and the medium becomes more steep (for a recent review see Karnaky, 1991).

Contact of glucose solutions with the apical region of intestinal cells produces focal separations between junctional strands and increases paracellular permeability (Madara and Pappenheimer, 1987). Interestingly, these effects of glucose are inhibited by agents such as phlorizin, a specific blocker of NA^+/glucose cotransporters which are *not* located along the paracellular route, but in the apical membrane. This indicates that the sugar penetrating into the cytoplasm of the cells, acts indirectly on the TJ. Even under the blockade of Na^+/glucose cotransporters elicited by phlorizin, the effects on the TJ and paracellular permeability may be elicited by the activation of other cotransporters, such as the one translocating Na^+/alanine. According to Madara (1991), Na^+ and glucose penetrate across the apical membrane from the lumen into the cytoplasm during the absorption that follows a meal; Na^+,K^+-ATPases located at the basolateral membrane extrude this Na^+ toward the intercellular space, thus developing an osmotic gradient between the solution in the intestinal lumen and the one in the intercellular space of the epithelium, i.e., across the TJ. In parallel with this process, a cascade of intracellular events provokes a modification of the cytoskeleton that would in turn decrease the restriction offered by the TJ resulting in a high solvent drag of glucose along the intercellular space. Madara and Pappenheimer (1987) and Pappenheimer and Reiss (1987) suggest that above 250 mM glucose in the lumen, this flux of glucose through the paracellular route is comparatively larger than the one crossing through the cytoplasm via cotransporters. On the basis that the concentration of glucose in the lumen after a meal is about twice that figure, these authors suggested that solvent drag through the paracellular pathway is the principal route for intestinal translocation of glucose. However, Diamond (1991) pointed out that there must be an error in the reasoning by which these authors reached their conclusion, because (1) the poor discrimination of the paracellular route would allow toxic substances of small molecular size to pass and intoxicate the organism; (2) *in vitro* studies of the small intestine indicate that most of the glucose translocation occurs through highly specific carriers; (3) glucose absorption is virtually absent in human infants with congenital defective glucose carriers.

The permeability of the paracellular pathway is affected by the status of the cytoskeleton. Cytochalasin B, a drug that affects actin, blocks junction formation in monolayers of MDCK cells (González-Mariscal *et al.*, 1985) and opens already sealed TJs (Meza *et al.*, 1980). Electrical studies by Meza *et al.* (1982) demonstrate that these effects are not due to cell damage, but to an enhancement of paracellular permeability. Cytochalasin B also affects the *Necturus* gallbladder epithelium in a complex way: below 1.5 μM it increases the TER, but at higher concentrations it has an opposite effect (Bentzel *et al.*, 1980). Cytochalasin D increases the paracellular flux of small inert hydrophilic solutes in the intestine (Madara *et al.*, 1986). It also decreases the TER in monolayers of T_{84} cells. Phalloidin, another actin-affecting drug, produces a rearrangement of hepatocyte TJs and increases permeation to La^{3+} (Elias *et al.*, 1980). In keeping with these effects, immunofluorescence studies of the distribution of actin microfilaments show that they form a continuous ring that circles the membrane on the cytoplasmic side, in close contact with the adherent junctions

(Hirano *et al.*, 1987). Upon addition of cytochalasin D, the actinomyosin ring contracts and this effect induces the opening of the paracellular pathway (Madara *et al.*, 1986).

On the contrary, microtubules do not seem to participate in the modulation of the paracellular pathway. Thus, immunofluorescence studies of the distribution of tubulin indicate that microtubules are located toward the nuclear region, far from the TJ. Colchicine, a poison that completely distorts microtubules, fails to produce any appreciable effect on the permeability of the paracellular pathway (Meza *et al.*, 1980).

The permeability of the paracellular pathway is also sensitive to drugs that affect the intracellular concentration of cAMP (see Bentzel *et al.*, 1980, 1991; Duffey *et al.*, 1981; Balda, 1991; Balda *et al.*, 1991).

Chapter 10 provides a detailed discussion on the sieving properties of the TJ, i.e., according to size, shape, and charge selectivity.

5. INTRACELLULAR STRUCTURES AND CHEMICAL SIGNALS THAT CONTROL THE ASSEMBLY AND SEALING OF THE TIGHT JUNCTION, AND THAT REGULATE PERMEATION THROUGH THE PARACELLULAR ROUTE

There are several indications that TJs and the paracellular pathway are under strict, albeit delicate cellular control: (1) Their structure and tightness in different epithelia exhibit a wide range of variation so as to adapt the epithelium to specific roles; (2) these variations also may be observed in the same epithelium, but in two different physiological situations; (3) as mentioned above, TJs are known to respond to changes in Ca^{2+} concentration, cAMP, osmolarity, pH, cytochalasin B and D, perturbation of CAM molecules located far from the TJ itself, and so forth; (4) TJs occupy a very precise position in these cells at the exact apical/basolateral limit, suggesting that it is the result of complex machinery for the synthesis, assembly, and vectorial delivery of its components to the cell membrane.

TJ formation may be easily studied in monolayers of MDCK cells plated at confluence on a permeable support. Junction formation does not require the synthesis of new mRNA (Cereijido *et al.*, 1978) unless the cells have been harvested from sparse cultures (i.e., cells without cell–cell contacts) (Griepp *et al.*, 1983). It does require the synthesis of proteins (Cereijido *et al.*, 1978). This synthesis may proceed even if Ca^{2+} is kept at a low concentration (below 50 μM). However, at these low Ca^{2+} concentrations, junctional components are not delivered to the plasma membrane, but are retained in an intracellular compartment located beyond the Golgi cisternae (González-Mariscal *et al.*, 1985). Addition of Ca^{2+} under such conditions triggers a process of exocytosis that transfers the components to the membrane surface (González-Mariscal *et al.*, 1990). Yet this ion does not seem to act, at least initially, on an intracellular target, but at the level of the plasma membrane, where it activates Ca^{2+}-dependent cell-attaching molecules. This requires the presence of transducing steps that transmit the contact signal to intracellular effectors that are in charge of delivering and assembling specific molecules to build the several types of junctions spanning the intercellular spaces (Contreras *et al.*, 1992a,b).

Balda *et al.* (1991) and Balda (1991) have shown that Ca^{2+} addition to MDCK cells preincubated in Ca^{2+}-free medium stimulates at least one G-protein in the plasma membrane, and triggers a cascade of signals involving phospholipase C, protein kinase C, and calmodulin (Fig. 6).

Synthesis, assembly, and sealing of the TJ also require the participation of other junction-associated proteins, such as ZO-1, a 210-kDa phosphoprotein (Stevenson *et al.*, 1986; Anderson *et al.*, 1989), cingulin, a 140-kDa peptide (Citi *et al.*, 1988), and a 192-kDa peptide (Chapman and Eddy, 1989). Although these peptides are exclusively associated with the TJs, their precise role is not understood.

6. RELATIONSHIP BETWEEN TIGHT JUNCTIONS AND APICAL/ BASOLATERAL POLARITY

Vectorial transport of substances across epithelia requires an asymmetric distribution of mechanisms in the apical or in the basolateral poles of the cells (Koefoed-Johnson and Ussing, 1953; see Cereijido *et al.*, 1989b). Anatomically, these two domains of the plasma membrane usually differ in area, density of intramembrane particles, presence of microvilli, gap junctions, desmosomes, and so forth (Galli *et al.*, 1976; (Cereijido *et al.*, 1980a; González-Mariscal *et al.*, 1985). They also differ in the composition of their lipid matrix (van Meer *et al.*, 1987), proteins (Caplan *et al.*, 1986; Simons and Fuller, 1985; Bartles *et al.*, 1987), glycolipid molecules (Lisanti *et al.*, 1988, 1990), and the types of ion channels expressed (Ponce *et al.*, 1991; Ponce and Cereijido, 1991). Once these differences between the two poles of epithelial cells are established by sorting processes (see Rodríguez-Boulan and Nelson, 1989), they do not remain static, but are subject to intense recycling and may be modified by a number of physiological signals and experimental manipulations (Cereijido *et al.*, 1989b).

For a long time, it was supposed that polarity was the result of a fence constituted by the TJ at the apical/basolateral limit (see Cereijido *et al.*, 1988). Yet even when the TJ may act as a fence to prevent the randomization of lipids (Dragsten *et al.*, 1981; van Meer and Simons, 1986; van Meer *et al.*, 1986) and of proteins that are not anchored to the cytoskeleton, and are therefore free to diffuse in the outer leaflet of the plasma membrane (Nelson and Hammerton, 1989; Nelson *et al.*, 1990), it cannot *sort* membrane components. There is now ample evidence that polarity may be achieved in the absence of TJs, and may even be modified in its presence (see Cereijido *et al.*, 1989b; Vega-Salas *et al.*, 1987).

As a consequence of the polarized distribution of transporting mechanisms, the

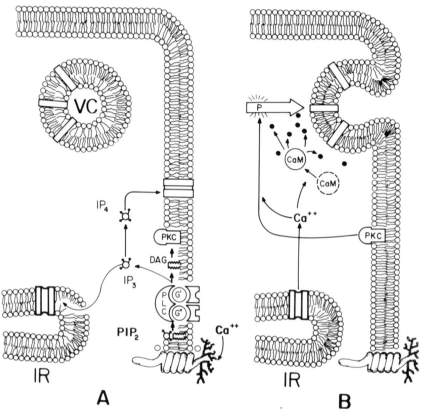

Figure 6. Schematic view of the effect of Ca^{2+} (A) Cells incubated for 20 hr in Ca^{2+}-free media have a cytoplasmic vesicular compartment (VC) where junctional components might be stored. Extracellular Ca^{2+} activates a highly glycosylated uvomorulin-like molecule and permits the attachment of neighboring cells. This contact would stimulate PLC, which is connected to membrane receptors via two G-proteins (G' and G'') and converts PIP_2 into IP_3 and DAG. IP_3 mobilizes Ca^{2+} from an internal reservoir (IR), IP_4 decreases Ca^{2+} permeability, and DAG activates PKC. (B) This cascade of reactions provokes the phosphorylation and incorporation of junctional components through exocytic fusion of the VC, and activation of calmodulin (CaM). (C) Activation of CaM causes actin filaments to organize into a continuous ring that circles the cell (represented by the two groups of solid circles). TJs and other type of junctions are established (gaps, adherens, etc., not shown), the paracellular space is sealed, and TER develops. (Taken with permission from Contreras *et al.*, 1992b, and *News in Physiological Sciences*.)

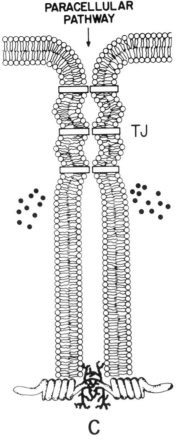

PARACELLULAR
PATHWAY

TJ

C

Figure 6. (*Continued*)

plasma membrane surrounding the paracellular pathway has a very special set of pumps, carriers, and channels. Thus, in most epithelia, the walls of the paracellular route are rich in NA^+,K^+-ATPases. In this way, the paracellular pathway truly acts as an assembly line that removes or adds specific ions and molecules to the intercellular space as the flux proceeds from the apical to the basolateral end. In this respect, the paracellular pathway may be considered a small version of the nephron.

Uvomorulin, a molecule that participates in the assembly and sealing of TJs, also plays a role as inducer of cell polarity, for its transfection to fibroblasts is sufficient to induce a polarized distribution of NA^+,K^+-ATPase, even when these cells have no TJs (McNeill *et al.*, 1990). Therefore, in principle, a variation of paracellular permeability, that is certainly due to modifications in the tightness of the TJ, may in fact reflect a change in the degree of attachment of other intercellular elements.

7. CONCLUDING REMARKS

Cells from transporting epithelia have two basic characteristics: (1) they form TJs that confer to the layer of cells their ability to act as diffusion barriers; and (2) their plasma membrane polarizes into an apical and a basolateral domain, an asymmetry that enables them to transport substances vectorially. These cells are able to regulate the degree of attachent between themselves and control the amount and nature of the substances allowed to permeate the epithelium, either by traversing their cytoplasm or by crossing through the paracellular pathway. When the two fluid compartments separated by the epithelium have similar compositions and the volume to be trans-located is large, the cells, by controlling the permeability of the TJ, allow an easy passage of substances through the paracellular pathway. Furthermore, the structure and permeability of the TJ and the paracellular pathway may be drastically changed in response to a variety of signals, so as to adapt the epithelium to physiological requirements. From the teleological point of view, all of these properties are quite understandable and have a paramount importance. Yet we do not know at this moment how the cell "senses" the difference in concentrations between the two sides, and how this information triggers the mechanisms in charge of synthesizing, assembling, and sealing a TJ, or how this structure is regulated.

One may foresee that in the near future, manipulation of permeation of peptides and even proteins through the paracellular route will become a crucial medical resource. This expectancy is based on: (1) the use of cultured monolayers of epithelial and endothelial cells (Cereijido *et al.*, 1978) to provide crucial information on the molecular structure of the TJ; (2) the use of these models to provide essential information on intracellular mechanisms and messengers that control the TJ, such as cytosolic Ca^{2+}, microfilaments, G-proteins, calmodulin (Meza *et al.*, 1980; Balda, 1991; Balda *et al.*, 1991); (3) the availability of suitable *in vitro* preparations and detailed knowledge on mechanisms that control TJs, stimulating intensive research on pharmacological agents that may affect them. Since epithelia and endothelia not only constitute continuous sheets lining major biological compartments, but also carry out and regulate the exchange of substances between them, these pharmacological agents would act as keys allowing the selective access of antibiotics and antibodies to the desired region of the organism.

In summary, paracellular spaces are potential routes for the transfer of almost anything, ranging in size from a proton to a leukocyte, from one biological compartment to another. We should develop suitable chemicals that act as keys to open such routes.

ACKNOWLEDGMENTS. We wish to acknowledge the pleasant and efficient help of Elizabeth Del Oso and Maricarmen De Lorenz. This work was supported by research grants from COSNET, the National Research Council of Mexico (CONACYT), and the Cystic Fibrosis Foundation of the United States.

REFERENCES

Anderson, J. M., Van Itallie, G. M., Peterson, M. D., Stevenson, B. R., Carew, E. A., and Mooseker, M. S., 1989, ZO-1 mRNA and protein expression during tight junction assembly in Caco-2 cells, *J. Cell Biol.* **109**:1047–1056.

Balda, M. S., 1991, Intracellular signals and the tight junction, in: *The Tight Junction* (M. Cereijido, ed.), CRC Press, Boca Raton, Fla., pp. 121–137.

Balda, M. S., González-Mariscal, L., Contreras, R. G., and Cereijido, M., 1991, The assembly and sealing of tight junctions: Participation of G-proteins, phospholipase C, protein kinase C and calmodulin, *J. Membr. Biol.* **122**:193–202.

Bartles, J. R., Feracci, H. M., Stieger, B., and Hubbard, A. L., 1987, Biogenesis of the rat hepatocyte plasma membrane in vivo: Comparison of the pathways taken by apical and basolateral proteins using subcellular fractionation, *J. Cell Biol.* **105**:1241–1251.

Bentzel, C. J., Hainau, B., Ho, S., Huis, W., Edelman, A., Anagnostopoulus, T., and Benedetti, E. L., 1980, Cytoplasmic regulation of tight-junction permeability: Effect of plant cytokinins, *Am. J. Physiol.* **239**:C75–C89.

Bentzel, C. J., Palant, C. E., and Fromm, M., 1991, Physiological and pathological factors affecting the tight junction, in: *The Tight Junction* (M. Cereijido, ed.), CRC Press, Boca Raton, Fla., pp. 151–173.

Boller, K., Vestweber, D., and Kemler, R., 1985, Cell-adhesion molecule uvomorulin is located in the intermediate junctions of adult intestinal epithelial cells, *J. Cell Biol.* **100**:327–332.

Boulpaep, E. L., and Seely, J. F., 1971, Electrophysiology of proximal and distal tubules in the autoperfused dog kidney, *Am. J. Physiol.* **221**:1084–1096.

Caplan, M. J., Anderson, H. C., Palade, G. E., and Jamieson, J. D., 1986, Intracellular sorting and polarized cell surface delivery of (NA^+, K^+) ATPase, an endogenous component of MDCK cell basolateral plasma membranes, *Cell* **46**:623–631.

Cereijido, M. (ed.), 1991, *Tight Junctions*, CRC Press, Boca Raton, Fla.

Cereijido, M., Rotunno, C. A., Robbins, E. S., and Sabatini, D. D., 1978, Polarized epithelial membranes produced in vitro, in: *Membrane Transport Processes* (J. F. Hoffman, ed.), Raven Press, New York, pp. 443–461.

Cereijido, M., Ehrenfeld, J., Meza, I., and Martínez-Palomo, A., 1980a, Structural and functional membrane polarity in cultured monolayers of MDCK cells, *J. Membr. Biol.* **52**:147–159.

Cereijido, M., Stefani, E., and Martínez-Palomo, A., 1980b, Occluding junctions in a cultured transporting epithelium: Structural and functional heterogeneity, *J. Membr. Biol.* **53**:19–32.

Cereijido, M., Meza, I., and Martínez-Palomo, A., 1981, Occluding junctions in cultured epithelial monolayers, *Am. J. Physiol.* **240**:C96–C102.

Cereijido, M., González-Mariscal, L, Avila, G., and Contreras, R. G., 1988, Tight junctions, *CRC Press Rev. Anat. Sci.* **1**:171–192.

Cereijido, M., González-Mariscal, L., and Contreras, R. G., 1989a, Tight junction: Barrier between higher organisms and environment. *News Physiol. Sci.* **4**:72–75.

Cereijido, M., Ponce, A., and González-Mariscal, L., 1989b, Tight junctions and apical/ basolateral polarity, *J. Membr. Biol.* **110**:1–9.

Chapman, L. M., and Eddy, E. M., 1989, A protein associated with the mouse and rat hepatocyte junctional complex, *Cell Tissue Res.* **257**:333–341.

Citi, S., Sabanay, H., Jakes, R., Geiger, B., and Kendrich-Jones, J, 1988, Cingulin, a new peripheral component of tight junctions, *Nature* **333**:272–276.

Civan, M. M., and Frazier, H. S., 1968, The site of the stimulatory action of vasopressin on the sodium transport in toad bladder, *J. Gen. Physiol.* **51**:589–605.

Claude, P., 1978, Morphological factors influencing transepithelial permeability: A model for the resistance of the zonula occludens, *J. Membr. Biol.* **39**:219–232.

Claude, P., and Goodenough, D. A., 1973, Fracture faces of zonulae occludentes from "tight" and "leaky" epithelia, *J. Cell Biol.* **58**:390–400.

Contreras, R. G., Ponce, A., and Bolívar, J. J., 1991, Calcium and tight junction, in: *The Tight Junction* (M. Cereijido, ed.), CRC Press, Boca Raton, Fla., pp. 139–149.

Contreras, R. G., Miller, J. H., Zamora, M., González-Mariscal, L., and Cereijido, M., 1992a, Interaction of calcium with the plasma membrane of epithelial (MDCK) cells during junction formation, *Am. J. Physiol.* **263**:C313–318.

Contreras, R. G., González-Mariscal, L., Balda, M. S., and Cereijido, M., 1992b, The role of calcium in the making of a transporting epithelium, *News Physiol. Sci.* **7**:105–108.

Curran, P. F., and McIntosh, J. R., 1962, A model system for biological water transport, *Nature* **193**:347–348.

Diamond, J. M., 1962, The mechanism of solute transport by the gall-bladder, *J. Physiol. (London)* **161**:474–502.

Diamond, J. M., 1991, Evolutionary design of intestinal nutrient absorption: Enough but not too much, *News Physiol Sci.* **6**:92–96.

Dragsten, P. R., Blumenthal, R., and Handler, J. S., 1981, Membrane asymmetry in epithelia: Is the tight junction a barrier to diffusion in plasma membrane? *Nature* **294**:718–722.

Duffey, M. E., Hainau, B., Ho, S., and Bentzel, C. J., 1981, Regulation of epithelial tight junction permeability by cyclic AMP, *Nature* **294**:451–453.

Elias, E., Hruban, Z., Wade, J. B., and Boyer, J. L., 1980, Phalloidin-induced cholestasis: A microfilament-mediated change in junctional complex permeability, *Proc. Natl. Acad. Sci. USA* **77**:2229–2230.

Frömter, E., and Diamond, J., 1972, Route of passive ion permeation in epithelia, *Nature* **235**: 9–13.

Galli, P., Brenna, A., De Camilli, P., and Meldolesi, J., 1976, Extracellular calcium and the organization of tight junctions in pancreatic acinar cells, *Exp. Cell Res.* **99**:178–183.

González-Mariscal, L., Chávez de Ramírez, B., and Cereijido, M., 1984, The effect of temperature on the occluding junctions of monolayers of epithelioid cells (MDCK), *J. Membr. Biol.* **79**:175–184.

González-Mariscal, L., Chávez de Ramírez, B., and Cereijido, M., 1985, Tight junction formation in cultured epithelial cells (MDCK), *J. Membr. Biol.* **86**:113–125.

González-Mariscal, L., Chávez de Ramírez, B., Lázaro, A., and Cereijido, M., 1989,

Establishment of tight junctions between cells from different animal species and different sealing capacities, *J. Membr. Biol.* **107**:43–56.

González-Mariscal, L., Contreras, R. G., Bolívar, J. J., Ponce, A., Chávez de Ramírez, B., and Cereijido, M., 1990, Role of calcium in tight junction formation between epithelial cells, *Am. J. Physiol.* **259**:C978–C986.

Griepp, E. B., Dolan, W. J., Robbins, E. S., and Sabatini, D. D., 1983, Participation of plasma membrane proteins in the formation of tight junctions by cultured epithelial cells, *J. Cell Biol.* **96**:693–702.

Gumbiner, B., Stevenson, B., and Grimaldi, A., 1988, The role of the cell adhesion molecule uvomorulin in the formation and maintenance of the epithelial junctional complex, *J. Cell Biol.* **107**:1575–1587.

Hirano, S., Nose, A., Hatta, K., Kawakami, A., and Takeichi, M., 1987, Calcium-dependent cell–cell adhesion molecules (cadherins): Subclass specificities and possible involvement of actin bundles, *J. Cell Biol.* **105**:2501–2510.

Kachar, B., and Reese, T., 1982, Evidence for the lipidic nature of tight junction strands, *Nature* **296**:464–466.

Karnaky, K., 1991, Teleost osmoregulation: Changes in the tight junction in response to the salinity of the environment, in: *The Tight Junction* (M. Cereijido, ed.), CRC Press, Boca Raton, Fla., pp. 175–185.

Kaye, G. I., and Pappas, G. D., 1962, Studies on the cornea. I. The fine structure of the rabbit cornea and the uptake and transport of colloidal particles by the cornea in vivo, *J. Cell Biol.* **12**:457–480.

Kaye, G. I., Pappas, G. D., Don, A., and Mallett, N., 1962, Studies on the cornea. II. The uptake and transport of colloidal particles by the living rabbit cornea in vitro, *J. Cell Biol.* **12**:481–501.

Kluge, H., Hartmann, W., Mertins, B., and Wieczorek, V., 1986, Correlation between protein data in normal lumbar CSF and morphological findings of choroid plexus epithelium: A biochemical corroboration of barrier transport via tight junction pores, *J. Neurol.* **233**:195–199.

Koefoed-Johnson, V., and Ussing, H. H., 1953, The contributions of diffusion and flow to the passage of D_2O through living membranes, *Acta Physiol. Scand.* **28**:60–76.

Lindemann, B., and Solomon, A. K., 1962, Permeability of luminal surface of intestinal mucosal cells, *J. Gen. Physiol.* **45**:801–810.

Lisanti, M., Sargiacomo, M., Graeve, L., Saltiel, A., and Rodriguez-Boulan, E., 1988, Polarized apical distribution of glycosyl phosphatidylinositol anchored proteins in a renal epithelial line, *Proc. Natl. Acad. Sci. USA* **85**:9557–9561.

Lisanti, M. P., Le Bivic, A., Saltiel, A., and Rodriguez-Boulan, E., 1990, Preferred apical distribution of glycosyl-phosphatidylinositol (GPI) anchored proteins: A highly conserved feature of the polarized epithelial cell phenotype, *J. Membr. Biol.* **113**:155–167.

McNeill, H., Ozawa, M., Kemler, R., and Nelson, W. J., 1990, Novel function of the cell adhesion molecule uvomorulin as an inducer of cell surface polarity, *Cell* **62**:309–316.

Madara, J., 1991, Relationships between the tight junction and the cytoskeleton, in: *The Tight Junction* (M. Cereijido, ed.), CRC Press, Boca Raton, Fla., pp. 105–119.

Madara, J. L., and Dharmsathaphorn, K., 1985, Occluding junction structure–function relationship in cultured epithelia monolayers, *J. Cell Biol.* **101**:2124–2133.

Madara, J. L., and Pappenheimer, J. R., 1987, Structural basis for physiological regulation of paracellular pathways in intestinal epithelia, *J. Membr. Biol.* **100**:149–164.

Madara, J. L., Berenberg, D., and Carlson, S., 1986, Effects of cytochalasin D on occluding junctions of intestinal absorptive cells: Further evidence that the cytoskeleton may influence paracellular permeability and junctional charge selectivity, *J. Cell Biol.* **102**: 2125–2136.

Marcial, M. A., Carlson, S. L., and Madara, J. L., 1984, Partitioning of paracellular conductance along the ileal crypt–villus axis: A hypothesis based on structural analysis with detailed consideration of tight junction structure function relationships, *J. Membr. Biol.* **80**:59–70.

Meza, I., Ibarra, G., Sabanero, M., Martínez-Palomo, A., and Cereijido, M., 1980, Occluding junctions and cytoskeletal components in a cultured transporting epithelium, *J. Cell Biol.* **87**:746–754.

Meza, I., Sabanero, M., Stefani, E., and Cereijido, M., 1982, Occluding junctions in MDCK cells: Modulation of transepithelial permeability by the cytoskeleton, *J. Cell Biochem.* **18**: 407–421.

Miller, F., 1960, Hemoglobin absorption by the cell of the proximal convoluted tubules in mouse kidney, *J. Biophys. Biochem. Cytol.* **8**:689–718.

Nelson, W. J., and Hammerton, R. W., 1989, A membrane–cytoskeletal complex containing NA$^+$,K$^+$-ATPase, ankyrin, and fodrin in Madin–Darby canine kidney (MDCK) cells: Implications for the biogenesis of epithelial cell polarity, *J. Cell Biol.* **108**:893–902.

Nelson, W. J., Shore, E. S., Wang, A. Z., and Hammerton, R. W., 1990, Identification of a membrane–cytoskeletal complex containing the cell adhesion molecule uvomorulin (E-cadherin), ankyrin, and fodrin in Madin–Darby canine kidney epithelial cells, *J. Cell Biol.* **110**:349–357.

O'Keefe, E. J., Briggaman, R. A., and Herman, B., 1987, Calcium-induced assembly of adherens junctions in keratinocytes, *J. Cell Biol.* **105**:908–917.

Paganelli, C. V., and Solomon, A. K., 1957, The rate of exchange of tritiated water across the human red cell membrane, *J. Gen. Physiol.* **41**:259–277.

Pappenheimer, J. R., and Reiss, K. Z., 1987, Contribution of solvent drag through intercellular junctions to absorption of nutrients by the small intestine of the rat, *J. Membr. Biol.* **100**: 123–136.

Pasdar, M., and Nelson, W. J., 1988a, Kinetics of desmosome assembly in Madin Darby canine kidney epithelial cells: temporal and spatial regulation of desmoplakin organization and stabilization upon cell–cell contact. I. Biochemical analysis, *J. Cell Biol.* **106**:677–685.

Pasdar, M., and Nelson, W. J., 1988b, Kinetics of desmosome assembly in Madin Darby canine kidney epithelial cells: Temporal and spatial regulation of desmoplakin organization and stabilization upon cell–cell contact. II. Morphological analysis, *J. Cell Biol.* **106**: 687–695.

Pinto da Silva, P., and Kachar, B., 1982, On tight junction structure, *Cell* **28**:441–450.

Polak-Charcon, S., 1991, Proteases and the tight junction, in: *Tight Junctions* (M. Cereijido, ed.), CRC Press, Boca Raton, Fla., pp. 257–277.

Ponce, A., and Cereijido, M., 1991, Polarized distribution of cation channels in epithelial cells, *Cell Physiol. Biochem.* **1**:13–23.

Ponce, A., Contreras, R. G., and Cereijido, M., 1991, Polarized distribution of chloride channels in epithelial cells (MDCK), *Cell Physiol. Biochem.* **1**:60–69.

Renkin, E. M., and Gilmore, J. P., 1973, Glomerular filtration, in: *Handbook of Physiology* (J. Orloff, R. W. Berliner, and S. R. Geiger, eds.), American Physiological Society, Washington, D.C.

Reuss, L., 1991, Tight junction permeability to ions and water, in: *Tight Junctions* (M. Cereijido, ed.), CRC Press, Boca Raton, Fla., pp. 49–66.

Rodríguez-Boulan, E., and Nelson, W. J., 1989, Morphogenesis of the polarized epithelial cell phenotype, *Science* **245**:718–725.

Sidel, V. W., and Solomon, A. K., 1957, Entrance of water into human red cells under an osmotic pressure gradient, *J. Gen. Physiol.* **41**:243–257.

Simons, K., and Fuller, S. D., 1985, Cell surface polarity in epithelia, *Annu. Rev. Cell Biol.* **1**: 243–288.

Stefani, E., and Cereijido, M., 1983, Electrical properties of MDCK cells, *Fed. Proc.* **73**: 177–184.

Stevenson, B. R., Siliciano, J. D., Mooseker, M. S., and Goodenough, D. A., 1986, Identification of ZO-1: a high molecular weight polypeptide associated with the tight junction (zonula occludens) in a variety of epithelia. *J. Cell Biol.* **103**:755–766.

Stevenson, B., Anderson, J. M., Braun, I. D., and Mooseker, M. S., 1989a, Phosphorylation of the tight-junction protein ZO-1 in two strains of Madin–Darby canine kidney cells which differ in transepithelial resistance, *Biochem. J.* **263**:597–599.

Stevenson, B. R., Heintzelman, M. B., Anderson, J. M., Citi, S., and Mooseker, M., 1989b, ZO-1 and cingulin: Tight junction proteins with distinct identities and localization, *Am. J. Physiol.* **257**:C621–C628.

van Meer, G., and Simons, K., 1986, The function of tight junctions in maintaining differences in lipid composition between the apical and the basolateral cell surface domains of MDCK cells, *EMBO J.* **5**:1455–1466.

van Meer, G., Gumbiner, B., and Simons, K., 1986, The tight junction does not allow lipid molecules to diffuse from one epithelial cell to the next, *Nature* **322**:639–641.

van Meer, G., Stelzer, E. H. K., Wijnaendts-van Resandt, R., and Simons, K., 1987, Sorting of sphingolipids in epithelial (Madin–Darby canine kidney) cells, *J. Cell Biol.* **105**:1623–1635.

Vega-Salas, D. E., Salas, P., Gundersen, D., and Rodriguez-Boulan, E., 1987, Formation of the apical pole of epithelial (MDCK) cells: Polarity of an apical protein is independent of tight junctions while segregation of a basolateral marker requires cell–cell interactions, *J. Cell Biol.* **105**:905–916.

Wang, A. Z., Ojakian, G. K., and Nelson, W. J., 1990, Steps in the morphogenesis of a polarized epithelium. I. Uncoupling the roles of cell–cell and cell–substratum contact in establishing plasma membrane polarity in multicellular epithelial (MDCK) cysts, *J. Cell Sci.* **95**:137–151.

Chapter 2

Ectopeptidases

Nigel M. Hooper

1. INTRODUCTION

The surfaces of a wide variety of mammalian cell types are rich in a group of enzymes that are responsible for the hydrolysis of a range of biologically active peptides. These ectopeptidases are integral proteins of the plasma membrane, asymmetrically oriented, with the catalytic site exposed at the extracytoplasmic surface (see Fig. 1). For the majority of those ectopeptidases thus far characterized, the protein is anchored in the membrane by a sequence of hydrophobic amino acid residues, probably in an α-helical conformation (Fig. 1a). Recently, though, it has become apparent that at least two peptidases are anchored by a complex glycosyl–phosphatidylinositol structure covalently linked to the C-terminal amino acid of the polypeptide chain (Fig. 1b) (see Cross, 1990; Turner and Hooper, 1990). Regardless of the mode of membrane attachment, the most important feature of an ectopeptidase is that its catalytic site faces the extracellular space where it can act on peptide substrates. Thus, peptidases that have a predominantly, if not exclusively, cytosolic location, such as endopeptidase-24.15 (Pierotti *et al.*, 1990), will not *in vivo* have access to circulating peptides and will not be discussed further.

The substrate specificity of ectopeptidases is limited to small peptides (di-, tri-, and oligopeptides) up to a maximum of approximately 30 residues. Thus, ectopeptidases will not act directly on large proteins or polypeptides. The classification of ectopeptidases is based on their site of action in a susceptible substrate (Table I) (Barrett and McDonald, 1980; McDonald and Barrett, 1986). Those peptidases that

Nigel M. Hooper • Department of Biochemistry and Molecular Biology, University of Leeds, Leeds LS2 9JT, United Kingdom.

Biological Barriers to Protein Delivery, edited by Kenneth L. Audus and Thomas J. Raub. Plenum Press, New York, 1993.

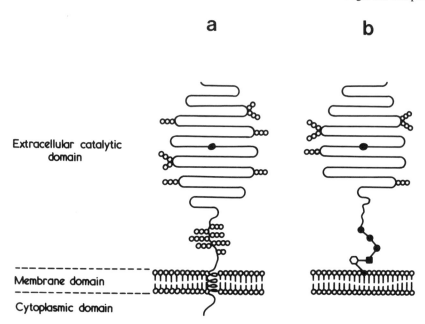

Figure 1. Membrane topology of ectopeptidases. The two modes of membrane anchorage of mammalian ectopeptidases are shown: (a) a transmembrane sequence of hydrophobic amino acids located near either the N- or C-terminus; (b) a C-terminal glycosyl–phosphatidylinositol anchor. The catalytic site (●) and the N-linked glycans (ooooo) are also shown.

are capable of cleaving the internal peptide bonds of a substrate are designated as endopeptidases. For larger peptides or those with blocked ends, e.g., an N-terminal pyroglutamate residue or an amidated C-terminal amino acid, it is the endopeptidases that will be required to initiate hydrolysis. The rest of the peptidases are classified as exopeptidases, removing one or more residues from the ends of peptides; e.g., aminopeptidases, carboxypeptidases, etc. (Table I). Because each of the peptidases has a wide substrate specificity, this relatively small (14) complement of enzymes is capable of metabolizing a vast number of biologically active peptides. The membrane topology and wide tissue distribution of ectopeptidases mean that both natural peptides and those delivered to mammalian organs or tissues are faced, at the cell surface, by an extremely effective barrier.

In this chapter, I briefly describe the properties of the mammalian ectopeptidases that have been characterized in some detail. I then examine their tissue distribution and how they may act in concert when a membrane is presented with a susceptible peptide substrate. Finally, I consider the physiological functions of ectopeptidases with details of a few selected systems.

Table I
Properties of the Ectopeptidases Described in This Chapter

Class	Enzyme	Subunit size (kDa)	Specificity	Active site	cDNA sequence available
Endopeptidases	Endopeptidase-24.11 EC 3.4.24.11	90	↓ -o-o-●-o-o- (hydrophobic)	Zn²⁺	Yes
	Endopeptidase-2	100	↓/↓ -o-o-●-o-o- (aromatic)	Zn²⁺	—
Aminopeptidases	Aminopeptidase N EC 3.4.11.2	160	↓ ●-o-o-o- (nonspecific)	Zn²⁺	Yes
	Aminopeptidase A EC 3.4.11.7	170	↓ ●-o-o-o- (Glu/Asp)	?Ca²⁺	Yes
	Aminopeptidase P EC 3.4.11.9	90	↓ o-●-o-o (Pro)	Zn²⁺	—
	Aminopeptidase W EC 3.4.11.16	130	↓ o-●-o-o (Trp)	Zn²⁺	—
Carboxypeptidases	Carboxypeptidase M EC 3.4.17.-	62	↓ -o-o-o-● (Arg/Lys)	Zn²⁺	Yes
	Carboxypeptidase P EC 3.4.17.-	135	↓ -o-o-●-o (Pro, Ala, Gly)	Zn²⁺	—
Dipeptidyl peptidase	Dipeptidyl peptidase IV EC 3.4.14.5	110	↓ o-●-o-o- (Pro, Ala)	Ser	Yes
Peptidyl dipeptidase	Angiotensin-converting enzyme EC 3.4.15.1	175 100	↓ -o-o-●-● (nonspecific)	Zn²⁺	Yes
Dipeptidases	Membrane dipeptidase EC 3.4.13.11	59	↓ ●-● (nonspecific)	Zn²⁺	Yes
	N-Acetylated-α-linked acidic dipeptidase	94	↓ ●-● (acidic)	Me²⁺	—
Omega peptidases	γ-Glutamyltranspeptidase EC 2.3.2.2	50, 30	↓ ●-o-o-o- (γ-Glu)	—	Yes
	Pyroglutamyl peptidase II EC 3.4.19.-	230	↓ Glp-His-ProNH₂	Me²⁺	—

2. PROPERTIES OF MAMMALIAN ECTOPEPTIDASES

2.1. Endopeptidases

2.1.1. ENDOPEPTIDASE-24.11

Endopeptidase-24.11 (enkephalinase; neutral metallo-endopeptidase) is probably one of the best characterized ectopeptidases. The cDNA sequences for the human (Malfroy *et al.*, 1988), rabbit (Devault *et al.*, 1987), and rat (Malfroy *et al.*, 1987) enzymes have been elucidated and show a high degree of similarity with each other. As with nearly all of the ectopeptidases, endopeptidase-24.11 is glycosylated, and the slight variations in size of the enzyme between tissues is due to differences in the extent of glycosylation (Relton *et al.*, 1983). The catalytic site of endopeptidase-24.11 contains one atom of zinc which is essential for activity (Kerr and Kenny, 1974). The specificity of endopeptidase-24.11 is directed toward peptide bonds involving the amino groups of hydrophobic residues (Table I), provided this residue is not C-terminal or penultimate at the N-terminus. However, there are a few irregu-

Table II
Peptides Hydrolyzed by Endopeptidase-24.11[a]

Peptide	k_{cat}/K_m (min^{-1}μM^{-1})	Peptide	k_{cat}/K_m (min^{-1}μM^{-1})
Substance P (deamidated)	203	Atrial natriuretic peptide	—
Neurokinin A	191	Brain natriuretic peptide	—
Substance P	159	Dynorphin 1–13	—
[Met^5]-enkephalin-Arg^6-Phe^7	117	β-Endorphin	—
Physalaemin	70	γ-Endorphin	—
Bradykinin	69	[Leu^5]-enkephalin-Arg^6	—
Chemotactic peptide	62	[Met^5]-enkephalin-Arg^6-Gly^7-Leu^8	—
Endothelin-1	57	FMRF amide	—
[Met^5]-enkephalin-Arg^6	54	Gastrin	—
[Leu^5]-enkephalin	44	Gastrin releasing peptide	—
[Met^5]-enkephalin	42	Insulin B chain	—
Cholecystokinin-8	29	Interleukin 1β	—
Dynorphin 1–9	24	β-Lipotropin 61–69	—
Endothelin-2	21	α-Neoendorphin	—
Neurotensin	16	β-Neoendorphin	—
Endothelin-3	6	Neurokinin B	—
[Leu^5]-enkephalinamide	2	Oxytocin	—
Luliberin (LH-RH)	1	Safratoxin-b	—
Angiotensin I	—	Somatostatin	—
Angiotensin II	—	Vasoactive intestinal polypeptide	—
Angiotensin III	—		

[a]Data from Turner *et al.* (1987), Sakurada *et al.* (1990), Sokolovsky *et al.* (1990), and Vijayaraghaven *et al.* (1990).

larities, e.g., an Arg–Ser bond is cleaved in atrial natriuretic peptide (Vanneste *et al.*, 1988). The catalytic site of endopeptidase-24.11 has an extended substrate binding site (Hersh and Morihara, 1986), such that residues distant from the scissile bond can have a marked influence on the rates of hydrolysis of substrates. Endopeptidase-24.11 can hydrolyze a wide range of natural peptides (Table II), although only a few of these peptides (atrial and brain natriuretic peptides, bradykinin, chemotactic peptide, enkephalins, gastrin, neurotensin, and substance P) have been shown to be cleaved *in vivo* by the enzyme (see Erdos and Skidgel, 1989; Section 5.1).

2.1.2. ENDOPEPTIDASE-2

Endopeptidase-2 (PABA peptide hydrolase; meprin) is the only other true endopeptidase that has been identified in human tissues (Sterchi *et al.*, 1988). This enzyme is closely related if not identical to meprin in mice (Beynon *et al.*, 1981) and endopeptidase-2 in rats (Kenny and Ingram, 1987). Like endopeptidase-24.11, endopeptidase-2 has an extended substrate binding site and has an atom of zinc which is essential for activity. Endopeptidase-2 is capable of hydrolyzing several natural peptides including angiotensins I and II, bradykinin, [Met^5]-enkephalin-Arg^6-Phe^7, luliberin, neurotensin, oxytocin, and substance P.

2.2. Aminopeptidases

2.2.1. AMINOPEPTIDASE N

Aminopeptidase N (aminopeptidase M) is the best characterized of this class of peptidases. The cDNAs coding for the human (Olsen *et al.*, 1988), pig (Olsen *et al.*, 1989), and rat (Malfroy *et al.*, 1989) enzymes have been sequenced and display a high degree of similarity. Interestingly, there is also considerable homology with the soluble aminopeptidase N from *E. coli*, especially in the catalytic zinc binding domain (Olsen *et al.*, 1988). Aminopeptidase N has a broad substrate specificity, releasing the N-terminal amino acid from unblocked peptides. Substrate hydrolysis is enhanced with increasing chain length of the peptide. Hydrolysis is most rapid when the N-terminal amino acid is Ala, although Phe, Leu, Tyr, and Arg are hydrolyzed at intermediate rates. Glu, Asp, and Pro are hydrolyzed very slowly (McDonald and Barrett, 1986). Aminopeptidase N is the major activity releasing the N-terminal Tyr residue from the enkephalins (Matsas *et al.*, 1985), and the enzyme may also play a role in the metabolism of cholecystokinin-8 and neurokinin A (Matsas *et al.*, 1984; Nau *et al.*, 1986).

2.2.2. AMINOPEPTIDASE A

Until recently, aminopeptidase A had only been purified from pig kidney (Danielsen *et al.*, 1980). However, molecular cloning of the murine BP-1/6C3 antigen revealed that it was a member of the zinc metallopeptidase family (Wu *et al.*, 1990) and it was subsequently shown to possess aminopeptidase A activity (Wu *et al.*, 1991) (see Section 6). This peptidase is unusual compared with the other metallopeptidases in possibly containing one atom of calcium at its catalytic site. Aminopeptidase A hydrolyzes the acidic residues Asp and Glu from the N-terminus of peptide substrates, and may be involved in the conversion of angiotensin II to angiotensin III.

2.2.3. AMINOPEPTIDASE P

Until recently, aminopeptidase P had not been purified or characterized to any extent due to difficulties encountered in solubilizing the enzyme from other membrane components. However, the observation that aminopeptidase P was anchored by a covalently attached glycosyl–phosphatidylinositol moiety (Hooper and Turner, 1988; see Section 1) means that the enzyme can be readily released from the membrane in a soluble form by bacterial phosphatidylinositol-specific phospholipase C (Hooper *et al.*, 1990a). Aminopeptidase P contains one atom of zinc at its catalytic site and appears to be highly specific for peptides with a penultimate Pro residue at the N-terminus (Dehm and Nordwig, 1970; Hooper, unpublished). Thus, the enzyme may play a role in metabolizing bradykinin, substance P, and peptides derived from the breakdown of collagen (Dehm and Nordwig, 1970; Orawski *et al.*, 1987). Unlike the other aminopeptidases, aminopeptidase P is not inhibited by actinonin, amastatin, or bestatin. Surprisingly, however, and somewhat disturbingly, the clinically used inhibitors of angiotensin-converting enzyme (Section 2.5) are also potent inhibitors of aminopeptidase P (Hooper and Turner, unpublished).

2.2.4. AMINOPEPTIDASE W

Aminopeptidase W is unique among the ectopeptidases in being discovered by affinity chromatography using a monoclonal antibody generated to pig kidney microvillar membranes (Gee and Kenny, 1985). Aminopeptidase W prefers short peptides, and exhibits maximal rates toward dipeptides, in which the penultimate residue at the N-terminus is aromatic (Gee and Kenny, 1987).

2.3. Carboxypeptidases

2.3.1. CARBOXYPEPTIDASE M

Carboxypeptidase M was originally believed to be a membrane-bound form of the soluble carboxypeptidase N (EC 3.4.17.3) present in plasma. However, these two enzymes are quite distinct and are coded for by separate genes (Gebhard *et al.*, 1989;

Tan *et al.*, 1989). In certain cell types, carboxypeptidase M appears to be anchored by a glycosyl–phosphatidylinositol structure (Deddish *et al.*, 1990). Carboxypeptidase M cleaves C-terminal basic residues from dipeptides and oligopeptides, and has been shown to hydrolyze the natural peptides bradykinin, dynorphin A(1–13), [Met[5]]-enkephalin-Lys[6], [Met[5]]-enkephalin-Arg[6], and [Leu[5]]-enkephalin-Arg[6] (Skidgel *et al.*, 1989).

2.3.2. CARBOXYPEPTIDASE P

Carboxypeptidase P is a poorly characterized enzyme, cleaving C-terminal amino acids from peptides, but not dipeptides, when the penultimate residue is Pro, although Ala and Gly are also tolerated (Hedeager-Sorensen and Kenny, 1985). Peptides such as angiotensins II and III are potential substrates for this enzyme.

2.4. Dipeptidyl Peptidase

Dipeptidyl peptidase IV is alone among the mammalian ectopeptidases in containing an active-site Ser residue (Kenny *et al.*, 1976; Table I). The cDNA for the rat liver enzyme has been cloned and sequenced (Ogata *et al.*, 1989). Dipeptidyl peptidase IV releases dipeptides from the N-terminus of susceptible peptides when the penultimate residue is Pro or Ala (Bella *et al.*, 1982). Substance P is the only natural peptide that has been shown to be hydrolyzed by dipeptidyl peptidase IV (Heymann and Mentlein, 1978).

2.5. Peptidyl Dipeptidase

Angiotensin-converting enzyme (peptidyl dipeptidase A) is one of the best characterized ectopeptidases due, primarily, to its central role in blood pressure regulation (reviewed in Valloton,1987; and see Section 5.2). The cDNA sequences for the human (Soubrier *et al.*, 1988) and mouse (Bernstein *et al.*, 1989) endothelial forms (175 kDa) of angiotensin-converting enzyme have been elucidated. The most striking feature of the amino acid sequence is a high degree of internal similarity between two large domains and the presence of two zinc binding sites, and therefore, possibly two functional catalytic sites (Fig. 2) (reviewed in Hooper, 1991). As with endopeptidase-24.11, the variations in size between tissues appear to be due to differences in the extent of glycosylation (Hooper and Turner, 1987; Williams *et al.*, 1991), except in the testis where there is a distinct mRNA which encodes a smaller protein (100 kDa) (Lattion *et al.*, 1989; Ehlers *et al.*, 1989). The testicular enzyme, with only one catalytic site (Fig. 2), corresponds to the C-terminal domain of

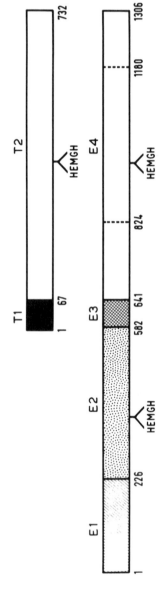

Figure 2. Schematic comparison of human endothelial and testicular angiotensin-converting enzyme. The structure of human testicular ACE (Lattion *et al.*, 1989) is shown at the top and that of human endothelial ACE (Soubrier *et al.*, 1988) at the bottom. T, testicular; E, endothelial. The amino acids at the junction of the domains are numbered. Domains T2 (residues 67–732) and E4 (residues 641–1306) have 100% identity, E2 (residues 226–582) and E4 insert (residues 824–1180) 67.7% identity. T1, E1, and E3 are unique to the respective isozymes. The characteristic zinc binding ligands (HEMGH) are labeled, indicating that endothelial angiotensin-converting enzyme has two putative catalytic domains.

endothelial angiotensin-converting enzyme, suggesting that the endothelial form has resulted from gene duplication while the testicular form corresponds to the ancestral, nonduplicated form of the gene.

Angiotensin-converting enzyme was originally classified as a peptidyl dipeptidase due to its action in removing C-terminal dipeptides from angiotensin I and bradykinin. Although this is its preferred action, with certain substrates angiotensin-converting enzyme displays peptidyl tripeptidase and/or endopeptidase activities (Table III). Angiotensin-converting enzyme is known to hydrolyze several natural peptides (Table III).

Table III
Substrate Specificity and Peptides Hydrolyzed by Angiotensin-Converting Enzyme[a]

Peptide structure	Cleavage product	Examples	k_{cat}/K_m (min$^{-1}\mu$M^{-1})
Free C-terminus	C-terminal dipeptide	Angiotensin I	125
		Bradykinin	1250
	-V-W$\overset{\downarrow}{-}$X-Y unless X = Pro or Y = Asp, Glu	BzGly-His-Leu	3.6
		Dynorphin 1–6	—
		Dynorphin 1–8	—
		f-Met-Leu-Phe	—
		[Leu⁵]-enkephalin	0.7
		[Met⁵]-enkephalin-Arg-Phe	3.5
		[Met⁵]-enkephalin-Arg-Gly-Leu	—
		β-Neoendorphin	—
		Neurotensin	2.4
		Substance P (deamidated)	637
	C-terminal tripeptide	Angiotensin II	—
		des-Arg⁹-bradykinin	—
	-V$\overset{\downarrow}{-}$W-X-Y when X = Pro or Trp		
Amidated C-terminus	C-terminal dipeptide amide	Substance P	9
		[Leu⁵]-enkephalinamide	—
		[Met⁵]-enkephalinamide	—
		Cholecystokinin-8	0.3
	-V-W$\overset{\downarrow}{-}$X-YNH₂	[Leu¹⁵]-gastrin-11–17	0.3
	C-terminal tripeptide amide	Substance P	9
		Luliberin (LH-RH)	1.3
	-V$\overset{\downarrow}{-}$W-X-YNH₂		
Blocked N-terminus	N-terminal tripeptide	Luliberin (LH-RH)	1.3
	V-W-X$\overset{\downarrow}{-}$Y-		

[a]Data from Erdos and Skidgel (1985, 1987), Skidgel and Erdos (1985), and Dubreuil *et al.* (1989).

2.6. Dipeptidases

2.6.1. MEMBRANE DIPEPTIDASE

Membrane dipeptidase (renal dipeptidase; microsomal dipeptidase; dehydro-peptidase I) was the first ectopeptidase identified as possessing a glycosyl–phosphatidylinositol membrane anchor (Hooper et al., 1987b; Littlewood et al., 1989; Hooper and Turner, 1989). Recently, the cDNA sequences for both the human (Adachi et al., 1990) and pig (Rached et al., 1990) enzymes have been deduced and shown to display a high degree of similarity. The large (14 kDa) size difference between the human and pig forms is due exclusively to differences in the extent of N-linked glycosylation (Hooper et al., 1990b). Membrane dipeptidase is a disulfide-linked dimer, each subunit containing one atom of zinc. However, the enzyme does not possess the characteristic zinc-binding signature (His-Glu-Xaa-Xaa-His) present in all of the other zinc metallopeptidases sequenced to date (Jongeneel et al., 1989a). Membrane dipeptidase hydrolyzes dipeptides including those with a C-terminal D-amino acid (Campbell, 1970). The enzyme may also play a role in the metabolism of leukotriene D_4 (Kozak and Tate, 1982) and β-lactam antibiotics such as imipenem (Kropp et al., 1982; see Section 5.3).

2.6.2. N-ACETYLATED-α-LINKED ACIDIC DIPEPTIDASE

N-Acetylated-α-linked acidic dipeptidase was first identified as the activity that metabolized the natural acidic dipeptide, N-acetyl-Asp-Glu, subsequent to its release from nerve endings (Blakely et al., 1986). The peptidase has recently been purified from rat brain synaptosomal membranes and located by immunocytochemistry in cerebellum, kidney, and testis (Slusher et al., 1990). N-Acetylated-α-linked acidic dipeptidase appears to be a metallopeptidase, and will also hydrolyze non-N-acetylated dipeptides containing two acidic amino acids, e.g., Asp-Glu (Serval et al., 1990).

2.7. Omega Peptidases

2.7.1. γ-GLUTAMYL TRANSPEPTIDASE

γ-Glutamyl transpeptidase is unusual in existing in the membrane as a glycosy-lated heterodimer (Table I). The cDNA sequences for the enzymes from human (Rajpert-De Meyts et al., 1988), pig (Papandrilkopoulou et al., 1989), and rat (Laperche et al., 1986) have been elucidated and show a high degree of similarity. Recently, an alternatively processed mRNA for γ-glutamyl transpeptidase has been isolated which may account for the inactive form of the enzyme identified in human tissues (Pawlak *et al.*, 1990). γ-Glutamyl transpeptidase cleaves γ-Glu from the N-terminus of peptides, its principal natural substrate being glutathione (Tate and Meister, 1974; see Section 5.4).

2.7.2. PYROGLUTAMYL PEPTIDASE II

Pyroglutamyl peptidase II (thyroliberinase) is a metallopeptidase and is thus distinct from the cytosolic cysteine enzyme, pyroglutamyl peptidase I (EC 3.4.19.3) (O'Connor and O'Cuinn, 1984). Pyroglutamyl peptidase II has a restricted specificity in hydrolyzing only the Glp-His bond of thyroliberin and related synthetic peptides (Wilk, 1986). The same bond in luliberin is not cleaved; however, it is probably premature to describe this enzyme as a peptide-specific peptidase.

3. ORGAN, TISSUE, AND CELLULAR DISTRIBUTION OF ECTOPEPTIDASES

The majority of the ectopeptidases are present on a wide range of cell types in a variety of different organs throughout the body (Table IV). However, the data for some of the ectopeptidases, especially in human tissues, are somewhat scarce. The most detailed surveys have been carried out on angiotensin-converting enzyme and endopeptidase-24.11. Primarily, immunohistochemical techniques (immunofluorescence and immunoperoxidase staining) by light microscopy have been used to detail the tissue and cellular localization of these two enzymes (Caldwell *et al.*, 1976; Gee *et al.*, 1983). In addition, a sensitive immunoradiometric assay has been used to locate endopeptidase-24.11 in a variety of pig tissues (Gee *et al.*, 1985), and autoradiography of tissue sections with radiolabeled inhibitors has been used to study the distribution of angiotensin-converting enzyme, especially in the CNS (Mendelsohn, 1984; Strittmatter *et al.*, 1984). Recently, the immunogold method with electron microscopy has allowed a more detailed subcellular localization of both angiotensin-converting enzyme (Schulz *et al.*, 1988) and endopeptidase-24.11 (Barnes *et al.*, 1988).

Although such detailed surveys have yet to be carried out on most of the other peptidases, what is clearly evident is that the major epithelial surfaces, e.g., in the intestine, placenta, choroid plexus, and kidney, contain a battery of ectopeptidase activities. Thus, together these enzymes form an extremely effective barrier, preventing the access of a wide range of peptides to the underlying tissues (see Section 4).

Some of the ectopeptidases have been identified in soluble forms in certain body fluids, such as blood, urine, CSF, and amniotic fluid (Table IV), albeit at very low levels. The origin of the soluble forms of the ectopeptidases is uncertain. They most probably derive from the membrane-bound forms by proteolytic action (Stewart

Table IV
Distribution of Ectopeptidases[a]

Location	Cell type	EP-24.11	E-2	AP-N	AP-A	AP-P	AP-W	CP-M	CP-P	DPPIV	ACE	MDP	AAD	γ-GT	PPII
Blood vessels															
Peripheral tissues	Vascular endothelium	×[b]	×	✓	—	—	×	✓	—	✓	✓	—	—	—	—
CNS	Microvessels	×	×	✓	(✓)	—	×	—	—	✓	✓	—	—	✓	—
Epithelial															
Choroid plexus	Epithelium	✓	—	✓	—	—	×	—	—	—	✓	—	—	—	—
Kidney	Proximal tubule epithelial cells	✓	✓	✓	✓	✓	✓	✓	(✓)	✓	✓	✓	✓	✓	—
Liver	Hepatocytes	×	×	✓	—	—	—	—	—	✓	—	—	—	✓	✓
Lung	Acinal cells	✓	✓	—	—	—	—	✓	—	—	✓	—	—	—	✓
	Mucous glands	✓	—	—	—	—	—	—	—	—	—	—	—	—	—
	Duct epithelium	✓	×	✓	✓	—	—	✓	—	✓	✓	—	—	—	—
Placenta	Syncytial trophoblast	✓	—	—	—	—	—	—	—	✓	✓	—	—	✓	—
Reproductive (male and female)	Epithelium	✓	—	✓	✓	✓	✓	—	—	—	✓	—	—	—	—
Small intestine	Enterocyte	✓	✓	✓	✓	✓	✓	—	(✓)	✓	✓	—	—	✓	—

Neuropil															
Cerebellum		×	×	×	—	×	—	—	×	—	—	×	—	—	✓
Cortex	?Neuronal	×	×	✓	—	✓	—	—	✓	—	✓	✓	—	✓	✓
Striatum		✓	×	✓	—	✓	—	—	✓	—	✓	✓	—	✓	—
Cervical spinal cord		✓	×	×	—	×	—	—	×	—	×	×	—	—	—
Other															
Lymph nodes	Reticular cells	✓	×	×	—	×	—	—	×	—	×	×	—	—	—
	Macrophages	×	✓	✓	—	✓	—	✓	✓	—	✓	✓	—	—	✓
Blood	Plasma	✓	—	—	—	—	✓	—	—	—	✓	✓	—	—	—
	Polymorphonuclear leukocytes	✓	—	—	—	—	—	—	—	—	×	×	—	—	—

[a]Ectopeptidases: E-24.11, endopeptidase-24.11; E-2, endopeptidase-2; AP-N, aminopeptidase N; AP-A, aminopeptidase A; AP-P, aminopeptidase P; AP-W, aminopeptidase W; CP-M, carboxypeptidase M; CP-P, carboxypeptidase P; DPPIV, dipeptidyl peptidase IV; ACE, angiotensin-converting enzyme; MDP, membrane dipeptidase; AAD, N-acetylated-α-linked acidic dipeptidase; γ-GT, γ-glutamyl transpeptidase; PPII, pyroglutamyl peptidase II.

[b]Key: ✓, positively identified by immunocytochemistry; (✓), identification is tentative and based only on activity measurements; ×, not present; —, data not available.

et al., 1981; Hooper *et al.*, 1987a) or, in the case of those anchored by a glycosyl–phosphatidylinositol structure, by the action of phospholipases (Low, 1989). In certain disease states, the soluble forms of the ectopeptidases can be markedly increased, e.g., angiotensin-converting enzyme in sarcoidosis, Gaucher's disease, leprosy, and hyperthyroidism (Erdos and Skidgel, 1987), and endopeptidase-24.11 in adult respiratory distress syndrome and in patients with end-stage renal failure (Almenoff *et al.*, 1984; Johnson *et al.*, 1985). The role of these circulating forms of angiotensin-converting enzyme and endopeptidase-24.11 in these pathophysiological situations is unclear, but they are likely to make an important contribution to the metabolism of natural and administered peptides.

4. CONCERTED ACTION OF ECTOPEPTIDASES

From the preceding section it is apparent that most cell types contain a large complement of ectopeptidases, and thus a peptide substrate is likely to be acted on by several of the enzymes either simultaneously or in quick succession. Thus, although the information obtained from detailed studies on purified enzymes is important in identifying a potential substrate and delineating the bond(s) hydrolyzed, physiological conditions are far removed from such artificial situations. In addition, the concentration of the peptide substrate *in vivo* will be several orders of magnitude lower than those necessary for *in vitro* determinations. A membrane preparation rich in the ectopeptidases, such as the renal or intestinal brush border, can be used to investigate the effect of multiple activities on a particular peptide substrate. In the presence of such a battery of peptidases, most peptides rapidly undergo virtually complete hydrolysis. However, under appropriate conditions it is possible to stop the hydrolysis at an early stage so that the initial peptide fragments can be identified. Specific inhibitors are then also used to confirm the identity of the ectopeptidase initiating the attack. Most of the ectopeptidases detailed in this chapter can be selectively inhibited (Table V), although it may be necessary to assess the effects of more than one inhibitor in order to unequivocally identify the ectopeptidase involved.

An example of this approach is illustrated in Fig. 3 where a number of biologically active peptides have been incubated with pig kidney microvillar membranes (Stephenson and Kenny, 1987a). The hydrolysis of angiotensins I, II, and III, bradykinin, and oxytocin is initiated by endopeptidase-24.11. The degradation of all of the peptides was markedly inhibited by the specific inhibitor of endopeptidase-24.11, phosphoramidon (Table V). Angiotensin-converting enzyme has a minor role in hydrolyzing the Pro-Phe bond of bradykinin and in releasing His-Leu from angiotensin I, although most of the His-Leu arises from the combined attack of endopeptidase-24.11 (generating Phe-His-Leu) and aminopeptidase N (removing Phe). The C-terminal Pro-Phe bond in angiotensins II and III is most probably cleaved by carboxypeptidase P. A similar experimental approach has revealed that the initial

Table V

Specific Inhibitors of Ectopeptidases[a]

Peptidase	Inhibitor	K_i (nM)	I_{50} (μM)	Recommended concentration (μM)	Ref.[b]
Endopeptidase-24.11	Phosphoramidon	2	0.013	1	1
	Thiorphan	4.7	0.013	1	2
	Retrothiorphan	6.0	—	1	3
	Kelatorphan	1.4	0.002	1	4
Aminopeptidase N	Actinonin	—	1	100	5
	Amastatin	19	0.05	1[a]	6
	Bestatin	4100	3.03	100	6
	Kelatorphan	7000	0.71	100	4
Aminopeptidase A	Amastatin	250	—	10	7
Aminopeptidase P	Enalaprilat	—	2.17	100	8
	L155, 212	—	0.33	10	8
	Ramiprilat	—	12.0	100	8
Aminopeptidase W	Amastatin	—	2	100[a]	9
	Bestatin	—	6	100[a]	9
Carboxypeptidase M	MGTA	—	0.3	10	10
	GEMSA	—	—	100	10
Dipeptidyl peptidase IV	Di-isopropyl fluoro-phosphate	—	10	100[a]	11
	Diprotin A	—	3.2	100	12
	N-Ala-Pro-o-(4-nitro-benzoyl) hydroxylamine	—	—	—	13
Angiotensin-converting enzyme	Captopril	1.7	0.021	1	14
	Enalaprilat	0.2	0.003	1	15
	Lisinopril	0.1	0.011	1	15
	L155, 212	—	0.003	1	15
Membrane dipeptidase	Cilastatin	700	0.11	10	16
N-Acetylated-α-linked-acidic dipeptidase	Quisqualate	—	0.48	100	17
γ-Glutamyl transpeptidase	AT-125	0.6 mM	—	—	18

[a]Preincubation (30–60 min) required for maximal inhibition. No specific inhibitors of endopeptidase-2, carboxypeptidase P, or pyroglutamyl peptidase II have yet been identified.

[b]References: 1, Kenny (1977); 2, Fulcher et al. (1982); 3, Roques et al. (1983); 4, Fournie-Zaluski et al. (1985); 5, Takeuchi (1985); 6, Rich et al. (1984); 7, Aoyagi et al. (1978); 8, Hooper and Turner (unpublished data); 9, Gee and Kenny (1987); 10, Skidgel et al. (1989); 11, Kenny et al. (1976); 12, Umezawa et al. (1984); 13, Demuth et al. (1989); 14, Cushman et al. (1977); 15, Hooper and Turner (1987); 16, Campbell et al. (1984); 17, Robinson et al. (1987); 18, Gardel and Tate (1980).

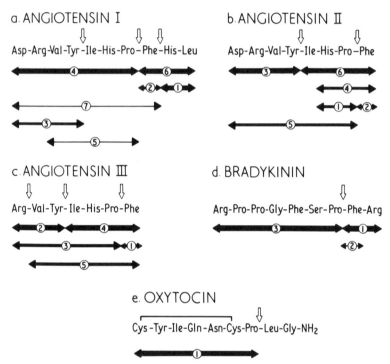

Figure 3. Metabolism of peptides by pig kidney microvillar membranes. The figure shows the initial points of attack when the peptides were incubated with pig kidney microvillar membranes. The horizontal arrows define the peptide fragments characterized by amino acid analyses, after separation by HPLC, and the thickness of each arrow relates to the yield of that peptide. The vertical arrows indicate the bonds hydrolyzed. By including specific peptidase inhibitors in parallel incubations, it was possible to assess the relative contributions of the principal peptidases. (See text for details.) (Reproduced from Stephenson and Kenny, 1987a, with permission from the Biochemical Society, Portland Press, London.)

attack on atrial natriuretic peptide and brain natriuretic peptide is also due to endopeptidase-24.11 (Stephenson and Kenny, 1987b; Bourne and Kenny, 1990). With the natriuretic peptides, as with most of the other peptides studied, the initial attack on the peptide is also the inactivating step. It should be noted that as well as tissues having different complements of the ectopeptidases, they may also have different amounts of the individual enzymes relative to one another, and this may lead to tissue-specific differences in the hydrolysis pattern of a peptide.

To overcome possible problems associated with the high concentration (50–250 μM) of substrate used in the above *in vitro* experiments revealing only peptidases exhibiting high K_m values, similar incubations were performed with 20 nM [^{3}H]brady-kinin. The main labeled product had the same retention time as bradykinin 1–7 and the formation of this peptide was inhibited by phosphoramidon, indicating that the initial, inactivating attack is due to endopeptidase-24.11 (Stephenson and Kenny, 1987a). This argues against the existence of other ectopeptidases, with lower K_m values, which might be dominant in physiological conditions. These experiments demonstrate the crucial role of endopeptidase-24.11, rather than angiotensin-converting enzyme, in the inactivation of bradykinin by renal microvillar membranes. However, this does not imply that angiotensin-converting enzyme lacks this role *in vivo*, as on the surfaces of the endothelium of the pulmonary vasculature it is the only enzyme present (Table IV) capable of removing the C-terminal dipeptide.

5. PHYSIOLOGICAL FUNCTIONS OF ECTOPEPTIDASES

Obviously the location of an ectopeptidase is going to have a significant bearing on its physiological role. In general terms, three distinct types of role can be envisaged for ectopeptidases. (1) The scavenging of peptides for nutritional purposes, for example in the intestine, where the ectopeptidases lining the lumen will act on the small peptide products generated from the breakdown of ingested protein. The resultant amino acids can then be efficiently transported into the epithelial cells. (2) The protection of tissues against the unwanted actions of peptides that might otherwise cause effects at a remote location. The ectopeptidases lining the lumen of the renal proximal tubule and those on the epithelial surface of the choroid plexus, being bathed by the CSF, may well fall into this category. (3) The termination (or modification) of an intercellular signal, where the regulatory molecule is a peptide. In this mode, ectopeptidases are acting in a fashion analogous to acetylcholinesterase in hydrolyzing and inactivating acetylcholine. However, ectopeptidases are not confined to the nervous system and will also be involved in the inactivation of bioactive peptides derived from endocrine and paracrine secretions.

A few of the more well-characterized systems have been selected below to demonstrate the variety of roles that ectopeptidases can play.

5.1. Endopeptidase-24.11

Endopeptidase-24.11 has twice been the focus of intensive research by the pharmaceutical industry. The first arose with the observation that exogenous [Met[5]]-enkephalin was rapidly hydrolyzed and inactivated by brain preparations (Craves *et al.*, 1978). After much research, it became evident that the ectopeptidase responsible for this action was the enzyme previously characterized from kidney, endopeptidase-24.11 (Matsas *et al.*, 1983; Relton *et al.*, 1983; Almenoff and Orlowski, 1984). Inhibitors of the enzyme were found to increase the recovery of [Met[5]]-enkephalin when brain slices were depolarized by K^+, suggesting an *in vivo* role for endopeptidase-24.11 in the metabolism of enkephalins (Chaillet *et al.*, 1983). This then raised the possibility that inhibitors of endopeptidase-24.11 might provide a new class of analgesics (reviewed in Erdos and Skidgel, 1989). However, the aminopeptidase inhibitor, bestatin, also augmented the recovery of [Met[5]]-enkephalin, pointing to a contribution from aminopeptidases. The recovery of substance P from brain slices is also increased in the presence of inhibitors of endopeptidase-24.11 (Mauborgne et al., 1987; Littlewood et al., 1988). These findings reiterate that ectopeptidases are not peptide-specific, and that any one peptide is likely to be metabolized by more than one peptidase.

The second wave of interest in endopeptidase-24.11 is linked with atrial natriuretic peptide. This peptide is involved in the homeostasis of fluid balance and blood pressure, raising the possibility that atrial natriuretic peptide could be administered to treat certain cardiovascular disorders. However, the administered peptide is rapidly inactivated, the critical enzyme being endopeptidase-24.11 (Bergey et al., 1987; Stephenson and Kenny, 1987b; Lecomte et al., 1990). Inhibitors of endopeptidase-24.11 markedly increase the half-life of the hormone in plasma, and may represent a novel class of therapeutic agents with potential applications in congestive heart failure, essential hypertension, and various sodium-retaining states (Schwartz et al., 1990). The related hormone, brain natriuretic peptide, which displays similar pharmacological effects as atrial natriuretic peptide, but which has a distinct distribution, is also inactivated by endopeptidase-24.11 (Bourne and Kenny, 1990; Vanneste et al., 1990).

In both of these cases, however, the pharmaceutical industry is working with a flawed hypothesis, that endopeptidase-24.11 is, to all practical purposes, peptide-specific. The wide range of cell types expressing this ectopeptidase (Table IV) and the large repertoire of susceptible peptides (Table II) raise serious doubts about the capacity of inhibitors of endopeptidase-24.11 to manipulate selectively a single (patho-) physiological process.

5.2. Angiotensin-Converting Enzyme

Angiotensin-converting enzyme is probably best known for its role in the renin–angiotensin system (reviewed in Valloton, 1987). The lung was known for some years (Ng and Vane, 1968) to be the major site in the body for the conversion of angiotensin I to angiotensin II, a potent vasoconstrictor that also stimulates the release of aldosterone from the adrenal cortex leading to sodium ion retention. Angiotensin-converting enzyme was then observed to have an endothelial location in this tissue (Caldwell et al., 1976), and, with its ectoenzyme topology, is able to act on circulating angiotensin I. Independently, angiotensin-converting enzyme was also shown to inactivate the vasodilator, bradykinin (Erdos and Yang, 1967). Thus, the same enzyme has a dual role in the maintenance of blood pressure and fluid and electrolyte homeostasis. It is because of this role that inhibitors of angiotensin-converting enzyme have been so successful in the treatment of hypertension and congestive heart failure (Cohen, 1985; Johnston, 1988; Unger et al., 1990).

However, angiotensin-converting enzyme has a wide tissue and cellular distribution (Table IV) and is not always colocalized with other components of the renin–angiotensin system (Ehlers and Riordan, 1989). Together with its broad substrate specificity (Table III), this has led to angiotensin-converting enzyme being implicated in a number of other physiological processes such as immunity, reproduction, and neuropeptide metabolism (Erdos and Skidgel, 1987; Ehlers and Riordan, 1989). Thus, some of the observed effects (or side effects) of administration of angiotensin-converting enzyme inhibitors may be due to inhibition of the enzyme in other processes. In addition, some of the side effects noted with the clinical use of these inhibitors (Edwards and Padfield, 1985; Gavras and Gavras, 1988) may be due to inhibition of aminopeptidase P (see Section 2.2.3).

5.3. Membrane Dipeptidase

This example highlights the problem encountered when a compound has been developed to treat a disorder and is found to be inactivated upon administration. During the search for a β-lactam antibiotic with activity against a broad spectrum of bacteria, a natural product of streptomyces, thienamycin, was discovered. This compound was, unfortunately, rather unstable in concentrated solution or as a solid, so an amidine derivative, imipenem, was developed, which has substantially improved antibacterial properties. However, when imipenem was administered to humans and other animals, although the serum half-life was very high, urinary recoveries were extremely low, thus rendering the antibiotic ineffective against urinary tract infections. Further investigation revealed that imipenem was rapidly metabolized in the kidneys, and the enzyme responsible was found to be an ectopeptidase located on the surface of the proximal tubules, membrane dipeptidase (Kropp et al., 1982). This observation led to the development of cilastatin to inhibit membrane dipeptidase, and when coadministered with imipenem, high urinary recovery of the antibiotic was observed (Kahan et al., 1983). Thus, membrane dipeptidase appears to play a physiological role in metabolizing β-lactam compounds (Campbell et al., 1984) and is the first example of a mammalian β-lactamase.

5.4. γ-Glutamyl Transpeptidase

Glutathione (γ-Glu-Cys-Gly) is the most abundant peptide and the major nonprotein thiol present within mammalian cells (Meister and Anderson, 1983). Glutathione is resistant to attack by intracellular aminopeptidases and is metabolized in the kidney subsequent to its release from the cell into the blood. γ-Glutamyl transpeptidase lining the renal proximal tubule initiates the degradation of gluta-thione. This is the only enzyme known to catalyze the cleavage of the γ-glutamyl bond in glutathione as evidenced from the pronounced glutathionemia and glutathion-uria that result from the genetic deficiency or selective inhibition of γ-glutamyl transpeptidase (Schulman et al., 1976; Griffith and Meister, 1979). This ectopepti-dase can also catalyze the transfer of the γ-glutamyl group to other amino acids and dipeptides. The cysteinyl-glycine generated by γ-glutamyl transpeptidase is then acted on by either of two other ectopeptidases, aminopeptidase N (Section 2.2.1) or membrane dipeptidase (Section 2.6.1).

6. CONCLUSIONS AND FUTURE PERSPECTIVES

In this chapter, I have attempted to describe those ectopeptidases that may have a role in metabolizing biologically active peptides and related compounds. It is evident that certain of these peptidases, endopeptidase-24.11 and angiotensin-converting enzyme, are extremely well characterized in their structural and catalytic properties, and they have been identified as playing central roles in some defined physiological functions. However, for the majority of the ectopeptidases described in this chapter, precise physiological functions have yet to be described, mainly because we are still ignorant of their natural substrates. What is clearly apparent, though, is that there is a limited number of ectopeptidases that are neither peptide- nor tissue-specific. This limited number of enzymes suffices to inactivate a large number of biologically active peptides. In addition, because the ectopeptidases are abundant on various epithelial and endothelial surfaces, an exogenously administered peptide has to surmount a virtually impregnable wall of enzymes in order to enter tissues and cells.

Recently, following the isolation of the cDNA sequences for certain of the ectopeptidases, it has been shown that (1) endopeptidase-24.11 is identical with the common acute lymphocytic leukemia antigen (CALLA) or cluster differentiation (CD) 10 antigen (Letarte et al., 1988; Jongeneel et al., 1989b; Shipp et al., 1989); (2) aminopeptidase N is identical with CD13 (Look et al., 1989), (3) aminopeptidase A is identical with the murine β-lymphocyte differentiation antigen BP-1/6C3 (Wu et al., 1991); and (4) dipeptidyl peptidase IV is identical with CD26 or gp110 (Ogata et al., 1989). These observations would be consistent with a role for the ectopepti-dases in terminating peptide signals affecting the proliferation, in either a stimulatory or an inhibitory way, of transformed and normal cells (Kenny et al., 1989). Thus, inhibitors of ectopeptidases may alter the growth and differentiation of different cell types. Also, the overexpression of ectopeptidases on transformed cells may present problems to those administering peptide-like drugs to such patients.

The mammalian ectopeptidases present a substantial barrier to the delivery of peptides and related compounds. In the future, the development and use of more specific and selective inhibitors and antibodies and the use of cDNA probes will undoubtedly further our understanding of the physiological functions of ectopeptidases, and thus aid in the design of peptidase-resistant drugs.

REFERENCES

Adachi, H., Tawaragi, Y., Inuzuka, C., Kubota, I., Tsujimoto,M., Nishihara, T., and Nakazato, H., 1990, Primary structure of human microsomal dipeptidase deduced from molecular cloning, *J. Biol. Chem.* **265:**3992–3995.

Almenoff, J., and Orlowski, M., 1984, Biochemical and immunological properties of a membrane-bound metalloendopeptidase: Comparison with thermolysin-like kidney neutral metalloendopeptidase, *J. Neurochem.* **42:**151–157.

Almenoff, J., Teirstein, A. S., Thornton, J. C., and Orlowski, M., 1984, Identification of a thermolysin-like metallopeptidase in serum: Activity in normal subjects and in patients with sarcoidosis, *J. Lab. Clin. Med.* **103:**420–431.

Aoyagi, T., Tobe, H., Kojima, F., Hamada, M., Takeuchi, T., and Umezawa, H., 1978, Amastatin, an inhibitor of aminopeptidase A, produced by actinomycetes, *J. Antibiot.* **31:** 636–638.

Barnes, K., Turner, A. J., and Kenny, A. J., 1988, Electronmicroscopic immunocytochemistry of pig brain shows that endopeptidase-24.11 is localized in neuronal membranes, *Neurosci. Lett.* **94:**64–69.

Barrett, A. J., and McDonald, J. K., 1980, *Mammalian Proteases: A Glossary and Bibliography*, Volume 1, Academic Press, New York.

Bella, A. M., Erickson, R. M., and Kim, Y. S., 1982, Rat intestinal brush border membrane dipeptidyl aminopeptidase IV: Kinetic properties and substrate specificities of the purified enzyme, *Arch. Biochem. Biophys.* **218:**156–162.

Bergey, J. L., Kotler, D., Delaney, N. G., and Cushman, D. W., 1987, Factors affecting the vasorelaxant activity of atrial natriuretic peptides, *Fed. Proc. Fed. Am. Soc. Exp. Biol.* **46:**1296.

Bernstein, K. E., Martin, B. M., Edwards, A. S., and Bernstein, E. A., 1989, Mouse angiotensin-converting enzyme is a protein composed of two homologous domains. *J. Biol. Chem.* **264:**11945–11951.

Beynon, R. J., Shannon, J. D., and Bond, J. S., 1981, Purification and characterization of metalloendopeptidase from mouse kidney, *Biochem. J.* **199:**591–598.

Blakely, R. D., Ory-Lavollee, L., Thompson, R. C., and Coyle, J. T., 1986, Synaptosomal transport of radiolabel from N-acetyl-aspartyl-[^{3}H]-glutamate suggests a mechanism of inactivation of an excitatory neuropeptide, *J. Neurochem.* **47:**1013–1019.

Bourne, A., and Kenny, A. J., 1990, The hydrolysis of brain and atrial natriuretic peptides by porcine choroid plexus is attributable to endopeptidase-24.11, *Biochem. J.* **271:**381–385.

Caldwell, P. R. B., Seegar, B. C., Hsu, K. C., Das, M., and Soffer, R. L., 1976, Angiotensin-converting enzyme: Vascular endothelial localizations, *Science* **191:**1050–1051.

Campbell, B. J., 1970, Renal dipeptidase, *Methods Enzymol.* **19:**722–729.

Campbell, B. J., Forrester, L. J., Zahler, W. L., and Burks, M., 1984, β-Lactamase activity of purified and partially characterized human renal dipeptidase, *J. Biol. Chem.* **259:**14586–14590.

Chaillet, P., Marcus-Collado, H., Costentin, J., Yi, C. C., de la Baume, S., and Schwartz, J. C., 1983, Inhibition of enkephalin metabolism by, and antinociceptive activity of, bestatin, an aminopeptidase inhibitor. *Eur. J. Pharmacol.* **86**:329–336.

Cohen, M. L., 1985, Synthetic and fermentation-derived angiotensin-converting enzyme inhibitors, *Annu. Rev. Pharmacol. Toxicol.* **25**:307–323.

Craves, F. B., Law, P. Y., Hunt, C. A., and Loh, H. H., 1978, The metabolic disposition of radiolabeled enkephalins *in vitro* and *in situ*. *J. Pharmacol. Exp. Ther.* **206**:492–506.

Cross, G. A. M., 1990, Glycolipid anchoring of plasma membrane proteins, *Annu. Rev. Cell Biol.* **6**:1–39.

Cushman, D. W., Cheung, H. S., Sabo, E. F., and Ondetti, M. A., 1977, Design of potent competitive inhibitors of angiotensin converting enzyme. Carboxyalkanoyl and mercaptoalkanoyl amino acids, *Biochemistry* **16**:5485–5491.

Danielsen, E. M., Noren, O., Sjostrom, H., Ingram, J., and Kenny, A. J., 1980, Proteins of the kidney microvillar membrane. Aspartate aminopeptidase: Purification by immunoadsorbent chromatography and properties of the detergent- and proteinase-solubilized forms, *Biochem. J.* **189**:591–603.

Deddish, P. A., Skidgel, R. A., Kriho, V. B., Li, X.-Y., Becker, R. P., and Erdos, E. G., 1990, Carboxypeptidase M in Madin–Darby canine kidney cells. Evidence that carboxypeptidase M has a phosphatidylinositol glycan anchor, *J. Biol. Chem.* **265**:15083–15089.

Dehm, P., and Nordwig, A., 1970, The cleavage of prolyl peptides by kidney peptidases. Partial purification of a "X-prolyl-aminopeptidase" from swine kidney microsomes, *Eur. J. Biochem.* **17**:364–371.

Demuth, H.-V., Neumann, U., and Barth, A., 1989, Reactions between dipeptidyl peptidase IV and diacyl hydroxylamines: Mechanistic investigations, *J. Enzyme Inhib.* **2**:239–248.

Devault, A., Lazure, C., Nault, C., Moual, H. L., Seidah, N. G., Chretien, M., Kahn, P., Powell, J., Mallet, J., Beaumont, A., Roques, B. P., Crine, P., and Boileau, G., 1987, Amino acid sequence of rabbit kidney neutral endopeptidase-24.11 (enkephalinase), *EMBO J.* **6**:1317–1322.

Dubreuil, P., Fulcrand, P., Rodriguez, M., Fulcrand, H., Laur, J., and Martinez, J., 1989, Novel activity of angiotensin converting enzyme: Hydrolysis of cholecystokinin and gastrin analogues with release of the amidated COOH-terminal dipeptide, *Biochem. J.* **262**:125–130.

Edwards, C. R. W., and Padfield, P. L., 1985, Angiotensin-converting enzyme inhibitors: Past, present, and bright future. *Lancet* **1**:30–34.

Ehlers, M. R. W., and Riordan J. F., 1989, Angiotensin-converting enzyme: New concepts concerning its biological role, *Biochemistry* **28**:5311–5318.

Ehlers, M. R. W., Fox, E. A., Strydom, D. J., and Riordan, J. F., 1989, Molecular cloning of human testicular angiotensin-converting enzyme: The testis isozyme is identical to the C-terminal half of endothelial ACE, *Proc. Natl. Acad. Sci. USA* **86**:7741–7745.

Erdos, E. G., and Skidgel, R. A., 1985, Structure and functions of human angiotensin I converting enzyme (kininase II), *Biochem. Soc. Trans.* **13**:42–44.

Erdos, E. G., and Skidgel, R. A., 1987, The angiotensin I-converting enzyme, *Lab. Invest.* **56**: 345–348.

Erdos, E. G., and Skidgel, R. A., 1989, Neutral endopeptidase-24.11 (enkephalinase) and related regulators of peptide hormones, *FASEB J.* **3**:145–151.

Erdos, E. G., and Yang, H. Y. T., 1967, An enzyme in microsomal fraction of kidney that inactivates bradykinin, *Life Sci.* **6**:569–574.

Fournie-Zaluski, M.-C., Coulaud, A., Bouboutou, R., Chaillet, P., Devin, J., Waksman, G., Costentin, J., and Roques, B. P., 1985, New bidentates as full inhibitors of enkephalin-degrading enzymes: Synthesis and analgesic properties, *J. Med. Chem.* **28:**1158–1169.

Fulcher, I. S., Matsas, R., Turner, A. J., and Kenny, A. J., 1982, Kidney neutral endopeptidase and the hydrolysis of enkephalin by synaptic membranes show similar sensitivity to inhibitors, *Biochem. J.* **203:**519–522.

Gardel, S. J., and Tate, S. S., 1980, Affinity labeling of γ-glutamyl transpeptidase by glutamine antagonists, *FEBS Lett.* **122:**171–174.

Gavras, H., and Gavras, I., 1988, Angiotensin converting enzyme inhibitors. Properties and side effects, *Hypertension* **11**(Suppl. II):37–41.

Gebhard, W., Schube, M., and Eulitz, M., 1989, cDNA cloning and complete primary structure of the small, active subunit of human carboxypeptidase N (kininase I), *Eur. J. Biochem.* **178:**603–607.

Gee, N. S., and Kenny, A. J., 1985, Proteins of the kidney microvillar membrane. The 130 kDa protein in pig kidney, recognized by monoclonal antibody GK5Cl, is an ectoenzyme with aminopeptidase activity. *Biochem. J.* **230:**753–764.

Gee, N. S., and Kenny, A. J., 1987, Proteins of the kidney microvillar membrane. Enzymic and molecular properties of aminopeptidase W, *Biochem. J.* **246:**97–102.

Gee, N. S., Bowes, M. A., Buck, P., and Kenny, A. J., 1985, An immunoradiometric assay for endopeptidase-24.11 shows it to be a widely distributed enzyme in pig tissues, *Biochem. J.* **228:**119–126.

Gee, N. S., Matsas, R., and Kenny, A. J., 1983, A monoclonal antibody to kidney endopepti-dase-24.11, *Biochem. J.* **214:**377–386.

Griffith, O. W., and Meister, A., 1979, Translocation of intracellular glutathione to membrane-bound γ-glutamyl transpeptidase as a discrete step in the γ-glutamyl cycle: Glutathionuria after inhibition of transpeptidase, *Proc. Natl. Acad. Sci. USA* **76:**268–272.

Hedeager-Sorensen, S., and Kenny, A. J., 1985, Proteins of the kidney microvillar membrane. Purification and properties of carboxypeptidase P from pig kidneys, *Biochem. J.* **229:** 251–257.

Hersh, L. B., and Morihara, K., 1986, Comparison of the subsite specificity of the mammalian neutral endopeptidase-24.11 (enkephalinase) to the bacterial neutral endopeptidase ther-molysin, *J. Biol. Chem.* **261:**6433–6437.

Heymann, E., and Mentlein, R., 1978, Liver dipeptidyl aminopeptidase IV hydrolyses substance P, *FEBS Lett.* **91:**360–364.

Hooper, N. M., 1991, Mini review. Angiotensin converting enzyme: Implications from molecular biology for its physiological functions, *Int. J. Biochem.* **23:**641–647.

Hooper, N. M., and Turner, A. J., 1987, Isolation of two differentially glycosylated forms of peptidyl dipeptidase A (angiotensin converting enzyme) from pig brain: A re-evaluation of their role in neuropeptide metabolism, *Biochem. J.* **241:**625–633.

Hooper, N. M., and Turner, A. J., 1988, Ectoenzymes of the kidney microvillar membrane. Aminopeptidase P is anchored by a glycosylphosphatidylinositol moiety, *FEBS Lett.* **229:** 340–344.

Hooper, N. M., and Turner, A. J., 1989, Ectoenzymes of the kidney microvillar membrane. Isolation and characterization of the amphipathic form of renal dipeptidase and hydrolysis of its glycosyl-phosphatidylinositol anchor by an activity in plasma, *Biochem. J.* **261:** 811–818.

Hooper, N. M., Keen, J., Pappin, D. J. C., and Turner, A. J., 1987a, Pig kidney angiotensin converting enzyme. Purification and characterization of amphipathic and hydrophilic forms of the enzyme establishes C-terminal anchorage to the plasma membrane, *Biochem. J.* **247:**85–93.

Hooper, N. M., Low, M. G., and Turner, A. J., 1987b, Renal dipeptidase is one of the membrane proteins released by phosphatidylinositol-specific phospholipase C, *Biochem. J.* **244:**465–469.

Hooper, N. M., Hryszko, J., and Turner, A. J., 1990a, Purification and characterization of pig kidney aminopeptidase P. A glycosylphosphatidylinositol anchored ectoenzyme, *Biochem. J.* **267:**509–515.

Hooper, N. M., Keen, J. N., and Turner, A. J., 1990b, Characterization of the glycosyl-phosphatidylinositol anchored human renal dipeptidase reveals that it is more extensively glycosylated than the pig enzyme, *Biochem. J.* **265:**429–433.

Johnson, A. R., Coalson, J. J., Ashton, J., Larumbide, M., and Erdos, E. G., 1985, Neutral endopeptidase in serum samples from patients with adult respiratory distress syndrome. Comparison with angiotensin-converting enzyme, *Am. Rev. Respir. Dis.* **132:**1262–1267.

Johnston, C. I., 1988, Angiotensin converting enzyme inhibitors—The balance sheet, *Med. J. Aust.* **148:**488–489.

Jongeneel, C. V., Bouvier, J., and Bairoch, A., 1989a, A unique signature identifies a family of zinc-dependent metallopeptidases, *FEBS Lett.* **242:**211–214.

Jongeneel, C. V., Quackenbush, E. J., Ronco, P., Verroust, P., Carrel, S., and Letarte, M., 1989b, Common acute lymphoblastic leukemia antigen expressed on leukemia and melanoma cell lines has neutral endopeptidase activity, *J. Clin. Invest.* **83:**713–717.

Kahan, F. M., Kropp, H., Sundelof, J. G., and Birnbaum, J., 1983, Thienamycin development of imipenem-cilastatin, *J. Antimicrob. Chemother.* **12**(Suppl. D):1–35.

Kenny, A. J., 1977, Proteinases associated with cell membranes, in: *Proteinases in Mammalian Cells and Tissues* (A. J. Barrett, ed.), Elsevier/North-Holland, Amsterdam, pp. 393–444.

Kenny, A. J., and Ingram, J., 1987, Proteins of the kidney microvillar membrane. Purification and properties of the phosphoramidon-insensitive endopeptidase ('endopeptidase-2') from rat kidney, *Biochem. J.* **245:**515–524.

Kenny, A. J., Booth, A. G., George, S. C., Ingram, J., Kershaw, D., Wood, E. J., and Young, A. T., 1976, Dipeptidyl peptidase IV, a kidney brush border serine peptidase. *Biochem. J.* **157:**169–182.

Kenny, A. J., O'Hare, M. J., and Gusterson, B. A., 1989, Cell-surface peptidases as modulators of growth and differentiation, *Lancet* **2:**785–787.

Kerr, M. A., and Kenny, A. J., 1974, The purification and specificity of a neutral endopeptidase from rabbit kidney brush border, *Biochem. J.* **137:**477–488.

Kozak, E. M., and Tate, S. S., 1982, Glutathione-degrading enzymes of microvillus membrane, *J. Biol. Chem.* **257:**6322–6327.

Kropp, H., Sundelof, J. S., Hajdu, R., and Kahan, F. M., 1982, Metabolism of thienamycin and related carbapenem antibiotics by the renal dipeptidase, dehydropeptidase-I, *Antimicrob. Agents Chemother.* **22:**62–70.

Laperche, Y., Bulle, F., Aissani, T., Chobert, N.-N., Aggerbeck, M., Hanoune, J., and Guellaen, G., 1986, Molecular cloning and nucleotide sequence of rat kidney γ-glutamyl transpeptidase cDNA, *Proc. Natl. Acad. Sci. USA* **83:**937–941.

Lattion, A.-L., Soubrier, F., Allegrini, J., Hubert, C., Corvol, P., and Alhenc-Gelas, F., 1989,

The testicular transcript of the angiotensin I-converting enzyme encodes for the ancestral, non-duplicated form of the enzyme, *FEBS Lett.* **252**:99–104.

Lecomte, J.-M., Baumer, P., Lim, C., Duchier, J., Cournot, A., Dussaule, J.-C., Ardaillou, R., Gros, C., Chaignon, B., Souque, A., and Schwartz, J.-C., 1990, Stereoselective protection of exogenous and endogenous atrial natriuretic factor by enkephalinase inhibitors in mice and humans. *Eur. J. Pharmacol.* **179**:65–73.

Letarte, M., Vera, S., Tran, R., Addis, J. B. L., Onizuka, R. J., Quackenbush, E. J., Jongeneel, C. V., and McInnes, R. R., 1988, Common acute lymphocytic leukemia antigen is identical to neutral endopeptidase, *J. Exp. Med.* **168**:1247–1253.

Littlewood, G. M., Iversen, L. L., and Turner, A. J., 1988, Neuropeptides and their peptidases: Functional considerations, *Neurochem. Int.* **12**:383–389.

Littlewood, G. M., Hooper, N. M., and Turner, A. J., 1989, Ectoenzymes of the kidney microvillar membrane. Affinity purification, characterization and localization of the phospholipase C-solubilized form of renal dipeptidase, *Biochem. J.* **257**:361–367.

Look, A. T., Ashmun, R. A., Shapiro, L. H., and Peiper, S. C., 1989, Human myeloid plasma membrane glycoprotein CD13 (gp150) is identical to aminopeptidase N, *J. Clin. Invest.* **83**:1299–1307.

Low, M. G., 1989, The glycosyl-phosphatidylinositol anchor of membrane proteins, *Biochim. Biophys. Acta* **988**:427–454.

McDonald, J. K., and Barrett, A. J., 1986, *Mammalian Proteases: A Glossary and Bibliography*, Volume 2, Academic Press, New York.

Malfroy, B., Schofield, P. R., Kuang, W.-J., Seeburg, P. H., Mason, A. J., and Henzel, W. J., 1987, Molecular cloning and amino acid sequence of rat enkephalinase, *Biochem. Biophys. Res. Commun.* **144**:59–66.

Malfroy, B., Kuang, W.-J., Seeburg, P. H., Mason, A. J, and Schofield, P. R., 1988, Molecular cloning and amino acid sequence of human enkephalinase (neutral endopeptidase), *FEBS Lett.* **229**:206–210.

Malfroy, B., Kado-Fong, H., Gros, C., Giros, B., Schwartz, J.-C., and Hellmiss, R., 1989, Molecular cloning and amino acid sequence of rat kidney aminopeptidase M: A member of a super family of zinc-metallohydrolases, *Biochem. Biophys. Res. Commun.* **161**:236–241.

Matsas, R., Fulcher, I. S., Kenny, A. J., and Turner, A. J., 1983, Substance P and [Leu]enkephalin are hydrolyzed by an enzyme in pig caudate synaptic membranes that is identical with the endopeptidase of kidney microvilli, *Proc. Natl. Acad. Sci. USA* **80**:3111–3115.

Matsas, R., Turner, A. J., and Kenny, A. J., 1984, Endopeptidase-24.11 and aminopeptidase activity in brain synaptic membranes are jointly responsible for the hydrolysis of cholecystokinin octapeptide (CCK-8), *FEBS Lett.* **175**:124–128.

Matsas, R., Stephenson, S. L., Hryszko, J., Kenny, A. J., and Turner, A. J., 1985, The metabolism of neuropeptides: Phase separation of synaptic membrane preparations with Triton X-114 reveals the presence of aminopeptidase N, *Biochem. J.* **231**:445–449.

Mauborgne, A., Bourgoin, S., Bendiel, J. J., Hirsch, M., Berthier, J. L., Haman, M., and Cesselin, F., 1987, Enkephalinase is involved in the degradation of endogenous substance P released from slices of rat substantia nigra, *J. Pharmacol. Exp. Ther.* **243**:674–680.

Meister, A., and Anderson, M. E., 1983, Glutathione, *Annu. Rev. Biochem.* **52**:711–760.

Mendelsohn, F. A. O., 1984, Localization of angiotensin converting enzyme in rat forebrain and other tissues by *in vitro* autoradiography using [^{125}I]-labelled MK351A, *Clin. Exp. Pharmacol. Physiol.* **11**:431–436.

Nau, R, Schafer, G., Deacon, C. F., Cole, T., Agoston, D. V., and Conlon, J. M., 1986, Proteolytic inactivation of substance P and neurokinin A in the longitudinal muscle layer of guinea pig small intestine, *J. Neurochem.* **47**:856–864.

Ng, K. K. F., and Vane, J. R., 1968, Fate of angiotensin I in the circulation, *Nature* **218**: 144–150.

O'Connor, B., and O'Cuinn, G., 1984, Localization of a narrow specificity thyroliberin hydrolyzing pyroglutamate aminopeptidase in synaptosomal membranes of guinea-pig brain. *Eur. J. Biochem.* **144**:271–278.

Ogata, S., Misumi, Y., and Ikehara, Y., 1989, Primary structure of rat liver dipeptidyl peptidase IV deduced from its cDNA and identification of the NH_2-terminal signal sequence as the membrane-anchoring domain, *J. Biol. Chem.* **264**:3596–3601.

Olsen, J., Cowell, G. M., Konigshofer, E., Danielsen, M., Moller, J., Laustsen, L., Hansen, O. C., Welinder, K. G., Engberg, J., Hunziker, W., Spiess, M., Sjostrom, H., and Noren, O., 1988, Complete amino acid sequence of human intestinal aminopeptidase N as deduced from cloned cDNA, *FEBS Lett.* **238**:307–314.

Olsen, J., Sjostrom, H., and Noren, O., 1989, Cloning of the pig aminopeptidase N gene. Identification of possible regulatory elements and the exon distribution in relation to the membrane-spanning region, *FEBS Lett.* **251**:275–281.

Orawski, A. T., Susz, J. P., and Simmons, W. H., 1987, Aminopeptidase P from bovine lung: Solubilization, properties, and potential role in bradykinin degradation, *Mol. Cell. Biochem.* **75**:123–132.

Papandrilkopoulou, A., Frey, A., and Gassen, H. G., 1989, Cloning and expression of γ-glutamyl transpeptidase from isolated porcine brain capillaries, *Eur. J. Biochem.* **183**: 693–698.

Pawlak, A., Cohen, E. H., Octave, J.-N., Schweickhardt, R., Wu, S.-J., Bulle, F., Chikki, N., Baik, J.-H., Siegrist, S., and Guellaen, G., 1990, An alternatively processed mRNA specific for γ-glutamyl transpeptidase in human tissues, *J. Biol. Chem.* **265**:3256–3262.

Pierotti, A., Dong, K.-W., Glucksman, M. J., Orlowski, M., and Roberts, J. L., 1990, Molecular cloning and primary structure of rat testes metallopeptidase EC 3.4.24.15, *Biochemistry* **29**:10323–10329.

Rached, E., Hooper, N. M., James, P., Semenza, G., Turner, A. J., and Mantei, N., 1990, cDNA cloning and expression in *Xenopus laevis* oocytes of pig renal dipeptidase, a glycosyl-phosphatidylinositol-anchored ectoenzyme, *Biochem. J.* **271**:755–760.

Rajpert-De Meyts, E., Heisterkamp, N., and Groffen, J., 1988, Cloning and nucleotide sequence of human γ-glutamyl transpeptidase, *Proc. Natl. Acad. Sci. USA* **85**:8840–8844.

Relton, J. M., Gee, N. S., Matsas, R., Turner, A. J., and Kenny, A. J., 1983, Purification of endopeptidase-24.11 ('enkephalinase') from pig brain by immunoadsorbent chromatography, *Biochem. J.* **215**:519–523.

Rich, D. H., Moon, B. J., and Harbeson, S., 1984, Inhibition of aminopeptidases by amastatin and bestatin derivatives. Effect of inhibitor structure on slow-binding processes, *J. Med. Chem.* **27**:417–422.

Robinson, M. B., Blakely, R. D., Couts, R., and Coyle, J. T., 1987, Hydrolysis of the brain dipeptide N-acetyl-L-aspartyl-L-glutamate. Identification and characterization of a novel N-acetylated α-linked acidic dipeptidase activity from rat brain, *J. Biol. Chem.* **262**: 14498–14506.

Roques, B. P., Lucas-Soroca, E., Chaillet, P., Costentin, J., and Fournie-Zaluski, M. C., 1983,

Complete differentiation between enkephalinase and angiotensin-converting enzyme inhibition by retrothiorphan, *Proc. Natl. Acad. Sci. USA* **80**:3178–3182.

Sakurada, C., Yokosawa, H., and Ishii, S.-I., 1990, The degradation of somatostatin by synaptic membrane of rat hippocampus is initiated by endopeptidase-24.11, *Peptides* **11**: 287–292.

Schulman, J. D., Goodman, S. J., Mace, J. W., Patrick, A. D., Tietze, F., and Butler, E. J., 1976, L-Glutathionuria: Inborn error of metabolism due to deficiency of γ-glutamyl transpeptidase, *Biochem. Biophys. Res. Commun.* **65**:68–74.

Schulz, W. W., Hagler, H. K., Buja, L. M., and Erdos, E. G., 1988, Ultrastructural localization of angiotensin I-converting enzyme and neutral metalloendopeptidase in the proximal tubule of the human kidney, *Lab. Invest.* **59**:789–797.

Schwartz, J.-C., Gros, C., Lecomte, J.-M., and Bralet, J., 1990, Enkephalinase (EC 3.4.24.11) inhibitors: Protection of endogenous atrial natriuretic factor against inactivation and potential therapeutic applications, *Life Sci.* **47**:1279–1297.

Serval, V., Barbeito, L., Pittaluga, A., Cheramy, A., Lavielle, S., and Glowinski, J., 1990, Competitive inhibition of N-acetylated α-linked acidic dipeptidase activity by N-acetyl-L-aspartyl-β-linked L-glutamate, *J. Neurochem.* **55**:39–46.

Shipp, M. A., Vijayaraghavan, J., Schmidt, E. V., Masteller, E. L., d'Adamio, L., Hersh, L. B., and Reinherz, E. L., 1989, Common acute lymphoblastic leukemia antigen (CALLA) is active neutral endopeptidase 24.11 ("enkephalinase"): Direct evidence by cDNA transfection analysis, *Proc. Natl. Acad. Sci. USA* **86**:297–301.

Skidgel, R. A., and Erdos, E. G., 1985, Novel activity of human angiotensin I converting enzyme: Release of the NH_2- and COOH-terminal tripeptides from the luteinizing hormone-releasing hormone, *Proc. Natl. Acad. Sci. USA* **82**:1025–1029.

Skidgel, R. A., Davis, R. M., and Tan, F., 1989, Human carboxypeptidase M. Purification and characterization of a membrane-bound carboxypeptidase that cleaves peptide hormones, *J. Biol. Chem.* **264**:2236–2241.

Slusher, B. S., Robinson, M. B., Tsai, G., Simmons, M. L., Richards, S. S., and Coyle, J. T., 1990, Rat brain N-acetylated α-linked acidic dipeptidase activity. Purification and immunological characterization, *J. Biol. Chem.* **265**:21297–21301.

Sokolovsky, M., Galron, R., Kloog, Y., Bdolah, A., Indig, F. E., Blumberg, S., and Fleminger, G., 1990, Endothelins are more sensitive than safratoxins to neutral endopeptidase: Possible physiological significance, *Proc. Natl. Acad. Sci. USA* **87**:4702–4706.

Soubrier, F., Alhenc-Gelas, F., Hubert, C., Allegrini, J., John, M., Tregear, G., and Corvol, P., 1988, Two putative active centres in human angiotensin I-converting enzyme revealed by molecular cloning, *Proc. Natl. Acad. Sci. USA* **85**:9386–9390.

Stephenson, S. L., and Kenny, A. J., 1987a, Metabolism of neuropeptides. Hydrolysis of the angiotensins, bradykinin, substance P and oxytocin by pig kidney microvillar membranes, *Biochem. J.* **241**:237–247.

Stephenson, S. L., and Kenny, A. J., 1987b, The hydrolysis of α-human atrial natriuretic peptide by pig kidney microvillar membranes is initiated by endopeptidase-24.11, *Biochem. J.* **243**:183–187.

Sterchi, E. E., Naim, H. Y., Lentze, M. J., Hauri, H.-P., and Fransen, J. A. M., 1988, N-benzoyl-L-tryosyl-p-aminobenzoic acid hydrolase: A metalloendopeptidase of the human intestinal microvillus membrane which degrades biologically active peptides, *Arch. Biochem. Biophys.* **265**:105–118.

Stewart, T. A., Weare, J. A., and Erdos, E. G., 1981, Purification and characterization of human converting enzyme (kininase II), *Peptidases* **2**:145–152.

Strittmatter, S. M., Lo, M. M. S., Javitch, J. A., and Snyder, S. H., 1984, Autoradiographic localization of angiotensin-converting enzyme in rat brain with [^{3}H]captopril. Localization to a striatonigral pathway, *Proc. Natl. Acad. Sci. USA* **81**:1599–1603.

Takeuchi, T., 1985, Production of actinonin, an inhibitor of aminopeptidase M_1, by actinomycetes, *J. Antibiot.* **38**:1629–1630.

Tan, F., Chan, S.-J., Steiner, D. F., Schilling, J. W., and Skidgel, R. A., 1989, Molecular cloning and sequencing of the cDNA for human membrane-bound carboxypeptidase M. Comparison with carboxypeptidases A, B, H and N, *J. Biol. Chem.* **264**:13165–13170.

Tate, S. S., and Meister, A., 1974, Interaction of γ-glutamyl transpeptidase with amino acids, dipeptides and derivatives and analogs of glutathione, *J. Biol. Chem.* **249**:7593–7602.

Turner, A. J., and Hooper, N. M., 1990, The membrane anchors of microvillar hydrolases, in: *Molecular and Cell Biology of Membrane Proteins: Glycolipid Anchors of Cell-Surface Proteins* (A. J. Turner, ed.), Ellis Horwood, Chichester, pp. 129–150.

Turner, A. J., Hooper, N. M., and Kenny, A. J., 1987, Metabolism of neuropeptides, in: *Mammalian Ectoenzymes* (A. J. Turner and A. J. Kenny, eds.), Elsevier, Amsterdam, pp. 211–248.

Umezawa, H., Aoyagi, T., Ogawa, K., Naganawa, H., Hamada, M., and Takeuchi, T., 1984, Diprotins A and B, inhibitors of dipeptidyl aminopeptidase IV, produced by bacteria, *J. Antibiot.* **37**:422–425.

Unger, T., Gohlke, P., and Gruber, M.-G., 1990, Converting enzyme inhibitors, in: *Pharmacology of Anti-hypertensive Therapeutics* (D. Ganten and P. J. Mulrow, eds.), Springer-Verlag, Berlin, pp. 379–481.

Valloton, M. B., 1987, The renin-angiotensin system, *Trends Pharmacol. Sci.* **8**:69–74.

Vanneste, Y., Michel, A., Dimaline, R., Najdovski, T., and Deschodt-Lanckman, M., 1988, Hydrolysis of α-human atrial natriuretic peptide *in vitro* by human kidney membranes and purified endopeptidase-24.11. Evidence for a novel cleavage site, *Biochem. J.* **254**:531–537.

Vanneste, Y., Pauwels, S., Lambotte, L., and Deschodt-Lanckman, M., 1990, *In vivo* metabolism of brain natriuretic peptide in the rat involves endopeptidase-24.11 and angiotensin converting enzyme, *Biochem. Biophys. Res. Commun.* **173**:265–271.

Vijayaraghavan, J., Scicli, A. G., Carretero, O. A., Slaughter, C., Moomaw, C., and Hersh, L. B., 1990, The hydrolysis of endothelins by neutral endopeptidase-24.11 (enkephalinase), *J. Biol. Chem.* **265**:14150–14155.

Wilk, S., 1986, Neuropeptide-specific peptidases: Does brain contain a specific TRH-degrading enzyme? *Life Sci.* **39**:1487–1492.

Williams, T. A., Hooper, N. M., and Turner, A. J., 1991, Characterization of neuronal and endothelial forms of angiotensin converting enzyme in pig brain, *J. Neurochem.* **57**:193–199.

Wu, Q., Lahti, J. M., Air, G. M., Burrows, P. D., and Cooper, M. D., 1990, Molecular cloning of the murine BP-1/6C3 antigen: A member of the zinc-dependent metallopeptidase family, *Proc. Natl. Acad. Sci. USA* **87**:993–997.

Wu, Q., Li, L., Cooper, M. D., Pierres, M., and Gorvel, J. P., 1991, Aminopeptidase A activity of the murine B-lymphocyte differentiation antigen BP-1/6C3, *Proc. Natl. Acad. Sci. USA* **88**:676–680.

Chapter 3

Endosomal and Lysosomal Hydrolases

Sandra A. Brockman and Robert F. Murphy

1. INTRODUCTION

Intracellular hydrolases degrade all types of biological polymers. Acting in concert, they are capable of degrading protein, polysaccharide, lipid, DNA, and RNA. In addition, sulfate and phosphate groups can be specifically removed from these polymers. Cytochemistry, immunolocalization, subcellular fractionation, and analysis of the kinetics of hydrolysis of endocytosed substrates have all been used to localize and identify the intracellular compartments in which these enzymes are contained (for reviews see Kornfeld, 1987; Glaumann and Ballard, 1987; Storrie, 1988; Holtzman, 1989; Kornfeld and Mellman, 1989). Most acid hydrolases are targeted to lysosomes by the addition of mannose 6-phosphate (M6P), which is recognized by mannose 6-phosphate receptors (MPR) that are thought to cycle between the Golgi apparatus and endosomes or lysosomes (for reviews see von Figura and Hasilik, 1986; Kornfeld and Mellman, 1989). The pH optima of many of these enzymes are less than 5 (for reviews see Barrett and McDonald, 1980; McDonald and Barrett, 1986) consistent with their acting in lysosomes, whose pH is in this vicinity (Ohkuma and Poole, 1978). However, some enzymes have a higher pH optimum, and activity has been detected in early endosomal compartments (Storrie *et al.*, 1984; Diment and Stahl, 1985; Roederer *et al.*, 1987) that have a pH near 6 (Murphy *et al.*, 1984; Roederer and Murphy, 1986; Kielian *et al.*, 1986; Sipe and Murphy, 1987). Therefore, it appears that some of these enzymes may be present (perhaps in varying amounts) in many, if not all, compartments of the endocytic pathway, and that pH

Sandra A. Brockman and Robert F. Murphy • Department of Biological Sciences, Carnegie Mellon University, Pittsburgh, Pennsylvania 15213.

Biological Barriers to Protein Delivery, edited by Kenneth L. Audus and Thomas J. Raub. Plenum Press, New York, 1993.

51

may play an important role in regulating their activity (Murphy, 1988; Yamashiro and Maxfield, 1988). In this chapter we discuss the endocytic pathway in various cell types, the regulation of pH within its compartments, and the enzymatic components of the pathway. In addition, we discuss the processing of known physiological ligands and implications for drug targeting.

1.1. Endosome

Endosomes are normally defined as acidic, prelysosomal compartments containing endocytosed material. Support for the concept of an early acidic compartment came from studies of virus and toxin entry into mammalian cells and direct measurements of pH using fluorescence methods (for review see Mellman et al., 1986). Two temporally related types of endosomes have been detected using cell fractionation methods (Galloway et al., 1983; Murphy, 1985; Schmid et al., 1988). The first compartment, the "early endosome," is accessible to endocytosed ligands within 2 min, has a lower buoyant density than later compartments, and is apparently the mildly acidic (pH 6) compartment characterized by measurements of ligand acidification (see Section 2). By electron microscopy, this compartment does not appear to contain significant amounts of MPR or lysosomal membrane glycoproteins (LGPs; see below). It does, however, contain cathepsin D (Diment and Stahl, 1985; Geuze et al., 1985), cathepsin B (Roederer et al., 1987; Harding et al., 1990), and acid phosphatase (Storrie et al., 1984; Braun et al., 1989; Bowser and Murphy, 1990) activities.

Based on in vitro assays involving horseradish peroxidase-induced density shifts (Ajioka and Kaplan,1987) or avidin–biotin interaction between different endocytic ligands (Braell, 1987), it appears that virtually all endocytosed molecules (fluid-phase and receptor-mediated) pass through the same early endocytic compartment where recycled molecules are separated from those to be retained and/or degraded.

A second class of endosomal compartments, variously referred to as late endosomes, dense endosomes, multivesicular bodies, light lysosomes, or prelysosomes, has been identified by a variety of criteria. These include higher buoyant density, more anodal deflection in free-flow electrophoresis, larger amounts of hydrolases, and decreased accessibility to endocytosed material at 20°C relative to early endosomes. These compartments contain a significant amount of MPR and a small amount of LGPs. [LGPs are highly charged membrane proteins that are mainly found in lysosomes and may play a role in protecting the lysosomal membrane from degradation; see Kornfeld and Mellman (1989)]. Most pre-1985 models of lysosome biogenesis postulated that intravesicular hydrolases were delivered directly to lysosomes after being sorted in the trans-Golgi network. However, Griffiths et al. (1988) showed that the receptor for many of these hydrolases, MPR, is found primarily in an acidic prelysosomal compartment. This, along with the observation of hydrolase activity in endosomes (discussed above), raised the possibility that M6P-containing hydrolases are delivered earlier in the endosome-to-lysosome pathway than had previously been thought. Since significant amounts of MPR were not found in morphologically mature lysosomes, Griffiths *et al.* have proposed that this late endosome is the site of MPR/ligand entry into the pathway. They proposed that MPR/ligand-containing vesicles are delivered to late endosomes, that ligand dissociates from the receptor at their acidic pH and continues to lysosomes along with soluble contents, and that MPR is excluded from the pathway to lysosomes so that it can be reused for another round of hydrolase transport from the Golgi. The precise location (or locations) at which newly synthesized and preexisting hydrolytic enzymes are delivered to the endocytic pathway remains unclear (see Storrie, 1988, for review). However, it is likely that even material in the process of being recycled to the plasma membrane (e.g., receptors, transferrin) may be exposed to active hydrolases.

1.2. Lysosome

The concept of a lysosome has evolved substantially from the initial demonstration of a membrane-enclosed set of hydrolases. Most current definitions emphasize an active role in degradation of endocytosed material, a relative absence of ligands that are normally recycled, a dense appearance in light and electron microscopy, and high concentrations of LGPs. Density centrifugation has been used to show that this compartment has the highest density of all of the endocytic compartments (see Storrie, 1988, for review).

2. ACIDIFICATION OF ENDOSOMES AND LYSOSOMES

2.1. pH Measurements

Both endosomes and lysosomes have been shown to have an acidic internal pH. The major function of endosomes, receptor–ligand segregation, requires an acidic pH to dissociate ligands and receptors, while the major function of lysosomes, macromolecular degradation, requires acidic pH to maximally activate hydrolases. While studies of endocytosed viruses and toxins were crucial in demonstrating the acidic nature of early endocytic compartments, fluorescence methods have played a major role in determining the pH of these compartments. These methods are based on the pH dependence of fluorescein fluorescence—the fluorescence emission of fluorescein conjugates is quenched at acidic pH. This property was used to show that mature lysosomes have a pH below 5 (Ohkuma and Poole, 1978), and to demonstrate that endocytosed material is acidified rapidly after endocytosis (Murphy *et al.*, 1982;

Tycko and Maxfield, 1982), presumably in endosomes. Measurements of the kinetics of acidification of fluorescent conjugates of a number of probes revealed that the pH of early endosomes is significantly higher (approximately pH 6) than that of late endosomes and lysosomes (Murphy et al., 1984; Roederer and Murphy, 1986; Roederer et al., 1987; Sipe and Murphy, 1987; Yamashiro and Maxfield, 1987). These results are supported by measurements of the kinetics of infection of a mutant Semliki Forest virus that requires a significantly lower pH for infection than wild-type virus (Kielian et al., 1986). In most cell types studied, all internalized molecules are rapidly acidified to pH 6 in early endosomes. At this point, molecules that are recycled (such as transferrin) are alkalinized to neutral pH after segregation for transit back to the plasma membrane. In contrast, molecules that are not recycled (such as epidermal growth factor) are further acidified to near pH 5. This biphasic acidification pattern has been observed in mouse 3T3 fibroblasts, Chinese hamster ovary cells, and the A549 human epidermoid cell line. However, at least one cell type shows altered acidification. In K562, a human erythroleukemia cell line, Tf is initially acidified with the same kinetics as described for recycled molecules above. However, rather than being realkalinized upon recycling, Tf is further acidified to pH 5.4, much like molecules that are to be degraded (van Renswoude et al., 1982; Sipe et al., 1991). The overall half-times for uptake and release of Tf remain the same as in the other cell types (5–15 min cycle time). A similar pattern of acidification has been observed in mouse Friend erythroleukemia cells (D. M. Sipe, R. F. Murphy, and P. Kulakosky, unpublished observations) and in chicken HD3 erythroblasts (Killisch et al., 1992). The function of this further acidification remains unclear. It may be an adaptation to ensure efficient extraction of iron from Tf for hematopoiesis or to facilitate transport of iron into the cytoplasm. Alternatively, it may be a consequence of the transformed nature of erythroleukemia cell lines.

2.2. pH Regulation

In order to explain the difference between endosomal and lysosomal acidification observed both in living cells and in subcellular fractions, Fuchs et al. (1989) proposed that the Na^+,K^+-ATPase might act to limit proton pumping in endosomes. The Na^+,K^+-ATPase is an electrogenic ion pump, i.e., its action leads to a change in charge distribution across the membrane containing it. If found in endosomes (either transiently or stably) in its normal orientation, it would be predicted to generate an interior-positive membrane potential (pumping three Na^+ into the endosome in exchange for two K^+ pumped out). This membrane potential was proposed to inhibit the proton-translocating ATPase, and ultimately to limit the pH of any vesicle containing both the Na^+,K^+-ATPase and the H^+-ATPase. Treatment of isolated endosomes (Fuchs et al., 1989) or living cells (Cain et al., 1989) with inhibitors of the Na^+,K^+-ATPase causes the pH within endosomes to decrease to near that of lysosomes. These treatments do not affect the pH of lysosomes. Similarly, treatment of K562 cells with ouabain does not decrease their already low endosomal pH (Sipe *et al.*, 1991). These results suggest that the Na^+,K^+-ATPase is either not present or not involved in pH regulation in lysosomes (and endosomes of K562 cells). The inhibitory effect of membrane potential on intravesicular acidification may be relieved by the presence of channels for anions such as chloride (Van Dyke, 1988; Bae and Verkman, 1990).

3. TRANSIENT AND STABLE COMPARTMENT MODELS FOR ENDOSOMES AND LYSOSOMES

One of the aspects of endocytic trafficking that is the subject of much current interest is the means by which endocytosed materials appear in compartments of different characteristics. By analogy with models debated for the Golgi apparatus some years earlier, Helenius *et al.* (1983) distinguished two classes of models for endocytic processing. The first, the vesicle shuttle model, postulates that endosomes and lysosomes are stable compartments and that communication between them occurs through transport vesicles. In the second, the maturation model, endosomes are proposed to be transient and individual endosomes, or portions thereof, are proposed to undergo a remodeling process that converts them into lysosomes. The arguments for and against each of these models have been recently summarized (Griffiths and Gruenberg, 1991; Murphy, 1991).

4. ENZYME ACTIVITIES OF ENDOSOMES AND LYSOSOMES

The major classes of enzyme activities found in lysosomes are presented below, along with information on the properties and localization of specific enzymes. The major mammalian proteases were catalogued in 1980 (endoproteases) and 1986 (exoproteases) (Barrett and McDonald, 1980; McDonald and Barrett, 1986). Additional reviews of lysosomal hydrolases may be found in Glaumann and Ballard (1987), Storrie (1988), and Holtzman (1989).

4.1. Distribution of Hydrolases between Endosomes and Lysosomes

Self-forming Percoll density gradients have been used extensively to distinguish endosomes from lysosomes (Merion and Poretz, 1981; Merion and Sly, 1983). Most hydrolases are found in two peaks when such gradients (typically 27% Percoll) are used to fractionate postnuclear supernatants. The lighter peak is also labeled with Tf and other markers in a short pulse. While the denser peak is not accessible to Tf, it is accessible to nonrecycled molecules when labeled for longer periods of time. However, when the lighter peak is collected and rerun on a 17% Percoll gradient, two peaks of endocytosed markers are again seen, presumably corresponding to early and late endosomes.

The distribution of individual hydrolases between the low- and high-density peaks varies for different enzymes. The distribution of individual enzymes has also been observed to vary between cells in active and quiescent states (Chu and Olden, 1984; Roederer *et al.*, 1989). Since the endosomal region of Percoll gradients also contains elements of the endoplasmic reticulum, Golgi apparatus, and plasma membrane, hydrolase activity observed in this region has frequently been assigned to these compartments rather than endosomes. However, a number of studies have demonstrated that endosomes contain a repertoire of hydrolases. Whether the presence of specific hydrolases has been demonstrated in early endocytic compartments is noted below. It has been proposed that the activity of hydrolases present in early endosomes may be limited by the higher endosomal pH and/or by interaction with inhibitory proteins (Murphy, 1988).

4.2. Proteases

The cysteine proteases, cathepsins B, H, and L, together with the aspartate protease, cathepsin D, are thought to be the most active intravesicular proteases (Shaw and Dean, 1980). Not only do these proteases have acidic pH optima, but cathepsins B, H, and L are irreversibly inactivated at mildly alkaline pH (Barrett, 1973). The cysteine proteases also require a free sulfhydryl group for optimal activity. *In vitro*, this is supplied as either cysteine, dithiothreitol, or β-mercaptoethanol. *In vivo*, cysteine may serve this function.

The specificity of these enzymes varies depending on pH and substrate size (Barrett and Kirschke, 1981). Cathepsin B cleaves after arginine residues in short peptide substrates, yet the bond between arginine-23 and glycine-24 of oxidized insulin B chain is not cleaved. In addition, cathepsin B possesses a rather nonspecific carboxy-terminal dipeptidyl peptidase activity. Cathepsin H has endopeptidase activity similar to that of cathepsin B on short peptide substrates, and also has significant aminopeptidase activity. Cathepsin L is an endopeptidase that cleaves after hydrophobic dipeptides. Finally, cathepsin D is an endopeptidase that cleaves after aromatic- and long-side-chain-containing amino acids. It is more active against larger peptides (Keilova, 1971).

Cathepsin D activity has been demonstrated to be present in macrophage endosomes (Diment and Stahl, 1985) and cathepsin B activity (which could also have been due to the action of cathepsin H or L) has been shown to be present in early compartments of a variety of cell types (Roederer *et al.*, 1987; Bowser and Murphy, 1990). Precursor forms of these enzymes may also be present in endosomes. Cathepsin B activity has also been detected in plasma membrane fractions from both normal and transformed bovine lymphoid cells (Sloane *et al.*, 1986). In this study, increasing amounts of cathepsin B and N-acetyl-β-glucosaminidase activities were found in a plasma membrane fraction in cell lines with increasing metastatic potential. In addition to being found in endosome/plasma membrane fractions, both precursor and mature lysosomal enzymes are released in small amounts by a number of normal cell lines (Lemansky *et al.*, 1985; Hanewinkel *et al.*, 1987). In addition, Moloney murine leukemia virus-transformed BALB/c 3T3 fibroblasts show increased secretion of cathepsin B precursor (Achkar *et al.*, 1990).

The remaining proteases are classified as exopeptidases. Before discussing them, the convention regarding name and function will be discussed (McDonald and Barrett, 1986). Aminopeptidases remove one amino acid at a time from the amino (N) terminus of a protein. Dipeptidyl peptidases remove dipeptides from the amino terminus of a protein. Tripeptidyl peptidases remove tripeptides from the amino terminus. On the other end, carboxypeptidases remove one amino acid from the carboxy (C) terminus, and peptidyl dipeptidases remove dipeptides from the carboxy terminus. Furthermore, dipeptidases hydrolyze dipeptides into single amino acids, and tripeptidases remove one amino acid from either end (not specified) of tripeptides.

Dipeptidyl peptidase I (DPPI) was originally named cathepsin C. It is found at high levels in bovine spleen, and in the lysosomes of peripheral blood lymphocytes (McDonald and Barrett, 1986; Thiele and Lipsky, 1990). More specifically, natural killer cells detected by the presence of the surface molecule, CD16, showed a 10-fold higher level of activity than CD4-positive T-helper cells, and a 20-fold higher level of DPPI activity than CD19-positive B cells (Thiele and Lipsky, 1990). This indicates that DPPI may serve a specific role in the cytotoxic immune response. The activity of this enzyme is greatly affected by pH. N-terminal dipeptides are removed at acidic pH, while at neutral to alkaline pH, DPPI polymerizes dipeptide esters or amides (Thiele and Lipsky, 1990). Since amino acids and dipeptide esters can accumulate in lysosomes in a similar manner to other substituted amines and might raise lysosomal pH if present at sufficient concentrations, it is possible that DPPI can be induced to carry out transpeptidation even in the lysosome. This enzyme appears to be responsible for the cytotoxic effect of Leu-Leu-OMe on cytotoxic lymphocytes due to the polymerization of Leu_{4+} chains, which causes cell lysis (Thiele and Lipsky, 1990).

In contrast to DPPI, which displays broad specificity and removes dipeptides from the N-termini of proteins of various lengths, dipeptidyl peptidase II is only active against tripeptides, cleaving prolyl bonds in the Gly-Pro-X sequence (McDonald *et al.*, 1968; Fukasawa *et al.*, 1983). This sequence is present at high frequency in collagen.

Tripeptidyl peptidase removes tripeptides from the N-terminus of bovine growth hormone. This enzyme has an acidic pH optimum (between 4.3 and 5.0 depending on the arylamide derivative used for assaying its activity), and is capable of depolymerizing poly(Gly-Pro-Ala-) at acidic pH. In conjunction with dipeptidyl peptidase II, tripeptidyl peptidase reduces poly(Gly-Pro-Ala) to Gly-Pro and free Ala, indicating that together, these enzymes are capable of degrading collagen (McDonald *et al.*, 1985).

The "protective protein" is an enzyme with carboxypeptidase activity (Tranchemontagne *et al.*, 1990) (optimal at pH 5.5) that was originally characterized because of its absence in galactosialidosis. This protein copurifies with β-galactosidase and neuraminidase activity. The deduced amino acid sequence from the cloned protective protein gene has homology to carboxypeptidase Y and the KEX1 gene product from yeast. At elevated pH, protective protein also has esterase and deamidase activities, which are also reduced in cells from galactosialidosis patients (Kase *et al.*, 1990). In light of the recent purification of an enzyme with esterase, peptidase, and deamidase activities from platelets, and the fact that the N-terminal 25 residues of the platelet-derived enzyme are identical to the deduced sequence of protective protein, it seems likely that protective protein is responsible for all three activities. Overexpression of protective protein increases cathepsin A-like activity, and cells from galactosialidosis patients have reduced cathepsin A activity. In addition, affinity-purified antibodies raised against purified recombinant protective protein remove cathepsin A activity from normal human fibroblast extracts. Protective protein serves both as a hydrolase and as a stabilizer for other hydrolases. Interestingly, inactivation of the enzymatic activity by site-directed mutagenesis does not impair the protective function. *In vitro*, protective protein has deamidase and carboxypeptidase activity against substance P, bradykinin, angiotensin I, and oxytocin (Jackman *et al.*, 1990). Expression of protective protein mRNA is high in mouse kidney, brain, and placenta (Galjart *et al.*, 1990).

Other lysomal proteases include carboxypeptidases A and B, prolyl carboxypeptidase, tyrosine carboxypeptidase, and dipeptidases I and II.

4.3. Disulfide Bond Reduction

It is currently unclear whether a reduction of intrapeptide disulfide bonds occurs in endosomes and/or lysosomes. A large body of evidence indicates that disulfide reduction enhances protein degradation in lysosomes, yet no reductase has been found. Mego (1984) showed that degradation of serum albumin in murine kidney and liver lysosomes (tritosomes) is stimulated by disulfide-reducing compounds like urea and cysteine. This stimulation was eliminated if the serum albumin disulfide bonds were reduced and alkylated. The pH optimum for degradation of serum albumin by cathepsin D is 3–4.4 but shifts up to 5.0 if the protein is denatured. Highly purified cathepsins B, D, H,and L were used to show that the presence of thiol compounds stimulates degradation of insulin, serum albumin, and denatured serum albumin by these enzymes (Kooistra *et al.*, 1982). Preincubation of substrates with reducing agent did not affect the rate of degradation and the disulfide bonds within the substrates were still intact. Therefore, it appears that proteins are partially degraded, then disulfide bonds are reduced, in a combination of proteolysis and reduction. Although thiols enhanced the intralysosomal degradation in isolated lysosomes, studies of cystinotic patients indicate that disulfide bonds may not be reduced in all lysosomes (Schulman et al., 1969; Thoene et al., 1977). In these patients, cystine accumulates in fibroblast or leukocyte lysosomes. The defect appears to be an impairment of cystine transport from lysosomes, suggesting that disulfide bonds or cystine products of proteolysis are not reduced in these lysosomes. Since the study of cystinosis revealed that cystine leaves normal lysosomes, without prior reduction to cysteine (Jonas et al., 1982; Steinherz et al., 1982; Gahl et al., 1982), Lloyd (1986) proposed that cysteine itself acts as the hydrogen donor in the reduction of intrachain disulfide bonds. Because cysteine freely crosses lipid bilayers, including the lysosomal membrane, it is possible that once inside the lysosome, cysteine may substitute for one partner in an intrapeptide disulfide bond. A second molecule of cysteine may then react with the other polypeptide cysteine, making re-formation of the original disulfide bond unfavored. The protein would then be more susceptible to unfolding and degradation. The result of full degradation would be two molecules of cystine per original disulfide bond; this cystine would be transported across the lysosomal membrane and into the cytoplasm (where it would presumably be reduced). This hypothesis provides for a simple mechanism of disulfide bond reduction without requiring a lysosomal reducing enzyme. Cysteine-specific uptake into lysosomes has since been reported (Pisoni et al., 1990).

In a more recent paper, Feener et al., (1990) presented evidence that disulfide bonds are cleaved in compartments other than endosomes or lysosomes. They proposed the Gogli apparatus as the site of disulfide bond reduction. Reduction of a [^{125}I]tyramine–poly-D-lysine conjugate, internalized by nonspecific adsorptive endocytosis, was monitored. Reduction of the probe was observed to begin much earlier after internalization than degradation of poly-L-lysine, suggesting that the site of reduction was earlier in the pathway than lysosomes. Little reduction of the probe in lysosomes was observed, at least at the times tested.

The mechanism of disulfide bond reduction in the endosomal pathway is especially important in light of the number of disulfide-containing toxins that enter the cell via endocytosis. Previous reports indicated that the disulfide bond in ricin must be reduced before the toxin is capable of inhibiting protein synthesis. Lewis and Youle (1986) demonstrated that the disulfide bond is only required to hold the two subunits together at low concentrations of ricin; if the complex is reduced before addition to cells, it is no longer toxic. They proposed that the disulfide bond is reduced in the cytoplasm.

Diphtheria toxin enters cells via receptor-mediated endocytosis and enters the cytoplasm after exposure to low pH in endosomes (Sandvig and Olsnes, 1980; Draper and Simon, 1980). The toxin undergoes a conformational change when exposed to the lower pH in endosomes, revealing hydrophobic domains. These domains may insert themselves into the lipid bilayer, facilitating the toxin's entry into the cytoplasm (see Chapter 4). The A fragment of diphtheria toxin is not fully active until the interfragment disulfide bond is reduced (Pappenheimer, 1977). Moskaug *et al.* (1987) examined the effect of low pH on entry of surface-bound nicked toxin into monensin-treated cells. It was found that exposure of surface-bound nicked toxin to pH less than 5.5 caused a reduction in protein synthesis, indicating that reduction of toxin could occur even when the toxin was artificially introduced into the cytoplasm. While this result does not prove that reduction cannot occur in endosomes or lysosomes, Moskaug *et al.* believe that the data in their paper and others indicate that disulfide bond reduction occurs in the cytoplasm.

4.4. Glycosidases

As discussed above, significant fractions of most protease activities are found in the endosomal region of density gradients. However, much lower fractions of glycosidase activities are found in this region (e.g., Diment and Stahl, 1985). In addition to implying differential delivery of these two types of enzymes into the endocytic apparatus, this observation suggests that removal of sugar residues occurs primarily in lysosomes.

α-Galactosidase is found in two different forms. α-Galactosidase A catalyzes the hydrolysis of terminal α-galactose from both water-soluble and lipid compounds. The pH optimum for the hydrolysis of glycolipid substrates is 4.1 while the pH optimum for the hydrolysis of a water-soluble fluorogenic substrate is higher (4.6) (Dean and Sweeley, 1979a). α-Galactosidase B has α-N-acetylgalactosamino hydrolase activity (see Dean and Sweeley, 1979b).

β-Galactosidase A hydrolyzes G_{M1}-ganglioside, N-acetyllactosamine, and asialofetuin (see Norden *et al.*, 1974). The saccharide side chains of these molecules are of the type found on many glycoproteins; therefore, β-galactosidase may play a significant role in the degradation of glycoproteins. In fact, glycopeptides and oligosaccharides accumulate within the lysosomes of individuals with missing or defective β-galactosidase A (G_{M1}-gangliosidosis).

β-Glucocerebrosidase is a membrane-bound enzyme found in lysosomes. It catalyzes the hydrolysis of glucocerebroside to glucose and ceramide with a pH optimum of less than 5.5 in the presence of Triton X-100 and taurocholate, and a pH of less than 5.0 in the absence of detergents (Aerts *et al.*, 1985). Deficiency in glucocerebrosidase causes Gaucher's disease (Brady *et al.*, 1965). There appear to be two forms of this enzyme, although they might actually be coded by the same gene. One form binds concanavalin A with higher affinity (type I). Type I is present at 15% of the normal level in Gaucher's disease, while type II is present at anywhere between 25 and 50% (Aerts *et al.*, 1985). Both types appear to have the same K_M against the artificial substrate, β-glucoside.

α-L-Iduronidase is a lysosomal enzyme that hydrolyzes α-L-iduronic acid residues from the glycosaminoglycans dermatan sulfate, heparan sulfate, and heparin. Two forms were originally found in human urine, termed the "high"- and "low"-uptake forms. The low-uptake form is metabolically processed from the high-uptake form. Myerowitz and Neufeld (1981) showed that the smaller form, 66 kDa, was secreted, while the larger form, 76 kDa, was intracellular. They also showed that the 66-kDa form had low uptake properties and the 76-kDa form had high uptake properties. The mature 66-kDa form is synthesized as a 75-kDa precursor form that is processed to 72 kDa and then to 66 kDa in five days.

α-Mannosidase is found in liver, fibroblasts, and other tissues in humans (Carroll *et al.*, 1972). The two predominant forms, A and B, can be separated by ion-exchange chromatography. The lysosomal storage disease, mannosidosis, shows loss of both forms. Both forms have a small subunit of 26 kDa, connected via a disulfide bond to a larger subunit. The large subunits bind to concanavalin A, indicating the presence of high-mannose oligosaccharides (Cheng *et al.*, 1986). There are two types of larger subunits, 58 and 62 kDa. The 58-kDa subunit has a lower pI, and contains less high-mannose oligosaccharides, but may contain more phosphate than the 62-kDa subunit. The A form consists of one small subunit and one 62-kDa subunit, whereas the B form consists of one small subunit and a mixture of the two large subunits. The B form has a higher affinity for MPR, indicating that it contains more M6P. This is supported by the fact that much less of form A is taken up by enzyme-deficient fibroblasts.

4.5. Sulfatases

α-Glucosaminide *N*-acetyltransferase is a membrane-bound lysosomal enzyme that catalyzes the acetylation of terminal α-linked glucosamine residues. This acetylation is one of four steps in the degradation of heparan sulfate in the lysosome. The pH optimum for this enzyme is above 5.5 (see Bame and Rome, 1985). It is supplied with acetyl-coenzyme A from the cytoplasm. Acetyl-CoA itself is not taken up by the lysosome; rather, the enzyme is acetylated on the cytoplasmic side, and the acetyl group is brought into the lysosome by a conformational change in the enzyme.

Arylsulfatases A and B are found in rat brain lysosomes. Selmi *et al.* (1989) found a high level of arylsulfatase A in pig thyroid lysosomes, up to one-third of the protein, suggesting that arylsulfatase A may play a role in the generation of thyroid hormones. Both A and B forms hydrolyze the same synthetic substrate, but their natural substrates are distinct. Arylsulfatase B degrades sulfatides while arylsulfatase A degrades sulfated mucopolysaccharides. Both forms from human tissues are soluble lysomal enzymes.

Other sulfatases include sulfatases A and B, chondroitin 6-sulfatase, heparin sulfamatase, and iduronosulfatase.

4.6. Phosphatases

Lysosomal acid phosphatase (LAP) is a glycoprotein with two identical subunits of 48–52 kDa. While most lysosomal enzymes are thought to be directed to the lysosome via MPR, acid phosphatase is one of the few enzymes that does not appear to use that system (Waheed *et al.*, 1988; Gottschalk *et al.*, 1989; Peters *et al.*, 1990). While the other non-M6P-containing enzymes are tightly bound to the membrane, acid phosphatase is partially soluble. Lemansky *et al.* (1985) found that two-thirds of newly synthesized acid phosphatase is secreted from I-cells, while only 10% is secreted in normal human skin fibroblasts (suggesting some role for the MPR in acid phosphatase targeting). Other lines of evidence indicate that at least some acid phosphatase can be transported to lysosomes via MPR; endocytosis of LAP is inhibited by M6P, and anti-MPR antibodies increase the secretion of LAP. However, other reports indicate that the enzyme is transported in a membrane form (Gottschalk *et al.*, 1989; Peters *et al.*, 1990; Waheed *et al.*, 1988). Significant differences exist between the systems used in the conflicting reports, making it difficult to draw a conclusion. The reports indicating membrane-bound transport used an interspecies expression system where human LAP was overexpressed (70-fold higher activity) in baby hamster kidney cells. However, in a follow-up paper, the disruption of proper targeting by a single amino acid change in the cytoplasmic tail of LAP provides compelling evidence that the membrane-bound form plays a major role in the delivery of LAP to lysosomes (Peters *et al.*, 1990).

While acid phosphatase has traditionally been used as a histochemical marker for lysosomes, a significant fraction of acid phosphatase activity is present in low-buoyant-density compartments (as much as 60% in actively growing 3T3 fibroblasts; Roederer *et al.*, 1989). Significant acid phosphatase activity has also been detected in endosomes using electron microscopic (Storrie *et al.*, 1984), biochemical (Braun *et al.*, 1989), and fluorometric (Bowser and Murphy, 1990) assays. While the hypothesis that endosomal acid phosphatase is entirely membrane-associated while the lysosomal activity is entirely soluble is attractive, it remains to be rigorously proven.

4.7. Lipases

Robinson and Waite (1983) examined the substrate specificity of purified phospholipase A. They found that phosphatidylethanolamine (PE) was the preferred substrate, phosphatidylcholine (PC) was second, with a hydrolysis rate one-fifth that for PE, while phosphatidylinositol (PI), phosphatidylglycerol (PG), and phosphatidylserine (PS) were degraded very slowly. Hydrolysis of PE was inhibited by the presence of Triton WR 1339, while hydrolysis of PC, PI, PS, and PG was stimulated.

At a 6:1 Triton/phospholipid ratio, PG was the preferred substrate. The authors found that fatty acid chain in position 1 affected the rate of PE hydrolysis, palmitic and oleic acids being preferred. They also found a requirement for negative surface charge for hydrolysis of PC and PE. The presence of Ca^{2+} stimulated hydrolysis of PI, PS, and PG, while inhibiting the hydrolysis of PE. In the initial purification from rat liver lysosomes, phospholipase A was recovered from the soluble fraction. The enzyme was most active at pH 4.0. However, in a report from a different group, multiple forms of phospholipase A were found. They found that PG was the preferred substrate, and the activity was not inhibited by the presence of positive charge (Ca^{2+} or Na^+).

Lysosomes also contain triacylglycerol lipase, phospholipase A_2, phosphatidate phosphatase, acylsphingosine deacylase, and sphingomyelin phosphodiesterase activities.

5. PROCESSING OF PHYSIOLOGICAL LIGANDS

As mentioned above, cathepsin D activity has been detected in early endosomes of macrophages (Diment and Stahl, 1985). Endosomal cathepsin D is likely to play an important role in parathyroid metabolism in macrophages. Using bovine parathyroid hormone (PTH) and rabbit alveolar macrophages, Diment et al. (1989) demonstrated that PTH is internalized, cleaved into fragments, including the bioactive 1–34 fragment, then released by exocytosis. They concluded that PTH is never delivered to lysosomes. The half-time of metabolism is 10–15 min, which is consistent with the half-time of peptide 1–34 generation in vivo. It has already been shown that Kupffer cells are the major site of PTH metabolism; kidney plays a lesser role, and bone tissue does not make a significant contribution. Like intact PTH, fragments from the N-terminus can be taken up by bone tissue, where they stimulate resorption of bone mineral. Liver metabolism of PTH may be an important source of bioactive PTH fragments which act in the skeleton, though the significance of intact and partially metabolized PTH remains unclear.

In another study, Yamaguchi et al. (1989) found that human PTH is degraded in UMR-106 cells (characteristic of mature osteoblasts) by chymotrypsin-like activity. Unlike Diment et al., they found no contribution by cathepsin D activity. They also found no contribution by cathepsin B or metalloendoproteases, nor was the degradation inhibited by treatment with ammonium chloride, chloroquine, or monensin. This indicates that lysosomal compartments are not involved in PTH metabolism in this osteoblast-like cell line though degradation appeared to occur intracellularly. The fragments generated by these cells strongly resemble fragments generated by purified chymotrypsin, yet these peptides have not been found circulating in the blood. The authors point out that PTH fragments generated by parathyroid slices and intact parathyroid cells are different from fragments generated by cathepsins B and D from parathyroid homogenate. It appears, then, that while osteoblast-like cells and parathyroid tissue are capable of degrading PTH, they are not responsible for generating the partially metabolized forms found in the blood.

6. ANTIGEN PRESENTATION

Proteases of the endosomal pathway play an important role in the humoral immune response. Antigens are internalized by either macrophages, B cells, or dendritic cells [collectively referred to as antigen-presenting cells (APC)] and proteolytically processed in an internal compartment (for reviews see Unanue, 1984; Schwartz, 1985; Allen, 1987; Lanzavecchia, 1988). The resulting peptide fragments interact with class II major histocompatibility molecules which are then expressed on the surface of the APC. The complex is recognized by T-helper cells via specific T-cell receptors, stimulating the T-helper cell to secrete various lymphokines which in turn stimulate the T cell and APC to divide and differentiate. In this manner, clonal expansion of the appropriate T cells and APCs is thought to occur.

The antigen presentation process requires time, temperatures above 20°C, an acidic intravesicular environment, and active cysteine proteases. These requirements have been interpreted as implicating endosomes as the site of antigen processing. Additional evidence for the role of endosomes in antigen processing is the rapid colocalization of internalized surface immunoglobulins with proteolytic enzymes and major histocompatibility molecules in endosome-like compartments (Guagliardi et al., 1990). While specific roles for cathepsins B and D in this process have been proposed (Diment, 1990), it is unclear how internalized antigens manage to escape complete degradation, especially by cells with high proteolytic potential, such as macrophages. Harding et al. (1991) have demonstrated, using liposome-encapsulated antigens, that at least some antigen fragments can escape from lysosomes and be presented to T cells.

7. CONCLUSIONS

The results reviewed above indicate the complexity of the enzyme activities to which endocytosed material is exposed in the endocytic system. Some of the characteristics of endosomes and lysosomes have been exploited in model systems for drug targeting. These include acidic pH (Shen and Ryser, 1981; Straubinger et al., 1983), proteases (Monsigny et al., 1980), and reducing activity (Wan et al., 1990). The next few years are anticipated to provide a dramatic increase in information about the mechanisms of regulation of enzyme activity and membrane traffic in this system. The availability of this information may permit exploitation of the highly specialized characteristics of endosomes and lysosomes to achieve particular pharmacological goals.

REFERENCES

Achkar, C., Gong, Q., Frankfater, A., and Bajowski, A. S., 1990, Differences in targeting and secretion of cathepsins B and L by BALB/c 3T3 fibroblasts and Moloney murine sarcoma virus-transformed BALB/c 3T3 fibroblasts, *J. Biol. Chem.* **265**:13650–13654.

Aerts, G. M. F. G., Donker-Koopman, W. E., van der Vliet, M. K., Jonsson, L. M. V., Ginns, E. I., Murray, G. J., Barranger, J. A., Tager, J. M., and Schram, A. W., 1985, The occurrence of two immunologically distinguishable β-glucocerebrosidases in human spleen, *Eur. J. Biochem.* **150**:565–574.

Ajioka, R. S., and Kaplan, J., 1987, Characterization of endocytotic compartments using the horseradish peroxidase–diaminobenzidine density shift technique, *J. Cell Biol.* **104**:77–85.

Allen, P. M., 1987, Antigen processing at the molecular level, *Immunol. Today* **8**:270.

Bae, H.-R., and Verkman, A. S., 1990, Protein kinase A regulates chloride conductance in endocytic vesicles from proximal tubules, *Nature* **348**:637–639.

Bame, K. J., and Rome, L. H., 1985, Acetyl coenzyme A:α-glucosaminide *N*-acetyltransferase. Evidence for a transmembrane acetylation mechanism, *J. Biol. Chem.* **260**:11293–11299.

Barrett, A. J., 1973, Human cathepsin B1, *Biochem. J.* **131**:809–822.

Barrett, A. J., and Kirschke, H., 1981, Cathepsin B, cathepsin H, and cathepsin L, *Methods Enzymol.* **80**:535–561.

Barrett, A. J., and McDonald, J. K., 1980, *Mammalian Proteases: A Glossary and Bibliography*, Academic Press, New York.

Bowser, R., and Murphy, R. F., 1990, Kinetics of hydrolysis of endocytosed substrates by mammalian cultured cells: Early introduction of lysosomal enzymes into the endocytic pathway, *J. Cell Physiol.* **143**:110–117.

Brady, R. O., Kanfer, J. N., and Shapiro, D., 1965, Metabolism of glucocerebrosides. II. Evidence of an enzymatic deficiency in Gaucher's disease, *Biochem. Biophys. Res. Commun.* **18**:221–225.

Braell, W. A., 1987, Fusion between endocytic vesicles in a cell-free system, *Proc. Natl. Acad. Sci. USA* **84**:1137–1141.

Braun, M., Waheed, A., and von Figura, K., 1989, Lysosomal acid phosphatase is transported to lysosomes via the cell surface, *EMBO J.* **8**:3633–3640.

Cain, C. C., Sipe, D. M., and Murphy, R. F., 1989, Regulation of endocytic pH by the Na^+,K^+-ATPase in living cells, *Proc. Natl. Acad. Sci. USA* **86**:544–548.

Carroll, M., Dance, N., Masson, P. K., Robinson, D., and Winchester, B. G., 1972, Human mannosidosis—The enzyme defect, *Biochem. Biophys. Res. Commun.* **49**:579–583.

Cheng, S. H., Malcolm, S., Pemble, S., and Winchester, B., 1986, Purification and comparison of the structures of human liver acidic α-D-mannosidases A and B, *Biochem. J.* **233**:65–72.

Chu, F.-F., and Olden, K., 1984, Distribution of acid hydrolases in subcellular fractions of proliferating vs non-proliferating fibroblasts, *Exp. Cell Res.* **154**:606–612.

Dean, K. J., and Sweeley, C. C., 1979a, Studies on human liver α-galactosidases. I. Purification of α-galactosidase A and its enzymatic properties with glycolipid and oligosaccharide substrates, *J. Biol. Chem.* **254**:9994–10000.

Dean, K. J., and Sweeley, C. C. 1979b, Studies on human liver α-galactosidases. II. Purification and enzymatic properties of α-galactosidase B (α-*N*-acetylgalactosaminidase), *J. Biol. Chem.* **254**:10001–10005.

Diment, S., 1990, Different roles for thiol and aspartyl proteases in antigen presentation of ovalbumin, *J. Immunol.* **145**:417–422.

Diment, S., and Stahl, P., 1985, Macrophage endosomes contain proteases which degrade endocytosed protein ligands, *J. Biol. Chem.* **260**:15311–15317.

Diment, S., Martin, K, J., and Stahl, P. D., 1989, Cleavage of parathyroid hormone in macrophage endosomes illustrates a novel pathway for intracellular processing of proteins, *J. Biol. Chem.* **264**:13403–13406.

Draper, R. K., and Simon, M. I., 1980, The entry of diphtheria toxin into the mammalian cell cytoplasm: Evidence for lysosomal involvement, *J. Cell Biol.* **87**:849–854.

Feener, E. P., Shen, W.-C., and Ryser, H. J.-P., 1990, Cleavage of disulfide bonds in endocytosed macromolecules, *J. Biol. Chem.* **265**:18780–18785.

Fuchs, R., Schmid, S., and Mellman, I., 1989, A possible role for the Na^+,K^+-ATPase in regulating ATP-dependent endosome acidification, *Proc. Natl. Acad. Sci. USA* **86**:539–543.

Fukasawa, K., Fukasawa, K. M., Hiraoka, B. Y., and Harada, M., 1983, Purification and properties of dipeptidyl peptidase II from rat kidney, *Biochim. Biophys. Acta* **745**:6–11.

Gahl, W. A., Tietze, R., Bashan, N., Steinherz, R., and Schulman, J. D., 1982, Defective cystine exodus from isolated lysosome-rich fractions of cystinotic leucocytes, *J. Biol. Chem.* **257**:9470–9475.

Galjart, N. J., Gillemans, N., Meijer, D., and d'Azzo, A., 1990, Mouse "protective protein" cDNA cloning, sequence comparison, and expression, *J. Biol. Chem.* **265**:4678–4684.

Galloway, C. J., Dean, G. E., Marsh, M., Rudnick, G., and Mellman, I., 1983, Acidification of macrophage and fibroblast endocytotic vesicles in vitro, *Proc. Natl. Acad. Sci. USA* **80**:3334–3338.

Geuze, H. J., Slot, J. W., Strous, G. J. A. M., Hasilik, A., and von Figura, K., 1985, Possible pathways for lysosomal enzyme delivery, *J. Cell Biol.* **101**:2253–2262.

Glaumann, H., and Ballard, F. J. (eds.), 1987, *Lysosomes: Their Role in Protein Breakdown*, Academic Press, New York.

Gottschalk, S., Waheed, A., and von Figura, K., 1980, Targeting of lysosomal acid phosphatase with altered carbohydrate, *Biol. Chem. Hoppe-Seyler* **370**:75–80.

Griffiths, D., Hoflack, B., Simons, K., Mellman, I., and Kornfeld, S., 1988, The mannose 6-phosphate receptor and the biogenesis of lysosomes, *Cell* **52**:329–341.

Griffiths, G., and Gruenberg, J., 1991, The arguments for pre-existing early and late endosomes, *Trends Cell Biol.* **1**:5–9.

Guagliardi, L. E., Kopppleman, B., Blum, J. S., Marks, M. S., Cresswell, P., and Brodsky, F. M., 1990, Co-localization of molecules involved in antigen processing and presentation in an early endocytic compartment, *Nature* **343**:133–139.

Hanewinkel, H., Glossl, J., and Kresse, H., 1987, Biosynthesis of cathepsin B in cultured normal and I-cell fibroblasts, *J. Biol. Chem.* **262**:12351–12355.

Harding, C. V., Unanue, E. R., Slot, J. W., Schwartz, A. L., and Geuze, H. J., 1990, Functional and ultrastructural evidence for intracellular formation of major histocompatibility complex class II–peptide complexes during antigen processing, *Proc. Natl. Acad. Sci. USA* **87**:5553–5557.

Harding, C. V., Collins, D. S., Slot, J. W., Geuze, H. J., and Unanue, E. R., 1991, Liposome-encapsulated antigens are processed in lysosomes, recycled, and presented to T cells, *Cell* **64**:393–401.

Helenius, A., Mellman, I., Wall, D., and Hubbard, A., 1983, Endosomes, *Trends Biochem. Sci.* **8:**245–249.

Holtzman, E., 1989, *Lysosomes*, Plenum Press, New York.

Jackman, H. L., Tan, F., Tamei, H., Beurling-Harbury, C., Li, X.-Y., Skidgel, R. A., and Erdos, E. G., 1990, A peptidase in human platelets that deamidates tachykinins, *J. Biol. Chem.* **265:**11265–11272.

Jonas, A. J., Greene, A. A., Smith, M. L., and Schneider, J. A., 1982, Cystine accumulation and loss in normal, heterozygous and cystinotic fibroblasts, *Proc. Natl. Acad. Sci. USA* **79:**4442–4445.

Kase, R., Itoh, K., Takiyama, N., Oshima, A., Sakuraba, H., and Suzuki, Y., 1990, Galactosialidosis: Simultaneous deficiency of esterase, carboxy-terminal deamidase and acid carboxypeptidase activities, *Biochem. Biophys. Res. Commun.* **172:**1175–1179.

Keilova, H., 1971, On the specificity an inhibition of cathepsins D and B, in: *Tissue Proteinases* (A. J. Barrett and J. T. Dingle, eds.), North-Holland, Amsterdam, pp. 45–65.

Kielian, M. C., Marsh, M., and Helenius, A., 1986, Kinetics of endosome acidification detected by mutant and wild-type Semliki Forest virus, *EMBO J.* **5:**3103–3109.

Killisch, I., Steinlein, P., Römisch, K., Hollinshead, R., Beug, H., and Griffiths, G., 1992, Characterization of early and late endocytic compartments of the transferrin cycle: Transferrin receptor antibody blocks erythroid differentiation by trapping the receptor in the early endosome, *J. Cell Sci.* **103:**211–232.

Kooistra, T., Millard, P. C., and Lloyd, J. B., 1982, Role of thiols in degradation of proteins by cathepsins, *Biochem. J.* **204:**471–477.

Kornfeld, S., 1987, Trafficking of lysosomal enzymes, *FASEB J.* **1:**462–468.

Kornfeld, S., and Mellman, I., 1989, The biogenesis of lysosomes, *Annu. Rev. Cell Biol.* **5:** 483–525.

Lanzavecchia, A., 1988, Clonal sketches of the immune response. *EMBO J.* **7:**2945.

Lemansky, P., Gieselmann, V., Hasilik, A., and von Figura, K., 1985, Synthesis and transport of lysosomal acid phosphatase in normal and I-cell fibroblasts, *J. Biol. Chem.* **260:**9023–9030.

Lewis, M. S., and Youle, R. J., 1986, Ricin subunit association, *J. Biol. Chem.* **261:**11571–11577.

Lloyd, J. B., 1986, Disulphide reduction in lysosomes, *Biochem. J.* **237:**271–272.

McDonald, J. K., and Barrett, A. J., 1986, *Mammalian Proteases: A Glossary and Bibliography*, Academic Press, New York.

McDonald, J. K., Leibach, F. H., Grindeland, R. E., and Ellis, S., 1968, Purification of dipeptidyl aminopeptidase II (dipeptidyl arylamidase II) of the anterior pituitary gland, *J. Biol. Chem.* **243:**4143–4150.

McDonald, J. K., Hoisington, A. R., and Eisenhauer, D. A., 1985, Partial purification and characterization of an ovarian tripeptidyl peptidase: A lysosomal exopeptidase that sequentially releases collagen-related (Gly-Pro-X) triplets, *Biochem. Biophys. Res. Commun.* **126:**63–71.

Mego, J. L., 1984, Role of thiols, pH and cathepsin D in the lysosomal catabolism of serum albumin, *Biochem. J.* **218:**775–783.

Mellman, I., Fuchs, R., and Helenius, A, 1986, Acidification of the endocytic and exocytic pathways, *Annu. Rev. Biochem.* **55:**663–700.

Merion, M., and Poretz, D. R., 1981, The resolution of two populations of lysosomal organelles containing endocytosed Wisteria floribunda agglutinin from murine fibroblasts, *J. Supramol. Struct. Cell Biochem.* **17:**337–346.

Merion, M., and Sly, W. S., 1983, The role of intermediate vesicles in the adsorptive endocytosis and transport of ligand to lysosomes by human fibroblasts, *J. Cell Biol.* **96:** 644–650.

Monsigny, M., Kieda, C., Roche, A.-C., and Delmotte, F., 1980, Preparation and biological properties of a covalent antitumor drug-arm-carrier (DAC) conjugate, *FEBS Lett.* **119:** 181–186.

Moskaug, J. O., Sandvig, K., and Olsnes, S., 1987, Cell-mediated reductions of the interfragment disulfide in nicked diphtheria toxin, *J. Biol. Chem.* **262:**10339–10345.

Murphy, R. F., 1985, Analysis and isolation of endocytic vesicles by flow cytometry and sorting: Demonstration of three kinetically distinct compartments involved in fluid-phase endocytosis, *Proc. Natl. Acad. Sci. USA* **82:**8523–8526.

Murphy, R. F., 1988, Processing of endocytosed material, *Adv. Cell Biol.* **2:**159–180.

Murphy, R. F., 1991, Maturation models for endosome and lysosome biogenesis, *Trends Cell Biol.* **1:**77–82.

Murphy, R. F., Jorgensen, E. D., and Cantor, C. R., 1982, Kinetics of histone endocytosis in Chinese hamster ovary cells, *J. Biol. Chem.* **257:**1695–1701.

Murphy, R. F., Powers, S., and Cantor, C. R., 1984, Endosomal pH measured in single cells by dual fluorescence flow cytometry: Rapid acidification of insulin to pH 6, *J. Cell Biol.* **98:**1757–1762.

Myerowitz, R., and Neufeld, E. F., 1981, Maturation of α-L-iduronidase in cultured human fibroblasts, *J. Biol. Chem.* **256:**3044–3048.

Norden, A. G. W., Tennant, L. L., and O'Brien, J. S., 1974, G_{M1} ganglioside β-galactosidase A, *J. Biol. Chem.* **249:**7969–7976.

Ohkuma, S., and Poole, B., 1978, Fluorescence probe measurement of the intralysosomal pH in living cells and the perturbation of pH by various agents, *Proc. Natl. Acad. Sci. USA* **75:**3327–3331.

Pappenheimer, A. M., Jr., 1977, Diphtheria toxin, *Annu. Rev. Biochem.* **46:**69–94.

Peters, C., Braun, M., Weber, B., Wendland, M., Schmidt, B., Pohlmann, R., Waheed, A., and von Figura, K., 1990, Targeting of a lysosomal membrane protein: A tyrosine-containing endocytosis signal in the cytoplasmic tail of lysosomal acid phosphatase is necessary and sufficient for targeting to lysosomes, *EMBO J.* **9:**3497–3506.

Pisoni, R. L., Acker, T. L., Lisowski, K. M., Lemons, R. M., and Thoene, J. C., 1990, A cysteine-specific lysosomal transport system provides a major route for the delivery of thiol to human fibroblast lysosomes: Possible role in supporting lysosomal proteolysis, *J. Cell Biol.* **110:**327–335.

Robinson, M., and Waite, M., 1983, Physical–chemical requirements for the catalysis of substrates by lysosomal phospholipase A_1, *J. Biol. Chem.* **258:**14371–14378.

Roederer, M., and Murphy, R. F., 1986, Cell-by-cell autofluorescence correction for low signal-to-noise systems: Application to EGF endocytosis by 3T3 fibroblasts, *Cytometry* **7:**558–565.

Roederer, M., Browser, R., and Murphy, R. F., 1987, Kinetics and temperature dependence of exposure of endocytosed material to proteolytic enzymes and low pH: Evidence for a maturation model for the formation of lysosomes, *J. Cell Physiol.* **131:**200–209,

Roederer, M., Mays, R. W., and Murphy, R. F., 1989, Effect of confluence on endocytosis by

3T3 fibroblasts: Increased pinocytosis and accumulation of residual bodies, *Eur. J. Cell Biol.* **48**:37–44.

Sandvig, K., and Olsnes, S., 1980, Diphtheria toxin entry into cells is facilitated by low pH, *J. Cell Biol.* **87**:828–832.

Schmid, S. L., Fuchs, R., Male, P., and Mellman, I., 1988, Two distinct subpopulations of endosomes involved in membrane recycling and transport to lysosomes, *Cell* **52**:73–83.

Schulman, J. D., Bradley, K. H., and Seegmiller, J. E., 1969, Cystine: Compartmentalization within lysosomes in cystinotic leukocytes, *Science* **166**:1152–1154.

Schwartz, R. H., 1985, T-lymphocyte recognition of antigen in association with gene products of the major histocompatibility complex, *Annu. Rev. Immunol.* **3**:237–261.

Selmi, S., Maire, I., and Rousset, B., 1989, Evidence for the presence of a very high concentration of arylsulfatase A in the pig thyroid: Identification of arylsulfatase A subunits as the two major glycoproteins in purified thyroid lysosomes, *Arch. Biochem. Biophys.* **273**:170–179.

Shaw, E., and Dean, R. T., 1980, The inhibition of macrophage protein turnover by a selective inhibitor of thiol proteinases, *Biochem. J.* **186**:385–390.

Shen, W.-C., and Ryser, H. J.-P., 1981, cis-Aconityl spacer between daunomycin and macromolecular carriers: A model of pH-sensitive linkage releasing drug from a lysosomotropic conjugate, *Biochem. Biophys. Res. Commun.* **102**:1048–1054.

Sipe, D. M., and Murphy, R. F., 1987, High resolution kinetics of transferrin acidification in BALB/c 3T3 cells: Exposure to pH 6 followed by temperature-sensitive alkalinization during recycling, *Proc. Natl. Acad. Sci. USA* **84**:7119–7123.

Sipe, D. M., Jesurum, A., and Murphy, R. F., 1991, Absence of NA$^+$,K$^+$-ATPase regulation of endosomal acidification in K562 erythroleukemia cells, *J. Biol. Chem.* **266**:3469–3474.

Sloane, B. F., Rozhin, J., Johnson, K., Taylor, H., Crissman, J. D., and Honn, K. V., 1986, Cathepsin B: Association with plasma membrane in metastatic tumors, *Proc. Natl. Acad. Sci. USA* **83**:2483–2487.

Steinherz, R., Tietze, F., Gahl, W. A. Triche, T. J., Chiang, H., Modesti, A., and Schulman, J., 1982, Cystine accumulation and clearance by normal and cystinotic leukocytes exposed to cystine dimethyl ester, *Proc. Natl. Acad. Sci. USA* **79**:4446–4450.

Storrie, B., 1988, Assembly of lysosomes: Perspectives from comparative molecular cell biology, *Int. Rev. Cytol.* **111**:52–105.

Storrie, B., Pool, R. R., Sachdeva, M., Maurey, K. M., and Oliver, C., 1984, Evidence for both prelysosomal and lysosomal intermediates in endocytic pathways, *J. Cell Biol.* **98**:108–115.

Straubinger, R. M., Hong, K., Friend, D. S., and Papahadjopoulos, D., 1983, Endocytosis of liposomes and intracellular fate of encapsulated molecules: Encounter with a low pH compartment after internalization in coated vesicles, *Cell* **32**:1069–1079.

Thiele, D. L., and Lipsky, P. E., 1990, Mechanism of L-leucyl-L-leucine methyl ester-mediated killing of cytotoxic lymphocytes: Dependence on a lysosomal thiol protease, dipeptidyl peptidase I, that is enriched in these cells, *Proc. Natl. Acad. Sci. USA* **87**:83–87.

Thoene, J. G., Oshima, R. G., Ritchie, D. G., and Schneider, J. A., 1977, Cystinotic fibroblasts accumulate cystine from intracellular protein degradation, *Proc. Natl. Acad. Sci. USA* **74**:4505–4507.

Tranchemontagne, J., Michaud, L., and Potier, M., 1990, Deficient lysosomal carboxypeptidase activity in galactosialidosis, *Biochem. Biophys. Res. Commun.* **168**:22–29.

Tycko, B., and Maxfield, F. R., 1982, Rapid acidification of endocytic vesicles containing alpha-2-macroglobulin, *Cell* **28:**643–651.

Unanue, E. R., 1984, Antigen-presenting function of the macrophage, *Annu. Rev. Immunol.* **2:**395–428.

Van Dyke, R. W., 1988, Proton pump-generated electrochemical gradients in rat liver multivesicular bodies, *J. Biol. Chem.* **263:**2603–2611.

van Renswoude, J., Bridges, K. R., Harford, J. B., and Klausner, R. D., 1982, Receptor-mediated endocytosis of transferrin and the uptake of Fe in K652 cells: Identification of a nonlysosomal acidic compartment, *Proc. Natl. Acad. Sci. USA* **79:**6186–6190.

von Figura, K., and Hasilik, A., 1986, Lysosomal enzymes and their receptors, *Annu. Rev. Biochem.* **55:**167–193.

Waheed, A., Gottschalk, S., Hille, A., Krentler, C., Pohlmann, R., Braulke, T., Hauser, H., Geuze, H., and von Figura, K., 1988, Human lysosomal acid phosphatase is transported as a transmembrane protein to lysosomes in transfected baby hamster kidney cells, *EMBO J.* **7:**2351–2358.

Wan, J., Persiani, S., and Shen, W.-C., 1990, Transcellular processing of disulfide- and thioether-linked peroxidase–polylysine conjugates in cultured MDCK epithelial cells, *J. Cell Physiol.* **145:**9–15.

Yamaguchi, T., Fukase, M., Nishikawa, M., Fujimi, T., and Fujita, T., 1989, Parathyroid hormone degradation by chymotrypsin-like endopeptidase in the clonal osteogenic UMR-106 cell, *Biochim. Biophys. Acta* **1010:**177–183.

Yamashiro, D. J., and Maxfield, F. R., 1987, Kinetics of endosome acidification in mutant and wild-type Chinese hamster ovary cells, *J. Cell Biol.* **105:**2713–2721.

Yamashiro, D. J., and Maxfield, F. R., 1988, Regulation of endocytic processes by pH, *Trends Pharmacol. Sci.* **9:**190–193.

Chapter 4

Protein Uptake and Cytoplasmic Access in Animal Cells

Bo van Deurs, Steen H. Hansen, Sjur Olsnes, and Kirsten Sandvig

1. INTRODUCTION

Cell membranes are the major barriers to protein delivery into cells. As a great deal of pharmaceutical and biotechnological research attempts to find ways of delivering drugs into cells—for instance, with the purpose of irreversibly inhibiting the protein synthesis machinery of cancer cells—mechanisms by which various protein ligands are internalized by cells and subsequently translocated across the membrane of intracellular compartments are coming into focus.

A protein can enter a cell in two ways, either directly by translocation across the plasma membrane into the cytosol, or by endocytosis. Following endocytosis, the internalized protein is still separated from the cytosol by a membrane (for instance, the endosome membrane). The protein can then either follow an endocytic pathway which leads back to the cell surface (recycling and transcytosis) or to the Golgi complex, or follow one leading to proteolytic destruction in lysosomes (Fig. 1). Alternatively, the protein can be translocated across the membrane in an enzymatically active form to reach the cytosol from, in principle, any station on the endocytic pathway. Once in the cytosol, the translocated protein may interact with the

Bo van Deurs and Steen H. Hansen • Structural Cell Biology Unit, Department of Anatomy, The Panum Institute, University of Copenhagen, DK-2200 Copenhagen N, Denmark. *Sjur Olsnes and Kirsten Sandvig* • Institute for Cancer Research, Norwegian Radium Hospital, Montebello, 0310 Oslo 3, Norway.

Biological Barriers to Protein Delivery, edited by Kenneth L. Audus and Thomas J. Raub. Plenum Press, New York, 1993.

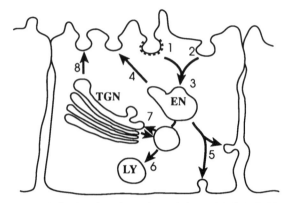

Figure 1. Schematic model of endocytic pathways in a polarized epithelial cell. Plasma membrane with receptors and bound ligands as well as solutes may be internalized by clathrin-coated vesicles (1) or noncoated vesicles (2) to reach (3) the endosomal compartment (EN). From the endosome, internalized molecules are sorted to their next or final destination. There is a recycling pathway (4) back to the cell surface from where the internalization took place, a transcytotic pathway (5) to the opposite surface, and a pathway (6) to lysosomes (LY). The transfer of molecules to lysosomes occurs via an intermediate stage which communicates (7) with the *trans*-Golgi network (TGN). Internalized molecules reaching the TGN may be transported back to the cell surface together with newly synthesized secretory and membrane proteins (8).

protein synthesis machinery, an aspect of great relevance in relation to the application of various plant and bacterial toxins for constructing immunotoxins (Olsnes *et al.*, 1989; van Deurs *et al.*, 1990b).

2. ENTRY OF PROTEINS INTO THE CYTOSOL

A number of plant and bacterial toxins are able to enter the cytosol of cells and inhibit the protein synthesis in these cells enzymatically. These toxins include the plant toxins abrin, ricin, modeccin, viscumin, and volkensin, as well as the bacterial toxins diphtheria toxin, Pseudomonas exotoxin A, Shiga toxin, and Shiga-like toxins (for review, see Jackson, 1990; Olsnes and Sandvig, 1988). Also, a number of bacterial toxins with other targets than the protein synthesis machinery seem to enter cells. Cholera toxin, pertussis toxin, and *Escherichia coli* heat-labile toxin all modify G-proteins (Moss and Vaughan, 1988), while *Clostridium botulinum* C2 toxin and the three immunologically related toxins. *C. perfringens* E iota toxin, *C. spiroforme* toxin, and one of the toxins produced by *C. difficile* all modify actin and lead to depolymerization of actin and eventually cell lysis (Aktories and Wegner, 1989). *C. difficile* also produces two other toxins, called toxin A and toxin B, which have a cytotoxic effect and which seem to enter the cytosol after uptake by endocytosis

(Henriques *et al.*, 1987). Also, botulinum and tetanus neurotoxins seem to exert their toxic effect after entry into the cytosol. However, in the case of these two toxins the intracellular target is still unknown (Ahnert-Hilger *et al.*, 1989; de Paiva and Dolly, 1990; Mochida *et al.*, 1989; Stecher *et al.*, 1989). Some bacteria (*Bacillus anthracis* and *Bordetella pertussis*) produce invasive adenylate cyclase, which also enters cells and intoxicates them (Donovan and Storm, 1990; Gordon *et al.*, 1989). In addition to the adenylate cyclase, *Bacillus anthracis* produces another protein which is highly toxic to some cells, and which also seems to enter the cytosol to exert its effect (Singh *et al.*, 1989).

As described above, there are a large variety of toxins which are able to gain access to the cytosol. Most of the toxins described here consist of a moiety that binds the toxin to cell surface receptors, and another, enzymatically active moiety that acts in the cytosol after translocation through the membrane. Relatively little is known about how most of these proteins cross the membrane. The results obtained so far suggest that many of them have to be endocytosed, in some cases modified, and that the transport to the cytosol then occurs across the membrane of an intracellular compartment (see Section 5). Recent data suggest that the same is the case with the man-made immunotoxins that are constructed to specifically kill certain cell types (Olsnes *et al.*, 1989).

In spite of the structural similarity between some of the naturally occurring toxins, their entry mechanisms seem to be quite different. Some of the toxins may have to be routed to the trans-Golgi network before entry into the cytosol (see Sections 4 and 5), whereas diphtheria toxin normally seems to enter from early acidic endosomes (Draper and Simon, 1980; Sandvig and Olsnes, 1980, 1981), and a few of the toxins may be able to enter directly from the plasma membrane [pertussis invasive adenylate cyclase (Donovan and Storm, 1990), cholera toxin, pertussis toxin (Moss and Vaughan, 1988)]. Also, when diphtheria toxin is bound to the cell surface, direct entry can be induced by exposing the cells to medium with low pH, thus mimicking the conditions in the endosome (see Section 4). Upon exposure to low pH, the conformation of the toxin is changed so that hydrophobic regions are exposed (Sandvig and Olsnes, 1981), the binding moiety is inserted into the plasma membrane (Moskaug *et al.*, 1988), there is formation of cation-selective channels (Sandvig and Olsnes, 1988), and the enzymatically active fragment is translocated (Fig. 2). Recent experiments have shown that one can, using diphtheria toxin as a vehicle, get translocation of other proteins into the cytosol (Stenmark *et al.*, 1991). Using this model system, the entry of diphtheria toxin has been studied in more detail than the entry of other toxins, and the conditions required for translocation of diphtheria toxin have thus been characterized. The experiments strongly suggest that the binding moiety is not only important for the binding of the toxin to the cell surface, but that both this protein and the receptor to which it binds play an important role in the translocation process (Stenmark *et al.*, 1988). Such information is important in connection with construction of immunotoxins or other molecules which one wants to bring into the cytosol. For this purpose it is important to know whether or not a

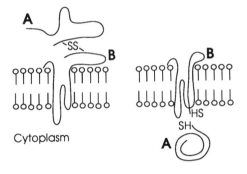

Cytoplasm

Figure 2. Model for translocation of diphtheria toxin. When cells with receptor-bound toxin are exposed to low pH, the B-fragment inserts into the membrane, and the A-fragment is translocated to the cyto-plasm.

toxin has to be internalized by endocytosis before translocation and to which destination in the cell it has to be transported. One would also like to know the requirements for translocation across the membrane.

3. ENDOCYTIC UPTAKE

According to classical definitions (see van Deurs *et al.*, 1989), *endocytosis* is subdivided into *phagocytosis* and *pinocytosis*. Phagocytosis is carried out by certain specialized cell types (e.g., macrophages) which actively encircle, for instance, invading bacteria and other particles followed by the formation of an intracellular vacuole. Pinocytosis, on the other hand, is uptake of ligands and solutes from the external environment, and occurs in all cell types. Endocytosis is generally used synonymously with pinocytosis, and in the following we analyze how cells endocytose various molecules, without taking phagocytosis into consideration. Some molecules are endocytosed in the *fluid phase*, that is, without any binding to the cell surface prior to internalization. In most cases, however, molecules bind to plasma membrane constituents and the subsequent internalization is referred to as *adsorptive endocytosis*. In its most elaborate form, adsorptive endocytosis is mediated by ligand-specific receptor molecules such as the low-density lipoprotein (LDL) receptor, the transferrin receptor, or the insulin receptor, and the process is referred to as *receptor-mediated endocytosis*. Many nonphysiological ligands such as viral, plant, and bacterial toxins (opportunistic ligands) utilize various cell surface molecules or specific carbohydrate moieties to become efficiently internalized, thereby imitating receptor-mediated endocytosis.

It has become evident from studies on various cell lines that cells have at least two different endocytic mechanisms, one depending on the clathrin molecule and being thoroughly studied, the other clathrin-independent and much less explored (Brodsky, 1988; Gruenberg and Howell, 1989; Hubbard, 1989; Pearse and Robinson,

1990; Rodman *et al.*, 1990; van Deurs *et al.*, 1989). In the first case, endocytosis takes place from specialized membrane domains, the coated pits, at the cell surface (Figs. 3–5). Whether these pits are almost flat (early stages in vesicle formation) or deeply invaginated (late stages, just before the coated vesicle pinches off), they are easily recognized in the electron microscope due to the characteristic clathrin coat on the cytoplasmic face of the membrane. Even though the coated pits occupy less than 2% of the cell surface, they are involved in a very efficient internalization of a large number of physiological ligands (receptor-mediated endocytosis) (Goldstein *et al.*, 1985). In some cases, the receptors (like those for LDL and transferrin) become clustered in coated pits also in the absence of ligand, and are internalized and recycled several times during their lifetime (constitutive endocytosis). In other cases, receptors [like the epidermal growth factor (EGF) receptor and others with tryosine

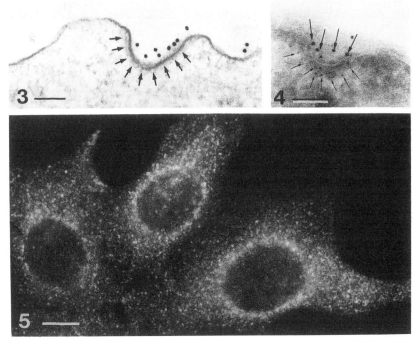

Figures 3–5. Micrographs showing clathrin-coated pits in HEp-2 cells. In Fig. 3, which is an electron micrograph of a section of Epon-embedded cells, the clathrin coat is indicated by arrows. The coated pit contains transferrin receptors as visualized by preembedding immunogold labeling on fixed cells. The electron micrograph in Fig. 4 is from an ultracryosection, where transferrin receptors are immunogold labeled with 10-nm gold particles (large arrows) while clathrin is labeled with 5-nm gold particles (small arrows). In Fig. 5, coated pits are visualized by immunofluorescence using the same anticlathrin antibody as in Fig. 4. Figures 3 and 4, bar = 100 nm; Fig. 5, bar = 10 μm.

kinase activity] only aggregate in coated pits after ligand binding (induced endocytosis).

It has been shown for the receptors of LDL (Davis and Czech, 1986; Lehrman *et al.*, 1985), polymeric immunoglobulin (poly-Ig) (Mostov *et al.*, 1986), EGF (Prywes *et al.*, 1986), transferrin (Iacopetta *et al.*, 1988; Rothenberger *et al.*, 1987), the Fc fragment of immunoglobulins (Miettinen *et al.*, 1989), and mannose 6-phosphate (M6P) (Lobel *et al.*, 1989) that a signal located in the cytoplasmic tail of these transmembrane glycoproteins is crucial for clustering into coated pits and efficient internalization. Although heterogeneity in the internalization sequence between different receptors exists, the structural determinants required for high-efficiency internalization appear, in general, to be found within a stretch of four to six amino acids containing a crucial tyrosine or at least an aromatic residue which must be separated from the transmembrane domain by some residues (Alvarez *et al.*, 1990; Breitfeld *et al.*, 1990; Chen, *et al.*, 1990; Collawn *et al.*, 1990; Davis *et al.*, 1986, 1987; Jing *et al.*, 1990; Ktistakis *et al.*, 1990; Lazarovits and Roth, 1988; Lobel *et al.*, 1989; McGraw and Maxfield, 1990). These data have been generated from studies employing site-directed mutagenesis to alter the cytoplasmic domain not only of receptors internalized by coated pits under physiological conditions (Alvarez *et al.*, 1990; Breitfeld *et al.*, 1990; Chen *et al.*, 1990; Collawn *et al.*, 1990; Davis *et al.*, 1986; Jing *et al.*, 1990; Lobel *et al.*, 1989; McGraw and Maxfield, 1990), but also of plasma membrane proteins such as the influenza virus hemagglutinin that are normally excluded from coated pits (Ktistakis *et al.*, 1990; Lazarovits and Roth, 1988). In the first case, receptor-mediated endocytosis is perturbed by mutations in the internalization sequence, whereas in the second, high-efficiency endocytosis is generated following insertion of one (a tyrosine) or more residues at specific positions in the cytoplasmic tail. Further evidence to support the idea that the internalization sequence operates at least to some extent by mediating receptor clustering in coated pits has come from studies on isolated receptors and plasma membrane proteins. These studies have shown that the HA-2 adaptor, a component of coated pits (see below), binds specifically to the cytoplasmic tail of the LDL receptor, the M6P receptor (M6PR), the poly-Ig receptor, and a point-mutated influenza virus hemagglutinin, where cysteine-543 in the cytoplasmic tail has been replaced by a tyrosine residue (Glickman *et al.*, 1989; Pearse, 1988).

In addition to the receptors mentioned above, which are all transmembrane glycoproteins, at least one example exists for endocytosis via coated pits of a membrane glycolipid, namely the receptor or binding site for Shiga toxin (from *Shigella dysenteriae*) (Sandvig *et al.*, 1989a, 1991a). The clustering of this glycolipid in coated pits apparently depends on ligand (Shiga toxin) binding, and it is still unclear which "signal" in the absence of a cytoplasmic tail leads to the clustering.

Even though efficient receptor concentration in coated pits is mediated by a signal, it is uncertain how efficient this clustering actually is. It often appears from the literature that the majority ($\sim$ 70%) of receptors, for example, for LDL and transferrin are present in coated pits (Anderson *et al.*, 1977; Hopkins, 1983; Hopkins and Trowbridge, 1983), although lower values have been reported in some cases (Anderson et al., 1981; Gal et al., 1982, Harding et al., 1983; Iacopetta et al., 1988; Watts, 1985). Recently, we have measured the efficiency of transferrin receptor clustering in coated pits of various human cell lines by using a combination of immunogold labeling for the receptor and biochemical measurements of uptake of [^{125}I]transferrin (Hansen et al., 1992). For all cell lines, we found in confluent cultures that only about 10% of the transferrin receptors were localized in coated pits (including all stages from flat to deeply invaginated ones). The rate of transferrin uptake was about 10% per min (for the first 2–4 min). These data confirm previous estimates of the lifetime of coated pits of approx. 1 min (Griffiths et al., 1989; Marsh and Helenius, 1980). In order to link these results to the data obtained with mutated transferrin receptors, summarized above, we measured the relative cell surface area occupied by coated pits in HEp-2 cells and found a value of 1.2% in confluent HEp-2. This level seems to be quite general, as previous studies have revealed 1.4% in fibroblasts (Anderson et al., 1976), and 1.6% in BHK cells (Griffiths et al., 1989). Deletion of the internalization motif of the transferrin receptor reduces the efficiency of transferrin uptake to 10–20% of the wild-type level (Collawn et al., 1990; Jing et al., 1990), an effect which would be expected if the mutation results in a decrease from 10% transferrin receptor clustering in coated pits to random distribution (1.2–1.6% of transferrin receptors).

Coated pits act as molecular filters, since some membrane molecules are excluded. This is, for instance, true of the phosphatidylinositol-anchored proteins like Thy-1 (Bretscher et al., 1980). This protein is endocytosed, however, although only at a rate of a few percent per hour (Lemansky et al., 1990), and similarly Thy-1 immunotoxins (Marsh, 1988) and the decay accelerating factor in leukocytes (Tausk et al., 1989) are endocytosed, possibly in a clathrin-independent way (see below).

Two classes of coat molecules are involved in the receptor clustering, generation of pit curvature, and subsequent formation of coated vesicles from the plasma membrane: the HA-2 adaptors and clathrin. The adaptor is a heterotetramer consisting of α- and β-adaptins and two smaller polypeptides, and in involved in the interaction between the cytoplasmic tail of the receptor molecules and the clathrin coat. This coat consists of a polyhedral lattice of clathrin triskelions, each made of three clathrin heavy chains and three clathrin light chains (Brodsky, 1988; Keen, 1990; Pearse and Robinson, 1990). Changes in the clathrin lattice geometry (occurrence of pentagons and hexagons) are thought to be responsible for the curvature eventually leading to formation of a free, coated vesicle (Heuser and Evans, 1980; Larkin et al., 1986).

The structure and function of coated pits can be experimentally modified although the underlying molecular mechanisms have not been clarified. Hence, Larkin et al. (1983) showed that K$^+$ depletion of cells in combination with a hypotonic shock removes coated pits from the cell surface. Apparently, the disappearance of coated pits in K$^+$-depleted cells as well as in cells incubated with hypertonic media is due to an abnormal clathrin polymerization into empty microcages (Heuser and

Anderson, 1989). Upon addition of KCl to K^+-depleted cells coated pits rapidly reassemble, beginning as flat clathrin lattices (Larkin *et al.*, 1986). Analysis of the (re)assembly of coated pits has also been carried out in an *in vitro* system where fibroblast membrane adhering to a substrate was first depleted of coated pits, then allowed to form new coated pits by adding cytosol from other cells (Mahaffey *et al.*, 1989; Moore et al., 1987). The coated pit formation in this system turned out to be rapid at both 4° and 37°C, independent of ATP, and restricted to a limited number of assembly sites. In contrast to the effect of K^+ depletion, acidification of the cytosol (Sandvig *et al.*, 1989b) leads to a paralysis of coated pits at the cell surface (Sandvig *et al.*, 1987). Using his freeze-dry replica technique, Heuser (1989) could further show that these paralyzed pits were associated with some clathrin microcage formation, suggesting that the effect of the various treatments may be somehow related.

Huet *et al.* (1980) studied cell surface distribution and endocytosis of MHC class I molecules (HLA antigens) in cultured human fibroblasts by cross-linking the molecules with ferritin-conjugated antibody against β_2-microglobulin. They found no labeling of coated pits, but the probe was internalized. Although this study is difficult to interpret in terms of the physiological role of endocytosis of class I molecules because of the cross-linking, it indicated that clathrin-independent endocytosis can take place. Morphological studies on endocytosis of tetanus and cholera toxin using gold-labeled probes similarly led to the conclusion that the absence of any labeling of coated pits, endocytosis occurs from noncoated membrane (Montesano *et al.*, 1982; Tran *et al.*, 1987).

Additional support for clathrin-independent endocytosis derives from studies where the coated pit pathway has been experimentally blocked. Moya *et al.* (1985) and Madshus *et al.* (1987) found that in K^+-depleted cells where the uptake of transferrig via coated pits was abolished, ricin was still endocytosed. Similarly, human rhinovirus type 2 was endocytosed in K^+-depleted cells although poliovirus (presumably taken up by coated pits) and transferrin were not (Madshus *et al.*, 1987). Hence, endocytosis takes place in situations where no coated pits are present at the cell surface. Moreover, following acidification of the cytosol, where uptake of transferrin is inhibited, ricin and the fluid-phase markers lucifer yellow (Sandvig *et al.*, 1987) and HRP (West *et al.*, 1989) are still internalized. Quantitative analysis of experiments with K^+ depletion and acidification indicates that clathrin-independent endocytosis may account for roughly (or slightly more than) 50% of the internalization of fluid-phase markers and ligands (such as ricin) not selectively taken up by coated pits (Madshus *et al.*, 1987; Moya *et al.*, 1985; Sandvig *et al.*, 1987). Important in this context is that using acidification to block endocytosis from coated pits, it was recently shown that a mistletoe lectin A-chain immunotoxin directed against mouse leukemia cells seemed to exert most of its toxic effect after entry from non-clathrin-coated membrane (Wiedlocha *et al.*, 1991). Also, it is important to stress that not only can endocytosis take place when the coated pit pathway is blocked, but endocytosis of ricin and fluid-phase markers can be modified experimentally in Vero and A431 cells without any changes in transferrin uptake via coated pits (Sandvig and van Deurs, 1990; and see below). Recently, Wang *et al.* (1990) found that fluid-phase uptake in CHO mutant cells was reduced by 50% without changes in transferrin uptake.

Since criticism can be raised against data obtained under nonphysiological conditions like K^+ depletion or acidification, and since the structural aspects of clathrin-independent endocytosis were unclear, we recently analyzed whether there actually exist distinct preendosomal (primary endocytic) vesicle populations under nonperturbing conditions (Hansen *et al.*, 1991). We took an ultrastructural approach with pulse-labeling of the cell surface at 4°C with concanavalin A (Con A)–gold and very short chase periods at 37°C (30 or 60 sec), followed by fixation and detection of surface-connected structures by an anti-Con A–HRP incubation. Con A–gold that has been internalized is inaccessible to the anti-Con A–HPR antibody and will thus reveal endocytic vesicles by their absence of HRP reaction product. With this technique, two distinct preendosomal vesicle populations were revealed with approximately the same frequency. One population consisted of the typical coated vesicles with an average diameter of 110 nm; the other comprised noncoated vesicles with an average diameter of 95 nm (Figs. 6–10). Following K^+ depletion, no coated endocytic vesicles were observed, whereas noncoated vesicles of the same size range as detected under nonperturbing conditions were still present. Additional control experiments with an antitransferrin receptor antibody as a marker for coated pits ruled out that noncoated vesicles could arise to any significant degree from uncoating of coated vesicles (Hansen *et al.*, 1991).

Which mechanism(s) could be involved in the formation of endocytic vesicles without clathrin, i.e., be responsible for membrane invagination and final membrane fusion–fission (pinching off)? Although the answer is uncertain, several lines of available information should be considered. First, we would like to stress that our term "noncoated" is used to distinguish between coated vesicles and vesicles which in routine preparations for the electron microscope are not covered by the characteristic clathrin coat. It may well be that some kind of protein coat, which could be involved in generating the needed forces for membrane invagination, exists, although it is not visible in the preparation. Hence, surface pits or caveolae of smooth muscle cells have been reported to exhibit a special striation on their cytoplasmic aspect (Prescott and Brightman, 1976; Somlyo *et al.*, 1971). A striped bipolar surface structure of endothelial plasmalemmal vesicles was described by Peters *et al.* (1985) using high-resolution scanning electron microscopy. Brown and co-workers (Brown and Orci, 1986; Brown *et al.*, 1987) found in rat kidney collecting duct cells that vesicles, with a prominent cytoplasmic coat that is immunocytochemically and morphologically distinct from clathrin, were involved in endocytosis of HRP. Also, Orci *et al.* (1986) have described vesicles with a characteristic nonclathrin coat which seem to be involved in the transport between individual Golgi cisternae.

Second, experimental data document that growth factors and cytoskeletal elements are involved in changes of cell surface geometry which could, in fact, lead to clathrin-independent endocytosis. Thus, insulin, insulin-like growth factor-I,

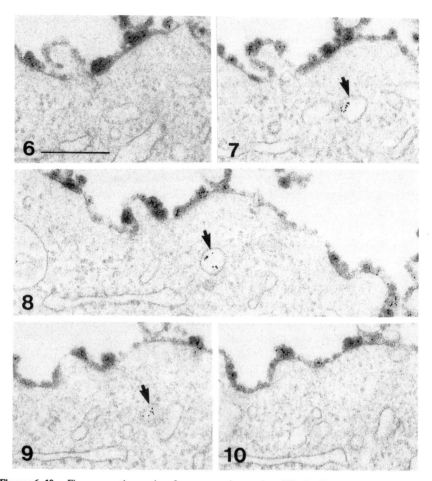

Figures 6–10. Five consecutive sections from an experiment where HEp-2 cells were labeled at 4°C with Con A–gold, and then warmed to 37°C to allow endocytosis and fixed within 60 sec. Following fixation the cells were incubated with anti-Con A/HRP to detect cell surface-associated Con A. With this approach it is possible to distinguish between surface-connected structures and free vesicles in the cytoplasm. It is evident that the vesicle profile appearing in Figs. 7 through 9 (arrows) is noncoated and truly endocytic, since it contains only Con A–gold, but not anti-Con A/HRP. Bar = 0.25 μm.

and EGF, as well as the tumor promoter 12-*O*-tetradecanoyl-phorbol-13-acetate (TPA), induce changes in surface architecture (ruffling) as well as increased endocytic activity (Brunk *et al.*, 1976; Chinkers *et al.*, 1979; Gibbs *et al.*, 1986; Haigler *et al.*, 1979a,b; Miyata *et al.*, 1988, 1989; Phaire-Washington *et al.*, 1980a,b; West *et al.*, 1989). The cytoskeleton is involved in these processes (Miyata *et al.*, 1988, 1989; Phaire-Washington *et al.*, 1980a,b) and cytoskeletal inhibitors can reduce endocytosis (Goldman, 1976; Phaire-Washington *et al.*, 1980a,b; Pratten and Lloyd, 1979; Wagner *et al.*, 1971). Recently, we found that in Vero cells cytochalasin D reduces endocytosis of ricin and of the fluid-phase marker sucrose without reducing the uptake of transferrin. Moreover, colchicine had a similar effect (Sandvig and van Deurs, 1990). In contrast to these cytoskeletal inhibitors, EGF and TPA stimulated clathrin-independent endocytosis in A431 cells (Sandvig and van Deurs, 1990; West *et al.*, 1989). Thus, the formation of noncoated endocytic vesicles described above could somehow involve cytoskeletal elements. However, it is still an open question whether several types of endocytosis not dependent on clathrin exist, since it has been suggested that the ruffling of membrane observed after addition of TPA and EGF gives rise to large endocytic vesicles (Haigler *et al.*, 1979b).

4. INTRACELLULAR SORTING AND TRANSPORT

4.1. Endosomes as the Site of Intracellular Sorting

Following endocytosis, internalized molecules are delivered within minutes to endosomes (Figs. 11–15). These are highly pleiomorphic structures with vacuolar and

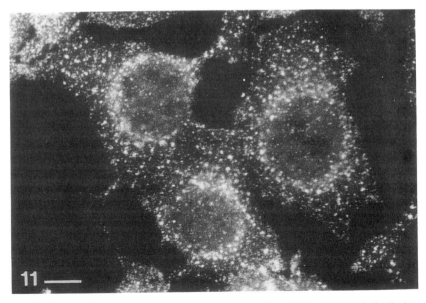

Figure 11. Immunofluorescence micrograph giving an impression of the amount and distribution of endosomes in HEp-2 cells, using endocytosed transferrin as a marker of this compartment. Bar = 10 μm.

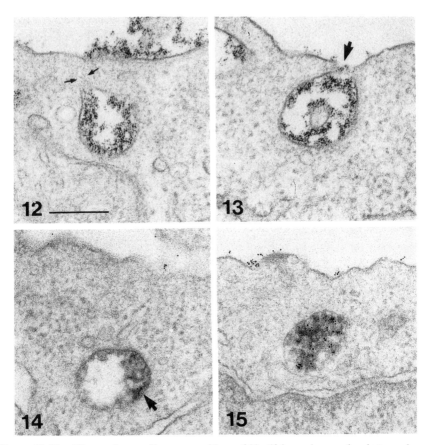

Figures 12–15. Micrographs revealing some problems of identifying endosomes by electron micros-copy. The vacuolar structure containing cationized ferritin in Fig. 12 may represent an endosome, or it may be connected to the cell surface by a "neck"—suggested by small arrows—not visible in this section. In Fig. 13 is shown an example of an endosome-like structure which, however, is clearly surface-connected (arrow). Both pictures are from T47D cells. Figures 14 and 15 (HEp-2 cells) show true endosomes. Following fluid-phase endocytosis of anti-Con A/HRP, the cells have been washed carefully, labeled at 4°C with Con A–gold, and then warmed to 37°C for 1 min (Fig 14) or 15 min (Fig 15) before fixation. The endosomes contain both anti-Con A/HRP and Con A–gold (indicated by the arrow in Fig. 14). Bar = 0.25 μm.

tubulovesicular segments (Geuze *et al.*, 1983, 1984; Griffiths *et al.*, 1989; Marsh *et al.*, 1986; van Deurs *et al.*, 1987). Recent studies using confocal microscopy and video recording led to the conclusion that endosomes may form an extensive, continuous network of vacuoles and tubular cisterns with swellings corresponding to the multivesicular bodies known from electron microscopy (Hopkins *et al.*, 1990). It is generally believed that the endosomal system represents the major sorting station on the endocytic pathway (Goldstein *et al.*, 1985; van Deurs *et al.*, 1989; Wileman *et al.*, 1985). Hence, from the endosomes internalized membrane and ligand may be delivered to intracellular proteolysis in lysosomes, pinched off in small vesicles and transported to the cell surface (recycling or transcytosis) or to the Golgi complex. LDL, for instance, dissociates from its receptor in the endosome, and while the receptor recycles back to the cell surface, LDL is delivered to lysosomes. In contrast, transferrin follows its receptor into the endosome and back to the cell surface; iron bound to transferrin, however, dissociates in the endosome. Some molecules like EGF and insulin may follow the receptor into lysosomes (receptor downregulation) and others like polymeric IgA may be transported across the cell together with the receptor (transcytosis; see below) (Goldstein *et al.*, 1985; Mostov and Simister, 1985; Wileman *et al.*, 1985). Some molecules, for instance the toxic protein ricin, are routed from endosomes to the Golgi complex (van Deurs *et al.*, 1989; see below). Although the low pH in endosomes is definitely involved in processing of internalized molecules, e.g., dissociation of some ligands from their receptors (Mellman *et al.*, 1986), the molecular signals and sorting mechanisms behind the endosomal function are largely unknown.

4.2. Transport to Lysosomes

From the endosomal compartment a major endocytic pathway leads to lysosomes (Figs. 16 and 17). There are, however, problems with generating a satisfactory model for this pathway fitting all experimental data. Our current thinking of how endosomes are "connected" to lysosomes is strongly influenced by two models of lysosome biogenesis (Hubbard, 1989). According to the first model, the *maturation model*, endosomes gradually mature, receive lysosomal enzymes, and finally become lysosomes. The other model, the *vesicle shuttle model*, assumes that early and late endosomes, prelysosomes, and lysosomes are stationary compartments connected by transport vesicles. Data in favor of the vesicle shuttle model were reported, for instance, by Schmid *et al.* (1988). Using free-flow electrophoresis, they found subpopulations of endosomes that were functionally distinct. In a study of intracellular pathways followed by a plasma membrane protein (LEP 100), Lippincott-Schwartz and Fambrough (1987) showed that membrane shuttling takes place between the plasma membrane and lysosomes via endosomes. Griffiths *et al.* (1988) found in NRK cells a distinct prelysosomal compartment characteristically enriched with M6PR, also supporting the vesicle shuttle model (Figs. 16 and 17). Recently, a membrane glycoprotein (plgp57) has been reported to be specific for prelysosomes (Park *et al.*, 1991). In contrast, Croze *et al.* (1989), characterizing a membrane protein (endolyn-78) present in both endosomes and lysosomes in NRK cells, interpreted their data in favor of the maturation model. Data obtained by Diment and Stahl (1985) that proteolysis by a cathepsin D-like protease takes place in macrophage endosomes,

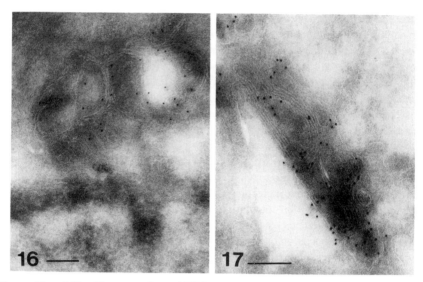

Figures 16 and 17. Ultracryosections of T47D cells showing late endosomes or lysosomes with characteristic internal membrane structures, in Fig. 16 immunogold-labeled for endocytosed ricin, in Fig. 17 for the M6PR. Bars = 0.25 μm.

and by Roederer *et al.* (1987) showing that hydrolysis of internalized molecules can take place very early in the endocytic pathway, and can occur below 20°C, also seem to favor the maturation model. It may be that the correct answer to the puzzle contains elements of both models. For instance, there might be an anterograde transport into lysosomes by maturation and a concomitant retrograde, vesicular backflow through earlier "maturation steps" to the cell surface. An obvious complication would be if the functional organization of the endocytic apparatus varies between cell types, depending, for instance, on the nature of specialization and the degree of differentiation.

Moreover, the transport of the various lysosomal membrane glycoproteins and soluble matrix proteins from their site of synthesis is quite complicated. Thus, whereas soluble lysosomal proteins are sorted from other proteins in the Golgi complex due to the M6P signal and are delivered directly to endosomes–lysosomes (Kornfeld and Mellman, 1989; von Figura and Hasilik, 1986; and see below), lysosomal membrane proteins are routed from the Golgi complex to lysosomes independently of the M6PR. Thus, lysosomal acid phosphatase, which is synthesized and handled as a transmembrane glycoprotein before final proteolytic release into the lysosomal matrix, is first transported from the Golgi complex to the cell surface. Here it is endocytosed and recycled between endosomes and the cell surface for 5– 6 hr before it is finally transferred to lysosomes (Braun *et al.*, 1989). Efficient endocytosis and further targeting to lysosomes depend on a tyrosine signal in the cytoplasmic tail of the lysosomal acid phosphatase. Deletion of the cytoplasmic tail or substitution of the tyrosine with phenylalanine reduces the uptake efficiency and delivery to lysosomes (Peters *et al.*, 1990). Possibly other lysosomal membrane proteins have similar tryosine-containing signals as well, as suggested by studies on the human lysosome membrane glycoprotein h-lamp-1 (Williams and Fukuda, 1990).

Although it is evident that plasma membrane and ligands can be internalized by at least two distinct vesicle types, it is uncertain whether there also exist two (or more) distinct intracellular pathways for endocytosed molecules. Hence, one could imagine that molecules endocytosed by clathrin-dependent and -independent mechanisms, respectively, remain separated intracellularly in different sets of endosomes and lysosomes. Alternatively, all internalized membrane and ligand may reach the same intracellular compartment immediately after internalization and thereafter follow the same routes to the various possible stations of the endocytic pathway. Using cholera toxin–gold as a marker for endocytosis by noncoated vesicles and α_2-macroglobulin–gold as a marker for endocytosis by coated vesicles, Tran *et al.* (1987) found that the internalized markers colocalized in the same endosomes and lysosomes. However, future studies will have to elucidate to what extent different endocytic pathways merge.

Recently, much interest has been paid to endocytic pathways and lysosome biogenesis in polarized epithelial cells. In simple, lining epithelia, discrete apical and basolateral membrane domains are separated by tight junctions that do not allow passive, intercellular diffusion of proteins (Simons and Fuller, 1985; see also Chapter 1). When cells are grown on permeable filters, protein tracers can be added from either the apical or the basolateral surface, or different tracers can be administered from the two sides simultaneously (Figs. 18 and 19). In this way, several recent studies have revealed that molecules taken up from the apical and basolateral surface domains of filter-grown MDCK cells (dog kidney epithelium) (Bomsel *et al.*, 1989; Parton *et al.*, 1989; van Deurs *et al.*, 1990a) and Caco-2 cells (human intestine epithelium) (Hughson and Hopkins, 1990) are initially separated in discrete populations of endosomes. Later, internalized molecules meet in apically localized late endosomes–prelysosomes (Fig. 20–23). Also, *in situ* studies on exocrine pancreas (Oliver, 1982), isolated rabbit proximal tubules (Nielsen *et al.*, 1985), and absorptive cells of suckling rat ileum (Fujita *et al.*, 1990) have provided evidence for such a meeting of apical and basolateral endocytic traffic.

4.3. Recycling

A large-scale recycling of membrane (Burgert and Thilo, 1983; Muller *et al.*, 1980; Schneider *et al.*, 1979) and internalized fluid and solutes (Adams *et al.*, 1982; Besterman *et al.*, 1981; Steinman *et al.*, 1976) from endocytic compartments to the surface of origin is well-established and mandatory to maintain a balance in the cell's

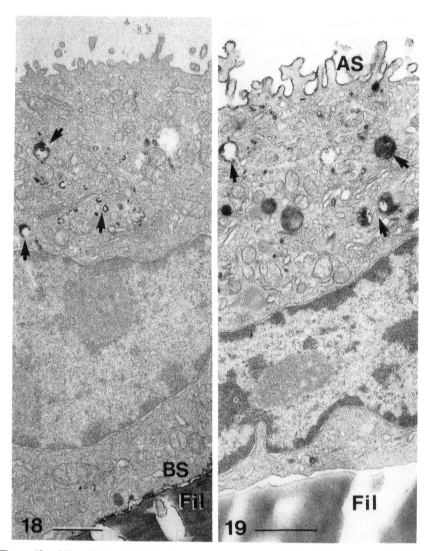

Figures 18 and 19.   Polarized MDCK cells grown on permeable filters (Fil). In Fig. 18 the epithelial cells have been exposed to ricin–HRP from the basal side (BS), whereas in Fig. 19 the cells were exposed from the apical side (AS). In either case, internalized ricin–HRP ends up in late endosomes/lysosomes in the apical cytoplasm (arrows). Bars = 1 μm.

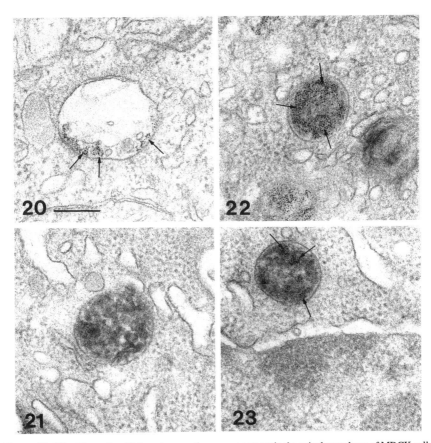

Figures 20–23. Examples of late endosomes/lysosomes present in the apical cytoplasm of MDCK cells. In Fig. 20 the cells have been incubated from the apical side for 1 hr with cationized ferritin (arrows). In Fig. 21 the cells have been incubated from the apical side for 1 hr with ricin–HRP. In Figs. 22 and 23 the cells have been incubated simultaneously with cationized ferritin (arrows) from the apical side and ricin–HRP from the basal side, and the two markers clearly colocalize. Bar =0.25 μm.

surface area and total membrane area of the endosome–lysosome system. About 50% of the fluid endocytosed by a cell, and certainly most membrane, will rapidly reappear at the cell surface. On the other hand, Burgert and Thilo (1983) calculated that only 3% of internalized membrane at any time will enter lysosomes in a macrophage cell line. This membrane is most likely degraded and slowly replaced by *de novo* synthesis as part of the cell's basal membrane turnover. [The reader specifically interested in kinetics of endocytosis and recycling should consult Besterman *et al.* (1981), Burgert and Thilo (1983), and Thilo (1985).]

The continuous inward–outward membrane traffic is a carrier mechanism utilized by numerous receptors like those for LDL and transferrin which, based on correct sorting in endosomes, cycle between the cell surface and the cell's interior repeatedly during their lifetime (Goldstein *et al.*, 1985; Mellman *et al.*, 1986; Steinman *et al.*, 1983). In addition to the LDL and transferrin receptors, the macrophage Fc receptor is a very interesting example in this context. Hence, as long as the Fc receptor is unoccupied or only monovalent Fab fragments of IgG molecules are bound, it is continuously recycled. However, when polyvalent IgG, which is the physiological ligand, binds to Fc receptors, these are transported to lysosomes (Mellman and Plutner, 1984; Mellman *et al.*, 1984). Immunogold labeling studies by Geuze and co-workers (Geuze *et al.*, 1983, 1984) showed that while internalized ligand was mainly found in the vacuolar portions of endosomes, receptors were concentrated in the tubulovesicular portions from where they most likely recycled to the cell surface. Ultrastructural double-labeling experiments and quantifications on L929 fibroblasts indicate that membrane recycling is mediated by numerous, very small vesicles (van Deurs and Nilausen, 1982), which presumably are derived from the tubulovesicular portions of the endosomes. Similarly, evidence has been obtained for a role in recycling of the small tubular vesicles in kidney proximal tubule epithelial cells *in situ* (Christensen, 1982).

4.4. Transcytosis

Instead of delivering internalized membrane and fluid from endosomes to the cell surface of origin as in recycling, *transcytosis* in polarized epithelia leads to transport from one pole (either the apical or the basolateral) to the opposite one. For instance, in liver cells and breast epithelial cells, polymeric immunoglobulins (poly-IgA and -IgM) bind to the basolaterally located poly-Ig receptor and subsequently become transcytosed to the apical surface and released together with a large, extracellular portion of the receptor, the secretory component, into the bile or milk (Mostov and Simister, 1985). It was found that after injection of [^{125}I]-poly-IgA in the saphenous vein of rats, 36% was transported intact to the bile within 3 hr (Hoppe *et al.*, 1985). In neonatal rat intestine, IgG is transcytosed in the opposite, apical-to-basolateral direction (Abrahamson and Rodewald, 1981), and prolactin is transported in the same direction in suckling rats (Gonnella *et al.*, 1989). Apical-to-basolateral transcytosis of thyroglobulin similarly takes place in the thyroid follicle cells (Herzog, 1983), and proteins in the cerebrospinal fluid can be removed by transcytosis from the apical surface of the choroid plexus epithelium (van Deurs *et al.*, 1981). Transcytosis is also of pathophysiological relevance. For instance, some viruses are transcytosed across intestinal epithelium to spread systemically (see Weltzin *et al.*, 1989).

At present, most of our knowledge on transcytosis (as is true for most other aspects of endocytosis as well) is based on studies of various cell lines, i.e., model systems which are easy to handle experimentally but do not necessarily reflect any specific *in situ* (organ-related) situation. Mostov and Deitcher (1986) expressed cDNA for the poly-Ig receptor in polarized MDCK cells and found that > 90% of the newly synthesized receptor was initially inserted into the basolateral membrane, that subsequent transcytosis of the receptor to the apical membrane was independent of ligand binding, and that bound ligand was transcytosed with a $t_{1/2}$ of 30 min. These experiments strongly indicated that the specific transport properties of the poly-Ig receptor depend on signals in the receptor, and that MDCK cells (which do not normally express poly-Ig receptors) are able to read and utilize these signals (Mostov and Deitcher, 1986). Similar results with respect to transcytosis of the poly-Ig receptor were obtained when the cDNA for the poly-Ig receptor was expressed in a rabbit mammary cell line (Schaerer *et al.*, 1990). In further studies with the cDNA expressed in MDCK cells, it was found that about 30% of basolaterally endocytosed poly-Ig is transcytosed to the apical surface, whereas 45% recycles to the basolateral membrane (Breitfeld *et al.*, 1989). In contrast, all apically endocytosed poly-Ig recycles to the apical surface, thus demonstrating that the transcytosis of poly-Ig is unidirectional. Recently, it was reported that phosphorylation of the poly-Ig receptor at serine-664 was required for efficient transcytosis (Casanova *et al.*, 1990).

A receptor-mediated, unidirectional basolateral-to-apical transcytosis in MDCK cells of EGF has also been reported (Maratos-Flier *et al.*, 1987). The opposite situation, a unidirectional, apical-to-basolateral transcytosis, was obtained by transfecting Fc receptors into MDCK cells (Hunziker and Mellman, 1989). The receptors were initially inserted into the apical membrane, and one isoform characterized by a high affinity for coated pits was then transcytosed to the basolateral membrane, while the other isotype largely remained on the apical membrane.

Although transcytosis of various physiological ligands and their receptors thus appears to be specifically unidirectional, the polarized epithelial cell clearly has the capability of bidirectional transcytosis. Thus, molecules endocytosed in the fluid phase, such as HRP and FITC–dextran, are transcytosed in both directions in MDCK cells (von Bonsdorff *et al.*, 1985). Moreover, bidirectional transcytosis of ricin in MDCK cells has also been shown (Figs. 24–26) (van Deurs *et al.*, 1990a). The transcytosis was found to be most efficient (measured in percent of endocytosed ricin) in the apical-to-basolateral direction. Transcytosed ricin could intoxicate other cells, suggesting that ricin can penetrate an epithelial barrier in intact form. Since ricin binds to various cell surface glycoproteins (and glycolipids) with terminal galactose, the observations suggested that endogenous glycoproteins could be bidirectionally transcytosed in MDCK cells. Recently, Brändli *et al.* (1990) used a ricin-resistant mutant of MDCK cells where surface glycoproteins could be efficiently labeled with [³H]galactose and found that whereas some groups of glycoproteins were transcytosed only unidirectionally, one group was actually transported bidirectionally. Studies on MDCK cells expressing Fc receptors and poly-Ig receptors revealed that whereas apical-to-basolateral transcytosis was not influenced by microtubule depolymerization with nocodazole, basolateral-to-apical transcytosis was blocked by this treatment (Hunziker *et al.*, 1990).

Taken together, experimental evidence suggests that although transcytosis can

Bo van Deurs *et al.*

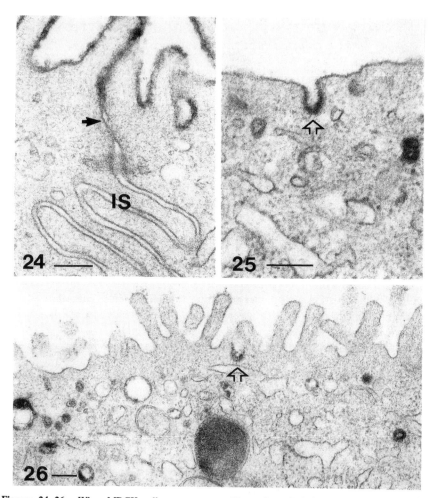

Figures 24–26. When MDCK cells are grown on filters, the apical tight junctions do not allow penetration of proteins, as shown in Fig. 24 where the cells have been incubated with ricin–HRP from the apical side. No labeling of the intercellular space (IS) is seen. Also, when the cells are incubated with ricin–HRP from the basal side, the tight junctions will not allow penetration onto the apical surface (not shown). In Figs. 25 and 26, cells have been incubated with ricin–HRP from the basal side. Labeled transcytotic vesicles in the process of releasing their content at the apical surface are shown with open arrows. Bars = 0.25 μm.

operate in both apical-to-basolateral and basolateral-to-apical directions, the process is clearly not an unselective "elevator mechanism" indiscriminately carrying membrane and content "up and down" in the epithelium. Signals determining whether a given membrane protein enters a transcytotic route and, subsequently, whether this will be strictly unidirectional or bidirectional, as well as a polarity in the epithelial cell's transport machinery, seem to be involved. Again, one may speculate to what extent variations exist between different types of epithelia *in situ*.

4.5. Transport to the Golgi Complex

Newly synthesized membrane proteins, lysosomal proteins, and proteins to be exocytosed from secretory vesicles all follow the same biosynthetic pathway in the cell (Dunphy and Rothman, 1985; Griffiths and Simons, 1986; Kornfeld and Kornfeld 1985; Pfeffer and Rothman, 1987). From the granular endoplasmic reticulum, the proteins are transported in vesicles to the *cis* compartment of the Golgi complex, possibly via an intermediate recycling compartment (Lippincott-Schwartz *et al.*, 1989, 1990). In the Golgi complex, from the *cis* to the *trans* side, a number of protein modifications take place, including terminal glycosylation. From the last Golgi compartment, the *trans*-Golgi network (TGN) (Griffiths and Simons, 1986), the proteins are delivered to their final destination (e.g., the plasma membrane, lysosomes, or secretory vesicles).

The endocytic pathway and the biosynthetic pathway intersect. Thus, newly synthesized hydrolytic enzymes destined for the endocytic pathway are provided with the M6P tag early in the Golgi complex, and in the TGN the M6PR selects the enzymes and mediates vesicular transport to the endosome–lysosome pathway (Griffiths *et al.*, 1988; Kornfeld and Mellman, 1989). Following dissociation of the lysosomal enzyme, the M6PR recycles back to the TGN. This means that there is a vesicular transport mechanism from prelysosomes to the Golgi complex, but other compartments on the endocytic route may also communicate with the Golgi complex (see above). Although highly efficient/selective for the M6PR, this vesicular transport mechanism also allows some other molecules on the endocytic pathway to reach the TGN (Goda and Pfeffer, 1988). Endocytosed ricin, for instance, was found to reach the Golgi complex when a monovalent ricin–HRP conjugate was used as tracer (Fig. 27), whereas polyvalent ricin–HRP as well as a ricin–colloidal gold conjugate, which is also polyvalent, did not. In parallel experiments, these findings were confirmed by detecting native (unconjugated) ricin by preembedding immunoperoxidase cytochemistry (van Deurs *et al.*, 1986). The delivery of internalized ricin to the Golgi complex was a discontinuous, temperature-sensitive process, since it could be completely abolished by incubating cells at 18–20°C (Sandvig *et al.*, 1986; van Deurs *et al.*, 1987). In these experiments, it was realized that it was formally unwarranted to assume that a ricin–HRP-labeled structure intimately associated with the Golgi stack

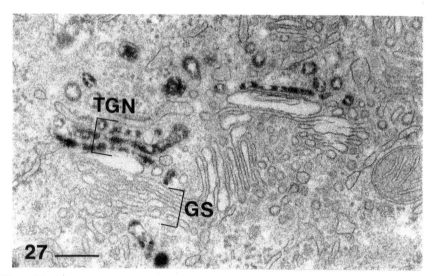

Figure 27.   Part of an MCF-7 cell incubated for 1 hr at 37°C with ricin–HRP. Marked labeling of the TGN is obtained whereas the Golgi stacks (GS) remain unlabeled. Bar = 0.25 μm.

actually represented a Golgi/TGN compartment, without further proof. We therefore applied the G-protein of a temperature-sensitive mutant of vesicular stomatitis virus (VSV tsO45) as a marker of the TGN (Griffiths *et al.*, 1985). Using quantitative analysis of immunogold labeling for ricin and VSV G-protein, it turned out that about 5% of the total amount of ricin internalized in 60 min had reached the TGN of BHK-21 cells (van Deurs *et al.*, 1988). The transport of internalized ricin to the TGN was confirmed in T47D cells using immunogold labeling for a newly synthesized membrane molecule normally occurring in these cells (MAM-6) as a Golgi/TGN marker (Hansen *et al.*, 1989). Similarly, we have found that internalized Shiga toxin–HRP reaches the Golgi/TGN of HeLa S3, Vero, and MDCK cells (Sandvig *et al.*, 1989a, 1991), and using cell fractionation, we recently found that after 60 min about 10% of the Shiga toxin was present in the Golgi fraction of MDCK cells (Sandvig *et al.*, 1991).

5. TRANSLOCATION ACROSS INTERNAL MEMBRANE SYSTEMS

Of the toxins transported into the cytosol from an intracellular compartment, most is known about ricin, abrin, Shiga toxin, and diphtheria toxin, and we will therefore concentrate on these toxins in this section. The delivery of internalized toxin to the TGN is a very important step as it appears necessary for translocation of ricin A-chain to the cytosol. Hence, Youle and Colombatti (1987) found that a hybridoma cell line producing antibodies against ricin was resistant to ricin. Most likely, internalized ricin meets the antibody in the TGN (or a pre-TGN compartment), before the ricin A-chain becomes translocated, strongly indicating that transport of ricin from the cell surface to endosomes–lysosomes does not fulfill the requirements for translocation. There is a lag time of about 45 min between ricin internalization and initial inhibition of protein synthesis, also suggesting that ricin has to penetrate rather deep into the cell along the endocytic pathway. At 18–20°C, when endocytosis is ongoing, although at a reduced rate, no ricin reaches the Golgi complex/TGN, and the toxic effect is abolished. Low concentrations of monensin with no effect on lysosomal and endosomal pH influence the Golgi complex functionally and structurally and sensitize cells to ricin (Sandvig and Olsnes, 1982), and NH_4Cl, which neutralizes endosomal and lysosomal pH, also increases the sensitivity to ricin (Sandvig et al., 1979). Similarly, both monensin and NH_4Cl are able to sensitize cells to some immunotoxins. In fact, a 30,000-fold increase in activity is observed with some immunotoxins (Olsnes et al., 1989). Drugs like swainsonine and tunicamycin, which inhibit various steps in glycosylation of newly synthesized proteins, and cycloheximide, which inhibits protein synthesis, all sensitive cells to ricin (Sandvig et al., 1986). Similar results have been obtained with abrin. These aspects are reviewed in greater detail elsewhere (van Deurs et al., 1989). Moreover, in recent experiments with the fungal drug Brefeldin A (BFA), which in some cells makes the Golgi complex disappear, apparently by a retrograde transport to the endoplasmic reticulum (Lippincott-Schwartz et al., 1989, 1990), we found that all cell lines in which BFA had a visible effect on the Golgi complex were protected against ricin, abrin, and Shiga toxin, whereas there was no protection against diphtheria toxin (Sandvig et al., 1991b). This is in agreement with the concept that diphtheria toxin enters the cytosol from early, acidic endosomes (see Section 2). In one cell line, MDCK, however, BFA had no effect on the structure of the Golgi complex and did not protect against ricin.

6. PERSPECTIVES

With our increasing knowledge about intracellular protein traffic and the signals that govern the routing of individual molecules, we may in the future be able to design immunotoxins and other cell-specific toxic molecules in such a way that they not only bind to the right cells, but that they are also routed to the correct intracellular compartments for translocation to occur. As more detailed information is obtained about the specific functions of the different domains in the toxin molecules and we learn more about the mechanisms for translocation of proteins through membranes, new toxic molecules may, with the molecular techniques available today, be constructed in such a way as to be not only highly toxic, but also more stable and perhaps smaller in size with a view to increasing their access to the cells we wish to eliminate.

ACKNOWLEDGMENTS. The work from our research group referred to in this review has been supported by the Danish and Norwegian Cancer Societies, the Danish Medical Research Council, the Norwegian Research Council for Science and Humanities, the NOVO Foundation, the Gangsted Foundation, the Madsen Foundation, the Leo Nielsen Foundation, the Wedell-Wedellsborg Foundation, the European Molecular Biology Organization (EMBO), and a NATO Collaborative Research Grant (CRG 900517).

NOTE ADDED IN PROOF. Recently we have shown that internalized shiga toxin in butyric acid-treated A431 cells can be transported to the endoplasmic reticulum. This seems necessary for the toxic effect (Sandvig *et al.*, 1992).

REFERENCES

Abrahamson, D. R., and Rodewald, R., 1981, Evidence for the sorting of endocytic vesicle contents during receptor-mediated transport of IgG across newborn rat intestine, *J. Cell Biol.* **91**:270–280.

Adams, C. J., Maurey, K. M., and Storrie, B., 1982, Exocytosis of pinocytic contents by Chinese hamster ovary cells, *J. Cell Biol.* **93**:632–637.

Ahnert-Hilger, G., Weller, U., Dauzenroth, M.-E., Habermann, E., and Gratzl, M., 1989, The tetanus toxin light chain inhibits exocytosis, *FEBS* **242**:245–248.

Aktories, K., and Wegner, A., 1989, ADP-ribosylation of actin by clostridal toxins, *J. Cell Biol.* **109**:1385–1387.

Alvarez, E., Girones, N., and Davis, R. J., 1990, Inhibition of the receptor-mediated endocytosis of diferric transferrin is associated with the covalent modification of the transferrin receptor with palmitic acid, *J. Biol. Chem.* **265**:16644–16655.

Anderson, R. G. W., Goldstein, J. L., and Brown, M. S., 1976, Localization of low density lipoprotein receptors on plasma membrane of normal human fibroblasts and their absence in cells from a familial hypercholesterolemia homozygote, *Proc. Natl. Acad. Sci. USA* **73**:2434–2438.

Anderson, R. G. W., Brown, M. S., and Goldstein, J. L., 1977, Role of coated endocytic vesicle in the uptake of receptor-bound low density lipoprotein in human fibroblasts, *Cell* **10**:351–364.

Anderson, R. G. W., Brown, M. S., and Goldstein, J. L., 1981, Inefficient internalization of receptor-bound low density lipoprotein in human carcinoma A-431 cells, *J. Cell Biol.* **88**:441–452.

Besterman, J. M., Airhart, J. A., Woodworth, R. C., and Low, R. B., 1981, Exocytosis of pinocytosed fluid in cultured cells: Kinetic evidence for rapid turnover and compartmentation, *J. Cell Biol.* **91**:716–727.

Bomsel, M., Prydz, K., Parton, R. G., Gruenberg, J., and Simons, K., 1989, Endocytosis in filter-grown Madin–Darby canine kidney cells, *J. Cell Biol.* **109**:3242–3258.

Brändli, A. W., Parton, R. G., and Simons, K., 1990, Transcytosis in MDCK cells: Identification of glycoproteins transported bidirectionally between both plasma membrane domains, *J. Cell Biol.* **111**:2909–2921.

Braum, M., Waheed, A., and von Figura, K., 1989, Lysosomal acid phosphatase is transported to lysosomes via the cell surface, *EMBO J.* **8**:3633–3640.

Breitfeld, P. P., Harris, J. M., and Mostov, K. E., 1989, Postendocytic sorting of the ligand for the polymeric immunoglobulin receptor in Madin–Darby canine kidney cells, *J. Cell Biol.* **109**:475–486.

Breitfeld, P. P., Casanova, J. E, McKinnon, W. C., and Mostov, K. E., 1990, Deletions in the cytoplasmic domain of the polymeric immunoglobulin receptor differentially affect endocytic rate and postendocytic traffic, *J. Biol. Chem.* **265**:13750–13757.

Bretscher, M. S., Thomson, J. N., and Pearse, B. M. F., 1980, Coated pits act as molecular filters, *Proc. Natl. Acad. Sci. USA* **77**:4156–4159.

Brodsky, F. M., 1988, Living with clathrin: Its role in intracellular membrane traffic, *Science* **242**:1396–1402.

Brown, D., and Orci, L., 1986, The "coat" of kidney intercalated cell tubulovesicles does not contain clathrin, *Am. J. Physiol.* **250**:C605–C608.

Brown, D., Weyer, P., and Orci, L., 1987, Nonclathrin-coated vesicles are involved in endocytosis in kidney collecting duct intercalated cells, *Anat. Rec.* **218**:237–242.

Brunk, U, Schellens, J., and Westermark, B., 1976, Influence of epidermal growth factor (EGF) on ruffling activity, pinocytosis and proliferation of cultivated human glia cells, *Exp. Cell Res.* **103**:295–302.

Burgert, H.-G., and Thilo, L., 1983, Internalization and recycling of plasma membrane glycoconjugates during pinocytosis in the macrophage cell line, P388D1, *Exp. Cell Res.* **144**:127–142.

Casanova, J. E., Breitfeld, P. P., Ross, S. A., and Mostov, K. E., 1990, Phosphorylation of the polymeric immunoglobulin receptor required for its efficient transcytosis, *Science* **248**: 742–745.

Chen, W.-J., Goldstein, J. L., and Brown, M. S., 1990, NPXY, a sequence often found in cytoplasmic tails, is required for coated pit-mediated internalization of low density lipoprotein receptor, *J. Biol. Chem.* **265**:3116–3123.

Chinkers, M., McKanna, J. A., and Cohen, S., 1979, Rapid induction of morphological changes in human carcinoma cells A-431 by epidermal growth factor, *J. Cell Biol.* **83**: 260–265.

Christensen, E. I., 1982, Rapid membrane recycling in renal proximal tubule cells. *Eur. J. Cell Biol.* **29**:43–49.

Collawn, J. F., Stangel, M., Kuhn, L. A., Esekogwu, V., Jing, S., Trowbridge, I. S., and Tainer, J. A., 1990, Transferrin receptor internalization sequence YXRF implicates a tight turn as the structural recognition motif for endocytosis, *Cell* **63**:1061–1072.

Croze, E., Ivanov, I. E., Kreibich, G., Adesnik, M., Sabatini, D. D., and Rosenfeld, M. G., 1989, Endolyn-78, a membrane glycoprotein present in morphologically diverse components of the endosomal and lysosomal compartments: Implications for lysosome biogenesis, *J. Cell Biol.* **108**:1597–1613.

Davis, C. G., Lehrman, M. A., Russell, D. W., Anderson, R. G. W., Brown, M. S., and Goldstein, J. L., 1986, The J. D. mutation in familial hypercholesterolemia: Amino acid substitution in cytoplasmic domain impedes internalization of LDL receptors, *Cell* **45**:15–24.

Davis, C. G., van Driel, I. R., Russel, D. W., Brown, M. S., and Goldstein, J. L., 1987, The low density lipoprotein receptor: Identification of amino acids in the cytoplasmic domain required for rapid endocytosis, *J. Biol. Chem.* **262**:4075–4082.

Davis, R. J., and Czech, M. P., 1986, Regulation of transferrin receptor expression at the cell surface by insulin-like growth factors, epidermal growth factor and platelet-derived growth factor, *EMBO J.* **5**:653–658.

de Paiva, A., and Dolly, J. O., 1990, Light chain of botulinum neurotoxin is active in mammalian motor nerve terminals when delivered via liposomes, *FEBS* **277**:171–174.

Diment, S., and Stahl, P., 1985, Macrophage endosomes contain proteases which degrade endocytosed protein ligands, *J. Biol. Chem.* **260**:15311–15317.

Donovan, M. G., and Storm, D. R., 1990, Evidence that the adenylate cyclase secreted from *Bordetella pertussis* does not enter animal cells by receptor-mediated endocytosis, *J. Cell Physiol.* **145**:444–449.

Draper, R. K., and Simon, M. I., 1980, The entry of diphtheria toxin into the mammalian cell cytoplasm: Evidence for lysosomal involvement, *J. Cell Biol.* **87**:849–854.

Dunphy, W. G., and Rothman, J. E., 1985, Compartmental organization of the Golgi stack, *Cell* **42**:13–21.

Fujita, M., Reinhart, F., and Neutra, M., 1990, Convergence of apical and basolateral endocytic pathways at apical late endosomes in absorptive cells of suckling rat ileum in vivo, *J. Cell Sci.* **97**:385–394.

Gal, D., Simpson, E. R., Porter, J. C., and Snyder, J. M., 1982, Defective internalization of low density lipoprotein in epidermoid cervical cancer cells, *J. Cell Biol.* **92**:597–603.

Geuze, H. J., Slot, J. W., Strous, G. J. A. M., Lodish, H. F., and Schwartz, A. L., 1983, Intracellular site of asialoglycoprotein receptor-ligand uncoupling: Double-label immuno-electron microscopy during receptor-mediated endocytosis, *Cell* **32**:277–287.

Geuze, H. J., Slot, J. W., Strous, G. J. A. M., Peppard, J., von Figura, K., Hasilik, A., and Schwartz, A. L., 1984, Intracellular receptor sorting during endocytosis: Comparative immunoelectron microscopy of multiple receptors in rat liver, *Cell* **37**:195–204.

Gibbs, E. M., Lienhard, G. E., Appleman, J. R., Lane, M. D., and Frost, S. C., 1986, Insulin stimulates fluid-phase endocytosis and exocytosis in 3T3-L1 adipocytes, *J. Biol. Chem.* **261**:3944–3951.

Glickman, J. N., Conibear, E., and Pearse, B. M. F., 1989, Specificity of binding of clathrin adaptors to signals on the mannose 6-phosphate/insulin-like growth factor II receptor, *EMBO J.* **8**:1041–1047.

Goda, Y., and Pfeffer, S. R., 1988, Selective recycling of the mannose 6-phosphate/IGF-II receptor to the trans Golgi network in vitro, *Cell* **55**:309–320.

Goldman, R., 1976, The effect of cytochalasin B and colchicine on concanavalin A induced vacuolation in mouse peritoneal macrophages, *Exp. Cell Res.* **99**:385–394.

Goldstein, J. L., Brown, M. S, Anderson, R. G. W., Russell, D. W., and Schneider, W. J., 1985, Receptor mediated endocytosis: Concepts emerging from the LDL receptor system, *Annu. Rev. Cell Biol.* **1**:1–39.

Gonnella, P. A., Harmatz, P., and Walker, W. A., 1989, Prolactin is transported across the epithelium of the jejunum and ileum of the suckling rat, *J. Cell Physiol.* **140**:138–149.

Gordon, V. M., Young, W. W., Lechler, S. M., Gray, M. C., Leppla, S. H., and Hewlett, E. L., 1989, Adenylate cyclase toxins from Bacillus anthracis and Bordella pertussis. Different processes for interaction with and entry into target cells, *J. Biol. Chem.* **264**:14792–14796.

Griffiths, G., and Simons, K., 1986, The trans Golgi network: Sorting at the exit site of the Golgi complex, *Science* **234**:438–443.

Griffiths, G., Pfeiffer, S., Simons, K., and Matlin, K., 1985, Exit of newly synthesized membrane proteins from the trans cisterna of the Golgi complex to the plasma membrane, *J. Cell Biol.* **101**:949–964.

Griffiths, G., Hoflack, B., Simons, K., Mellman, I., and Kornfeld, S., 1988, The mannose 6-phosphate receptor and the biogenesis of lysosomes, *Cell* **52**:329–341.

Griffiths, G., Back, R., Marsh, M., 1989, A quantitative analysis of the endocytic pathway in baby hamster kidney cells, *J. Cell Biol.* **109**:2703–2720.

Gruenberg, J., and Howell, K. E., 1989, Membrane traffic in endocytosis: Insights from cell-free assays, *Annu. Rev. Cell Biol.* **5**:453–481.

Haigler, H. T., McKanna, J. A., and Cohen, S., 1979a, Direct visualization of the binding and internalization of a ferritin conjugate of epidermal growth factor in human carcinoma cells A-431, *J. Cell Biol.* **81**:382–395.

Haigler, H. T., McKanna, J. A., and Cohen, S., 1979b, Rapid stimulation of pinocytosis in human carcinoma cells A-431 by epidermal growth factor, *J. Cell Biol.* **83**:82–90.

Hansen, S. H., Petersen, O. W., Sandvig, K., Olsnes, S., and van Deurs, B., 1989, Internalized ricin and the plasma membrane glycoprotein MAM-6 colocalize in the trans-Golgi network of T47D human breast carcinoma cells, *Exp. Cell Res.* **185**:373–386.

Hansen, S. H., Sandvig, K., and van Deurs, B., 1991, The preendosomal compartment comprises distinct coated and noncoated endocytic vesicle populations, *J. Cell Biol.* **113**: 731–741.

Hansen, S. H., Sandvig, K., and van Deurs, B., 1992, Internalization efficiency of the transferrin receptor, *Exp. Cell Res.* **199**:19–28.

Harding, C., Heuser, J., and Stahl, P., 1983, Receptor-mediated endocytosis of transferrin and recycling of the transferrin receptor in rat reticulocytes, *J. Cell Biol.* **97**:329–339.

Henriques, B., Florin, I., and Thelestam, M., 1987, Cellular internalisation of *Clostridium difficile* toxin A, *Microb. Pathogenesis* **2**:455–463.

Herzog, V., 1983, Transcytosis in thyroid follicle cells, *J. Cell Biol.* **97**:607–617.

Heuser, J., 1989, Effects of cytoplasmic acidification on clathrin lattice morphology, *J. Cell Biol.* **108**:401–411.

Heuser, J. E., and Anderson, R. G. W., 1989, Hypertonic media inhibit receptor-mediated endocytosis by blocking clathrin-coated pit formation, *J. Cell Biol.* **108**:389–400.

Heuser, J., and Evans, L., 1980, Three-dimensional visualization of coated vesicle formation in fibroblasts, *J. Cell Biol.* **84**:560–583.

Hopkins, C. R., 1983, Intracellular routing of transferrin and transferrin receptors in epidermoid carcinoma A431 cells, *Cell* **35**:321–330.

Hopkins, C. R., and Trowbridge, I. S., 1983, Internalization and processing of transferrin and the transferrin receptor in human carcinoma A431 cells, *J. Cell Biol.* **97**:508–521.

Hopkins, C. R., Gibson, A., Shipman, M., and Miller, K., 1990, Movement of internalized ligand–receptor complexes along a continuous endosomal reticulum, *Nature* **346**:335–339.

Hoppe, C. A., Connolly, T. P., and Hubbard, A. L., 1985, Transcellular transport of polymeric IGA in the rat hepatocyte: Biochemical and morphological characterization of the transport pathway, *J. Cell Biol.* **101**:2113–2123.

Hubbard, A, 1989, Endocytosis, *Curr. Opin. Cell Biol.* **1**:675–683.

Huet, C., Ash, J. F., and Singer, S. J., 1980, The antibody-induced clustering and endocytosis of HLA antigens on cultured human fibroblasts, *Cell* **21**:429–438.

Hughson, E. J., and Hopkins, C. R., 1990, Endocytic pathways in polarized Caco-2 cells: Identification of an endosomal compartment accessible from both apical and basolateral surfaces, *J. Cell Biol.* **110**:337–348.

Hunziker, W., and Mellman, I., 1989, Expression of macrophage–lymphocyte Fc receptors in Madin–Darby canine kidney cells: Polarity and transcytosis differ for isoforms with or without coated pit localization domains, *J. Cell Biol.* **109**:3291–3302.

Hunziker, W., Male, P., and Mellman, I., 1990, Differential microtubule requirements for transcytosis in MDCK cells, *EMBO J.* **9**:3515–3525.

Iacopetta, B. J., Rothenberger, S., and Kühn, L. C., 1988, A role for the cytoplasmic domain in transferrin receptor sorting and coated pit formation during endocytosis, *Cell* **54**:485–489.

Jackson, M. P., 1990, Structure–function analyses of Shiga toxin and the Shiga-like toxins, *Microb. Pathogenesis* **8**:235–242.

Jing, S., Spencer, T., Miller, K., Hopkins, C., and Trowbridge, I. S., 1990, Role of the human transferrin receptor cytoplasmic domain in endocytosis: Localization of a specific sequence for internalization, *J. Cell Biol.* **110**:283–294.

Keen, J. H., 1990, Clathrin and associated assembly and disassembly proteins, *Annu. Rev. Biochem.* **59**:415–438.

Kornfeld, R., and Kornfeld, S., 1985, Assembly of asparagine-linked oligosaccharides, *Annu. Rev. Biochem.* **54**:631–664.

Kornfeld, S., and Mellman, I., 1989, The biogenesis of lysosomes, *Annu. Rev. Cell Biol.* **5**:483–525.

Ktistakis, N. T., Thomas, D., and Roth, M. G., 1990, Characteristics of the tyrosine recognition signal for internalization of transmembrane surface glycoproteins, *J. Cell Biol.* **111**:1393–1407.

Larkin, J. M., Brown, M. S., Goldstein, J. L., and Anderson, R. G. W., 1983, Depletion of intracellular potassium arrests coated pit formation and receptor-mediated endocytosis in fibroblasts, *Cell* **33**:273–285.

Larkin, J. M., Donzell, W. C., and Anderson, R. G. W., 1986, Potassium-dependent assembly of coated pits: New coated pits form as planar clathrin lattices, *J. Cell Biol.* **103**:2619–2627.

Lazarovits, J., and Roth, M., 1988, A single amino acid mutation in the cytoplasmic domain allows the influenza virus hemagglutinin to be endocytosed through coated pits, *Cell* **53**:743–752.

Lehrman, M. A., Goldstein, J. L., Brown, M. S., Russell, D. W., and Schneider, W. J., 1985, Internalization-defective LDL receptors produced by genes with nonsense and frameshift mutations that truncate the cytoplasmic domain, *Cell* **41**:735–743.

Lemansky, P., Fatemi, S. H., Gorican, B., Meyale, S., Rossero, R., and Tartakoff, A. M., 1990, Dynamics and longevity of the glycolipid-anchored membrane protein, Thy-1, *J. Cell Biol.* **110**:1525–1531.

Lippincott-Schwartz, J., and Fambrough, D. M., 1987, Cycling of the integral membrane glycoprotein, LEP100, between plasma membrane and lysosomes: Kinetic and morphological analysis, *Cell* **49**:669–677.

Lippincott-Schwartz, J., Yuan, L. C., Bonifacino, J. S., and Klausner, R. D., 1989, Rapid redistribution of Golgi proteins into the ER in cells treated with Brefeldin A: Evidence for membrane cycling from Golgi to ER, *Cell* **56**:801–813.

Lippincott-Schwartz, J., Donaldson, J. G., Schweizer, A., Berger, E. G., Hauri, H.-P., Yuan, L. C., and Klausner, R. D., 1990, Microtubule-dependent retrograde transport of proteins into the ER in the presence of Brefeldin A suggests an ER recycling pathway, *Cell* **60**:821–836.

Lobel, P., Fujimoto, K., Ye, R. D., Griffiths, G., and Kornfeld, S., 1989, Mutations in the cytoplasmic domain of the 275 kd mannose 6-phosphate receptor differentially alter lysosomal sorting and endocytosis, *Cell* **57:**787–796.

McGraw, T. E., and Maxfield, F. R., 1990, Human transferrin receptor internalization is partially dependent upon an aromatic amino acid on the cytoplasmic domain, *Cell Regul.* **1:**369–377.

Madshus, I. H., Sandvig, K., Olsnes, S., and van Deurs, B., 1987, Effect of reduced endocytosis inducted by hypotonic shock and potassium depletion on the infection of HEp 2 cells by picornaviruses, *J. Cell Physiol.* **131:**14–22.

Mahaffey, D. T., Moore, M. S., Brodsky, F. M., and Anderson, R. G. W., 1989, Coat proteins isolated from clathrin coated vesicles can assemble into coated pits, *J. Cell Biol.* **108:**1615–1624.

Maratos-Flier, E., Yang Kao, C.-Y., Verdin, E. M., and King, G. L., 1987, Receptor-mediated vectorial transcytosis of epidermal growth factor by Madin–Darby canine kidney cells, *J. Cell Biol.* **105:**1595–1601.

Marsh, J. W., 1988, Antibody-mediated routing of diphtheria toxin in murine cells results in a highly efficacious immunotoxin, *J. Biol. Chem.* **263:**15993–15999.

Marsh, M., and Helenius, A., 1980, Adsorptive endocytosis of Semliki Forest virus, *J. Mol. Biol.* **142:**439–454.

Marsh, M., Griffiths, G., Dean, G. E., Mellman, I., and Helenius, A., 1986, Three-dimensional structure of endosomes in BHK-21 cells, *Proc. Natl. Acad. Sci. USA* **83:**2899–2903.

Mellman, I., and Plutner, H., 1984, Internalization and degradation of macrophage Fc receptors bound to polyvalent immune complexes, *J. Cell Biol.* **98:**1170–1177.

Mellman, I., Plutner, H., and Ukkonen, P., 1984, Internalization and rapid recycling of macrophage Fc receptors tagged with monovalent antireceptor antibody: Possible role of a prelysosomal compartment, *J. Cell Biol.* **98:**1163–1169.

Mellman, I., Fuchs, R., and Helenius, A., 1986, Acidification of the endocytic and exocytic pathways, *Annu. Rev. Biochem.* **55:**663–700.

Miettinen, H. M., Rose, J. K., and Mellman, I., 1989, Fc receptor isoforms exhibit distinct abilities for coated pit localization as a result of cytoplasmic domain heterogeneity, *Cell* **58:**317–327.

Miyata, Y., Hoshi, M., Koyasu, S., Kadowaki, T., Kasuga, M., Yahara, I., Nishida, E., and Sakai, H., 1988, Rapid stimulation of fluid-phase endocytosis and exocytosis by insulin, insulin-like growth factor-I, and epidermal growth factor in KB cells, *Exp. Cell Res.* **178:**73–83.

Miyata, Y., Nishida, E., Koyasu, S., Yahara, I., and Sakai, H., 1989, Regulation by intracellular Ca^{2+} and cyclic AMP of the growth factor-induced ruffling membrane formation and stimulation of fluid-phase endocytosis and exocytosis, *Exp. Cell Res.* **181:**454–462.

Mochida, S., Poulain, B., Weller, U., Habermann, E., and Tauc, L., 1989, Light chain of tetanus toxin intracellularly inhibits acetylcholine release at neuro-neuronal synapses, and its internalization is mediated by heavy chain, *FEBS* **253:**47–51.

Montesano, R., Roth, J., Robert, A., and Orci, L., 1982, Non-coated membrane invaginations are involved in binding and internalization of cholera and tetanus toxins, *Nature* **296:**651–653.

Moore, M. S., Mahaffey, D. T., Brodsky, F. M., and Anderson, R. G. W., 1987, Assembly of clathrin-coated pits onto purified plasma membranes, *Science* **236:**558–563.

Moskaug, J. Ø., Sandvig, K., and Olsnes, S., 1988, Low pH-induced release of diphtheria toxin A-fragment in Vero cells. Biochemical evidence for transfer to the cytosol, *J. Biol. Chem.* **263:**2518–2525.

Moss, J., and Vaughan, M., 1988, ADP-ribosylation of guanyl neucleotide-binding regulatory proteins by bacterial toxins, *Adv. Enzymol.* **61:**303–379.

Mostov, K. E., and Deitcher, D. L., 1986, Polymeric immunoglobulin receptor expressed in MDCK cells transcytoses IgA, *Cell* **46:**613–621.

Mostov, K. E., and Simister, N. E., 1985, Transcytosis, *Cell* **43:**389–390.

Mostov, K. E., de Bruyn Kops, A., and Deitcher, D. L., 1986, Deletion of the cytoplasmic domain of the polymeric immunoglobulin receptor prevents basolateral localization and endocytosis, *Cell* **47:**359–364.

Moya, M., Dautry-Varsat, A., Goud, B., Louvard, D., and Boquet, P., 1985, Inhibition of coated pit formation in HEp-2 cells blocks the cytotoxicity of diphtheria toxin but not that of ricin toxin, *J. Cell Biol.* **101:**548–559.

Muller, W. A., Steinman, R. M., and Cohn, Z. A., 1980, The membrane proteins of the vacuolar system. II. Directional flow between secondary lysosomes and plasma membrane, *J. Cell Biol.* **86:**304–314.

Nielsen, J. T., Nielsen, S., and Christensen, E. I., 1985, Transtubular transport of proteins in rabbit proximal tubules, *J. Ultrastruct. Res.* **92:**133–145.

Oliver, C., 1982, Endocytic pathways at the lateral and basal cell surfaces of exocrine acinar cells, *J. Cell Biol.* **95:**154–161.

Olsnes, S., and Sandvig, K., 1988, How protein toxins enter and kill cells, in: *Immunotoxins* (A. E. Frankel, ed.), Kluwer, Boston, pp. 39–73.

Olsnes, S., Sandvig, K., Petersen, O. W., and van Deurs, B., 1989, Immunotoxins—Entry into cells and mechanisms of action, *Immunol. Today* **10:**291–295.

Orci, L., Glick, B. S., and Rothman, J. E., 1986, A new type of coated vesicular carrier that appears not to contain clathrin: Its possible role in protein transport within the Golgi stack, *Cell* **46:**171–184.

Park, J. E., Lopez, J. M., Cluett, E. B., and Brown, W. J., 1991, Identification of a membrane glycoprotein found primarily in the prelysosomal endosome compartment, *J. Cell Biol.* **112:**245–255.

Parton, R. G., Prydz, K., Bomsel, M., Simons, K., and Griffiths, G., 1989, Meeting of the apical and basolateral endocytic pathways of the Madin–Darby canine kidney cell in late endosomes, *J. Cell Biol.* **109:**3259–3272.

Pearse, B. M. F., 1988, Receptors compete for adaptors found in plasma membrane coated pits, *EMBO J.* **7:**3331–3336.

Pearse, B. M. F., and Robinson, M. S., 1990, Clathrin, adaptors, and sorting. *Annu. Rev. Cell Biol.* **6:**151–171.

Peters, C. Braun, M., Weber, B., Wendland, M., Schmidt, B., Pohlmann, R., Waheed, A., and von Figura, K., 1990, Targeting of a lysosomal membrane protein: A tyrosine-containing endocytosis signal in the cytoplasmic tail of lysosomal acid phosphatase is necessary and sufficient for targeting to lysosomes, *EMBO J.* **9:**3497–3506.

Peters, K. R., Carley, W. W., and Palade, G. E., 1985, Endothelial plasmalemmal vesicles have a characteristic striped bipolar surface structure, *J. Cell Biol.* **101:**2233–2238.

Pfeffer, S. R., and Rothman, J. E., 1987, Biosynthetic protein transport and sorting by the endoplasmic reticulum and Golgi, *Annu. Rev. Biochem.* **56:**829–852.

Phaire-Washington, L., Silverstein, S. C., and Wang, E., 1980a, Phorbol myristate acetate stimulates microtubule and 10-nm filament extension and lysosome redistribution in mouse macrophages, *J. Cell Biol.* **86:**641–655.

Phaire-Washington, L., Wang, E., and Silverstein, S., 1980b, Phorbol myristate acetate stimulates pinocytosis and membrane spreading in mouse peritoneal macrophages, *J. Cell Biol.* **86:**634–640.

Pratten, M. K., and Lloyd, J. B., 1979, Effects of temperature, metabolic inhibitors and some other factors on fluid-phase and adsorptive pinocytosis by rat peritoneal macrophages, *Biochem. J.* **180:**567–671.

Prescott, L., and Brightman, M. W., 1976, The sarcolemma of aplysia smooth muscle in freeze-fracture preparations, *Tissue Cell* **8:**241–258.

Prywes, R., Livneh, E., Ullrich, A., and Schlessinger, J., 1986, Mutations in the cytoplasmic domain of EGF receptor affects EGF binding and receptor internalization, *EMBO J.* **5:** 2179–2190.

Rodman, J. S., Mercer, R. W., and Stahl, P. D., 1990, Endocytosis and transcytosis, *Curr. Opin. Cell Biol.* **2:**664–672.

Roederer, M., Bowser, R., and Murphy, R. F., 1987, Kinetics and temperature dependence of exposure of endocytosed material to proteolytic enzymes and low pH: Evidence for a maturation model for the formation of lysosomes, *J. Cell Physiol.* **131:**200–209.

Rothenberger, S., Iacopetta, B. J., and Kühn, L. C., 1987, Endocytosis of the transferrin receptor requires the cytoplasmic domain but not its phosphorylation site, *Cell* **49:** 423–431.

Sandvig, K., and Olsnes, S., 1980, Diphtheria toxin entry into cells is facilitated by low pH, *J. Cell Biol.* **87:**828–832.

Sandvig, K., and Olsnes, S., 1981, Rapid entry of nicked diphtheria toxin into cells at low pH. Characterization of the entry process and the effects of low pH on the toxin molecule, *J. Biol. Chem.* **256:**9068–9076.

Sandvig, K., and Olsnes, S., 1982, Entry of the toxic proteins abrin, modeccin, ricin and diphtheria toxin into cells. II. Effect of pH, metabolic inhibitors, and ionophores and evidence for toxin penetration from endocytic vesicles, *J. Biol. Chem.* **257:**7504–7513.

Sandvig, K., and Olsnes, S., 1988, Diphtheria toxin-induced channels in Vero cells selective for monovalent cations, *J. Biol. Chem.* **263:**12352–12359.

Sandvig, K., and van Deurs, B., 1990, Selective modulation of the endocytic uptake of ricin and fluid phase markers without alteration in transferrin endocytosis, *J. Biol. Chem.* **265:** 6382–6388.

Sandvig, K., Olsnes, S., and Pihl, A., 1979, Inhibitory effect of ammonium chloride and chloroquine on the entry of the toxic lectin modeccin into HeLa cells, *Biochem. Biophys. Res. Commun.* **90:**648–655.

Sandvig, K., Tønnessen, T. I., and Olsnes, S., 1986, Ability of inhibitors of glycosylation and protein synthesis to sensitize cells to abrin, ricin, Shigella toxin, and Pseudomonas toxin, *Cancer Res.* **46:**6418–6422.

Sandvig, K., Olsnes, S., Petersen, O. W., and van Deurs, B., 1987, Acidification of the cytosol inhibits endocytosis from coated pits, *J. Cell Biol.* **105:**679–689.

Sandvig, K., Olsnes, S., Brown, J. E., Petersen, O. W., and van Deurs, B., 1989a, Endocytosis from coated pits of Shiga toxin: A glycolipid-binding protein from *Shigella dysenteriae*, *J. Cell Biol.* **108**:1331–1343.

Sandvig, K., Olsnes, S., Petersen, O. W., and van Deurs, B., 1989b, Control of coated-pit function by cytoplasmic pH, *Methods Cell Biol.* **32**:365–382.

Sandvig, K., Prydz, K., Ryd, M., and van Deurs, B., 1991a, Endocytosis and intracellular transport of the glycolipid-binding ligand Shiga toxin in polarized MDCK cells, *J. Cell Biol.* **113**:553–562.

Sandvig, K., Prydz, K., Hansen, S. H., and van Deurs, B., 1991b, Ricin transport in Brefeldin A-treated cells: Correlation between Golgi structure and toxic effect, *J. Cell Biol.* **115**: 971–981.

Sandvig, K., Garred, Ø., Prydz, K., Kozlov, J. V., Hansen, S. H., and Van Deurs, B., 1992, Retrograde transport of endocytosed Shiga toxin to the endoplasmic reticulum, *Nature* **358**: 510–511.

Schaerer, E., Verrey, F., Racine, L., Tallichet, C., Reinhardt, M., and Kraehenbuhl, J.-P., 1990, Polarized transport of the polymeric immunoglobulin receptor in transfected rabbit mammary epithelial cells, *J. Cell Biol.* **101**:987–998.

Schmid, S. L., Fuchs, R., Male, P., and Mellman, I., 1988, Two distinct subpopulations of endosomes involved in membrane recycling and transport to lysosomes, *Cell* **52**:73–83.

Schneider, Y. J., Tulkens, P., deDuve, C., and Trouet, A., 1979, Fate of plasma membrane during endocytosis. I. Uptake and processing of anti-plasma membrane and control immunoglobulins by cultured fibroblasts, *J. Cell Biol.* **82**:449–465.

Simons, K., and Fuller, S. D., 1985, Cell surface polarity in epithelia, *Annu. Rev. Cell Biol.* **1**: 243–288.

Singh, Y., Leppla, S. H., Bhatnagar, R., and Friedlander, A. M., 1989, Internalization and processing of Bacillus anthracis lethal toxin by toxin-sensitive and -resistant cells, *J. Biol. Chem.* **164**:11099–11102.

Somlyo, A. P., Devine, C. E., Somlyo, A. V., and North, S. R., 1971, Sarcoplasmic reticulum and the temperature-dependent contraction of smooth muscle in calcium-free solutions, *J. Cell Biol.* **51**:722–741.

Stecher, B., Weller, U., Habermann, E., Gratzl, M., and Ahnert-Hilger, G., 1989, The light chain but not the heavy chain of botulinum A toxin inhibits exocytosis from permeabilized adrenal chromaffin cells, *FEBS* **255**:391–394.

Steinman, R. M., Brodie, S. E., and Cohn, Z. A., 1976, Membrane flow during pinocytosis. A stereologic analysis, *J. Cell Biol.* **68**:665–687.

Steinman, R. M., Mellman, I. S., Muller, W. A., and Cohn, Z. A., 1983, Endocytosis and the recycling of plasma membrane, *J. Cell Biol.* **96**:1–27.

Stenmark, H., Olsnes, S., and Sandvig, K., 1988, Requirement of specific receptors for efficient translocation of diphtheria toxin A fragment across the plasma membrane, *J. Biol. Chem.* **263**:13449–13455.

Stenmark, H., Moskaug, J.Ø., Madshus, I. H., Sandvig, K., and Olsnes, S., 1991, Peptides fused to the amino-terminal end of diphtheria toxin are translocated to the cytosol, *J. Cell Biol.* **113**:1025–1032.

Tausk, F., Fey, M., and Gigli, I., 1989, Endocytosis and shedding of the decay accelerating factor on human polymorphonuclear cells, *J. Immunol.* **143**:3295–3302.

Thilo, L., 1985, Quantification of endocytosis-derived membrane traffic, *Biochim. Biophys. Acta* **822**:243–266.

Tran, D., Carpentier, J.-L., Sawano, F., Gordon, P., and Orci, L., 1987, Ligands internalized through coated or noncoated invaginations follow a common intracellular pathway, *Proc. Natl. Acad. Sci. USA* **84:**7957–7961.

van Deurs, B., and Nilausen, K., 1982, Pinocytosis in mouse L-fibroblasts: Ultrastructural evidence for a direct membrane shuttle between the plasma membrane and the lysosomal compartment, *J. Cell Biol.* **94:**279–286.

van Deurs, B., von Bülow, F., and Møoller, M., 1981, Vesicular transport of cationized ferritin by the epithelium of the rat choroid plexus, *J. Cell Biol.* **89:**131–139.

van Deurs, B., Tønnessen, T. I., Petersen, O. W., Sandvig, K., and Olsnes, S., 1986, Routing of internalized ricin and ricin conjugates to the Golgi complex, *J. Cell Biol.* **102:**37–47.

van Deurs, B., Petersen, O. W., Olsnes, S., and Sandvig, K., 1987, Delivery of internalized ricin from endosomes to cisternal Golgi elements is a discontinuous, temperature-sensitive process, *Exp. Cell Res.* **171:**137–152.

van Deurs, B., Sandvig, K., Petersen, O. W., Olsnes, S., Simons, K., and Griffiths, G., 1988, Estimation of the amount of internalized ricin that reaches the *trans*-Golgi network, *J. Cell Biol.* **106:**253–267.

van Deurs, B., Petersen, O. W., Olsnes, S., and Sandvig, K., 1989, The ways of endocytosis, *Int. Rev. Cytol.* **117:**131–177.

van Deurs, B., Hansen, S. H., Petersen, O. W., Løkken Melby, E., and Sandvig, K., 1990a, Endocytosis, intracellular transport and transcytosis of the toxic protein ricin by a polarized epithelium, *Eur. J. Cell Biol.* **51:**96–109.

van Deurs, B., Sandvig, K., Petersen, O. W., and Olsnes, S., 1990b, Endocytosis and intracellular sorting of ricin, in: *Trafficking of Bacterial Toxins* (C. B. Saelinger, ed.), CRC Press, Boca Raton, Fla., pp. 91–119.

von Bonsdorff, C. H., Fuller, S. D., and Simons, K., 1985, Apical and basolateral endocytosis in Madin–Darby canine kidney (MDCK) cells grown on nitrocellulose filters, *EMBO J.* **4:** 2781–2792.

von Figura, K., and Hasilik, A., 1986, Lysosomal enzymes and their receptors, *Annu. Rev. Biochem.* **55:**167–193.

Wagner, R., Rosenberg, M., and Estensen, R., 1971, Endocytosis in Chang liver cells. Quantitation by sucrose-^{3}H uptake and inhibition by cytochalasin B., *J. Cell Biol.* **50:** 804–817.

Wang, R.-H., Colbaugh, P. A., Koa, C.-Y., Rutledge, E. A., and Draper, R. K., 1990, Impaired secretion and fluid-phase endocytosis in the End4 mutant of Chinese hamster ovary cells, *J. Biol. Chem.* **165:**20179–20187.

Watts, C., 1985, Rapid endocytosis of the transferrin receptor in the absence of bound transferrin, *J. Cell Biol.* **100:**633–637.

Weltzin, R., Lucia-Jandris, P., Michetti, P., Fields, N. B., Kraehenbuhl, J. P., and Neutra, M. R., 1989, Binding and transepithelial transport of immunoglobulins by intestinal M cells: Demonstration using monoclonal IgA antibodies against enteric viral proteins, *J. Cell Biol.* **108:**1673–1685.

West, M. A., Bretscher, M. S., and Watts, C., 1989, Distinct endocytotic pathways in epidermal growth factor-stimulated human carcinoma A431 cells, *J. Cell Biol.* **109:**2731–2739.

Wiedlocha, A., Sandvig, K., Walzel, H., Radzikowsky, C., and Olsnes, S., 1991, Internalization and action of an immunotoxin containing mistletoe lectin A-chain, *Cell Res.* **51:**916–920.

Wileman, T., Harding, C., and Stahl, P., 1985, Receptor-mediated endocytosis, *Biochem. J.* **232:**1–14.

Williams, M. A., and Fukuda, M., 1990, Accumulation of membrane glycoproteins in lysosomes requires a tyrosine residue at a particular position in the cytoplasmic tail, *J. Cell Biol.* **111:**955–966.

Youle, R. J., and Colombatti, M., 1987, Hybridoma cells containing intracellular anti-ricin antibodies show ricin meets secretory antibody before entering the cytosol, *J. Biol. Chem.* **262:**4676–4682.

II

Epithelial Barriers

Chapter 5

Transepithelial Transport of Proteins by Intestinal Epithelial Cells

Marian R. Neutra and Jean-Pierre Kraehenbuhl

1. INTRODUCTION

The mucosal surface of the digestive tract is a vast surface area covered by a monolayer of epithelial cells, joined by tight junctions that provide an effective barrier to proteins and peptides. Epithelial cells play important roles in nutrition and mucosal immune defense apart from their simple function as a barrier, however, and some of these functions require transepithelial vesicular transport of intact macromolecules (Neutra and Kraehenbuhl, 1992). For example, a minority population of epithelial cells (the M cells) are highly specialized for import of antigens to the cells of the mucosal immune system, while a major population of diverse epithelial and glandular cells selectively export polymeric immunoglobulins onto mucosal surfaces. In this chapter, we will focus on the basic mechanisms of membrane traffic and the epithelial cell specializations that allow the epithelium in the intestine to function as a gatekeeper. In addition to controlling transepithelial transport of proteins from the lumen to the interstitial tissue of the mucosa, the intestinal epithelium fulfills other roles such as digestion and absorption of nutrients and maintenance of a functional barrier.

Epithelial cell diversity in the intestine has important implications for protein

Marian R. Neutra • Gastrointestinal Cell Biology Laboratory, Children's Hospital and Department of Pediatrics, Harvard Medical School, Boston, Massachusetts 02115. *Jean-Pierre Kraehenbuhl* • Swiss Institute for Cancer Research and Institute of Biochemistry, University of Lausanne, CH-1066 Epalinges, Switzerland.

Biological Barriers to Protein Delivery, edited by Kenneth L. Audus and Thomas J. Raub. Plenum Press, New York, 1993.

transport. The intestinal absorptive cell or "enterocyte" is not a uniform cell type; its phenotype changes during intestinal development of the fetus and neonate, and during differentiation of individual cells in adults. Enterocyte phenotypes vary widely in their capacity for endocytosis and transcytosis. Dramatic changes in endocytic activity occur in the entire absorptive enterocyte population during fetal and postnatal development in mammals as they pass through progressive stages of digestive function (Neutra and Louvard, 1989). In addition, important changes in cell ultrastructure occur in the epithelium as cells migrate along the crypt–villus axis (Neutra, 1988; Madara and Trier, 1987). In adult villus cells, the endocytic capacity of individual absorptive enterocytes, and transport of antigens across the villus epithelial barrier, are generally very limited. Although only a very small proportion of endocytosed protein may enter transepithelial transport vesicles in these cells, the functional effect of such transport may be amplified by the immense numbers of cells involved (Smith *et al.*, 1992; Pusztai, 1989).

Over specific mucosal sites marked by the presence of organized lymphoid follicles, a unique epithelium occurs containing an unusual phenotype, the M cell. M cells represent an exceedingly small minority in the intestinal epithelium but their functional significance for mucosal immunity is amplified by their unique position over the organized lymphoid tissues, the inductive sites for mucosal immune responses (Neutra and Kraehenbuhl, 1992). Their potential significance for protein or peptide absorption is enhanced by their ability to transport antigens with great efficiency. In both M cells and enterocytes, the efficiency, selectivity, and consequences of protein uptake and transport depend on the directions of membrane traffic in the cell and the capacity of the intracellular pathways through which proteins are directed. These pathways function through the common mechanisms that govern membrane traffic in all cells.

2. TRANSEPITHELIAL TRANSPORT MECHANISMS

There is now a large body of information on receptor-mediated binding of macromolecular ligands to cell surfaces, and the cellular and molecular mechanisms whereby macromolecules and particles are internalized. Endocytic pathways and their component membrane compartments that carry macromolecules either toward the degradative lysosomal compartment or to other destinations, have been elucidated largely through studies on nonpolarized cells (Kornfeld and Mellman, 1989). Much recent attention has been focused on the endocytic compartments in polarized epithelial cells and the special function of transcytosis (Mostov and Simister, 1985; Schaerer *et al.*, 1991; Sztul et al., 1991; see Chapter 4). Because of its convenience, many studies have exploited the relatively simple model cell culture system of Madin–Darby canine kidney (MDCK) cells (Casanova *et al.*, 1990; Brandli *et al.*, 1990). Although the basic mechanisms operating in MDCK cells may be generalized to other cell types such as intestinal enterocytes, it is also clear that specific receptors and the activities of intracellular vesicular pathways differ among epithelial cell types. Thus, it is important that intestinal enterocyte-like cell lines derived from human colon adenocarcinomas (Caco-2, T84, and HT29) have also served as *in vitro* models (Hidalgo *et al.*, 1989; Heyman *et al.*, 1990; Lencer *et al.*, 1992). Thus, a few key findings derived from these *in vitro* models will be reviewed here.

2.1. Establishment and Maintenance of Cell Polarity

In cultured, polarized epithelial cells as well as in intestinal epithelial cells *in vivo*, tight junctions prevent diffusion of many ions, small molecules, and all macromolecules between cells (Madara, 1988) and lateral diffusion of glycolipids and proteins between apical and basolateral domains of the plasma membrane (Dragsten *et al.*, 1981). The apical domain of epithelial cells, including intestinal cells, is a differentiated structure containing components not present either in nonpolarized cell types, or in the same cells in culture prior to junction formation and polarization (Louvard *et al.*, 1985; Godefroy *et al.*, 1988; Rodriguez-Boulan and Nelson, 1989).

The basolateral cell surface is generally considered a single domain but in fact is divided into two major subdomains. The lateral subdomain is involved in cell–cell interactions via cell adhesion molecules (Nelson and Hammerton, 1989) and is enriched in NA^+,K^+-ATPase, poly Ig receptors, and other components in enterocytes (Slot and Geuze, 1984; Amerongen *et al.*, 1989). The basal subdomain is tethered to the basal lamina and is enriched in receptors that recognize extracellular matrix. Within both apical and lateral domains, there are specialized microdomains such as microvilli, cell–cell interdigitations, and other sites stabilized by submembrane cytoskeleton and clathrin-coated pits (Mooseker, 1985; Neutra *et al.*, 1988; Nelson and Hammerton, 1989).

Maintenance of the two major membrane domains in intestinal cells involves insertion of new membrane components via two routes. Vesicles derived from the trans-Golgi network and containing domain-specific proteins are transported directly to either the apical or basolateral domain in MDCK cells (Caplan *et al.*, 1986), but it was recently shown that in some cell types, specific components mis-inserted into one domain can be endocytosed and sent to the opposite domain by "corrective" transcytosis. In hepatocytes and intestinal enterocytes *in vivo* and in cell culture, certain proteins that at steady state are residents of the apical domain are consistently directed after synthesis form the trans-Golgi to the basolateral domain, but then are selectively withdrawn and carried to the apical domain in transcytotic vesicles (Massey *et al.*, 1987; Bartles *et al.*, 1987; Matter *et al.*, 1990). In MDCK cells, corrective transcytosis of viral proteins also occurs in the opposite direction (Matlin *et al.*, 1983) and if this also occurs in intestinal cells *in vivo*, it might provide a transcytotic pathway for uptake of foreign proteins from the lumen.

2.2. Endocytosis and Sorting into the Transcytotic Pathway

A prerequisite for transcytosis is the endocytic uptake of adsorbed or fluid-phase macromolecules via clathrin-coated or noncoated pits and vesicles (Anderson, 1991). In many cells types, incoming vesicles fuse to form "early endosomes" which are peripheral compartments consisting of clear vesicles with attached tubules (review: Kornfeld and Mellman, 1989; Chapter 4). This compartment in most cells acidifies to pH 6.0–6.2, and at the lowered pH certain ligands are released from their receptors (review: Maxfield and Yamashiro, 1991; Chapter 3). An important function of early endosomes is the physical segregation of ligands and receptors, and the segregation of various receptors from each other in the plane of the early endosome membrane. For example, in basolateral (sinusoidal) endosomes in hepatocytes, various receptors were localized by EM immunocytochemistry, and the images suggested that similar receptors cluster together and are removed from the endosomal tubules by selective budding of small vesicles destined for transport to different destinations: recycling to the same cell surface, transport to lysosomes, or transcytosis to the opposite membrane domain (Geuze et al., 1984, 1987).

The basic molecular mechanisms whereby sorting occurs in the plasma membrane or the endosome membrane are still unresolved. In addition, the exact site(s) at which transcytotic vesicles are formed is not established. There is recent evidence that recycling of membrane proteins and vesicle content can occur not only from early endosomes but also from late endosomes, also called multivesicular bodies, transport endosomes, and prelysosomal compartments (Hughson and Hopkins, 1990; Rabinowitz et al., 1992). This is consistent with earlier observations that vesicles all along the endosomal pathway have tubular extensions. It is not known, however, whether all of these extensions from early endosomes, late endosomes, and prelysosomes can give rise to transcytotic vesicles. This is important for intestinal cells since it would affect the degree of proteolytic processing that might occur during transcytosis.

Ultrastructural tracer studies in kidney tubular cell lines, intestinal cell lines, and enterocytes in vivo, have identified distinct sets of apical and basolateral early endosomes (Rodewald, 1980; Bomsel et al., 1989; Fujita et al., 1990; Hughson and Hopkins, 1990). Although both sets of endosomes have a tubulovesicular morphology and are capable of recycling to their respective cell surfaces, apical and basolateral early endosomes seem to be functionally and compositionally different. When isolated from MDCK cells, they were unable to fuse with each other in vitro (Bomsel et al., 1990), but both types of early endosomes nevertheless fused with common "late endosomes" isolated from the same cells (Gruenberg and Howell, 1989). These restricted fusion patterns correspond to the pathways of transfer of apical and basolateral tracer proteins from separate early endosomes to common late endosomes, observed in MDCK cells (Parton et al., 1989), and in intact intestinal epithelial cells in vivo (Fujita et al., 1990).

Endosomes were once thought to lack degradative enzymes because they lack traditional lysosomal enzyme markers such as acid phosphatase. It is now clear, however, that degradative enzymes are present even in early endosome compartments (Diment and Stahl, 1985; reviews: Courtoy, 1991; Chapter 3). Although lysosomal or endosomal enzymes can enter the endosome by receptor-mediated uptake from the outside in some cells, the major source of endosomal as well as lysosomal proteases seems to be receptor-mediated delivery in vesicles from the Golgi complex (Griffiths *et al.*, 1988; Kornfeld and Mellman, 1989). Hydrolases from the trans-Golgi are delivered into vesicles all along the endocytic pathway, but the major delivery site is the late endosome. It is not known whether the same enzymes are delivered to early apical and basolateral endosomes in polarized cells. In epithelial cells, uptake of enzymes from the external milieu may be limited to the basolateral side, since on MDCK cells mannose 6-phosphate receptors were found only basolaterally (Prydz *et al.*, 1990). Although lysosomes have been considered nonrecycling organelles, "resident" glycoproteins of the lysosome membrane are now known to cycle to the plasma membrane and back via the endocytic pathway (Lippincott-Schwartz and Fambrough, 1987). If this occurs in intestinal cells, partially digested protein fragments could be delivered to the interstitial tissue.

2.3. Mechanism of Transcytosis

Ultrastructural tracers representing both soluble proteins and specific ligands have been used to detect transcytosis in intact epithelial organs and epithelial cell culture systems, but since such tracers invariably enter multiple pathways, the identity of the transcytotic vesicles and the exact transcytotic route are not clear. A biochemical analysis of isolated transcytotic carrier vesicles, however, was achieved by immunoaffinity isolation of vesicles from hepatocytes (Sztul *et al.*, 1991). Access to apical as well as basolateral surfaces is restricted in most intact organs, and therefore the kinetics of internalization and the fates of endocytosed proteins have been analyzed primarily in cultured cell lines grown on permeable supports. Results derived from MDCK cells, and Caco-2 or HT29 human colon carcinoma cells, have revealed that there are striking differences among these models.

In MDCK cells, the amount of internalization of the fluid-phase protein horseradish peroxidase (HRP) from apical and basolateral surfaces was identical. Of the HRP taken up from the apical side, relatively small amounts were directed to lysosomes; most of the protein was released from the cells, half by recycling and half by transcytosis to the other side (Bomsel *et al.*, 1989). Of the HRP taken up basolaterally, only small amounts were transcytosed to the apical side. Certain membrane proteins were found to be transported bidirectionally across MDCK cells, indicating that single membrane vesicles may have made the entire transepithelial trip (Brandli *et al.*, 1990).

In Caco-2 cells, varying amounts of apical-to-basolateral transepithelial transport of HRP were clearly documented by some studies (Hidalgo *et al.*, 1989; Heyman *et al.*, 1990), although some investigators did not observe it (Hughson and Hopkins, 1990). These discrepancies may have been due to differences in the tracers or in the Caco-2 cell clones used. In a quantitative study (Heyman *et al.*, 1990), little intact HRP was transported from the apical to basal side, although significant amounts of breakdown products were released basally. In contrast, much larger amounts of intact HRP entered the basal-to-apical transcytotic pathway. The authors concluded that monolayers of this cell line may resemble intestinal crypt cells in their protein transport properties, and emphasized that the amount, rate, and direction of transcytosis in cultured epithelial cell lines provide useful information but cannot accurately reflect the special features of transport activity of all of the various enterocyte phenotypes present *in vivo* (Heyman *et al.*, 1990).

To date, only two receptor systems that mediate transcytosis *in vivo* have been analyzed at the molecular level (Breitfeld *et al.*, 1989a; Simister and Mostov, 1989; Apodaca *et al.*, 1991) and both of these receptor systems operate in the intestinal epithelium. The epithelial Fc receptor mediates apical-to-basolateral transcytosis of maternal immunoglobulins in the proximal small intestine of neonatal rodents, but is not present in neonatal humans (Rodewald, 1980; Simister and Mostov, 1989). The poly Ig receptor mediates basolateral-to-apical delivery of IgA in a wide variety of epithelial and glandular cells at all ages (Kuhn and Kraehenbuhl, 1982; Mostov *et al.*, 1984). Molecular signals mediate initial cell surface targeting, endocytosis, and trafficking of these immunoglobulin-like receptor molecules from the Golgi to the cell surface, through early endosomes, and into their respective transcytotic pathways to the opposite cell surface. These pathways have been studied in epithelial cell lines transfected with either wild-type or mutated receptor cDNAs encoding them (Breitfeld *et al.*, 1989a,b; Casanova *et al.*, 1990). Studies to date have shown that the major portion or perhaps all of the targeting information required for the complex itineraries of these proteins resides in the cytoplasmic tails of the receptors. "Sorting signals" may be contained both in specific amino acid phosphorylation sites and in motifs of secondary structure (Schaerer *et al.*, 1991; Apodaca *et al.*, 1991).

It has been assumed that transcytotic vesicles directly recognize and fuse with the contralateral cell surface, but studies on transfected MDCK cells expressing either the Fc receptor or the poly Ig receptor showed that transcytosed membrane proteins actually recycle between early endosomes and the adjacent cell surface. This may occur both at their original site of insertion and at the ultimate destination (Breitfeld *et al.*, 1989b; Hunziker and Mellman, 1989; Hunziker *et al.*, 1991a). Thus, transcytotic vesicles may be targeted not to plasma membrane but to "opposite" endosomes, and theoretically, a single population of transcytotic carrier vesicles could shuttle between the apical and basolateral early endocytic compartments (Schaerer *et al.*, 1991). These vesicles would have to be able to fuse with early endosomes at either cell pole. Since transcytotic vesicles from intestinal or MDCK cells have not yet been isolated, the *in vitro* experiments required to test this hypothesis have not been done.

2.4. Role of the Cytoskeleton in Vesicular Traffic

Microtubules play an important but controversial role in cell membrane polarity and transcytosis in epithelial cells. It is thought that microtubules serve primarily to move vesicles over "long" distances in cells, but that they are not required for endocytosis or exocytosis (Kelly, 1990). In epithelial cells, vesicles carrying newly synthesized cell surface and secretory materials move from the Golgi to the apical surface along a route corresponding to the microtubules that run parallel to the lateral cell membrane (Specian and Neutra, 1984; Eilers *et al.*, 1989; Achler *et al.*, 1989; Bacallao *et al.*, 1989; van der Sluijs *et al.*, 1990). The minus ends of microtubules have been shown to be oriented toward the apical cell surface of cultured epithelial cells (Bacallao *et al.*, 1989); therefore, polarized delivery of apically directed vesicles presumably involves a dynein-like motor (Schroer and Sheetz, 1989). Microtubules may also play a role in basolateral-to-apical transcytosis: they appeared to be involved in transcytosis of poly Ig receptors from the basolateral to the apical membrane in transfected MDCK cells (Hunziker *et al.*, 1990), and in transport of IgA into bile (Goldman *et al.*, 1983). In addition, movement of peripheral endosomes to more central locations and to lysosomes is mediated by microtubules in both polarized epithelial cells and nonpolarized cells (Gruenberg *et al.*, 1989; Hughson and Hopkins, 1990; Bomsel *et al.*, 1990). In this case, vesicles move toward the plus end of the microtubules, presumably via a kinesin-type motor (Schroer and Sheetz, 1989). Although microtubules were required for basolateral-to-apical transcytosis in transfected MDCK cells expressing both the poly Ig receptor and the epithelial Fc receptor, apical-to-basolateral transcytosis was microtubule independent (Hunziker *et al.*, 1990). It is possible that some other type of molecular motor and cytoskeletal "track" facilitates apical-to-basolateral vesicular transport.

Transport of proteins from one side of an epithelium to the other thus may not be accomplished by simple movement of a vesicle derived from one plasma membrane and fusion with the opposite membrane domain. Rather, it is likely that transcytosis requires a sequence of events including formation and fusion of endosomes, generation of transcytotic membrane vesicles, and perhaps participation of additional organelles and cytoskeleton. The complexity of these compartments was recently demonstrated by the dramatic alteration in basolateral endosomes that accompanied inhibition of transcytosis of the poly Ig receptor in MDCK cells by the antibiotic Brefeldin A (Hunziker *et al.*, 1991b). Current information about pathways of vesicular traffic must be considered in the interpretation of experiments exploring protein transport across intestinal epithelial cells *in vivo*.

3. MEMBRANE TRAFFIC IN INTESTINAL ENTEROCYTES

There is indirect evidence that in the intestine, antigens resulting from digestion in the lumen, and transepithelially transported or degraded intracellularly in enterocyte lysosomes, could associate with MHC II and be presented to lymphocytes within or below the epithelium. It has been suggested that this might suppress the systemic immune response to luminal antigens such as digestion products of food (Bland and Warren, 1986; Kaiserlain *et al.*, 1989; Mayer *et al.*, 1991). There is also evidence that food antigens can be transported across the intestinal epithelium in amounts sufficient to evoke an immune response (Pusztai, 1989; Smith *et al.*, 1992). The exact transepithelial transport pathways or the cell types that mediate these phenomena are not yet clear. Thus, it is useful to review what is currently known about membrane traffic in normal enterocytes.

3.1. Mechanisms of Antigen Exclusion by Enterocytes

Enterocytes of both villus and crypt epithelium *in vivo* are sealed by continuous tight junctions that permit charge-selective passage of certain ions, water, and perhaps small organic molecules such as glucose, but consistently exclude peptides or macromolecules (Madara, 1988). The apical surface of enterocytes is a highly differentiated structure covered by rigid, closely packed microvilli (Mooseker, 1985) that are coated with the glycocalyx, a thick layer of membrane-anchored and peripheral glycoproteins (Semenza, 1986). The glycocalyx impedes the passage of many viruses, bacteria, particles, and protein aggregates between microvilli and can prevent their contact with the plasma membrane of the microvillus (Gonnella and Neutra, 1985). Although the glycocalyx is rich in the many saccharides that serve as receptors for lectins or microorganisms on living enterocytes *in vivo*, some of these receptors are masked by the macromolecular assembly of the glycocalyx (Gonnella and Neutra, 1985; Neutra *et al.*, 1987). Others, however, serve as binding sites for lectins; indeed, certain lectins have been shown to induce damage to the enterocyte brush border (Weinman *et al.*, 1989). Crypt cell apical membranes lack a thick glycocalyx, but outward fluid flow inhibits passive entry of many macromolecules at this site (Phillips *et al.*, 1987). Normal enterocytes can also clear adherent molecules by "shedding" microvillus membrane or membrane components (Gonnella and Neutra, 1985). For example, microvillar vesicles were shed after cross-linking by lectins (Weinman *et al.*, 1989), and individual lipid-linked or integral membrane enzymes are routinely released (Alpers, 1975). Some enterocyte-like cell lines and MDCK cells in culture, whose apical surfaces may be less fully differentiated than enterocytes *in vivo*, can provide misleading information concerning protein adherence and endocytosis in the normal gut.

Specialized plasma membrane microdomains at the bases of microvilli are responsible for endocytosis (review: Neutra *et al.*, 1988). In adult enterocytes, the structure of the brush border and glycocalyx impedes access of many large molecules and particles to these microdomains. The clathrin-coated endocytic pits at this site in adult enterocytes are apparently intended for uptake of physiologic ligands such as intrinsic factor–cobalamin complexes (Levine *et al.*, 1984), but some endocytosis of adherent proteins such as lectins also occurs. Such proteins appear to be directed to lysosomes for degradation; there is also evidence that small amounts are transcytosed (Pusztai, 1989). In suckling rodents, these endocytic microdomains can be very large, apparently to provide for efficient uptake of maternal milk IgG in jejunum, and other milk macromolecules in ileum (Rodewald, 1980; Gonnella and Neutra, 1984; Neutra *et al.*, 1988). It has been assumed that such endocytic structures are present in neonatal human enterocytes, but they are not: these specialized cell types and expanded endocytic membrane domains are present in fetal human intestine only up to the midpoint of gestation (Moxey and Trier, 1979).

3.2. Membrane Traffic in Vacuolated Enterocytes

Enterocytes in the epithelium undergo two major bursts of cytodifferentiation during fetal and neonatal development of the mammalian intestine (Neutra and Louvard, 1989). The first dramatic change is the conversion of the undifferentiated fetal epithelium to a monolayer containing enterocytes that are specialized for active endocytosis of luminal contents. The "endocytic stage" of development varies widely among species (Kraehenbuhl *et al.*, 1979): as noted above, it begins and ends during the first half of human fetal life (Moxey and Trier, 1979). In sheep and cattle, however, it begins during fetal life but extends for a few days after birth (Trahair and Robinson, 1986, 1989). In rats, endocytic enterocytes appear a few days before birth and persist through the 3-week suckling period (Kraehenbuhl *et al.*, 1979).

In suckling rats, enterocytes in proximal intestine conduct active, receptor-mediated endocytosis and transcytosis of IgG from maternal milk (Rodewald, 1980). This highly selective IgG uptake is mediated by epithelial Fc receptors that show structural similarities to MHC I (Simister and Mostov, 1989). The enterocytes that conduct IgG uptake contain abundant apical endosomes into which both IgG–receptor complexes and soluble proteins from the lumen are delivered after uptake from apical clathrin-coated pits (Rodewald, 1980). Apical endosomes in these cells are a sorting compartment since proteins in the fluid content of the vesicles are directed to lysosomes, whereas IgG ligand–receptor complexes enter small transport vesicles that release their content by exocytosis at the lateral cell surface (Abrahamson and Rodewald, 1981).

Enterocytes in suckling rat distal intestine, like those throughout the intestines

Marian R. Neutra and Jean-Pierre Kraehenbuhl

of 10- to 20-week fetal humans and fetuses and newborns of other mammalian species, are a distinct phenotype known as "vacuolated enterocytes." These cells take up large amounts of luminal protein into an unusual apical endosomal system and hence into large lysosomal vacuoles (Gonnella and Neutra, 1984). It has been suggested, but not proven, that vacuolated enterocytes in the human fetus may conduct transfer of antibodies from amniotic fluid to the fetus (Israel et al., 1989). These endocytic compartments have been studied in detail in rats, where endocytosis of milk results in intracellular digestion of some proteins and transcytosis of others. Hormones and growth factors from milk including epidermal growth factor (EGF), nerve growth factor, and prolactin are sorted into a transepithelial transport pathway (Siminoski et al., 1986; Gonnella et al., 1987, 1989). Transcytosed proteins may encounter an intracellular protease either in endosomes or in transport vesicles since EGF appears to be reduced in molecular weight after transport (Gonnella et al., 1987).

There is evidence that positively charged peptides or lectins could also be transported across these cells. An adherent protein tracer, cationized ferritin, entered transepithelial transport vesicles in enterocytes of both proximal and distal intestine of rats (Siminoski et al., 1986; Abrahamson and Rodewald, 1981). In these cells, however, soluble proteins were consistently directed to lysosomes for degradation (Abrahamson and Rodewald, 1981; Gonnella and Neutra, 1984). In species where these cells function during fetal life, vacuolated enterocytes may transcytose proteins from amniotic fluid (Moxey and Trier, 1979), but the proteins transported or the nature of the transport vesicles are not known.

Human intestinal cell lines in culture and vacuolated enterocytes of suckling rats have served as models for mapping the interactions of apical and basolateral endocytic pathways. Apical endosomal markers such as the glycoprotein "endotubin" of rat ileum, and basolateral membrane receptors such as transferrin, cycle in and out of their separate plasma membrane domains and early endosome systems (Wilson et al., 1987; Fujita et al., 1990; Hughson and Hopkins, 1990; Godefroy et al., 1988). In enterocytes in vivo, soluble tracers that enter apical or basolateral early endosomes do not mix, but they do meet in late endosomes located in the apical cytoplasm (Fujita et al., 1990). Whether apical-to-basolateral transepithelial transport vesicles can transfer protein directly from apical to basolateral endosomes is not known. This could be relevant to antigen presentation by enterocytes, since delivery of transcytosed peptides into basolateral endosomes would provide an opportunity for interaction with newly synthesized, unoccupied MHC II.

3.3. Membrane Traffic in Adult Enterocytes

At about 20 weeks gestation in the human fetus, highly endocytic, vacuolated enterocytes are replaced by absorptive enterocytes that are analogous in most respects to adult enterocytes (Moxey and Trier, 1979). This second stage of cytodifferentiation occurs much later in other species: (at weaning in rodents and one or more days after birth in cattle, sheep, and pigs; Kraehenbuhl *et al.*, 1979). This change has been termed "closure," referring to the dramatic reduction in endocytosis, the halt in maternal immunoglobulin transfer, and the disappearance of specialized endocytic epithelial cells. In humans, maturational changes such as increased expression of brush border enzymes continue until after birth. Although there is indirect evidence for increased protein uptake in neonatal humans, the endocytic and transport capacities of individual enterocytes at this age have not been studied at a cellular level. It is important to note that monolayers of fully differentiated Caco-2 and HT29 enterocytes in culture are more analogous to enterocytes in human fetal colon during the second half of gestation than to enterocytes in human neonates or adults (Neutra and Louvard, 1989).

In the normal adult intestine, there is evidence for transepithelial transport of intact, biologically active insulin and ultrastructural evidence for a vesicular transport pathway carrying this hormone (Bendayan *et al.*, 1990). This transepithelial transport may be highly selective, but small proteins that can interact with insulin or other hormones, with their receptors, or with other components of endocytic membrane domains, could be carried along. Tracer proteins, including adherent and soluble molecules, have been found by most investigators to be endocytosed in very small amounts by adult enterocytes, and they appear to be directed entirely to apical lysosomes. If late endosome or lysosome membrane-derived vesicles are delivered to the basolateral endosomes or plasma membrane, however, intact or processed proteins would be transcytosed. Both lysosomal membrane proteins and mannose-6-phosphate receptors appear on basolateral membranes of MDCK cells (Nabi *et al.*, 1991). Lysosome–plasma membrane shuttles operate in nonpolarized cells (Braun *et al.*, 1989; Kornfeld and Mellman, 1989), but their existence or polarity in normal enterocytes is not established.

Certain adherent lectins and toxins may be particularly efficient in entering apical endocytic or transcytotic pathways. For example, cholera toxin binds with high affinity to membrane glycolipid GM1 and is efficiently endocytosed. In this case, endocytosis may have profound effects on enterocyte physiology. Lencer *et al.*, (1992) recently used confluent monolayers of polarized T84 cells in culture to demonstrate that cholera toxin is endocytosed and may be transported beyond early apical endosomes before it activates basolaterally located adenylate cyclase and evokes a chloride secretory response. Cholera toxin was shown to enter a transepithelial transport pathway in enterocytes *in vivo* (Hansson *et al.*, 1984). Thus, a vesicular transport in enterocytes may have multiple consequences; for example, this pathway could play a role in the unique modulatory action of cholera toxin in the mucosal immune system (Lycke and Holmgren, 1986; Czerkinsky *et al.*, 1989; Dertzbaugh and Elson, 1991). Cholera toxin has also been shown to enhance endocytic activity and if this occurs in enterocytes, uptake of bystander proteins would be enhanced.

4. MEMBRANE TRAFFIC IN INTESTINAL M CELLS

4.1. Induction of the M Cell Phenotype

The cellular epithelial barrier lining the intestine is organized around crypts, the centers of cell proliferation. A small clonal group of undifferentiated cells in each crypt proliferates and gives rise to several cell phenotypes that migrate in orderly columns onto the surrounding villi (Schmidt *et al.*, 1985). At sites of organized mucosal lymphoid tissue such as Peyer's patches, the crypts adjacent to the mucosal lymphoid follicles contribute cells to a villus on one side and to the follicle-associated epithelium on the other. Even deep in the crypt, the cells on the wall facing the follicle show a distinct phenotype lacking poly Ig receptors (Pappo and Owen, 1988), and these cells migrate onto the dome epithelium giving rise to M cells and dome absorptive cells (Bye *et al.*, 1984). On the other wall of the same crypt, a more conventional differentiating pattern is seen with goblet cells and absorptive enterocytes bearing poly Ig receptors. This pattern suggests that factors or cell contacts from the specialized cells of the mucosal lymphoid follicle induce the commitment of adjacent crypt cells to unusual phenotypes such as M cells of the follicle-associated epithelium. This idea was recently supported by observations in immunodeficient (SCID) mice that apparently lack mucosal lymphoid follicles and have no detectable follicle-associated epithelium or M cells. Injection of Peyer's patch cells from normal mice resulted in formation of new mucosal lymphoid follicles in the SCID mice, and this was accompanied by appearance of a follicle-associated epithelium with M cells (T. Savidge and M. Smith, personal communication).

It is also possible, however, that local transport of antigens by epithelial cells, perhaps reflecting local epithelial specializations, promotes the assembly of organized follicles in the mucosa. Newly differentiated M cells that emerge from the crypts (Bye *et al.*, 1984), as well as M cells in irradiated animals in which lymphocytes are depleted (Ermak *et al.*, 1989), conduct active endocytosis and transcytosis even though the intraepithelial pocket is not present as a morphological landmark. In addition, there is evidence that local infection or increased antigen load on epithelial surfaces (as in bacterial infection or gut stasis) induces formation of lymphoid follicles (Smith *et al.*, 1987).

4.2. Adherence of Macromolecules to M Cell Apical Membranes

Soluble protein tracers and inert particles are rapidly endocytosed and transcytosed by M cells (Owen, 1977; LeFevre *et al.*, 1978). This uptake is nonselective, but it is thought that the hydrophobicity of particle surfaces may enhance interaction with M cell surfaces. Binding of cationized ferritin to these cells suggests that simple electrostatic interactions could also play a role in binding of proteins, bacteria, and inert particles (Neutra *et al.*, 1987). This does not explain the M cell selectivity of binding often observed, however, since both M cell and enterocyte surfaces are rich in negatively charged carbohydrates.

There may be oligosaccharides unique to M cell surfaces but studies using specific lectins have so far failed to identify them (Owen and Bhalla, 1983; Neutra *et al.*, 1987). Since certain microorganisms bind to M cells selectively and with high efficiency, it seems likely that unique protein, glycoprotein, or glycolipid components are exposed on these cells. Adherence to M cells is of practical importance since particles, microbes, or macromolecules are effectively concentrated by adherence and may be transcytosed up to two orders of magnitude more efficiently than nonadherent materials (Neutra *et al.*, 1987). Adherent antigens tend to elicit strong secretory (and often systemic) immune responses (DeAizpurua and Russell-Jones, 1988) and these responses appear to be initiated in sites such as Peyer's patches. Thus, M cell adherence is thought to be important for induction of mucosal immunity. Transport of very small amounts of soluble antigens over time has been suggested to play a role in immune tolerance to soluble food antigens (Mayrhofer, 1984), but the role of M cell transport in mucosal or systemic tolerance is not clear.

We have also shown that M cells have binding sites for immunoglobulins on their apical membranes. It was previously observed that milk secretory IgA ingested by suckling rabbits accumulates on M cell surfaces (Roy and Varvayanis, 1987). We observed that monoclonal IgA or polyclonal sIgA as well as IgG antibodies bind selectively to M cells and compete with each other for binding sites (Weltzin *et al.*, 1989). Antibodies against epithelial and macrophage Fc receptors of other types, however, failed to recognize any M cell component. This binding mechanism allows M cells to endocytose and transport free IgA as well as IgA–antigen complexes. Such uptake could boost an existing secretory immune response or have other modulatory effects in the mucosal immune system (Kraehenbuhl and Neutra, 1992).

4.3. Adherence of Microorganisms to M Cells

Several gram-negative bacteria bind selectively to M cells including *Vibrio cholerae* (Owen *et al.*, 1986b; Winner *et al.*, 1991), some strains of *E. coli* (Inman and Cantey, 1983), *Salmonella* (Kohbata *et al.*, 1986) and *Shigella* (Wassef *et al.*, 1989). Since M cells might be expected to capture whole classes of pathogens for immunologic sampling, a common type of binding mechanism may be used. Many bacteria use lectin–carbohydrate recognition systems to interact with eukaryotic cells and these may also operate in M cell adherence. The participating lectin could be either a bacterial adhesin, or an M cell membrane protein. Because of the absence of a system for culture of polarized, functional M cells, these possibilities have not been explored.

Several pathogenic viruses also adhere selectively to M cells but the interacting viral and M cell surface molecules responsible for adherence have not been identified. An interesting example is the mouse pathogen, reovirus (Wolf et al., 1981). Processing of ingested reovirus by proteases in the intestinal lumen increases reovirus infectivity through cleavage of the major outer capsid protein sigma 3 and alteration of the viral hemagglutinin sigma 1 (Bass et al., 1988; Nibert et al., 1991). Helen Amerongen, in our laboratory, recently observed that proteolytic processing of the outer capsid is also required for M cell adherence since neither unprocessed virus nor capsid-less cores can bind. This implies that reovirus uses either the protease-resistant outer capsid protein mu1c or the extended sigma 1 protein to bind to M cells, presumably at a site not involved in serotype-specific cell tropism. Since sialylated glycoproteins serve as viral receptors on other cell types (Paul et al., 1989), and M cell surface components avidly bind wheat germ agglutinin (Neutra et al., 1987), M cell sialic acid residues might be involved. Such viral adhesins could eventually be used to elucidate the M cell molecules or oligosaccharides responsible for the interaction (Bass and Greenberg, 1992). Similar studies of other viruses that also adhere to M cells such as the neurotropic poliovirus (Sicinski et al., 1990) and the retrovirus HIV-1 (Amerongen et al., 1991) are needed to determine whether common molecular motifs are used for M cell binding.

The evidence gained so far does not prove the existence of unique M cell adhesins, since selective adherence could also be due to the absence of "blocking" molecules. Many of the complex stalked glycoprotein enzymes that cover the microvillar membranes of intestinal absorptive cells are not present on M cell apical membranes (Owen and Bhalla, 1983). In our laboratory, the "blocking" effect of the glycocalyx was demonstrated using cholera toxin. The cholera toxin receptor is glycolipid GM1 that is present on apical membranes of M cells as well as enterocytes. When rhodamine-labeled toxin was applied to Peyer's patch mucosa, it bound to all cell surfaces. When applied as polyvalent toxin complexes on 15-nm colloidal gold particles, however, the particles bound only to M cells, apparently because the thick glycocalyx of enterocytes prevented movement of the colloidal gold probe to microvillar membranes. Viruses and bacteria can also be considered particulate, polyvalent ligands, and thus some microbes may have access to receptors on M cells but not to the same receptors on enterocytes.

4.4. Transcytosis by M Cells

M cells endocytose adherent macromolecules such as cationized ferritin and lectins via clathrin-coated pits. These molecules are then delivered to tubulovesicular structures resembling early endosomes in the apical cytoplasm (Neutra et al., 1987). Lysosomes in M cells are scarce, and were not visualized in the apical cytoplasm using acid phosphatase as an enzymatic marker (Owen et al., 1986a). Viruses and bacteria are morphologically unaltered during transport (Amerongen *et al.*, 1991), but it is possible that proteins are altered in M cell vesicles since the "marker" glycoprotein lgp120 that is found in late endosomes and lysosomes (Lewis *et al.*, 1985) was recently detected in M cell apical vesicles (Allan *et al.*, 1992). If M cell apical endosomes contain endosomal hydrolases, immunogens may be processed by partial proteolysis in the transepithelial pathway and this could affect the specificity of the subsequent mucosal immune response. Transcytotic vesicle traffic in M cells is not directed toward the lateral cell surfaces (as in enterocytes); rather, all vesicles fuse with the modified, invaginated basolateral subdomain lining the intraepithelial pocket (Owen, 1977; Neutra *et al.*, 1987). The "pocket" membrane represents a distinct subdomain of the plasma membrane; it contains very low amounts of Na^+,K^+-ATPase (Neutra *et al.*, 1986), and it presumably contains adhesion molecules that recognize the subpopulation of lymphocytes that collect in the pocket (Ermak *et al.*, 1990).

Certain bacteria such as *Vibrio cholerae* are taken up by M cells by a process that resembles phagocytosis, with broad areas of close membrane interaction and assembly of M cell cytoplasmic actin (J. Mack, J. Mekalanos, and M. Neutra, unpublished). Presumably, specific M cell membrane components are recruited to the interaction site, but it is not known which microbial gene products of bacteria or viruses, or which M cell membrane components are required for binding and internalization. The fact that bacteria, viruses, and some macromolecules are released at the basolateral side after transport indicates that multiple, low-affinity interactions are involved that are reversed by a shift in ionic strength or pH in the transport vesicles, or in the intraepithelial pocket of the M cell.

5. CONCLUSION

There have been several attempts to measure and compare rates of transcytosis across Peyer's patch and nonpatch mucosa. While some investigators showed enhanced transport into Peyer's patch (Keljo and Hamilton, 1983; Ho *et al.*, 1990), others did not (Duroc *et al.*, 1983). It is clear that individual M cells conduct rapid and efficient transcytosis and that M cell transport can deliver immunologically important amounts of protein across the epithelium to the mucosal immune system. However, such proteins are efficiently endocytosed by the numerous macrophages and dendritic cells in the organized mucosal lymphoid tissue (Ho *et al.*, 1990). Thus, it is not clear to what extent proteins taken up by M cells enter the circulation. Furthermore, the relative contributions of M cells and enterocytes to the total amounts of luminal protein delivered across the entire intestinal epithelium and into the circulation remain undetermined. The answer is likely to be different for proteins of differing size, charge, and adherence properties because of the differing natures of the M cell and enterocyte membranes and transepithelial transport pathways.

REFERENCES

Abrahamson, D. R., and Rodewald, R., 1981, Evidence for the sorting of endocytic vesicle contents during receptor-mediated transport of IgG across newborn rat intestine, *J. Cell Biol.* **91:**270–280.

Achler, C., Filmer, D., Merte, C., and Drenckhahn, D., 1989, Role of microtubules in polarized delivery of apical membrane proteins to the brush border of the intestinal epithelium, *J. Cell Biol.* **109:**179–189.

Allan, C.H., Meyrick, D. L., and Trier, J. S., 1992, M cells contain acidic compartments and express class II MHC determinants, *Gastroenterology* **102:**A589.

Alpers, D. H., 1975, Protein turnover in intestinal mucosal villus and crypt brush border membranes, *Biochem. Biophys. Res. Commun.* **75:**130–135.

Amerongen, H. M., Mack, J. A., Wilson, J. M., and Neutra, M. R.,1989, Membrane domains of intestinal epithelial cells: Distribution of Na^+,K^+-ATPase and the membrane skeleton in adult rat intestine, during fetal development, and after epithelial isolation, *J. Cell Biol.* **109:**2129–2138.

Amerongen, H. M., Weltzin, R. A., Farnet, C. M., Michetti, P., Haseltine, W. A., and Neutra, M. R., 1991, Transepithelial transport of HIV-1 by intestinal M cells: A mechanism for transmission of AIDS, *J. Acquir. Immune Defic. Syndr.* **4:**760–765.

Anderson, R. G. W., 1991, Molecular motors that shape endocytic membrane, in: *Intracellular Trafficking of Proteins* (C. J. Steer and J. A. Hanover, eds.), Cambridge University Press, London, pp. 13–46.

Apodaca, G., Bomsel, M., Arden, J., Breitfeld, P. P., Tang, K. C., and Mostov, K. E., 1991, The polymeric immunoglobulin receptor. A model protein to study transcytosis, *J. Clin. Invest.* **87:**1877–1882.

Bacallao, R., Antony, C., Dotti, D., Karsenti, E., Stelzer, E. H., and Simons, K., 1989, The subcellular organization of Madin–Darby canine kidney cells during formation of a polarized epithelium, *J. Cell Biol.* **109:**2817–2832.

Bartles, J. R., Feracci, H. M., Steiger, B., and Hubbard, A. L., 1987, Biogenesis of the rat hepatocyte plasma membrane in vivo: Comparison of the pathways taken by apical and basolateral proteins using subcellular fractionation, *J. Cell Biol.* **105:**1241–1251.

Bass, D. M., and Greenberg, H. B., 1992, Strategies for the identification of icosahedral virus receptors, *J. Clin. Invest.* **89:**3–9.

Bass, D. M., Trier, J. S., Dambrauskas, R., and Wolf, J. L., 1988, Reovirus type I infection of small intestinal epithelium in suckling mice and its effect on M cells, *Lab. Invest.* **58:** 226–235.

Bendayan, M., Ziv, E., Ben-Sasson, R., Bar-On, H., and Kidron, M., 1990, Morphocytochemical and biochemical evidence for insulin absorption by the rat ileal epithelium, *Diabetologia* **33:**197–204.

Bland, P. W., and Warren, L. G., 1986, Antigen presentation by epithelial cells of the rat small intestine. II. Selective induction of suppressor T cells, *Immunology* **58:**9–14.

Bomsel, M., Prydz, K., Parton, R. G., Gruenberg, J., and Simons, K., 1989, Endocytosis in filter-grown Madin–Darby canine kidney cells, *J. Cell Biol.* **109:**3243–3258.

Bomsel, M., Parton, R., Kuznetsov, S. A., Schroer, T. A., and Gruenberg, J., 1990, Microtubule- and motor-dependent fusion *in vitro* between apical and basolateral endocytic vesicles from MDCK cells, *Cell* **62:**719–731.

Brandli, A. W., Parton, R. G., and Simons, K., 1990, Transcytosis in MDCK cells: Identification of glycoproteins transported bidirectionally between both plasma membrane domains, *J. Cell Biol.* **111:**2909–2921.

Braun, M., Waheed, A., and von Figura, K., 1989, Lysosomal acid phosphatase is transported to lysosomes via the cell surface, *EMBO J.* **8:**3633–3640.

Breitfeld, P. P., Casanova, J. E., Simister, N. E., Ross, S. A., McKinnon, W. C., and Mostov, K. E., 1989a, Sorting signals, *Curr. Opin. Cell Biol.* **1:**617–623.

Breitfeld, P. P., Harris, J. M., and Mostov, K. E., 1989b, Postendocytotic sorting of the ligand for the polymeric immunoglobulin receptor in Madin–Darby canine kidney cells, *J. Cell Biol.* **109:**475–486.

Bye, W. A., Allen, C. H., and Trier, J. S., 1984, Structure, distribution and origin of M cells in Peyer's patches of mouse ileum, *Gastroenterology* **86:**789–801.

Caplan, M. J., Anderson, H. C., Palade, G. E., and Jamieson, J. E., 1986, Intracellular sorting and polarized cell surface delivery of (Na+,K+)ATPase, an endogenous component of MDCK cell basolateral plasma membranes, *Cell* **46:**623–631.

Casanova, J. E., Breitfeld, P. P., Ross, S. A., and Mostov, K. E., 1990, Phosphorylation of the polymeric immunoglobulin receptor required for its efficient transcytosis, *Science* **248:** 742–745.

Courtoy, P. J., 1991, Dissection of endosomes, in: *Intracellular Trafficking of Proteins* (C. J. Steer and J. A. Hanover, eds.), Cambridge University Press, London, pp. 103–156.

Czerkinsky, C., Russell, M. W., Lycke, N., Lindblad, M., and Holmgren, J., 1989, Oral administration of a streptococcal antigen coupled to cholera toxin B subunit evokes strong antibody responses in salivary glands and extramucosal tissues, *Infect. Immun.* **57:**1072–1077.

DeAizpurua, H. J., and Russell-Jones, G. J., 1988, Oral vaccination: Identification of classes of proteins that provoke an immune response upon oral feeding, *J. Exp. Med.* **167:**440–451.

Dertzbaugh, M. T., and Elson, C.O., 1991, Cholera toxin as a mucosal adjuvant, in: *Topics in Vaccine Adjuvant Research* (D. R. Spriggs and W. C. Koff, eds.), CRC Press, Boca Raton, Fla., pp. 119–132.

Diment, S., and Stahl, P., 1985, Macrophage endosomes contain proteases which degrade endocytosed protein ligands, *J. Biol. Chem.* **260:**15311–15317.

Dragsten, P. R., Blumenthal, R., and Handler, J. S., 1981, Membrane asymmetry in epithelia: Is the tight junction a barrier to diffusion in the plasma membrane? *Nature* **294:**718–722.

Duroc, R., Heyman, M., Beaufrere, B., Morgat, J. L., and Desjeux, J. F., 1983, Horseradish peroxidase transport across rabbit jejunum and Peyer's patches *in vitro*, *Am. J. Physiol.* **245:**G54–G58.

Eilers, U., Klumperman, J., and Hauri, H. P., 1989, Nocodazole, a microtubule-active drug, interferes with apical protein delivery in cultured intestinal epithelial cells (Caco-2), *J. Cell Biol.* **108:**13–22.

Ermak, T. H., Steger, H. J., Strober, W., and Owen, L., 1989, M cells and granular mononuclear cells depleted of their lymphocytes by total lymphoid irradiation, *Am. J. Pathol.* **134:**529–537.

Ermak, T. H., Steger, H. J., and Pappo, J., 1990, Phenotypically distinct subpopulations of T cells in domes and M-cell pockets of rabbit gut-associated lymphoid tissues, *Immunology* **71:**530–537.

Fujita, M., Reinhart, F., and Neutra, M., 1990, Convergence of apical and basolateral endocytic pathways at apical late endosomes in absorptive cells of suckling rat ileum, *J. Cell Sci.* **97**:385–394.

Geuze, H. J., Slot, J. W., Strous, G. J. A. M., Peppard, J., von Figura, K., Hasilik, A., and Schwartz, A. L., 1984, Intracellular receptor sorting during endocytosis: Comparative immunoelectron microscopy of multiple receptors in rat liver, *Cell* **37**:195–204.

Geuze, H. J., Slot, J. W., and Schwartz, A. L., 1987, Membranes of sorting organelles display lateral heterogeneity in receptor distribution, *J. Cell Biol.* **104**:1715–1723.

Godefroy, O., Huet, C., Blair, L. A. C., Sahuquillo-Merino, C., and Louvard, D., 1988, Differentiation properties of a clone isolated from the HT29 cell line (a human colon carcinoma): Polarized differentiation of histocompatibility antigens (HLA) and of transferrin receptors, *Biol. Cell* **63**:41–56.

Goldman, I. S., Jones, A. L., Hradek, G. T., and Huling, S., 1983, Hepatocyte handling of immunoglobulin A in the rat: The role of microtubules, *Gastroenterology* **85**:130–140.

Gonnella, P. A., and Neutra, M. R., 1984, Membrane-bound and fluid-phase macromolecules enter separate prelysosomal compartments in absorptive cells of suckling rat ileum, *J. Cell Biol.* **99**:909–917.

Gonnella, P. A., and Neutra, M. R., 1985, Glycoconjugate distribution and mobility on apical membranes of absorptive cells of suckling rat ileum *in vivo*, *Anat. Rec.* **213**:520–528.

Gonnella, P. A., Simonski, K., Murphy, R. A., and Neutra, M. R., 1987, Transepithelial transport of epidermal growth factor by absorptive cells of suckling rat, *J. Clin. Invest.* **80**:22–32.

Gonnella, P. A., Harmatz, P., and Walker, W. A., 1989, Prolactin is transported across the epithelium of the jejunum and ileum of the suckling rat ileum, *J. Cell Physiol.* **140**:138–149.

Griffiths, G., Hoflack, B., Simons, K., Mellman, I., and Kornfeld, S., 1988, The mannose 6-phosphate receptor and the biogenesis of lysosomes, *Cell* **52**:329–341.

Gruenberg, J., and Howell, K. E., 1989, Membrane traffic in endocytosis: Insights from cell free assays, *Annu. Rev. Cell Biol.* **5**:453–481.

Gruenberg, J., Griffiths, G., and Howell, K. E., 1989, Characterization of the early endosome and putative endocytic carrier vesicles *in vivo* and with an assay of vesicle fusion *in vitro*, *J. Cell Biol.* **108**:1301–1316.

Hansson, H. A., Lange, S., and Lonnvoth, I., 1984, Internalization *in vivo* of cholera toxin in the small intestine of the rat, *Acta Pathol. Microbiol. Scand.* **92**:15–21.

Heyman, M., Crain-Denoyelle, A. M., Nath, S. K., and Desjeux, J. F., 1990, Quantification of protein transcytosis in the human colon carcinoma cell line Caco-2, *J. Cell Physiol.* **143**:391–395.

Hidalgo, I. J., Raub, T. J., and Borchardt, R. T., 1989, Characterization of the human colon carcinoma cell line (Caco-2) as a model system for intestinal epithelial permeability, *Gastroenterology* **96**:736–749.

Ho, N. F. H., Day, J. S., Barsuhn, C. L., Burton, P. S., and Raub, T. J., 1990, Biophysical model approaches to mechanistic transepithelial studies of peptides, *J. Controlled Release* **11**:3–24.

Hughson, E. J., and Hopkins, C. R., 1990, Endocytic pathways in polarized Caco-2 cells: Identification of an endosomal compartment accessible from both apical and basolateral surfaces, *J. Cell Biol.* **110**:337–348.

Hunziker, W., and Mellman, I., 1989, Expression of macrophage–lymphocyte Fc receptors in Madin–Darby canine kidney cells: Polarity and transcytosis differ for isoforms with or without coated pit localization domains, *J. Cell Biol.* **109**:3291–3302.

Hunziker, W., Male, P., and Mellman, I., 1990, Differential microtubule requirements for transcytosis in MDCK cells, *EMBO J.* **9**:3515–3525.

Hunziker, W., Harter, C., Matter, K., and Mellman, I., 1991a, Basolateral sorting in MDCK cells requires a distinct cytoplasmic domain determinant, *Cell* **66**:907–920.

Hunziker, W., Whitney, J. A., and Mellman, I., 1991b, Selective inhibition of transcytosis by Brefeldin A in MDCK cells, *Cell* **67**:617–628.

Inman, L. R., and Cantey, J. R., 1983, Specific adherence of Escherichia coli (strain RDEC-1) to membranous (M) cells of the Peyer's patch in Escherichia coli diarrhea in the rabbit, *J. Clin. Invest.* **71**:1–8.

Israel, E. J., Simister, N., Freiberg, E., Hendren, R., and Walker, W. A., 1989, Immunoglobulin G binding sites on the human fetal intestine, *Pediatr. Res.* **25**:116a.

Kaiserlain, D., Vidal, K., and Revillard, J. P., 1989, Murine enterocytes can present soluble antigen to specific class II restricted CD4+ T cells, *Eur. J. Immunol.* **19**:1513–1516.

Keljo, D. J., and Hamilton, J. R., 1983, Quantitative determination of macromolecular transport rate across intestinal Peyer's patches, *Am. J. Physiol.* **244**:G637–G644.

Kelly, R. B., 1990, Microtubules, membrane traffic, and cell organization, *Cell* **61**:5–7.

Kohbata, S., Yokobata, H., and Yabuuchi, E., 1986, Cytopathogenic effect of Salmonella typhi GIFU 10007 on M cells of murine ileal Peyer's patches in ligated ileal loops: An ultrastructural study, *Microbiol. Immunol.* **30**:1225–1237.

Kornfeld, S., and Mellman, I., 1989, The biogenesis of lysosomes, *Annu. Rev. Cell Biol.* **5**: 483–525.

Kraehenbuhl, J. P., and Neutra, M. R., 1992, Molecular and cellular basis of immune protection of mucosal surfaces, *Physiol. Rev.* **70**:853–879.

Kraehenbuhl, J. P., Bron, C., and Sordat, B., 1979, Transfer of humoral secretory and cellular immunity from mother to offspring, *Curr. Top. Pathol.* **66**:105–157.

Kuhn, L. C., and Kraehenbuhl, J. P., 1982, The sacrificial receptor—Translocation of polymeric IgA across epithelia, *Trends Biochem. Sci.* **7**:299–302.

LeFevre, M. E., Olivo, R., and Joel, D. D., 1978, Accumulation of latex particles in Peyer's patches and their subsequent appearance in villi and mesenteric lymph nodes, *Proc. Soc. Exp. Biol. Med.* **159**:298–302.

Lencer, W. E., Delp, C., Neutra, M. R., and Madara, J. L., 1992, Mechanism of cholera toxin action on a polarized human intestinal epithelial cell line: Role of vesicular traffic, *J. Cell Biol.* **117**:1197–1210.

Levine, J. S., Allen, R. H., Alpers, D. H., and Seetheram, B., 1984, Immunocytochemical location of intrinsic factor–cobalamin receptors in dog ileum, *J. Cell Biol.* **98**:1110–1117.

Lewis, V., Green, S. A., Marsh, M., Vihko, Helenius, A., and Mellman, I., 1985, Glycoproteins of the lysosomal membrane, *J. Cell Biol.* **100**:1839–1847.

Lippincott-Schwartz, J., and Fambrough, D. M., 1987, Cycling of the integral glycoprotein LEP 100 between plasma membrane and lysosomes: Kinetic and morphological analysis, *Cell* **49**:669–677.

Louvard, D., Godefroy, O., Huet, C., Sahuquillo-Merino, C., Robine, S., and Coudrier, E., 1985, Basolateral membrane proteins are expressed at the surface of immature intestinal cells whereas transport of apical proteins is abortive, in: *Current Communications in Molecular Biology, Protein Transport and Secretion* (M. J. Gething, ed.), Cold Spring Harbor Laboratory, Cold Spring Harbor, N.Y., pp. 168–173.

Lycke, N., and Holmgren, J., 1986, Strong adjuvant properties of cholera toxin on gut mucosal immune responses to orally presented antigens, *Immunology* **59**:301–308.

Madara, J. L., 1988, Tight junction dynamics: Is paracellular transport regulated? *Cell* **53**: 497–498.

Madara, J. L., and Trier, J. S., 1987, Functional morphology of the mucosa of the small intestine, in: *Physiology of the Gastrointestinal Tract* (L. R. Johnson, ed.), Raven Press, New York, pp. 1209–1250.

Massey, D., Feracci, H., Gorvel, J. P., Rigal, A., Soulie, J. M., and Maroux, S., 1987, Evidence for the transit of aminopeptidase N through the basolateral membrane before it reaches the brush border of enterocytes, *J. Membr. Biol.* **9**:19–25.

Matlin, K., Bainton, D. F., Pesonen, M., Louvard, D., Gent, N., and Simons, K., 1983, Transepithelial transport of a viral membrane glycoprotein implanted into the apical plasma membrane of MDCK cells. I. Morphological evidence, *J. Cell Biol.* **97**:627–637.

Matter, K., Brauchbar, M., Bucher, K., and Hauri, H. P., 1990, Sorting of endogenous plasma membrane proteins occurs from two sites in cultured human intestinal epithelial cells (Caco-2), *J. Cell Biol.* **60**:429–437.

Maxfield, F. R., and Yamashiro, D. J., 1991, Acidification of organelles and the intracellular sorting of proteins during endocytosis, in: *Intracellular Trafficking of Proteins* (C. J. Steer and J. A. Hanover, eds.), Cambridge University Press, London, pp. 157–182.

Mayer, L., Panja, A., and Li, Y., 1991, Antigen recognition in the gastrointestinal tract: Death to the dogma, *Immunol. Res.* **10**:356–359.

Mayrhofer, G., 1984, Physiology of the intestinal immune system, in: *Local Immune Responses of the Gut* (T. J. Newby and C. R. Stokes, eds.), CRC Press, Boca Raton, Fla., pp. 1–96.

Mooseker, M., 1985, Organization, chemistry and assembly of the cytoskeletal apparatus of the intestinal brush border, *Annu. Rev. Cell Biol.* **1**:209–241.

Mostov, K. E., and Simister, N. E., 1985, Transcytosis, *Cell* **43**:389–390.

Mostov, K. E., Friedlander, M., and Blobel, G., 1984, The receptor for transepithelial transport of IgA and IgM contains multiple immunoglobulin-like domains, *Nature* **308**:37–43.

Moxey, P. C., and Trier, J. S., 1979, Development of villus absorptive cells in the human fetal small intestine: A morphological and morphometric study, *Anat. Rec.* **195**:463–482.

Nabi, I. R., LeBivic, A., Fambrough, D., and Rodriguez-Boulan, E., 1991, An endogenous MDCK lysosomal membrane glycoprotein is targeted basolaterally before delivery to lysosomes, *J. Cell Biol.* **115**:1573–1584.

Nelson, W. J., and Hammerton, R. W., 1989, A membrane–cytoskeletal complex containing Na$^+$,K$^+$-ATPase, ankyrin, and fodrin in Madin–Darby canine kidney (MDCK) cells: Implications for the biogenesis of epithelial cell polarity, *J. Cell Biol.* **108**:893–902.

Neutra, M. R., 1988, The gastrointestinal tract, in: *Cell and Tissue Biology* (L. Weiss, ed.), Urban & Schwarzenberg, Munich, pp. 641–684.

Neutra, M. R., and Kraehenbuhl, J. P., 1992, Transepithelial transport and mucosal defense, *Trends Cell Biol.* **2**:134–138.

Neutra, M. R., and Louvard, D., 1989, Differentiation of intestinal cells *in vitro*, in: *Functional Epithelial Cells in Culture* (K. S. Matlin and J. D. Valentich, eds.), Liss, New York, pp. 363–398.

Neutra, M. R., Phillips, T. L., Fishkind, D. J., and Mack, J. A., 1986, Membrane domains of the intestinal M cell, *J. Cell Biol.* **103:**466a.

Neutra, M. R., Phillips, T. L., Mayer, E. L., and Fishkind, D. J., 1987, Transport of membrane-bound macromolecules by M cells in follicle-associated epithelium of rabbit Peyer's patch, *Cell Tissue Res.* **247:**537–546.

Neutra, M. R., Wilson, J. M., Weltzin, R. A., and Kraehenbuhl, J. P., 1988, Membrane domains and macromolecular transport in intestinal epithelial cells, *Am. Rev. Respir. Dis.* **138:**S10–S16.

Nibert, M. L., Furlong, D. B., and Fields, B. N., 1991, Mechanisms of viral pathogenesis. Distinct forms of reoviruses and their roles during replication in cells and host, *J. Clin. Invest.* **88:**727–734.

Owen, R. L., 1977, Sequential uptake of horseradish peroxidase by lymphoid follicle epithelium of Peyer's patch in the normal unobstructed mouse intestine: An ultrastructural study, *Gastroenterology* **72:**440–451.

Owen, R. L., and Bhalla, D. K., 1983, Cytochemical analysis of alkaline phosphatase and esterase activities and of lectin-binding and anionic sites in rat and mouse Peyer's patch M cells, *Am. J. Anat.* **168:**199–212.

Owen, R. L., Apple, R. T., and Bhalla, D. K., 1986a, Morphometric and cytochemical analysis of lysosomes in rat Peyer's patch follicle epithelium: Their reduction in volume fraction and acid phosphatase content in M cells compared to adjacent enterocytes, *Anat. Rec.* **216:** 521–527.

Owen, R. L., Pierce, N. F., Apple, R. T., and Cray, W. C., Jr., 1986b, M cell transport of *Vibrio cholerae* from the intestinal lumen into Peyer's patches: A mechanism for antigen sampling and for microbial transepithelial migration, *J. Infect. Dis.* **153:**1108–1118.

Pappo, J., and Owen, R. L., 1988, Absence of secretory component expression by epithelial cells overlying rabbit gut-associated lymphoid tissue, *Gastroenterology* **95:**1173–1177.

Parton, R. G., Prydz, K., Bomsel, M., Simons, K., and Griffiths, G., 1989, Meeting of the apical and basolateral endocytic pathways of the Madin–Darby canine kidney cell in late endosomes, *J. Cell Biol.* **109:**3259–3272.

Paul, R. W., Choi, A. H., and Lee, P. W., 1989, The alpha-anomeric form of sialic acid is the minimal receptor determinant recognized by reovirus, *Virology* **172:**382–385.

Phillips, T. E., Phillips, T. H., and Neutra, M. R., 1987, Macromolecules can pass through occluding junctions of rat ileal epithelium during cholinergic stimulation, *Cell Tissue Res.* **247:**547–554.

Prydz, K., Brandli, A. W., Bomsel, M., and Simons, K., 1990, Surface distribution of the mannose 6-phosphate receptors in epithelial Madin–Darby canine kidney cells, *J. Biol. Chem.* **265:**12629–12635.

Pusztai, A., 1989, Transport of proteins through the membranes of the adult gastrointestinal tract—A potential for drug delivery? *Adv. Drug Deliv. Rev.* **3:**215–228.

Rabinowitz, S., Horstmann, H., Gordon, S., and Griffiths, G., 1992, Immunocytochemical characterization of the endocytic and phagolysosomal compartments in peritoneal macrophages, *J. Cell Biol.* **116:**95–112.

Rodewald, R., 1980, Distribution of immunoglobulin G receptors in the small intestine of the young rat, *J. Cell Biol.* **85:**18–32.

Rodriguez-Boulan, E., and Nelson, W. J., 1989, Morphogenesis of the polarized epithelial cell phenotype, *Science* **245:**718–725.

Roy, M. J., and Varvayanis, M., 1987, Development of dome epithelium in gut-associated lymphoid tissues: Association of IgA with M cells, *Cell Tissue Res.* **248**:645–651.

Schaerer, E., Neutra, M. R., and Kraehenbuhl, J. P., 1991, Molecular and cellular mechanisms involved in transepithelial transport, *J. Membr. Biol.* **123**:93–103.

Schmidt, G. H., Wilkinson, M. M., and Ponder, B. A. J., 1985, Cell migration pathway in the intestinal epithelium: An *in situ* marker system using mouse aggregation chimeras, *Cell* **40**:425–429.

Schroer, T. A., and Sheetz, M. P., 1989, Role of kinesin and kinesin-associated proteins in organelle transport, in: *Cell Movement* (J. R. McIntosh and F. D. Warner, eds.), Liss, New York, pp. 295–306.

Semenza, G., 1986, Anchoring and biosynthesis of stalked brush border membrane glycoproteins, *Annu. Rev. Cell Biol.* **2**:255–314.

Sicinski, P., Rowinski, J., Warchol, J. B., Jarzcabek, Z., Gut, W., Szczygiel, B., Bielecki, K., and Koch, G., 1990, Poliovirus type 1 enters the human host through intestinal M cells, *Gastroenterology* **98**:56–58.

Siminoski, K., Gonnella, P., Bernanke, J., Owen, L., Neutra, M., and Murphy, R. A., 1986, Uptake and transepithelial transport of nerve growth factor in suckling rat ileum, *J. Cell Biol.* **103**:1979–1990.

Simister, N. E., and Mostov, K. E., 1989, An Fc receptor structurally related to MHC class I antigens, *Nature* **337**:184–187.

Slot, J. W., and Geuze, H. J., 1984, Transcytosis of IgA in duodenal epithelial cells observed by immunocytochemistry, *J. Cell Biol.* **99**:7a.

Smith, M. W., James, P. S., and Tivey, D. R., 1987, M cell numbers increase after transfer of SPF mice to a normal animal house environment, *Am. J. Pathol.* **128**:385.

Smith, P. L., Wall, D. A., Gochoco, C. H., and Wilson, G., 1992, Oral delivery of peptides and proteins, *Adv. Drug Deliv. Rev.* **8**:253–290.

Specian, R. D., and Neutra, M. R., 1984, The cytoskeleton of intestinal goblet cells, *Gastroenterology* **87**:1313–1325.

Sztul, E., Kaplin, A., Saucan, L., and Palade, G., 1991, Protein traffic between distinct plasma membrane domains: Isolation and characterization of vesicular carriers involved in transcytosis, *Cell* **64**:81–89.

Trahair, J. F., and Robinson, P. M., 1986, The development of the ovine small intestine, *Anat. Rec.* **214**:294–303.

Trahair, J. F., and Robinson, P. M., 1989, Enterocyte ultrastructure and uptake of immunoglobulins in the small intestine of the neonatal lamb, *J. Anat.* **166**:103–111.

van der Sluijs, P., Bennett, M. K., Antony, C., Simons, K., and Kreis, T. E., 1990, Binding of exocytic vesicles from MDCK cells to microtubules *in vitro*, *J. Cell Sci.* **95**:545–553.

Wassef, J. S., Keren, D. F., and Mailloux, J. L., 1989, Role of M cells in initial antigen uptake and in ulcer formation in the rabbit intestinal loop model of shigellosis, *Infect. Immun.* **57**: 858–863.

Weinman, M. D., Allan, C. H., Trier, J. S., and Hagen, S. J., 1989, Repair of microvilli in the rat small intestine after damage with lectins contained in the red kidney bean, *Gastroenterology* **97**:1193–1204.

Weltzin, R. A., Lucia-Jandris, P., Michetti, P., Fields, B. N., and Kraehenbuhl, J. P., 1989, Binding and transepithelial transport of immunoglobulins by intestinal M cells: Demonstration using monoclonal IgA antibodies against enteric viral proteins. *J. Cell Biol.* **108:** 1673–1685.

Wilson, J. M., Whitney, J. A., and Neutra, M. R., 1987, Identification of an endosomal antigen specific to absorptive cells of suckling rat ileum, *J. Cell Biol.* **105:**691–703.

Winner, L. S., III, Mack, J., Weltzin, R. A., Mekalanos, J. J., Kraehenbuhl, J. P., and Neutra, M. R., 1991, New model for analysis of mucosal immunity: Intestinal secretion of specific monoclonal immunoglobulin A from hybridoma tumors protects against *Vibrio cholerae* infection, *Infect. Immun.* **59:**977–982.

Wolf, J. L., Rubin, D. H., Finberg, R., Kauffman, R. S., Sharpe, A. H., Trier, J. S., and Fields, B. N., 1981, Intestinal M cells: A pathway for entry of reovirus into the host, *Science* **211:** 471–472.

Chapter 6

Intraoral Peptide Absorption

Hans P. Merkle and Gregor J. M. Wolany

1. INTRODUCTION

Among the different sites of administration for proteinaceous drugs, the intraoral route of delivery has recently gained considerable interest. In spite of its apparent disadvantages such as limited absorptive surface and moderate mucosal permeability, this route of administration might be of benefit especially for small- or medium-size peptide drugs. Potential advantages for intraoral delivery are the avoidance of gastrointestinal or liver first-pass effect, the feasibility of locally controlled absorption enhancement, the ease of administration and removal of an administered device, and the sustainment of drug action.

The aim of this chapter is to provide an overview on anatomical and biochemical features of the oral epithelium relevant for peptide and protein delivery. Focusing on the buccal epithelium, a critical literature survey, a discussion of possible mechanisms of drug absorption, and *in vivo* and *in vitro* approaches for investigating intraoral absorption will be given. Additionally, a brief description of intraoral dosage form design, of the current patent situation in the field, and a summary on chemical absorption enhancement will be presented.

Hans P. Merkle and Gregor J. M. Wolany • Department of Pharmacy, Swiss Federal Institute of Technology, CH-8092 Zurich, Switzerland.

Biological Barriers to Protein Delivery, edited by Kenneth L. Audus and Thomas J. Raub. Plenum Press, New York, 1993.

2. THE STRUCTURE OF THE ORAL EPITHELIUM

2.1. Anatomy of the Oral Cavity

The structure and function of the oral mucosa were comprehensively reviewed by Meyer *et al.* (1984). The oral cavity, functionally the initial part of the gastrointestinal tract, consists of lips, cheeks, floor of the mouth, gingivae, tongue, hard and soft palate (Thews *et al.*, 1989). The surface of the oral cavity is completely covered by a relatively thick and protective multilayered epithelium. The structure of this epithelium is not homogeneous throughout the entire oral cavity but reveals distinct histological differences relevant for drug absorption. The mucosae of the hard and soft palate, of the gingiva, and partly of the tongue are predominantly keratinized multilayered tissues. Due to their high degree of keratinization, these mucosal regions have been demonstrated to be less permeable for intraorally administered tracer compounds (Curatolo, 1987; Chen and Squier, 1984; Pimlott and Addy, 1985; Squier and Hall, 1985). Therefore, keratinized oral mucosae are regarded to be of limited value as an administration site for peptide and proteinaceous drugs and will not be discussed further. More promising intraoral tissues with or without minor keratinization only are the floor of the mouth and the buccal mucosae of the cheek and the lips.

Despite its small surface area, the floor of the mouth is, especially in combination with the intensively blood-perfused ground of the tongue, a possible site of absorption. This route of administration is commonly used for some small nonpeptide drugs; well-known examples are glycerol trinitrate and analogues for treatment in angina pectoris. For peptide drugs this intraoral site has received little attention. Laczi *et al.* (1980) investigated the sublingual absorption of vasopressin analogues. While 1-deamino-8-D-arginine-vasopressin (DDAVP) was able to provide sufficient antidiuretic effect in patients with diabetes insipidus, arginine-vasopressin failed to show sublingual absorption. The sublingual absorption of leuprolide in dogs was found to be limited (Vadnere *et al.*, 1990). Insulin was absorbed sublingually only upon coadministration of a vasodilatative agent (Earle, 1972); without enhancing agent, effects were negligible (Yokusaka *et al.*, 1977).

The nonkeratinized buccal epithelium of the cheeks covers one-third of the total intraoral surface of approximately 100 cm^2 (Ho and Higuchi, 1971). This site appears to be of special potential for the administration of systemic proteinaceous drugs and will therefore be described in more detail.

2.2. Histology of the Buccal Mucosa

The buccal mucosa consists principally of two components, the epithelium and the underlying connective tissue (Fig. 1). The interface between these two layers is

MUCUS

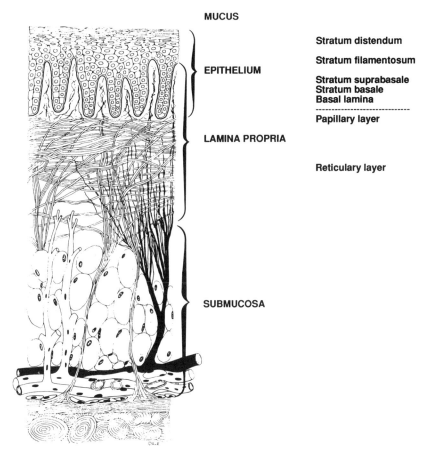

Stratum distendum

Stratum filamentosum

EPITHELIUM

Stratum suprabasale
Stratum basale
Basal lamina

Papillary layer

LAMINA PROPRIA

Reticulary layer

SUBMUCOSA

Figure 1.   Principal histology of human oral mucosa. After Squier and Hill (1989) (with permission from the copyright owner).

formed by the basal membrane. The human buccal epithelium is, similar to other epithelia like those of the esophagus, vagina, and anus, a multilayered and non-keratinized tissue. With regard to the skin, principal differences are the lack of both a keratinized stratum corneum and the mucus-containing salivary layer. The buccal epithelium has a mean thickness of 500–600 μm (Chen and Squier, 1984) and is extensively infiltrated by large, conical papillae of the underlying connective tissue, which reach up to 100 μm toward the outer surface (Sloan and Soames, 1984). Due to the connective tissue papillae, the interface between superficial epithelium and subepithelial connective tissue is greatly increased; thus, a broad anchorage of the epithelium and an enormous surface for intensive metabolic exchange are established. In total, the multilayered epithelium is formed by 40–50 successive cell layers; between the connective tissue papillae and the outer surface, 20 or fewer cell layers are found.

Due to mechanical abrasion in the course of mastication, the epithelial cells at the outer surface are continuously peeled off. Therefore, starting from nondifferentiated cells located above the basal membrane, fresh epithelial cells are permanently produced by mitosis. During their 5- to 8-day passage from the basal membrane to the outer surface, these epithelial cells undergo a process of maturation and change in form and size (Hill, 1984) and become progressively flatter. Macromolecules such as proteoglycans and glycogens are excreted by membrane-coated granules and form a complex intercellular matrix.

According to Schroeder (1981), the buccal epithelium can be subdivided into different histological sections (Fig. 1): The stratum basale is the inner nondifferentiated cell layer next to the basal membrane. In the following stratum suprabasale, the differentiation of the epithelial cells begins. The adjacent stratum filamentosum is characterized by multiple extracellular aggregates, the filaments. Finally, the superficial stratum distendum houses the fully differentiated cells and makes up to one-third of the complete thickness of the epithelium.

Between adjacent epithelial cells, three different types of intercellular junctions can occur, namely desmosomes, gap junctions, and tight junctions (Chen and Squier, 1984). Intercellular junctions serve either for communicative purposes or for maintenance of epithelial integrity (see Chapter 1). Desmosomes are one of the main features of oral epithelia and represent circular or oval regions of cell contact of about 0.2–0.5 μm diameter. Here a small intercellular zone 25–30 nm in diameter remains between the adjacent cells. Via this intercellular space the cells are fixed together by protein filaments serving as an intercellular contact layer. Such junctional complexes appear to be a minor barrier for penetrating tracer compounds or drug molecules. Gap junctions are, similar to desmosomes, regions of parallel plasma membranes between neighboring cells. Here only small intercellular spaces of 2–5 nm are left between the cells. Gap junctions present a higher resistance for penetrating molecules; they cannot be permeated by larger marker molecules, e.g., lanthanum tracer particles. Gap junctions are seen in the buccal epithelium upward from the stratum basale and are often associated with desmosomes. While these cell contact regions are rather common in keratinized mucosal tissues, e.g., hard palate, they are less frequent in buccal epithelia (Müller and Schroeder, 1980). Finally, tight junctions are the most intensive cell connections between neighboring cells (Alberts et al., 1989). They appear as though the two adjacent plasma membranes were fused so that no intercellular space remains. Such junctional complexes are common in the duodenal mucosa (Farquhar and Palade, 1963), where they play an important role in sealing the epithelial sheet. Interestingly, these highly occluding junctional complexes are rare in the buccal epithelium (Chen and Squier, 1984).

The interface between the epithelium and the underlying connective tissue is formed by the basal lamina, a complex layer about 1 μm thick (Chen and Squier, 1984). There is some evidence that the basal lamina may act as a barrier for relatively large, hydrophilic substances, e.g., immune complexes (Brandtzaeg and Tolo, 1977) or endotoxins (Alfano *et al.*, 1975). The underlying connective tissue can be further subdivided into the highly structuralized lamina propria, consisting of the papillary layer and the reticular layer, and the submucosa. The lamina propria contains a dense network of collagen serving for mechanical stability. The submucosa is rather variable in thickness and composition and may contain constituents such as adipose tissue salivary glands, nerve fibers, blood capillaries, and lymph vessels.

Blood supply to the buccal mucosa is derived primarily from the external carotid artery, which serves several smaller buccal blood vessels (Stablein and Meyer, 1984). Blood perfusion is high; e.g., the buccal blood flow in monkeys was found to be 21 ml/min per 100 g tissue (Nanny and Squier, 1982). A similarly diversified drainage of venous and lymphatic capillaries allows the exchange of metabolic products. Via the infiltrating connective tissue papillae, the arterial as well as the venous and the lymphatic capillaries reach far out into the depth of the multilayered epithelium and ensure an efficient supply of oxygen and nutrients.

2.3. Biochemistry of the Buccal Mucosa

An outstanding feature of the buccal epithelium is the high proportion of extracellular material. These macromolecular aggregates are excreted by the epithelial cells and can amount to a considerable part of the epithelium, e.g., filaments in the stratum filamentosum contribute more than 35% of the total volume (Schroeder, 1981). The extracellular matrix has multiple physiological functions, e.g., as permeability barrier, as gliding coat for the maturing epithelial cells, as elasticity factor or for maintaining tissue integrity.

The main components of the extracellular matrix were identified as glycosaminoglycans and multiple heterogeneous proteoglycans (Alberts *et al.*, 1989; Gerson and Harris, 1984). For simplicity, the extracellular matrix can be regarded as a hydrated polysaccharide gel, in which fibrous proteins are embedded. Soluble polysaccharides common in epithelia are hyaluronic acid, chondroitin sulfate, dermatan sulfate, heparan sulfate, and keratan sulfate (Alberts *et al.*, 1989). Except for hyaluronic acid, all of these polysaccharides are also bound covalently to proteins and form soluble proteoglycans. These macromolecular structures interact via multiple noncovalent and covalent forces and form a coherent gel structure of the extracellular matrix.

In this gel, serving as some sort of stabilizing backbone, fibrous protein components are embedded. These macromolecular aggregates consist of different collagens (type I–III) and elastins. Finally, these extracellular aggregates are linked to cellular membranes via connecting fibronectins. Fibronectins are dimer macromolecules of about 5000 amino acids and also seem to be involved in the migration process of the epithelial cells during maturation.

Lipids such as cholesterol, phospholipids, and free fatty acids are also part of the epithelium. Besides their ubiquitous occurrence in plasma membranes, they are to some extent part of the extracellular matrix (Gerson and Harris, 1984). In the nonkeratinized buccal epithelium, mostly polar lipids occur, such as phospholipids, cholesterol sulfate, and glycosylceramide. In contrast, the keratinized oral epithelium, very much comparable with the skin, contains predominantly neutral lipids like ceramides and triglycerides. The differences in types and distribution of lipids accord well with known permeability differences, which show that keratinized oral mucosae and epidermis have a similar impermeability to water and that nonkeratinized regions have higher permeability (Squier and Hall, 1985; Squier *et al.*, 1986).

Naturally, the proteolytic activity associated with the buccal mucosa is a relevant barrier for buccally administered peptides or proteins. Proteolytic activity of the buccal mucosa may include enzymes such as aminopeptidases, carboxypeptidases, and several endopeptidases (Garren and Repta, 1988; Lee, 1988) and may vary considerably between different species. So far only few and exemplary studies have dealt with the protease activity using buccal cell homogenates as test system (Lee, 1988; Cassidy and Quadros, 1988; Dodda-Kashi and Lee, 1986; Nakada et al., 1987). One paramount disadvantage of this experimental approach is its inability to discriminate between cytosolic, membrane-bound, and intercellular proteolytic activity. With respect to the possible intra- or paracellular routes of peptide or protein absorption (see Section 5), the exact localization of mucosal proteolytic activity is of crucial importance. Here new experimental techniques, e.g., the cell culture technique (see Section 3.3), will hopefully enlarge our knowledge.

Summing up, the buccal epithelium, like all other epithelia, possesses some proteolytic activity. So far the exact localization of these enzymes and their significance for the inactivation of buccally administered proteinaceous drugs are neither understood nor characterized in detail. As a whole, our knowledge on this topic must be regarded as insufficient and rudimentary.

2.4. Saliva and Mucus

The superficial salivary layer covering all oral epithelia can interfere with the penetration of drugs as well as with the adhesion of buccally administered delivery devices (see Section 6.1). The oral cavity is constantly washed by a stream of 0.5 to 2 liters of saliva daily (*Wissenschaftliche Tabellen*, 1977), produced mainly by the salivary glands. The glands are the three pairs of the parotid, the submaxillary, and the sublingual glands. The first are located under and in front of each ear, with ducts opening to the inner surface of the cheek. The submaxillary glands lie below the lower jaw releasing saliva through one duct on each side. Finally, the sublingual glands are located under the tongue with its ducts opening to the floor of the mouth. Minor salivary glands are situated in the buccal, palatal, and retromolar regions of the oral cavity. Additionally, multiple smaller salivary glands are located directly in the buccal epithelium. The pH of saliva is largely dependent on the person, age, and degree of stimulation and varies between pH 5.8 and 7.4 (Rettig, 1981). Saliva contains 4–6% solutes, mainly inorganic salts and mucus glycoproteins. Glandular saliva does not contain any peptidases (*Wissenschaftliche Tabellen*, 1977), but due to the complex intraoral bacterial microflora, peptidases might occur with considerable intra- and interindividual variability.

Structure forming macromolecules in saliva are the mucins, a group of charged high-molecular-weight glycoproteins found on the surfaces of all inner epithelia (Marriott and Hughes, 1990). Their mean molecular weight is 10^6–10^7 Da (Allen *et al.*, 1984). Mucins are water-soluble highly active gelling agents and form a coherent film 10–200 μm thick on the oral epithelia. Due to the multitude of charged functional groups in this macromolecule, the superficial mucus layer of the epithelium might act as a first barrier for buccally administered proteinaceous drugs. The physiological function of the mucus is as a protective layer for the viable tissue and as a lubricating agent for digested nutrients.

3. EXPERIMENTAL TECHNIQUES FOR STUDYING ORAL ABSORPTION

3.1. Human Assays

Human studies are indispensable for the clinical evaluation of a new (buccal) delivery devise or for the validation of an animal model. A number of different approaches may be used:

- Mouthwash procedure
- Application of mechanical absorption chambers
- Administration of buccal drug delivery devices

The most commonly described technique is the mouthwash procedure introduced by Beckett (Beckett and Hossie, 1971; Beckett and Triggs, 1967). In this test a volume of 25 ml of a drug solution of appropriate concentration and appropriate buffer, pH, and solvent composition is placed in a subject's mouth. Only by movement of the cheeks and the tongue the solution is then circulated 300–400 times around the mouth for 5 min. Care has to be taken that none of the solution is swallowed. After the required period the solution is put into a beaker, its volume recorded, and its pH measured. A subsequent mouth-rinse with 10 ml of water for 10 sec is designed to wash off nonabsorbed drug. Then the solutions are combined and analyzed for the drug remaining. In order to account for volume changes due to the dilution by saliva, a nonabsorbable marker may be added to the initial drug solution instead of recording its volume after the test. In spite of the somewhat imprecise-looking mouthwash procedure, the results obtained have been shown to be sufficiently reproducible, with respect to both intra- and intersubject variations. It should, however, be kept in mind that the amount of drug missing in the solution after the test cannot necessarily be set equal to the amount of drug absorbed during the test. Drug firmly bound to constituents of the mucin layer or to the cells is assumed as being absorbed. Also, permeability differences between different intraoral mucosal sites cannot be detected using this approach. Monitoring the degree of absorption after one given interval of mucosal contact only, e.g., 5 min, is also inadequate, since the amount absorbed in a given time period does not allow for a kinetic treatment of the data if absorption is a saturable process. Only under steady-state conditions is a full kinetic evaluation feasible. Modifications and improvements of the classical Beckett procedure were proposed by Dearden and Tomlinson (1971), Schürmann and Turner (1978), and, more recently, by Tucker (1988). These modifications stress the consideration of saliva secretion and protein binding, surface binding, accidental swallowing, and the sampling procedure.

Blood level monitoring upon buccal administration of drugs offers an alternative instrument for following buccal absorption. Hereby, the meaningfulness of the classical mouthwash procedure can be increased significantly. Such a setup may be regarded as satisfactory for most clinical purposes. Pharmacokinetic modeling of the plasma profiles and/or the urinary excretion data provides complete information about the overall amount absorbed and the rate of the process. However, a clear distinction among the different permeabilities for the various intraoral epithelia and a precise interpretation of the factors determining the absorption process at and in the mucosa are not possible with the mouthwash procedure even in this improved experimental setup. The same restrictions apply to the application of conventional nondirected intraoral dosage forms, e.g., chewable capsules or tablets.

Pimlott and Addy (1985) compared the mucosal uptake of a steroid from different intraoral regions, e.g., sublingual, buccal, and palatal mucosae. Here the drug solution was soaked up by a filter paper and this device was applied locally on a defined epithelial surface for 5 min. Analyzing the drug remaining in the devices after removal, the authors were able to show significant differences; the uptake of the steroid was found in the sequence: sublingual > buccal > palatal.

Buccal absorption experiments were also carried out using a Polytef absorption chamber (Anders, 1984; Anders et al., 1983). This design limits the contact area between the device and the buccal mucosa and, therefore, is able to more realistically simulate the absorption process for adhesive dosage forms. Similar chambers are used routinely for transdermal penetration and absorption studies, e.g., the Hill-Top Chamber. We recommend this experimental approach for exploratory studies in humans. More sophisticated absorption chambers offering the possibility for controlled perfusion of mucosal segments may allow even further insights into the factors governing the buccal absorption process. This approach is under evaluation in humans (Ho et al., 1990); in animals such perfusion chambers have already been used successfully (Veillard et al., 1987; Ho and Barsuhn, 1989; see Section 3.2).

Another technique for studying oral mucosal absorption is administration of drug-loaded adhesive patches (Anders, 1984; Merkle et al., 1986). In addition to pharmacokinetic or pharmacodynamic monitoring, the absorption process can be characterized by the amount of drug remaining in the patch after certain time intervals. Such an experimental setup combines the absorption kinetics with the *in vivo* liberation of a drug from an individually tailored delivery device and can therefore be regarded as a realistic approach.

3.2. Animal Models

An increasing number of absorption studies are being conducted in animals. Besides rodents such as hamster, rat, and rabbit, sheep and dog are used as experimental models. So far, information is scarce whether and to what extent these different animal models are representative of the nonkeratinized human buccal mucosa. The rat buccal epithelium, whose thickness of about 500 mm is rather comparable to the human buccal epithelium, is heavily keratinized and additionally covered by a dense and coherent keratin layer (Ebert *et al.*, 1986). Additionally, the state of hydration of the rat buccal mucosa appears to be lower than in humans. Concerning proteinaceous drugs, the buccal mucosa of rodents, therefore, appears to possess a more pronounced barrier function as compared with the nonkeratinized human buccal epithelium. Despite the relatively numerous buccal absorption experiments in rodents, no attempts to correlate these results with those of human studies have been made. Also, systematic studies correlating different animal models for possible extrapolation from one model to another are still missing.

Tanaka *et al.* (1980) were the first to study the *in vivo* absorption of sodium salicylate in the cheek pouch of the male golden hamster. Modifications of this technique were subsequently employed by Ishida *et al.* (1983), Eggerth *et al.* (1987), and Kurosaki *et al.* (1988, 1989a). Anesthetized rats were used by Aungst and co-workers (Aungst and Rogers, 1988, 1989; Aungst *et al.*, 1988), Merkle and co-workers (Anders, 1984; Wolany *et al.*, 1989), Nakada *et al.* (1988), and Paulesu *et al.* (1988). Rabbits were investigated by Oh and Ritschel (1988a,b). In sheep a microporous hollow fiber delivery system was evaluated for buccal use (Burnside *et al.*, 1989). Finally, dogs are a common animal model for buccal absorption studies (Ebert *et al.*, 1986; Ho *et al.*, 1990; Ishida *et al.*, 1981; Ritschel *et al.*, 1988; Veillard *et al.*, 1987; Wolany *et al.*, 1990a). The dog buccal mucosa is, like the human epithelium, a nonkeratinized tissue. Among all animal models focusing on buccal absorption, only for the beagle has a correlation of animal and human data been claimed (Cassidy *et al.*, 1989; Ebert *et al.*, 1986). In these studies the buccal absorption of the nonpeptide drug diclofenac sodium was evaluated. Whether these results are directly transferable to the more complex situation for peptides or proteins is not known. Nevertheless, the beagle has also been employed as an animal model for peptide drugs

(Ishida *et al.*, 1981; Ritschel *et al.*, 1988; Wolany *et al.*, 1990a) and can therefore be considered as the best characterized animal with regard to buccal absorption studies. In contrast to rodent experiments, the evaluations in dogs can be performed not only in anesthetized (Ebert *et al.*, 1986; Ho *et al.*, 1990) but also in conscious dogs (Ishida *et al.*, 1981; Wolany *et al.*, 1990a). This might be of interest with regard to possible changes in local blood perfusion upon anesthesia. It is our experience that conscious beagles show perfect compliance to bioadhesive dosage forms in case those devices are optimized for flexibility, smoothness, size, and bioadhesion. As far as the animal models mentioned above relate to the intraoral absorption of peptides or proteins, more data will be discussed in Section 4.

3.3. *In Vitro* Techniques

For *in vitro* studies on buccal absorption, excised mucosal preparations of rabbits were used by Galey *et al.* (1976), Siegel and Gordon (1985), and Robinson and co-workers (Banerjee and Robinson, 1988; Dowty *et al.*, 1990; Gandhi and Robinson, 1990; Harris and Robinson, 1990). Other authors investigated isolated dog frenulum (Siegel *et al.*, 1976), porcine buccal mucosa (De Vries *et al.*, 1990a,b), and hamster buccal mucosa (Kurosaki *et al.*, 1989a,b; Tanaka *et al.*, 1980) as *in vitro* test system. With respect to peptides and proteins, a critical and thorough evaluation of the significance of such *in vitro* studies has not been reported, although Tanaka *et al.* (1980) describe a good *in vivo/in vitro* correlation for their hamster experiments on sodium salicylate absorption.

In this context, two recent articles on the *in vitro* permeation and metabolism of an amino acid substrate (leucine-*p*-nitroanilide) in isolated hamster cheek pouch (Garren and Repta, 1989; Garren *et al.*, 1989) have to be mentioned. Such experimental setups should allow deeper insights into the so far poorly understood proteolytic processes which a buccally administered peptide or protein faces during its mucosal passage. The cell culture approach reported by Tavakoli-Saberi and Audus (1989) with cultivated hamster pouch buccal epithelium is anticipated to become another valuable tool to describe the enzymatic degradation of peptides or proteins.

4. LITERATURE SURVEY ON INTRAORAL PEPTIDE ABSORPTION

An integral bibliographic index covering the literature on buccal drug absorption has been published recently by Merkle *et al.* (1990b). In the past decade, our knowledge regarding buccal peptide absorption has expanded remarkably. A chronological survey is given in Table I.

Due to the outstanding therapeutic importance of insulin, much of the research

Table I

Chronological Survey of *in Vivo* Experiments
on Buccal Peptide and Protein Absorption

Author(s)	Year	Peptide (*in vivo* model)
Wieriks	1964	α-Amylase (guinea pig, rat)
Earle	1972	Insulin[a] (human)
Laczi *et al.*	1980	Vasopressin analogues[a] (human)
Ishida *et al.*	1981	Insulin (dog)
Anders *et al.*	1983	Protirelin (human)
Anders	1984	Protirelin, buserelin (rat)
Schurr *et al.*	1985	Protirelin (human)
Veillard *et al.*	1987	Model tripeptide (dog, hamster)
Aungst and Rogers	1988	Insulin (rat)
Aungst *et al.*	1988	Insulin (rat)
Nakada *et al.*	1988	Calcitonin (rat)
Oh and Ritschel	1988a,b	Insulin (rabbit)
Paulesu *et al.*	1988	Interferon-α_2 (rat)
Ritschel *et al.*	1988	Insulin (dog)
Aungst and Rogers	1989	Insulin (rat)
Burnside *et al.*	1989	Gonadorelin agonist (sheep)
Ho and Barsuhn	1989	Protirelin, oxytocin (dog)
Wolany *et al.*	1989	Octreotide (rat)
Dowty *et al.*	1990	Protirelin (dog, hamster)
Vadnere *et al.*	1990	Leuprolide[a] (dog)
Wolany *et al.*	1990a	Octreotide (dog)
Zhang *et al.*	1990	Insulin (dog)

[a]Sublingual administration.

interest has focused on it. Insulin is a relatively large peptide of 51 amino acids, and is able to pass the oral mucosa only in the presence of absorption enhancers (see Section 6.2) to some extent. Ishida *et al.* (1981) reported only a moderate bioavailability of 0.5% in dogs when administered together with sodium glycocholate in a cacao butter matrix. More recent studies in rats (Aungst and Rogers, 1988, 1989) report a remarkable buccal bioefficiency of 25–30% as compared with i.m. administration. In these studies, cholic acid derivatives, e.g., sodium glycocholate, were effective in enhancing buccal insulin absorption. Other additives, like sodium taurodihydrofusidate (STDHF), laureth-9, sodium dodecylsulfate, and palmitoylcarnitine, also demonstrated significant effects. In general, enhancer concentrations higher than 1% had to be employed to obtain significant increases in peptide bioefficacy. Principally similar and encouraging results on buccal insulin absorption were found in rabbits (Oh and Ritschel, 1988a,b) and dogs (Ritschel *et al.*, 1988; Zhang *et al.*, 1990). The buccal absorption of human calcitonin was investigated in anesthetized rats (Nakada *et al.*, 1988). In this study a decrease in plasma calcium levels could only be achieved by coadministration of absorption enhancers. Cholic acid derivatives and sodium dodecylsulfate were found to be most effective; sugar esters, e.g., sucrose monopalmitate, and quillajasaponine also enhanced the buccal calcitonin absorption. Possibilities and restrictions associated with absorption enhancement will be addressed in Section 6.2.

Interferon-α_2 (168 amino acids) was only absorbed in trace amounts via the oropharyngeal cavity of rats (Paulesu et al., 1988). For such complex and high-molecular-weight proteinaceous compounds, the buccal mucosa seems to be of limited value as an alternative site of administration as compared with parenteral administration.

Merkle and co-workers studied the buccal absorption of protirelin (TRH) in humans (Anders et al., 1983; Schurr et al., 1985). Here, sustained pharmacodynamic effects were observed relative to either intravenous or nasal administration. Conducting similar buccal experiments with an adhesive delivery device in rats (Anders, 1984), the author assumed the absolute buccal TRH bioavailability to be approximately 2%. However, the buccal absorption of buserelin, an LH-RH analogue, was variable or negligible, respectively.

The buccal absorption of the somatostatin analogue octreotide (Sandostatin®, Sandoz, CH-Basle), an enzymatically stable cyclic octapeptide (Bauer et al., 1982; Battershill and Clissold, 1989), was studied in some detail by Merkle and co-workers (Wolany, 1990. In rats, an experimental drug delivery device (see Section 6.1) was administered to a defined area of the buccal mucosa (Wolany et al., 1989). Blood samples were withdrawn at intervals by puncture of the jugular vein immediately before and for up to 10 hr after administration. Plasma was separated by centrifugation and analyzed for octreotide content by direct radioimmunoassay. As compared with an i.v. bolus injection, marked sustainment of peptide plasma levels with a moderate maximum 1 hr after application can be achieved upon buccal administration (Fig. 2). This indicates a depot function of the oral mucosa. In the range of 20–1600 μg octreotide per animal, octreotide plasma levels increase linearly with the applied dose (Fig. 3). However, the absolute buccal octreotide bioavailability in this animal model without absorption enhancer was found to be moderate with only 0.2–0.6%. In contrast to the above-mentioned significant effects of absorption enhancers with insulin, the effects of several absorption enhancers (administered in 3% aqueous solution together with the peptide drug) on buccal octreotide absorption were rather low or negligible (Fig. 4). Comparing the area under the curves (AUC 0–6 hr), only a combination of disodium-EDTA and sodium glycocholate raised the bioavailability to about 100%; a marked increase was also found with 1-dodecylazacycloheptan-2-one (Azone®, ethanolic solution). When administered alone with the peptide, disodium-EDTA failed to enhance buccal peptide absorption. Bacitracin increased the AUC 0–6 hr to approximately 60%, while the effects of the cholic acid derivatives sodium glycocholate, sodium taurocholate, sodium deoxycholate, and sodium taurodihydrofusidate were insignificant (AUC 0–6 hr increase < 20%). Based on these experiments, the heavily keratinized buccal mucosa of the rat appears to be a rather impermeable barrier.

Corresponding studies were conducted in beagle (Wolany et al., 1990a). Due

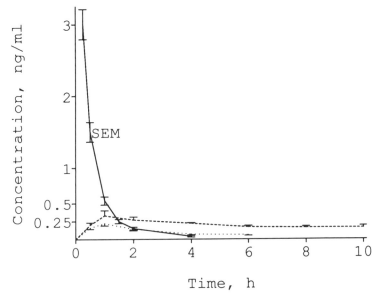

Figure 2. Octreotide plasma levels after intravenous and buccal administration in rats (N = 5–6). ——, i.v., 0.005 mg/rat; ·····, buccal, 80 mg/rat; ----, buccal, 80 mg/rat, plus 3% Na_2-EDTA + 3% sodium glycocholate.

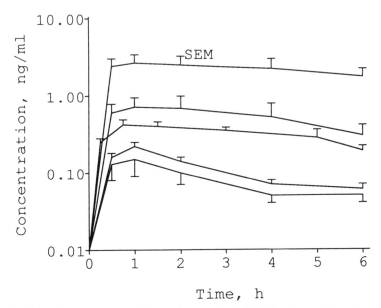

Figure 3. Octreotide plasma concentrations after buccal administration in rats (N = 5–6). Curves represent in ascending order: 50 mg, 80 mg, 200 mg, 400 mg, 1600 mg octreotide/rat.

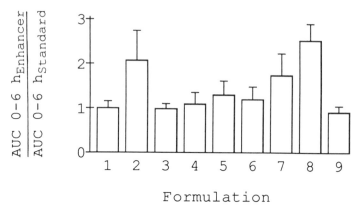

Figure 4. Relative efficacy of absorption enhancers (3% w/v) on the buccal absorption of octreotide in rats (N = 5–6). AUC 0–6 hr of reference experiment without enhancer = 1. Bars represent mean + SEM. 1, Standard, no additive; 2, Na_2-EDTA plus sodium glycocholate; 3, sodium taurocholate; 4, sodium glycocholate; 5, sodium taurodeoxycholate; 6, Na_2-EDTA; 7, bacitracin; 8, 1-dodecylazacycloheptan-2-one (ethanolic solution); 9, 1-dodecylazacycloheptan-2-one (aqueous formulation).

to an apparent minimum threshold concentration, the absolute buccal bioavailability of octreotide in this model is dose-dependent and varies from 0.3 to 2% (Fig. 5). Plasma levels remain sustained for an extended period of time relative to i.v. injection. However, the sustainment of peptide plasma profiles in the dog seems to be less pronounced than in the rat experiments. Among the absorption enhancers, 1-dodecylazacycloheptan-2-one (Fig. 6) and the cholic acid derivatives, i.e., sodium glycocholate (Fig. 7), had significant effects on buccal octreotide absorption with interesting octreotide bioavailabilities of 4–6%. Obviously, the permeability of the nonkeratinized buccal mucosa of the dog can be altered in a more pronounced way than can the heavily keratinized rat mucosa.

5. MECHANISMS OF PEPTIDE AND PROTEIN ABSORPTION

The mechanisms and quantitative relationships of buccal/sublingual absorption were the subject of a number of reviews and original papers (e.g., Beckett and Moffat, 1970; Dearden and Tomlinson, 1971; Dowty and Robinson, 1989; Ho and Higuchi, 1971; Schürmann and Turner, 1978; Siegel, 1984; Veillard, 1990; Wagner and Sedman, 1973). None of the literature, however, includes or explicitly covers peptides and proteins. Instead, their focus is on the common low-molecular-weight type of drugs or nutrients. From the information available, it seems to be premature to design a fully satisfying and realistic model for absorption of peptides in the oral cavity. But the following preliminary assumptions appear appropriate.

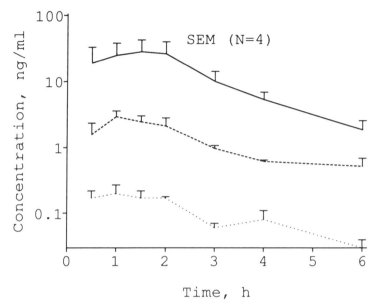

Figure 5. Octreotide plasma concentrations after buccal administration in beagle (*N* = 4). ——, 2.00 mg/kg; ----, 1.00 mg/kg; ·····, 0.20 mg/kg.

It is reasonable to suppose that the intraoral absorption of peptides and proteins is a passive transport mechanism. Endocytotic processes are not apparent in buccal epithelium (Siegel, 1984). Except for amino acids (Evered and Vadgama, 1981; Gandhi and Robinson, 1990), neither active nor carrier-mediated peptide transport systems are present in the buccal epithelium (Ho and Barsuhn, 1989). Such enzymatic transport systems are well-known to contribute to the intestinal absorption of dipeptides and tripeptides in the course of protein digestion.

In the case of simple passive diffusion, the two principal routes of drug transport are the transcellular and the paracellular pathway. The transcellular pathway via partitioning into lipid membranes is the well-established route of absorption for the majority of the usual low-molecular-weight type of compounds. This is confirmed by a multitude of successful correlations between pH, pK$_a$, lipophilicity, and permeability or absorption, respectively. It is difficult to conceive that peptides or proteins with predominantly hydrophilic behavior, having a relatively high molecular weight and being usually charged under physiological conditions, could be buccally absorbed by this mechanism. If cell membranes were permeable to molecules of comparable size (> 1000 Da), then ion concentration gradients would collapse and the living epithelial cells could not maintain a discrete internal environment. Therefore, other pathways must be considered to explain the reported biological responses of buccally administered peptides or proteins (see Section 4).

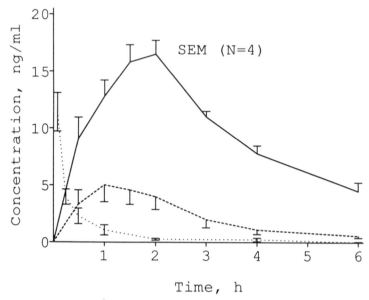

Figure 6.   Octeotide plasma levels after intravenous and buccal administration in beagle ($N = 4$). ·····,
i.v., 0.005 mg/kg; ——, buccal, 1.00 mg/kg, 3% dodecylazacycloheptan-2-one (ethanolic solution); ----,
buccal, 1.00 mg/kg, no additive.

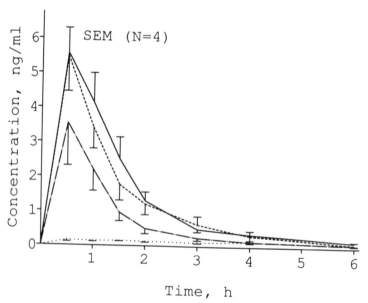

Figure 7.   Peptide plasma levels after buccal administration of 0.20 mg/kg octreotide in beagle—effect
of cholic acid derivatives (4% w/v) as absorption enhancing agents.   ——, Sodium glycocholate; ----,
sodium taurocholate plus Na_2-EDTA; –·–, sodium taurocholate; ·····, standard, no additive.

The paracellular route, also referred to as the "aqueous pore pathway" (Ho *et al.*, 1990; see Chapters 1 and 10), involves the passage of drug molecules through intercellular spaces. This pathway operates along cell–cell interfaces and through the different junctional complexes of adjacent epithelial cells. McMartin *et al.* (1987) propose this mechanism for the nasal penetration of oligopeptides. Despite the differences in structure of the various epithelia, such a route of absorption also appears to be most realistic for peptide or protein absorption in the buccal epithelium. In this context, the lack of excluding tight junctions in the buccal epithelium appears favorable for a paracellular transport pathway for peptides.

One might also speculate about peptides carrying lipophilic moieties. Such molecules could preferentially be adsorbed into the bilayer and then find transient holes to slip through the bilayer and thus into the cell. However, once inside an epithelial cell, the proteinaceous drug is exposed to the entire cytosolic and membrane-bound proteolytic activity and might, therefore, not survive such a route. Keeping in mind the multilayered structure of the buccal epithelium, at least 10–20 successive passages into and out of the epithelial cells are required until the basal lamina is reached at the interface to the underlying connective tissue.

Ion-pair formation may contribute to the permeation of charged peptides, e.g., cationic drug molecules may interact with ubiquitous steroids like cholesterol sulfate. Uncharged complexes may then diffuse directly into the hydrated extracellular matrix or pass into the depth of the epithelium via a more membrane-based lipid pathway. But again, no hard facts are available to firmly support or to exclude one or the other idea and further evidence is clearly needed to develop our understanding regarding the basic mechanisms of buccal absorption.

Fundamental research on this topic has been initiated by Ho and co-workers (Ho and Barsuhn, 1989; Ho *et al.*, 1990). The authors performed systematic *in vivo* and *in vitro* studies on buccal absorption employing a homologous series of small and enzymatically stable model peptides and their *t*-butyl-oxycarbonyl (BOC) derivatives. The peptide series (D-Phe)n-Gly, Ac(D-Phe)n, and Ac(N-CH$_3$) (D-Phe)$_3$NH$_2$ (with $n = 1$–3) were investigated. The results indicate that charged amino acids and their BOC derivatives tend to have membrane permeabilities which are smaller than structurally similar charged nonamino acids, although the effective partition coefficients (*n*-octanol/Krebs–Ringer buffer) were comparable. Terminal charges on zwitterionic peptides exert a negative effect on membrane permeability even though the effective partition coefficient of such compounds may be relatively high. The authors specify that, although the partition coefficient of the homologous peptide series studied may be increased by increasing the number of amino acids, the membrane permeability tends to decrease. Therefore, not only the molecular size, but also additional factors, e.g., the number of solvated amide bonds, may be rate-determining. In contrast to many conventional nonpeptide drugs, *n*-octanol/water partition coefficients seem to be inappropriate for predicting buccal peptide absorption.

To increase the buccal absorption of peptides, factors like molecular size, lipophilicity, charge, conformation, and stability of the specific compound have to be considered first. In a general sense, making peptides smaller and "less peptidic" seems to be a useful strategy for improving their permeability characteristics (Ho *et al.*, 1990). The available data base, however, is still not sufficient to predict absorption of proteinaceous compounds on the basis of the physical model designed by these authors. Other strategies for enhancing peptide absorption will be discussed in Section 6.

6. STRATEGIES FOR ENHANCING INTRAORAL PEPTIDE ABSORPTION

6.1. Dosage Form Design and Patent Review

As reviewed by Merkle *et al.* (1990a), conventional dosage forms for oral mucosal delivery include solutions, gels, lozenges, and buccal or sublingual tablets. Solutions in small quantities may be added to suitable capsules and the contents released upon chewing. Erodible buccal or sublingual tablets employ gradually dissolving excipients and binders. The disadvantage of the aforementioned conventional dosage forms is that the drug released is subject to continuous dilution by saliva and, therefore, to accidental loss of drug due to involuntary swallowing. Naturally, any controlled or locally restricted or directed release of the drug, respectively, is not within the scope of these formulations. Moreover, such dosage forms interfere with drinking and eating, and perhaps even talking. Their application time is thus rather limited.

Nevertheless, owing to the excellent accessibility of the oral cavity for drug delivery, adhesive dosage forms have been developed and appear to be of some commercial interest, as indicated by an increasing number of patents (Table II). Their formulation is predominantly based on the use of bioadhesive polymers. Such polymers have been extensively covered in recent textbooks and reviews (e.g., Lenaerts and Gurny, 1990; Gurny and Junginger, 1990). Their interaction with the glycoprotein calyx of the oral mucosae is mainly based on the interdiffusion of the mucosal mucin glycoproteins and the polymer chains resulting in more or less pronounced adhesive binding depending on the polymer applied and on the thermodynamics and kinetics of the interaction involved. Strong binding is usually associated with hydrogen bonds. Firm attachment to mucosal sites is an established fact (e.g., Anders and Merkle, 1989; Boddé *et al.*, 1990; Merkle *et al.*, 1990a; Wolany *et al.*, 1990b) and is normally well tolerated on the oral mucosae.

Polymers for mucosal bioadhesion are readily available and include erodible polymers like substituted cellulose derivatives, polyacrylates, polymethacrylates, or nonerodible cross-linked polymers, e.g., polyacrylic acid gels (polycarbophil type), gelatin, agarose, and others. Superior bioadhesion is demonstrated by ionic polymers like polyacrylic acid derivatives, mainly because of their extensive hydrogen bond-

Table II

Unidirectional Buccal Drug Delivery Systems (Chronological Patent Survery)

Inventor	Short title	Patent classification	Date
Sterwin	Sticking plaster which adheres to mucous membranes	GB 1384537	19.02.75
American Home Products	Buccal dosage form	US 3972995	04.08.76
Alza	Adhesive patch containing oxytocic drug	US 3699963	15.12.76
Teijin	Preparation adhering to mucous membrane of oral cavity	J 56100714	12.08.80
Nippon Soda	Film able to adhere to mucous membrane	WO 82201129	15.04.82
Nitto Electric	Mucosal preparation for oral application	J 58128314	30.07.83
	Adhesive oral bandage	EP 200508	10.12.86
Squibb, Yamanouchi	Adhesive medicament containing tape for oral mucosa	DE 3618553	08.01.87
Sato Seiyaku	Medical film composition for adhering to mucous membrane	J 62135417	08.06.87
Johnson & Johnson	Controlled release, extruded single or multilayer film	EP 250187	23.12.87
Nitto Electric	Bandage for oral cavity	J 63005757	11.01.88
Teikoku Seiyaku	Thin sustained release buccal delivery form	EP 262422	06.04.88
	Adhesive device for application to body, especially oral cavity	EP 275550	27.07.88
Ciba–Geigy	Pharmaceutical plasters for application to mucous membranes	EP 283434	21.09.88

ing. A comprehensive survey of mechanisms and experimental methods for evaluating adhesion was given by Peppas and Buri (1986) and by Gu *et al.* (1990).

The use of bioadhesive polymers for buccal delivery allows exact localization of the dosage forms on a defined area of the oral mucosae. Such dosage forms were reviewed by Merkle *et al.* (1990a). Adhesive tablets for buccal use were suggested, e.g., by Davis *et al.* (1982). Adhesive gels have been developed, among others, by Bremecker *et al.* (1983) and Ishida *et al.* (1983). More recently, adhesive patches were introduced by a number of researchers. The design and manufacture of such dosage forms is partly derived from polymer technologies. Schematic illustrations of different patches are given in Fig. 8. Two different approaches may be distinguished. One type uses face-adhesive layers (Merkle *et al.*, 1986) to ensure exact localization and contact at the site of administration; another option is to use peripheral adhesive rings (Wolany, 1990). The proteinaceous drug can be incorporated directly into the face-adhesive polymer layer (form A, Fig. 8) or into a specific drug matrix (forms B, C, and D), the matrices being, e.g., cocoa butter (form B) or a spongy gelatinous matrix (form D).

The maximum size of the above-mentioned patches is dependent on the desired

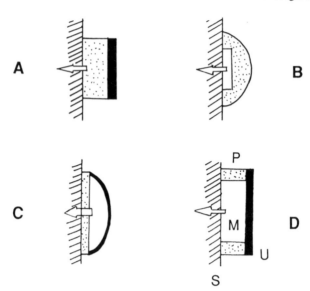

Figure 8.   Schematic illustrations of unidirectional buccal patches. A, Anders (1984); B, Ishida *et al.* (1981); C, Veillard *et al.* (1987); D, Wolany (1990). Explantaion of symbols: M, drug matrix; P, adhesive polymer; S, mucosa; U, impermeable backing layer.

site of administration. Large circular or oval patches of up to ca. 10 cm² may be exclusively localized to the buccal cheek surface. The sublingual area and the mucosal surface at the inner sides of the upper or lower lip are more restrictive and are suitable for patches of up to ca. 2 cm² only or even less. The optimization of patch composition, size, flexibility, and elasticity is a prerequisite for prefect adhesion and prevention of any local discomfort and for patient acceptance and compliance.

It appears that the maximum time span available for administration to the buccal cheek surface is naturally restricted to about 3 to 4 hr, although much longer periods are technically feasible. Otherwise, interference with normal meal schedules may occur, requiring detaching the patch. On the other hand, the mucosae of the upper and lower lip allow for much longer times of administration without regard to meals, since these areas are less involved in the masticatory process.

The use of impermeable backing layers to achieve unidirectional release (Merkle *et al.*, 1986) of the drug to the mucosa and not to the saliva, along with the exact localization of the patches, creates a rather unique situation: The absorption site is strictly confined to the site of administration and a large drug activity gradient is maintained during the entire time of administration. Additionally, the release of any absorption-enhancing compound from the patch takes place only to the mucosa underneath the patch and the agent does not spread across the whole surface. Thus, irritation, if any, is restricted to the site of application, and local stress may be limited by changing sites frequently. Nevertheless, localized high enhancer concentrations may also intensify adverse reactions and must, therefore, be critically evaluated in light of maintaining the health and appearance of the oral mucosa. However, as a rule, the buccal mucosa is certainly more robust than any other mucosal site under discussion for nonparenteral peptide absorption. No adverse reactions of the bioadhesive polymers mentioned have been reported.

A rather attractive aspect of mucoadhesive patches for buccal delivery may be the option of controlling release rates by formulation factors, e.g., by the rate of erosion of the bioadhesive polymers used (Anders and Merkle, 1989; Merkle *et al.*, 1990b) or by the effective diffusion coefficient within a cross-linked polymer gel. For buccal peptide delivery, however, immediate peptide availability seems to be crucial for achieving meaningful or even detectable levels of absorption (Merkle *et al.*, 1986; Wolany, 1990). Thus, any sustainment of release has no practical relevance for buccal peptide delivery at this point. Instead, immediate release of the peptide, e.g., from a porous polymer matrix, is highly advisable.

6.2. Chemical Strategies for Enhancing Buccal Peptide Absorption

Model studies on the efficacy of buccal absorption enhancement of proteinaceous drugs were presented in Section 4. The focus of this section is to highlight the underlying principal strategies and options.

Based on the intrinsically poor performance of mucosally administered proteinaceous compounds, recent research is turning toward the investigation of absorption-enhancing agents to improve peptide or protein bioavailability. As reviewed by Lee (1990), absorption enhancers are usually low-molecular-weight compounds which increase the drug penetration in, or the absorption through, a given mucosal surface. Five major classes may be distinguished (Lee, 1990):

• Chelators, e.g., disodium-EDTA
• Surfactants, e.g., sodium dodecylsulfate
• Bile salts, e.g., sodium glycocholate
• Fatty acids, e.g., caprylic acid
• Nonsurfactants, e.g., 1-dodecylazacycloheptan-2-one

The compounds so far evaluated for buccal use are surface-active compounds; a survey and classification according to their physicochemical or biochemical characteristics is given in Table III.

The above-mentioned enhancers may act by several mechanisms, and multiple mechanisms may occur with certain compounds. The following hypothetical mechanisms were postulated:

• Opening of calcium-dependent junctional complexes
• Intercalation and fluidization of phospholipid bilayers

- Peptidase or protease inhibition
- Ion-pair formation or micellar solubilization of the drug
- Solubilization of mucosal membrane lipids or proteins
- Inhibition of peptide aggregation

In a general sense, absorption enhancement may be achieved either by altering the thermodynamic activity of the proteinaceous drug or by interacting with the mucosal barrier. Surface-active agents are usually effective only above their critical micelle concentration (CMC), where micellar drug solubilization becomes possible. Additionally, the thermodynamic activity of the surfactant is then constant and provides a steady-state flux of the absorption enhancer through the buccal epithelium, an apparent requirement for altering membrane permeability throughout the depth of the multilayered epithelium (Ho *et al.*, 1990). It should be pointed out that the multilayered structure of the buccal epithelium would require higher doses of absorption enhancing agents for full alteration of the barrier function as compared with monolayered epithelia, e.g., the nasal mucosa.

In addition to absorption enhancers, protease inhibitors, e.g., aprotinin or bestatin, were also investigated. However, despite their reported effects on peptide stability in buccal tissue homogenates (Lee, 1988), these compounds lacked significant effects upon buccal administration *in vivo*. For example, Aungst and Rogers (1988, 1989) observed in rats no improvement of buccal insulin bioefficacy upon coadministration with either aprotinin or a peptidase-inhibiting pentapeptide (Z-Gly-Pro-Leu-Gly-Pro).

Robinson and co-workers (Knuth *et al.*, 1990) conducted initial *in vitro* studies

Table III

Exemplary Absorption Enhancers
under Investigation for Buccal Administration

Class	Exemplary compounds	Reference
Chelators	Ethylenediaminetetraacetic acid	De Vries *et al.* (1990a)
Surfactants	Benzalkonium chloride	Siegel and Gordon (1985)
	Brij 35	Oh and Ritschel (1988a)
	Laureth-9	Aungst and Rogers (1989)
	Sodium dodecylsulfate	Kurosaki *et al.* (1988)
Bile salts	Sodium deoxycholate	Nakada *et al.* (1988)
	Sodium glycocholate	Aungst *et al.* (1988, 1989)
Fatty acids	Sodium myristate	Nakada *et al.* (1988)
Peptidase inhibitors	Aprotinin	Aungst and Rogers (1988)
Miscellaneous	Chondroitinase ABC	Knuth *et al.* (1990)
	1-Dodecylazacycloheptan-2-one	Kurosaki *et al.* (1989a,b)
	Cyclodextrin	Nakada *et al.* (1988)
	Quillajasaponine	Nakada *et al.* (1988)
	Sodium salicylate	Aungst and Rogers (1988)

on several enzymes as a possible tool for specifically altering the paracellular pathway. Excised buccal mucosa of rabbits was investigated. Upon treatment with chondroitinase ABC, heparitinase, and trypsin, significant decreases in apparent resistances of the mucosal preparations were found. However, *in vivo* the enzymes chondroitinase and hyaluronidase were ineffective in enhancing buccal insulin absorption in anesthetized rats (Aungst and Rogers, 1989). Neither for protease inhibitors nor for topically administered enzymes have any studies on local compatibility and/or immunological safety been reported.

Despite the marked differences between various mucosal barriers, most studies on mucosal absorption enhancement have been performed without consideration of site-specific aspects, e.g., histological and biochemical features. So far, only Robinson and co-workers (Knuth *et al.*, 1990) address specific structures of the buccal mucosa as a target for absorption enhancement, i.e., the proteoglycan matrix in which the buccal epithelial cells are embedded. Although interesting in nature, this approach was not successful, as discussed above. Thus, the scientific degree of sophistication in the field of buccal absorption enhancement is rather low and needs further improvement and/or new strategies.

Despite the outstanding robustness of the buccal mucosa among the different epithelia currently under evaluation for alternative peptide and protein delivery, an elementary problem of absorption enhancement also applied to buccal delivery. While numerous studies on the efficacy of these agents were conducted, little or nothing, respectively, is known about the local toxicity of absorption enhancers. However, most of the characterized compounds possess at least some irritating activity. As peptide or protein therapy usually has to be frequent and long-term, the histocompatibility of these agents under realistic and long-term treatment conditions needs to be thoroughly evaluated. Therefore, suitable screening systems must be developed and validated. Not only evaluation of efficacy, but also safety considerations should guide our future research efforts in the field of absorption enhancement.

Nevertheless, keeping in mind the poor mucosal bioavailabilities of proteinaceous compounds, the coadministration of absorption enhancers appears to be a prerequisite for feasible buccal drug delivery at least with respect to larger peptides and proteins. Although bioavailability may be improved to some extent with the aid of these agents, it will presumably remain relatively low, reinforcing the fact that buccally delivered peptides or proteins of therapeutic relevance must be highly potent.

7. CONCLUSIONS

The intraoral route of peptide and protein drug delivery offers some unique advantages such as robustness of the epithelium, excellent accessibility, ease of administration, possibility to remove dosage form, and directed and localized drug delivery by dosage form design. Among the different intraoral epithelia, the buccal mucosa appears most promising due to its nonkeratinization and relatively large surface.

Relative to parenteral administration, buccal administration of peptides and proteins may allow a fundamental change of the pharmacokinetics and, therefore, of the pharmacodynamic profile of a peptide drugs. This is indicated, e.g., by a prolonged apparent half-life of octreotide after buccal administration versus i.v. administration, both in the rat and in the dog, as derived from our studies (Wolany *et al.*, 1989, 1990a). Obviously, the buccal mucosa acts as a tissue depot from which a sustained peptide drug release into the circulation can be achieved. Therefore, a practical use of intraoral peptide delivery seems to be feasible for drug candidates where a sustained and long-lasting pharmacokinetic and pharmacodynamic profile is desirable.

At this point, no satisfactory physical model for buccal absorption of protein-aceous compounds is available. This is mainly due to the fact that we do not fully understand the exact pathway taken by a peptide or protein from the oral cavity to the systemic circulation. There is a profound need for more information on the nature of the aqueous pores and how these pathways may be established by using suitable absorption enhancers.

Absorption enhancement is a prerequisite for obtaining peptide or protein bioavailabilities of therapeutic interest. With respect to long-term administration in humans, the histocompatibility of these excipients as of the administered devices will become a factor of paramount interest and, therefore, has to be evaluated in detail. There is a need for simple and meaningful *in vitro* techniques to supplement *in vivo* experiments, and especially on absorption mechanisms a body of work remains to be done.

The formulation of adequate dosage forms does not appear to be an essential problem in intraoral drug delivery. Well-known and widely characterized polymers of sufficient bioadhesion and proven compatibility are available, e.g., cellulose deriva-tives or polymethacrylates. By means of these classical excipients, the dosage form design as well as the release properties of the device can be varied in a broad manner and match virtually any demand of the individual proteinaceous compound and/or the site of intraoral administration.

Nevertheless, even under optimized conditions an intraoral route of peptide or protein delivery may not allow bioavailabilities as high as with other mucosal sites, e.g., the nasal mucosa. Thus, chances for buccal peptide delivery, if any, will be restricted to special cases and for special peptides of high permeability. In this instance, however, buccal delivery might be a preferred route of administra-tion, mainly due to the undisputed acceptance and compliance to oral dosage forms and due to the unmatched robustness of the epithelium.

ACKNOWLEDGMENTS. For the fruitful collaboration on buccal absorption of octreotide, the authors are indebted to Dr. T. Kissel, J. Munzer, and Dr. A. Rummelt (Drug Delivery Systems Department, Sandoz AG, CH-Basle). The support of Prof.

E. Mutschler (Department of Pharmacology, University of Frankfurt, D-Frankfurt/ Main) and Prof. Buéno (Institute National de la Recherche Agronomique, F-Toulouse) in the realization of the *in vivo* experiments is greatly acknowledged. For his most valuable help the authors are also indebted to Dr. N. F. H. Ho (Upjohn, USA-Kalamazoo, MI). This project was supported partly by the Deutsche Forschungsgemeinschaft (DFG, D-Bonn).

REFERENCES

Alberts, B., Bray, D., Lewis, J., Raff, M., Roberts, K., and Watson, J. D., 1989, *Molecular Biology of the Cell*, 2nd ed., Garland, New York, pp. 792–838.

Alfano, M. C., Drummond, J. F., and Miller, S. A., 1975, Localization of rate-limiting barrier to penetration of endotoxin through nonkeratinized oral mucosa in vitro, *J. Dent. Res.* **54:** 1143–1148.

Allen, A., Hutton, D. A., Pearson, J. P., and Sellars, L. A., 1984, Mucus and mucosa, *Ciba Found.* **109:**137.

Anders, R., 1984, Selbsthaftende Polymerfilme zur bukkalen Applikation von Peptiden, Ph.D. thesis, Universität Bonn.

Anders, R., Merkle, H. P., 1989, Evaluation of laminated muco-adhesive patches for buccal drug delivery. *Int. J. Pharm.* **49:**231–240.

Anders, R., Merkle, H. P., Schurr, W., and Ziegler, R., 1983, Buccal absorption of protirelin: An effective way to stimulate thyreotropin and prolactin, *J. Pharm. Sci.* **72:**1481–1483.

Aungst, B. J., and Rogers, N. J., 1988, Site dependence of absorption-promoting actions of Laureth-9, Na-salicylate, Na_2-EDTA, and aprotinin on rectal, nasal and buccal insulin delivery, *Pharm. Res.* **5:**305–308.

Aungst, B. J., and Rogers, N. J., 1989, Comparison of the effect of various transmucosal absorption promoters on buccal insulin delivery, *Int. J. Pharm.* **53:**227–235.

Aungst, B. J., Rogers, N. J., and Shefter, E., 1988, Comparison of nasal, rectal, buccal, sublingual, and intramuscular insulin efficiency and the effects of a bile salt absorption promoter, *J. Pharmacol. Exp. Ther.* **244:**23–27.

Banerjee, P. S., and Robinson, J. R., 1988, Electrophysiological characterization of rabbit buccal mucosa, *Pharm. Res.* **5**(Suppl.):185.

Battershill, P. E., and Clissold, S. P., 1989, Octeotide, *Drugs* **38:**658–702.

Bauer, W., Briner, U., and Doepfner, W., 1982, SMS 201–995: A very potent and selective octapeptide analogue of somatostatin with prolonged action, *Life Sci.* **31:**1133–1140.

Beckett, A. H., and Hossie, R. D., 1971, Buccal absorption of drugs, in: *Concepts in Biochemical Pharmacology* (B. B. Brodie, J. R. Gilette, and H. S. Ackerman, eds.), Springer, Berlin, pp. 25–46.

Beckett, A. H., and Moffat, A. C., 1970, Kinetics of buccal absorption of some carboxylic acids and the correlation of the rate constants and n-heptane:aqueous phase partition coefficients, *J. Pharm. Pharmacol.* **22:**15–19.

Beckett, A. H., and Triggs, E. R., 1967, Buccal absorption of basic drugs and its application as an in vivo model of passive drug transfer through lipid membranes, *J. Pharm. Pharmacol.* **19**(Suppl.):31–41.

Boddé, H. E., De Vries, M. E., and Junginger, H. E., 1990, Mucoadhesive polymers for the buccal delivery of peptides, structure–adhesiveness relationships, *J. Controlled Release* **13**:225–231.

Brandtzaeg, P., and Tolo, K., 1977, Mucosal penetrability enhanced by serum-derived antibodies, *Nature* **266**:262–263.

Bremecker, K.-D., Klein, G., Strempel, H., and Rübesamen-Vokuhl, A., 1983, Formulierung und klinische Erprobung einer neuartigen Schleimhauthaftsalbe, *Arzneim. Forsch./Drug Res.* **45**:591–594.

Burnside, B. A., Keith, A. D., and Snipes, W., 1989, Microporous hollow fibers as a peptide delivery system via the buccal cavity, *Proc. Int. Symp. Control. Rel. Bioact. Mater.* **16**: 93–94.

Cassidy, J., and Quadros, E., 1988, Buccal aminopeptidase activity, *Pharm. Res.* **5** (Suppl.):100.

Cassidy, J., Berner, B., Chan, K., John, V., Toon, S., Holt, B., and Rowland, M., 1989, Buccal delivery of diclofenac sodium in man using a prototype hydrogel drug delivery device, *Proc. Int. Symp. Control. Rel. Bioact. Mater.* **16**:91–92.

Chen, S.-Y., and Squier, C. A., 1984, The ultrastructure of the oral epithelium, in: *The Structure and Function of Oral Mucosa* (J. Meyer, C. A. Squier, and S. J. Gerson, eds.), Pergamon Press, Elmsford, N.Y., pp. 7–30.

Curatolo, W., 1987, The lipoidal permeability barriers of the skin and the alimentary tract, *Pharm. Res.* **4**:271–277.

Davis, S. S., Daly, P. B., Kennerly, J. W., Frier, M., Hardy, J. G., and Wilson, C. G., 1982, Design and evaluation of sustained release formulations for oral and buccal administration, in: *Advances in Pharmacotherapy* (G. Stille, W. Wagner, and W. M. Hermann, eds.), Karger, Basel, pp. 17–25.

Dearden, J. C., and Tomlinson, E., 1971, A new buccal absorption model, *J. Pharm. Pharmacol.* **23**(Suppl.):68–72.

De Vries, M. E., Boddé, H. E., Coos Verhoef, J., and Junginger, H. E., 1990a, Transport of the peptide desgylcinamide-arginine-vasopressin across buccal mucosa, *J. Controlled Release* **13**:316.

De Vries, M. E., Boddé, H. E., and Junginger, H. E., 1990b, The in vitro permeation of β-blocking agents through porcine buccal mucosa, *Proc. Int. Symp. Control. Rel. Mater.* **17**, abstract S 204.

Dodda-Kashi, S., and Lee, V. H. L., 1986, Enkephalin hydrolysis in homogenates of various absorptive mucosae in the albino rabbit: Similarities in rates and involvement of aminopeptidases, *Life Sci.* **38**:2019–2028.

Dowty, M. E., and Robinson, J. R., 1989, Drug delivery via the oral mucosa, in: *Drug Delivery to the Gastrointestinal Tract* (J. G. Hardy, S. S. Davis, and C. G. Wilson, eds.), Ellis Horwood, Chichester, pp. 123–131.

Dowty, M. E., Irons, B. K., Knuth, K. E., and Robinson, J. R., 1990, Characterization of transport pathways of thyreotropin releasing hormone in rabbit buccal mucosa, *Proc. Int. Symp. Control. Rel. Mater.* **17**, abstract D 342.

Earle, M. P., 1972, Experimental use of oral insulin, *Isr. J. Med. Sci.* **8**:899–900.

Ebert, C. D., John, V. A., Beall, P. T., and Rosenzweig, K. A., 1986, Transbuccal absorption of diclofenac in a dog model, in: *Controlled Release Technology* (P. I. Lee, ed.), Dekker, New York, pp. 310–321.

Eggerth, R. M., Rashidbaigi, Z. A., Mahjour, M., Goodhart, F. W., and Fawzi, M. B., 1987,

Evaluation of hamster cheek pouch as a model for buccal absorption, *Proc. Int. Symp. Control. Rel. Bioact. Mater.* **14**:180–181.

Evered, D. F., and Vadgama, J. V., 1981, Absorption of amino acids from the human buccal cavity, *Biochem. Soc. Trans.* **9**:132–133.

Farquhar, M. G., and Palade, G. E., 1963, Junctional complexes in various epithelia, *J. Cell Biol.* **17**:375–412.

Galey, W. R., Lonsdale, H. K., and Nacht, S., 1976, The in vitro permeability of skin and buccal mucosa to selected drugs and tritiated water, *J. Invest. Dermatol.* **67**:713–717.

Gandhi, R. B., and Robinson, J. R., 1990, Mechanism of transport of charged compounds across rabbit buccal mucosa, *Pharm. Res.* **7**(Suppl.):116.

Garren, K. W., and Repta, A. J., 1988, Buccal drug absorption. I. Comparative levels of esterase and peptidase activities in rat and hamster buccal and intestinal homogenates, *Int. J. Pharm.* **48**:189–194.

Garren, K. W., and Repta, A. J., 1989, Buccal drug absorption. II. In vitro diffusion across the hamster cheek pouch, *J. Pharm. Sci.* **78**:160–164.

Garren, K. W., Topp, E. M., and Repta, A. J., 1989, Buccal absorption. III. Simultaneous diffusion and metabolism of an aminopeptidase substrate in the hamster cheek pouch, *Pharm. Res.* **6**:966–970.

Gerson, S. J., and Harris, R. R., 1984, Biochemical features of oral epithelium, in: *The Structure and Function of Oral Mucosa* (J. Meyer, C. A. Squier, and S. J. Gerson, eds.), Pergamon Press, Elmsford, N.Y., pp. 31–52.

Gu, J. M., Robinson, J. R., and Leung, S. H. S., 1990, Binding of acrylic polymers to mucin/epithelial surfaces: Structure/property relationships, *CRC Crit. Rev. Ther. Drug Carrier Syst.* **5**:21–67.

Gurny, R., and Junginger, H. E. (eds.), 1990, *Bioadhesion—Possibilities and Future Trends*, WVG, Stuttgart.

Harris, D., and Robinson, J. R., 1990, Effects of inflammation on buccal permeability, *Proc. Int. Symp. Control. Rel. Mater.* **17**, abstract D 343.

Hill, M. W., 1984, Cell renewal in oral epithelia, in: *The Structure and Function of Oral Mucosa* (J. Meyer, C. A. Squier, and S. J. Gerson, eds.), Pergamon, Elmsford, N.Y., pp. 53–81.

Ho, N. F. H., and Barsuhn, C. L., 1989, Buccal delivery of drugs, *Proc. Int. Symp. Control. Rel. Bioact. Mater.* **16**:24.

Ho, N. F. H., and Higuchi, W. I., 1971, Quantitative interpretation of in vivo buccal absorption of n-alkanoic acids by the physical model approach, *J. Pharm. Sci.* **69**:537–541.

Ho, N. F. H., Barsuhn, C. L., Burton, P. S., and Merkle, H. P., 1990, Mechanistic insights to buccal delivery of proteinaceous substances, *Advanced Drug Delivery Rev.*

Ishida, M., Machida, Y., Nambu, N., and Nagai, T., 1981, New mucosal dosage form of insulin, *Chem. Pharm. Bull.* **29**:810–816.

Ishida, M., Nambu, N., and Nagai, T., 1983, Highly viscous gel ointment containing carbopol for application to the oral mucosa, *Chem. Pharm. Bull.* **31**:4561–4564.

Knuth, K. E., Dowty, M. E., and Robinson, J. R., 1990, Permeability enhancement of thyreotropin releasing hormone across excised rabbit buccal tissue, *Proc. Int. Symp. Control. Rel. Mater.* **17**, abstract S 205.

Kurosaki, Y., Hisaichi, S.-I., Hamada, C., Nakayama, T., and Kimura, T., 1988, Effects of surfactants on the absorption of salicylic acid from hamster cheek pouch as a model of keratinized oral mucosa, *Int. J. Pharm.* **47**:13–19.

Kurosaki, Y., Hisaichi, S.-I., Hong, L.-Z., Nakayama, T., and Kimura, T., 1989a, Enhanced permeability of keratinized oral mucosa to salicylic acid with 1-dodecylazacycloheptan-2-one (Azone). In vitro studies in hamster cheek pouch, *Int. J. Pharm.* **49:**47–55.

Kurosaki, Y., Hisaichi, S.-I., Nakayama, T., and Kimura, T., 1989b, Enhancing effect of 1-dodecylazacycloheptan-2-one (Azone®) on the absorption of salicylic acid from keratinized oral mucosa and the duration of enhancement in vivo, *Int. J. Pharm.* **51:**47–54.

Laczi, F., Mezei, G., and Laszlo, F. A., 1980, Effects of vasopressin analogues (DDAVP, DVDAVP) in the form of sublingual tablets in central diabetes insipidus, *Int. J. Clin. Pharmacol. Ther. Toxicol.* **18:**63–68.

Lee, V. H. L., 1988, Enzymatic barriers to peptide and protein absorption, *CRC Crit. Rev. Ther. Drug Delivery Syst.* **5:**69–97.

Lee, V. H. L., 1990, Protease inhibitors and penetration enhancers as approaches to modify peptide absorption, *J. Controlled Release* **13:**213–223.

Lenaerts, V., and Gurny, R. (eds.), 1990, *Bioadhesive Drug Delivery Systems*, CRC Press, Boca Raton, Fla.

McMartin, C., Hutchinson, L. E. F., Hyde, R., and Peters, G. E., 1987, Analysis of structural requirements for the absorption of drugs and macromolecules from the nasal cavity, *J. Pharm. Sci.* **76:**535–540.

Marriott, C., and Hughes, D. R. L., 1990, Mucus physiology and pathology, in: *Bioadhesion—Possibilities and Future Trends* (R. Gurny and H. E. Junginger, eds.), WVG, Stuttgart, pp. 29–43.

Merkle, H. P., Anders, R., Sandow, J., and Schurr, W., 1986, Drug delivery of peptides: The buccal route, in: *Delivery Systems for Peptide Drugs* (S. S. Davis, L. Illum, and E. Tomlinson, eds.), Plenum Press, New York, pp. 159–176.

Merkle, H. P., Anders, R., and Wermerskirchen, A., 1990a, Mucoadhesive buccal patches for peptide delivery, in: *Bioadhesive Drug Delivery Systems* (V. Lenaerts and R. Gurny, eds.), CRC Press, Boca Raton, Fla., pp. 105–136.

Merkle, H. P., Anders, R., Wermerskirchen, A., Raehs, S. C., and Wolany, G., 1990b, Buccal route of peptide and protein drug delivery, in: *Peptide and Protein Drug Delivery* (V. H. L. Lee, ed.), Plenum Press, New York.

Meyer, J., Squier, C. A., and Gerson, S. J. (eds.), 1984, *The Structure and Function of Oral Mucosa*, Pergamon Press, Elmsford, N.Y.

Müller, W., and Schroeder, H. E., 1980, Differentiation of the epithelium of the human hard palate, *Cell Tissue Res.* **209:**295–313.

Nakada, Y., Awata, N., Nakamichi, C., and Sugimoto, I., 1987, Stability of human calcitonin in the supernatant of the rat's oral mucosa homogenate, *Yakuzaigaku* **47:**217–223.

Nakada, Y., Awata, N., Nakamichi, C., and Sugimoto, I., 1988, The effect of additives on the oral mucosal absorption of human calcitonin, *J. Pharmacobio-Dyn.* **11:**394–401.

Nanny, D., and Squier, C. A., 1982, Blood flow in the oral mucosa of normal and atherosclerotic monkeys, *J. Dent. Res.* (Suppl.), abstract 465.

Oh, C. K., and Ritschel, W. A., 1988a, Biopharmaceutic aspects of buccal absorption of insulin in rabbits. I. Effect of dose size, pH, and sorption enhancers: in vivo–in vitro correlation, *Pharm. Res.* **5**(Suppl.):100.

Oh, C. K., and Ritschel, W. A., 1988b, Biopharmaceutic aspects of buccal absorption of insulin in rabbits. II. Absorption characteristics of insulin through the buccal mucosa, *Pharm. Res.* **5**(Suppl.):100.

Paulesu, L., Corradeschi, F., Nicoletti, C., and Bocci, V., 1988, Oral administration of human recombinant interferon-α_2 in rats, *Int. J. Pharm.* **46**:199–202.

Peppas, N. A., and Buri, P. A., 1986, Surface, interfacial and molecular aspects of polymer bioadhesion on soft tissues, in: *Advances in Drug Delivery Systems* (G. M. Anderson and S. W. Kim, eds.), Elsevier, Amsterdam, pp. 257–275.

Pimlott, S. J., and Addy, M., 1985, Evaluation of a method to study the uptake of prednisolone sodium phosphate from different oral mucosal sites, *Oral Surg. Oral Med. Oral Pathol.* **60**:35–37.

Rettig, H., 1981, Physiologische Transportvorgänge, in: *Biopharmazie* (J. Meier, H. Rettig, and H. Hess, eds.), Thieme, Stuttgart, pp. 93–124.

Ritschel, W. A., Forusz, H., and Kraeling, M., 1988, Buccal absorption of insulin in beagle dogs, *Pharm. Res.* **5**(Suppl.):108.

Schroeder, H. E., 1981, *Differentiation of Human Oral Stratified Epithelia*, Karger, Basel.

Schürmann, W., and Turner, P., 1978, A membrane model of the human oral mucosa as derived from buccal absorption and physicochemical properties of the β-blocking drugs atenolol and propranolol, *J. Pharm. Pharmacol.* **30**:137–147.

Schurr, W., Knoll, B., Ziegler, R., Anders, R., and Merkle, H. P., 1985, Comparative study of intravenous, nasal, oral, and buccal TRH administration among healthy subjects, *J. Endocrinol. Invest.* **8**:41–44.

Siegel, I. A., 1984, Permeability of the oral mucosa, in: *The Structure and Function of Oral Mucosa* (J. Meyer, C. A. Squier, and S. J. Gerson, eds.), Pergamon Press, Elmsford, N.Y., pp. 95–117.

Siegel, I. A., and Gordon, H. P., 1985, Effects of surfactants on the permeability of canine oral mucosa in vitro, *Toxicol. Lett.* **26**:153–157.

Siegel, I. A., Izutzu, K. T., and Burkhart, J., 1976, Transfer of alcohols and ureas across the oral mucosa using streaming potentials and radioisotopes. *J. Pharm. Sci.* **65**:129–133.

Sloan, P., and Soames, J. V., 1984, Microscopic anatomy and regional organization of the lamina propria, in: *The Structure and Function of Oral Mucosa* (J. Meyer, C. A. Squier, and S. J. Gerson, eds.), Pergamon Press, Elmsford, N.Y., pp. 141–157.

Squier, C. A., and Hall, B. H., 1985, The permeability of skin and oral mucosa to water and horseradish peroxidase as related to the thickness of permeability barrier, *J. Invest. Dermatol.* **84**:176–179.

Squier, C. A., and Hill, M. W., 1989, Oral mucosa, in: *Oral Histology: Development, Structure, and Function* (A. R. Ten Cate, ed.), Mosby, St. Louis, pp. 341–381.

Squier, C. A., Cox, P. S., Wertz, P. W., and Downing, D. T., 1986, The lipid composition of porcine epidermis and oral epithelium, *Arch. Oral Biol.* **31**:741–747.

Stablein, M. J., and Meyer, J., 1984, The vascular system and blood supply, in: *The Structure and Function of Oral Mucosa* (J. Meyer, C. A. Squier, and S. J. Gerson, eds.), Pergamon Press, Elmsford, N.Y., pp. 237–256.

Tanaka, M., Yanagibashi, N., Fukuda, H., and Nagai, T., 1980, Absorption of salicylic acid through the oral mucous membrane of hamster cheek pouch, *Chem. Pharm. Bull.* **28**:1056–1061.

Tavakoli-Saberi, M. R., and Audus, K. L., 1989, Cultured buccal epithelium: An in vitro model derived from the hamster pouch for studying drug transport and metabolism, *Pharm. Res.* **6**:160–166.

Thews, G., Mutschler, E., and Vaupel, P. (eds.), 1989, *Anatomie, Physiologie, Pathophysiologie*, 3rd ed., WVG, Stuttgart.

Tucker, I., 1988, A method to study the kinetics of oral mucosal drug absorption from solutions, *J. Pharm. Pharmacol.* **40:**679–683.

Vadnere, M., Adjei, A., Doyle, R., and Johnson, E., 1990, Evaluation of alternative routes for delivery of leuprolide, *J. Controlled Release* **13:**322.

Veillard, M., 1990, Buccal and gastrointestinal drug delivery systems, in: *Bioadhesion— Possibilities and Future Trends* (R. Gurny and H. E. Junginger, eds.), WVG, Stuttgart, pp. 124–139.

Veillard, M. M., Longer, M. A., Martens, T. W., and Robinson, J. R., 1987, Preliminary studies of oral mucosal delivery of peptides, *J. Controlled Release* **6:**123–131.

Wagner, J. G., and Sedman, A. J., 1973, Quantitation of rate of gastrointestinal and buccal absorption of acidic and basic drugs based on excretion theory, *J. Pharmacokinet. Biopharm.* **1:**23–50.

Wieriks, J., 1964, Resorption of α-amylase upon buccal application, *Arch. Int. Pharmacodyn.* **151:**127–135.

Wissenschaftliche Tabellen (Documenta Geigy), 1977, 8th ed., Ciba–Geigy, Basel.

Wolany, G. J. M., 1990, Zur bukkalen Applikation und Absorption des Oktapeptids Octreotid, Ph.D. thesis, ETH Zürich, No. 9293.

Wolany, G. J. M., Rummelt, A., and Merkle, H. P., 1989, Buccal absorption of Sandostatin in rats, Proceedings of the 49th International Congress of Pharmaceutical Sciences of F.I.P., Munich, abstract 146.

Wolany, G. J. M., Munzer, J., Rummelt, A., and Merkle, H. P., 1990a, Buccal absorption of Sandostatin (octreotide) in conscious beagle dogs, *Proc. Int. Symp. Control. Rel. Mater.* **17**, abstract D 341.

Wolany, G. J. M., Rummelt, A., and Merkle, H. P., 1990b, In vivo and in vitro studies on Eudispert as a bioadhesive for buccal drug delivery, *Acta Pharm. Technol.* **36**(Suppl.):30.

Yokusaka, T., Omori, Y., Hirata, Y., and Hirai, S., 1977, Nasal and sublingual administration of insulin in man, *J. Jpn. Diabet. Soc.* **20:**146–152.

Zhang, J., Ebert, C. D., Gijsman, H., McJames, S., and Stanley, T. H., 1990, Transbuccal mucosal delivery of insulin—An in vivo dog model, *Pharm. Res.* **7**(Suppl.):116.

Chapter 7

Macromolecular Transport across Nasal and Respiratory Epithelia

Larry G. Johnson and Richard C. Boucher

1. INTRODUCTION

The airway and alveolar epithelia in conjunction with the mucociliary escalator act as a barrier to prevent the translocation of inhaled macromolecules, e.g., protein and peptide antigens, across pulmonary surfaces. Nevertheless, some exogenous macromolecules appear to cross the epithelia antigenically intact through either the paracellular or the transcellular route. Endogenous macromolecules including albumin, immunoglobulin, lactoferrin, α_2-macroglobin, and α_1-antitrypsin have been measured in either bronchoalveolar lavage fluid or the liquid that lines airway surfaces, airway surface liquid (Mentz *et al.*, 1984; Bignon *et al.*, 1976). These proteins may have specialized methods for translocation into and out of the airway surface liquid and may also play significant roles in fluid homeostasis, particularly in the alveolar region. Yet the knowledge of the mechanisms responsible for the maintenance of protein gradients across respiratory epithelia is primitive in relation to similar knowledge regarding ion gradients.

Larry G. Johnson and Richard C. Boucher • Division of Pulmonary Diseases, Department of Medicine, The University of North Carolina at Chapel Hill, Chapel Hill, North Carolina 27599.

Biological Barriers to Protein Delivery, edited by Kenneth L. Audus and Thomas J. Raub. Plenum Press, New York, 1993.

2. ANATOMY OF THE RESPIRATORY EPITHELIA

2.1. Airway Epithelia

The respiratory epithelia consist of two primary divisions: the proximal conducting airways and the distal respiratory portion responsible for gas exchange. In addition, the nasal turbinates are lined by respiratory epithelia that are morphologically and functionally similar to the proximal conducting airways (Carson *et al.*, 1986). The conducting airways extend proximally from the trachea (~2 cm diameter) branching initially into the mainstem bronchi, then dichotomously ~16–20 times ending in the terminal bronchioles (~0.2 mm diameter). The epithelium lining the proximal (cartilaginous) conducting airways is pseudostratified columnar and consists of three major cell types: basal cells, ciliated cells, and goblet cells. Although the exact function of basal cells is not fully understood, they appear to serve either as a scaffolding cell or alternatively as a pluripotential cell. Ciliated cells are highly polarized columnar cells with cilia, microvilli, and a prominent glycocalyx located on the apical surface. Goblet cells, prominent in the trachea and proximal bronchi, progressively decrease in frequency with airway division in the normal lung until they disappear from epithelia of the more distal airway regions. These cells are primarily secretory and contain large numbers of periodic acid–Schiff (PAS)-stained granules, the contents of which can be released onto the apical surface of the epithelium. In the smaller airways, the Clara cell or nonciliated bronchiolar cell becomes one of the two dominant cell types, the other being the ciliated cell which becomes much more cuboidal with decreasing airway diameter. The Clara cell is a secretory cell with the ability to secrete apolipoprotein, phospholipids, high-molecular-weight glycoconjugates, and a 10-kDa protein homologous to uteroglobin (Patton *et al.*, 1986; Singh *et al.*, 1987). Accordingly it has been implicated as a potential progenitor cell for the goblet cell as well as a progenitor cell for the type II pneumocyte (Penney, 1988) and ciliated cell.

Superficial airway epithelial cells are connected by an interdigitated lateral intercellular space and gap junctions at the lateral surfaces and tight junctions at the apical surface. The tight junctions function at the cellular level to separate the apical and basolateral surfaces and at the epithelial level as a molecular sieve restricting the flow of large polar solutes across the airway mucosa. The goblet cell, ciliated cell, and type II pneumocyte have all been implicated as cell types involved in the transcellular transport of proteins and peptides across respiratory epithelia (Richardson *et al.*, 1976; Ranga and Kleinerman, 1982; Brown *et al.*, 1985; Sugahara *et al.*, 1987).

Proximal airways also have submucosal glands, which arise from invaginations of the surface epithelium into the submucosa during fetal development and consist of two principal cell types, serous cells and mucous cells. Both cell types secrete high-molecular-weight glycoconjugates and mucins. In addition, serous cells secrete lysozyme, lactoferrin, antileukoprotease, and possibly albumin (Bowes *et al.*, 1981; Jacquot *et al.*, 1988). Submucosal glands have also been shown to participate in IgA secretion in human bronchial epithelium (Goodman *et al.*, 1981). The ability of these glands to absorb proteins and peptides has not been studied. Similar to goblet cells, submucosal glands are not present in the distal airway regions.

The airway epithelium is lined by an airway surface liquid (ASL) ~10 μm in depth and is thought to consist of two layer: (1) a mucous or gel layer positioned atop the cilia presumably composed of mucus glycoproteins secreted from goblet and glandular cells, and (2) a serous periciliary fluid (sol) layer in which the cilia beat synchronously to propel the mucus containing trapped inhaled particles, cellular debris, or infectious agents toward the major bronchi, trachea, and then the pharynx where it can be either swallowed or expectorated. The composition and some of the protein constituents of the periciliary fluid have been identified and include albumin, immunoglobulins, and a variety of enzymes (Boucher *et al.*, 1981a; Mentz *et al.*, 1984). The regulation of the volume and contents of this periciliary fluid is only partially understood and is reviewed elsewhere (Boucher *et al.*, 1981a,b).

2.2. Alveolar Epithelia

The gas exchange portion of the lung in humans consists of two to four generations of respiratory bronchioles (0.15–0.2 μm diameter), alveolar ducts, and alveolar sacs, and alveoli. While the major conducting airways have a surface area of only ~2 m^2, the estimated 300 million alveoli present a surface area for gas exchange of ~143 m^2 (Penney, 1988). The major cell types in the respiratory portion of the lung are the type I pneumocyte or squamous alveolar cell, the type II pneumocyte, the type III pneumocyte or alveolar brush cell, and the alveolar macrophage. These cell types are all located within the alveolar ducts, alveolar sacs, and alveoli, and Clara cells particularly in the region of the respiratory bronchioles. With the exception of the alveolar macrophage, all of these cells rest on a basement membrane forming a complex epithelium for the intimate contact of air and blood over a large surface area.

The type I cell apposes the endothelial cells at the fusion of the two basement membranes to form an air–blood gas exchange barrier measuring 0.2–0.5 μm thick, separating the alveolar wall into the septal and alveolar surfaces. This cell, known for its extensive surface area and branching, also has limited phagocytic abilities to ingest particulates, e.g., chrysotile asbestos (Penny, 1988). The type II pneumocyte, the predominant cell type in the respiratory region of the lung, is a cuboidal cell characterized by numerous cytoplasmic lamellar bodies representing the storage granules for surfactant. This organelle-rich secretory cell replaces the type I cell in the event of injury, becoming attenuated and losing its organelles as it migrates down the basement membrane. This cell is rich in a basolateral Na^+,K^+-ATPase and rich in cytochome P-450 isozymes (Penney, 1988). The rare type III pneumocyte or alveolar brush cell is a cuboidal cell present in the region of the first alveolar duct bifurcation and is essentially absent in the more distal alveoli. This type III cell may play a role in detoxification of inhaled pollutants given their location in a region that is often the primary site of initial injury evoked by these agents.

As observed in airway epithelia, the alveolar cell types possess the ability to form tight junctions and generate spontaneous transepithelial potential differences both *in vivo* and *in vitro*. In fact, alveolar tight junctions may even be tighter than airway tight junctions as evidenced by freeze-fracture studies (Schneeberger, 1980; Olver *et al.*, 1981), increased transepithelial resistance in culture (Cheek *et al.*, 1989; Kim *et al.*, 1989), and the increased transepithelial resistance of excised bullfrog alveolar epithelia relative to airway tight junctions (Gatzy, 1982; Kim *et al.*, 1985). Alveoli, like the airways, are covered by a complex epithelial lining fluid (ELF) consisting of (1) an epiphase or surfactant layer comprised of a surface-active phospholipid layer at the air–liquid interface, and (2) a hypophase beneath this layer comprised of tubular myelin, apoprotein, IgG, albumin, phospholipid, and carbo-hydrates. Unlike the conducting airways, no cilia are located in this region to facilitate clearance. Moreover, no glands or goblet cells are present.

Macrophages are also located in the respiratory portion of the lung. The alveolar macrophage is a postmitotic cell with a half-life of several days residing in the alveolar lumen that may increase in number in response to inhaled pollutants. The septal macrophage, which retains the ability to divide, putatively resides in the alveolar septum for years, and is the immediate precursor cell for the alveolar macrophage. Both macrophages are obviously derived from peripheral blood mono-cytes. Although these cells are not of epithelial origin, they serve important lung defense functions and may potentially have an impact on the stability of proteins in the epithelial lining fluid.

3. EVIDENCE FOR PARACELLULAR FLOW OF PEPTIDES AND PROTEINS

Although respiratory epithelia serve as a major barrier to the flow of soluble macromolecules into the interstitium, some of the molecules cross the epithelium. This molecular translocation may occur either through transcellular routes or through small aqueous channels empirically defined as cylindrical pores in the paracellular path.

3.1. Diffusion through Pores

Evidence for functional pores comes from studies examining relative rates of egress of large polar solutes of differing molecular sizes in the fluid-filled lung (Taylor and Gaar, 1970; Normand *et al.*, 1971; Theodore *et al.*, 1975; Egan *et al.*, 1975). Using excised airway epithelia (Boucher, 1980) and excised bullfrog alveolar epithelia (Gatzy, 1982; Kim *et al.*, 1985) mounted in Ussing chambers, permeability coefficients for probes of differing molecular size were measured and referenced to the permeability coefficient of mannitol to calculate equivalent pore radii (Solomon, 1986; Normand *et al.*, 1971; Boucher, 1980; Gatzy, 1982). The relationship between solute molecular size and flow across the respiratory epithelium (shown for canine airway epithelia in Fig. 1) is linear for small solutes, but becomes nonlinear with increasing hydrated molecular radius. Accordingly, a two-pore model has been generated in canine airway epithelia: a small-pore population which might exist in either the cellular or paracellular paths and another population of larger pores (mean radius 7.5 nm) presumably too large to exist within a viable cell. This large-pore population, corresponding to the size of many water-soluble antigens (5–50 kDa), is presumably located in the paracellular path. Substantiation of this concept has been realized in flux studies of canine airway epithelia in which permeability coefficients for large polar solutes across the barrier were symmetric consistent with a diffusional process (Boucher, 1980). Alveolar epithelia exhibit a predominant population of small pores with radii of ~1 nm (Gatzy, 1982), although a few large pores of radius >8 nm may also be present (Theodore *et al.*, 1975). Unfortunately, the morphologic correlate of pores in the paracellular path has not been demonstrated satisfactorily to date.

3.2. Regulation of Paracellular Permeability

The rate of diffusion through the paracellular path may be regulable (see Chapter 1). A variety of physical factors have been shown to affect the paracellular permeability of airway epithelia. Lowering the pH of the luminal bathing fluid to 2.8 has been shown to increase paracellular permeability in isotopic tracer studies in airway epithelia (Boucher, 1981), possibly by displacing calcium from the tight junctions. Increased osmolality of the luminal fluid has been reported to increase mannitol permeability across canine trachea (Yankaskas *et al.*, 1987). Transalveolar pressures greater than 30 cm H_2O, which are known to occur during barotrauma, increase apparent equivalent pore radii, making the barrier less effective (Egan *et al.*, 1976; Egan, 1980). Finally, metabolic inhibition of airway epithelia, with either hypoxia or sodium cyanide, has been shown to increase paracellular permeability (Stutts *et al.*, 1988). Hence, a variety of factors may affect paracellular permeability to increase access of inhaled macromolecules to the submucosa or interstitium of respiratory epithelia. Similarly, these factors may facilitate increased transudation of serum proteins or locally produced proteins into the ASL or ELF.

Adjuvants or promoters that enhance paracellular permeability might be used to increase delivery of peptides and proteins across respiratory epithelia. Current research has largely focused on the nasal epithelia where a number of peptides are

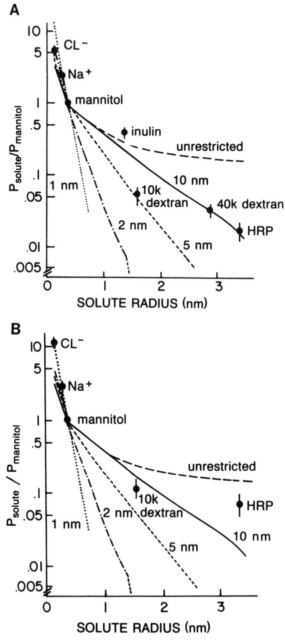

Figure 1. Relationship between solute permeability (P_{solute}) normalized to mannitol permeability ($P_{mannitol}$) and hydrated molecular radius of the solute across canine tracheal (A) and bronchial (B) epithelium. Also depicted are the plots for unrestricted diffusion and lines for predicted pores of 1 to 10 nm.

already available for intranasal delivery clinically in the United States and Europe. Among the list of approved peptides are oxytocin, desmopressin, luteinizing hormone-releasing hormone analogues and salmon calcitonin. The first three of these are small peptides of less than 10 amino acids which easily cross the epithelia presumably through pores. Larger peptides containing greater than 30 amino acids traverse the nasal epithelia very poorly. Some of these which have been more actively investigated include insulin, glucagon, growth hormone-releasing hormone, and corticotropin-releasing hormone, and calcitonin.

A variety of promoters have been used in an attempt to enhance intranasal absorption of these peptides and include chelating agents, surfactants, fatty acids, and bile salts (Salzman et al., 1985; Gordon et al., 1985; Aungst and Rogers, 1988; Pontiroli et al., 1989; Pontiroli and Pozza, 1990; Yamamoto et al., 1990). The mechanisms of action of these promoters have not been clearly delineated but focus mainly on changes in the paracellular path and the formation of hydrophilic pores or channels for transepithelial flow. Chelating agents such as disodium ethylenediamine tetraacetate (Na_2-EDTA), despite increasing the permeability of the tight junctions as consequence of removal of luminal Ca^{2+}, cause only a minimal increase in intranasal absorption of insulin (Aungst and Rogers, 1988). Simply dissolving insulin in dilute acid medium (pH 3.1) to increase paracellular permeability also enhances intranasal insulin absorption (Hirai et al., 1977). In contrast, surfactants act to enhance intranasal absorption of peptides by a different mechanism. These agents have been reported to bind to hydrophobic regions of membranes and tight junctions to form pores or hydrophilic channels for the transfer of peptides from the nasal lumen down a concentration gradient into the extracellular space (McMartin et al., 1987). The most frequently studied surfactant, laureth-9, has been shown to significantly enhance intranasal absorption of insulin in both animals and humans sufficiently to achieve desired hypoglycemic effects (Salzman et al., 1985; Aungst and Rogers, 1988). However, intranasal insulin remains only about 30% as efficacious as intramuscular insulin and may be associated with local toxicity at concentrations above 0.25% laureth-9 (Salzman et al., 1985; Aungst and Rogers, 1988). Like surfactants, bile salts such as sodium glycocholate and its derivatives may also enhance intranasal absorption of peptides by reverse micellar binding with subsequent formation of hydrophilic pores or channels in either cell membranes or tight junctions (Gordon et al., 1985; McMartin et al., 1987). Biles salts may also enhance absorption by binding Ca^{2+} (Kahn et al., 1982) to increase paracellular permeability (Alberts et al., 1989) and by inhibiting intranasal proteases to increase drug availability for absorption (Yamamoto et al., 1990). Enhancement of intranasal delivery using bile salts as a promoter has been reported for insulin, calcitonin, corticotropin-releasing hormone, and growth hormone-releasing hormone (Gordon et al., 1985; Pontiroli et al., 1989; Pontiroli and Pozza, 1990).

Hence, transudation of peptides and proteins across nasal and respiratory epithelia can be regulated using a variety of physical factors and promoters. The efficiency of delivery can perhaps be further increased through the use of protease inhibitors such as *p*-chloromercuriphenylsulfonic acid (PCPMS) and aprotinin (Aungst and Rogers, 1988; Yamamoto *et al.*, 1990). However, in the absence of physical factors or promoters, the contribution of protease inhibitors to enhancement of peptide absorption may be small (Aungst and Rogers, 1988).

4. EVIDENCE FOR TRANSCELLULAR FLOW OF PEPTIDES AND PROTEINS

4.1. Nasal Epithelia

In vivo studies of human nasal respiratory epithelium suggest that protein absorption is transcellular and regulable. Buckle and Cohen (1975) compared the rate of appearance of [^{125}I]albumin in venous blood over several intervals within a 30-min time period after intranasal application in normal subjects, patients with extrinsic asthma, and patients with atopic rhinitis. When the venous sampling period was limited to the 30-min time period that [^{125}I]albumin remained in contact with the nasal mucosa, 90% of the patients with atopic rhinitis had antigenically intact albumin present in their blood samples, versus 11% of patients with extrinsic asthma, and 33% of normal subjects. Dialysis of samples prior to immunoprecipitation, however, prevented the authors from determining whether degradation of albumin occurred during translocation across the epithelium. More recently, Svensson *et al.* (1989) have demonstrated rapidly reversible increases in albumin concentrations in nasal lavage fluid following nasal histamine challenge in normal subjects. The rapid return of albumin concentration to prehistamine nasal levels implicates an active transcellular process for absorption of this soluble protein against a concentration gradient from the airway surface liquid. Moreover, since histamine has been shown to have minimal effect on the paracellular or shunt resistance in cultured human nasal epithelia (L. C. Clarke, A. P. Paradiso, and R. C. Boucher, unpublished observations), the reversible changes in albumin concentrations in nasal lavage fluid suggest that transcellular movement of albumin has occurred in response to histamine.

4.2. Conducting Airway Epithelia

We have previously reported net albumin absorption across canine bronchial epithelium (Johnson *et al.*, 1989). In this series of experiments, permeability coefficients of ^{14}C-labeled canine albumin (P_{alb}) across excised canine bronchi in the absorptive or mucosal-to-serosal (M→S) direction, were significantly greater than in the serosal-to-mucosal (S→M) or secretory direction. The absence of asymmetry in transport rates of [^{3}H]inulin across the epithelia suggests that net absorption of albumin occurred through the transcellular path. Intracellular degradation of the [^{14}C]albumin also appears to have occurred since the majority of the label in the submucosal bath represented albumin fragments. Because no asymmetries in transport rates were seen in albumin fragments *per se*, isolated from spontaneous degradation of the tracer, it is likely that intact albumin was the substrate for transport. In addition, lowering the temperature to 4°C reduced the P_{alb} M→S to the same level as P_{alb} S→M consistent with the transcellular path as the route responsible for net absorption of albumin.

In contrast to our work in canine bronchial epithelia, Price *et al.* (1990) have reported active secretion of albumin into rabbit trachea *in vitro*. However, the authors failed to examine the rates of albumin transport in the absorptive direction and hence were unable to make statements about net direction of albumin transfer across rabbit airways. They also failed to test the effects of the various agonists cited in their study on the paracellular pathway. Nevertheless, their observation of increased flow of labeled albumin into the lumen in response to albuterol, which is generally felt to have minimal effect on the paracellular path, suggests that albumin moves across the airway epithelia by a transcellular route. The abolition of this response to albuterol by sodium cyanide, which inhibits vesicular transport but also modestly increases paracellular permeability in airway epithelia (Stutts *et al.*, 1988), is consistent with this notion. Hence, translocation of albumin into the fluid-filled lumen of rabbit trachea can occur via a transcellular route. As virtually no submucosal glands are found in rabbit trachea, secretion cannot be attributed to glands.

Similar findings have been reported by Webber and Widdicombe (1989) in ferret trachea which has abundant submucosal glands. Again, albuterol significantly increases the rate of albumin translocation into the lumen although no change in lysozyme, a marker of serous cell secretion, occurred. Nevertheless, a preliminary study by Jacquot *et al.* (1988) suggests that submucosal glands in culture may secrete an albumin-like substance.

4.3. Alveolar Epithelia

In vitro studies of isolated mammalian alveolar epithelia have been technically impossible. Many early studies have used the readily accessible planar sheet bullfrog alveolus to study both ion and macromolecular transport across alveolar epithelia. Using this model, Kim *et al.* (1985) reported a fourfold asymmetry in [^{14}C]albumin transport favoring absorption from lumen to the interstitium. As was seen by Johnson *et al.* (1989) in canine bronchial epithelia, no asymmetries in permeabilities of radiolabeled inulin across the bullfrog alveolar epithelia were detected, consistent with diffusion of inulin through the small pores of the paracellular pathway and active transcellular transport of albumin across the alveolar epithelia in the luminal-to-serosal direction. Moreover, gel electrophoretograms demonstrated that at least

50% of material absorbed was intact albumin. Kim *et al.* (1989) have also described net albumin absorption across cultured rat alveolar epithelial cells although the magnitude was less than observed in bullfrog lung. Hence, evidence for transcellular permeation of proteins and/or amino acids exists in nasal, conducting airway, and alveolar epithelia. Moreover, these processes may be regulable and saturable in nasal and conducting airway epithelia (Svensson *et al.*, 1989; Webber and Widdicombe, 1989; Price *et al.*, 1990).

5. POTENTIAL MECHANISMS OF TRANSCELLULAR FLOW

5.1. Receptor-Mediated Endocytosis in Airways

The potential mechanisms most often cited for protein and peptide translocation across respiratory epithelia involve the processes of endocytosis or transcytosis. A detailed description of these processes is found in Chapter 4.

Evidence for a transcytotic path in respiratory epithelia arises from the ultrastructural localization of IgA transport across the cell. IgA is sequentially localized in the basolateral plasmalemma, plasmalemmal invaginations, cytoplasmic vesicles, and gland lumina of human bronchial epithelial gland cells (Goodman *et al.*, 1981). These findings are consistent with the binding of polymeric IgA to the transmembrane receptor on the basolateral cell surface and transcytosis of IgA to the apical cell surface as has been shown in intestinal epithelia (Mostov and Simister, 1985; Breitfeld *et al.*, 1989). Notably, IgA transport could not be demonstrated across ciliated cells in this study. Further evidence for the existence of transcytosis comes from the work of Richardson *et al.* (1976) in which the exogenous tracer proteins horseradish peroxidase and ferritin were detected by electron microscopy in vesicles within the tracheal and bronchial epithelial cells that appeared to move toward the basal surface and be released into the extracellular space. At intervals of 5, 15, 30, and 60 min after instillation of these tracer proteins into guinea pig trachea, they could not be detected in the vicinity of the tight junctions of this epithelium, suggesting that the paracellular path was not the major route of protein flux across the epithelium.

Little is known of the mechanisms by which endogenous macromolecules other than IgA are transported across pulmonary epithelia. Precedents do exist for receptor-mediated endocytosis in other organ systems (see Chapter 5) as has been described for IgG in suckling rat ileum, human placenta, and rabbit yolk sac (Abrahamson and Rodewald, 1981; Mostov and Simister, 1985; Breitfeld *et al.*, 1989) and for albumin in capillary endothelium (Ghitescu *et al.*, 1986; see Chapter 10) and sinusoidal liver cell membranes (Horiuchi *et al.*, 1985). Park and Maack (1984) described both a low-affinity (K_m 18 μM) and a high-affinity (K_m 460 nM) system for albumin absorption across isolated perfused rabbit proximal renal tubules in which 80% of the albumin absorbed was catabolized to low-molecular-weight fragments. A more detailed discussion of other potential precedents is presented in other chapters of this volume. However, the contribution of receptor-mediated endocytosis to transcellular respiratory epithelial cell protein and peptide transport has not been determined.

The recycling of surfactant proteins is one example where receptor-mediated endocytosis may not participate in transcellular or transepithelial transfer of proteins (Possmayer, 1988; Hawgood and Shiffer, 1991; Weaver and Whitsett, 1991; Wright and Dobbs, 1991). The type II cell synthesizes both the lipid and protein components of surfactant and stores them in lamellar bodies until they are secreted by exocytosis into the alveolar space. Once secreted, the contents of the lamellar body form a lattice-like structure called tubular myelin which is believed to be the precursor to the final surface film in the alveoli enriched to dipalmitoylphosphatidylcholine. Type II cells have also been shown to recycle surfactant with the percentage reutilized ranging from 23 to 90% (Wright and Dobbs, 1991). The surfactant proteins SP-A, SP-B, and SP-C have been shown to stimulate uptake of surfactant phospholipids (Wright et al., 1987; Rice et al., 1989). SP-A also appears to inhibit phospholipid secretion by type II cells (Rice et al., 1987), implying a role for SP-A in the feedback regulation of surfactant pool size. SP-A-stimulated phospholipid uptake occurs at a time-, temperature-, and protein concentration-dependent manner (Wright et al., 1987). Binding of SP-A to the type II cell has been shown to be a high-affinity, saturable process with a K_d of ~5 × 10^{-10} M consistent with a receptor-mediated process (Kuroki et al., 1988; Wright et al., 1989). Subsequent internalization reveals sequential localization in coated pits, coated vesicles, endosomes, multivesicular bodies, and in close proximity to lamellar bodies without evidence for significant degradation consistent with recycling (Ryan et al., 1989). No apparent transcellular transport of intact protein occurs. The mechanisms by which other surfactant-associated proteins are recycled have not been clearly delineated but apparently do not involve receptor-mediated endocytosis. Whether or not recycling of exogenous peptides and proteins occurs in the alveolar region in vivo is unknown.

5.2. Adsorptive Endocytosis in Airways

The low-affinity system for albumin absorption in canine bronchial epithelium (Johnson et al., 1989), with the K_m (1.6 mM) being 100-fold greater than the K_m of the low-affinity system described by Park and Maack (1984), 10^4-fold greater than their high-affinity system, and 10^7-fold greater than the high-affinity system described by Ghitescu et al. (1986) in capillary endothelium, appears to be quite high for a pure receptor-mediated system. Adsorptive endocytosis may better describe this process. In adsorptive endocytosis, the asymmetric distribution of glycocalyceal components on nasal and conducting airway epithelia allows selective concentration of albumin and other soluble macromolecules on the apical cell surface that when coupled with endocytosis enhances absorption of macromolecules from the airway surface liquid in the absence of a specific cell surface receptor. Precedent for such a mechanism has been reported in studies of intestinal macromolecular absorption (Gonella and Neutra, 1984; Neutra *et al.*, 1987). The absence of a specific cell surface receptor containing a putative signal targeting endocytotic vesicles to the contralateral cell surface membrane (transcytosis) in its cytoplasmic domain (Davis *et al.*, 1987; Mostov *et al.*, 1989) or to endosomal sorting (Breitfeld *et al.*, 1989), might be expected to lead to more frequent fusion of endocytotic vesicles with lysosomes. The degradation of the majority of absorbed albumin in canine bronchial epithelium described by Johnson *et al.* (1989) is consistent with such interaction of macromolecules with the lysosomal compartment. However, transcytosis of endogenous surface glycoproteins by MDCK cells (Brandli *et al.*, 1990) suggests that other sorting signals for transcytosis may exist, permitting macromolecules absorbed via adsorptive endocytosis to undergo either lysosomal degradation or transcytosis to the contralateral cell membrane. In this scenario, one might expect to find that the material absorbed across the epithelium would consist of only a small amount of the intact macromolecule with a predominance of degraded macromolecular fragments.

5.3. Sodium-Dependent Amino Acid Cotransport in Alveolar Epithelia

The contribution of glycocalyceal components to alveolar epithelial macromolecular transport has not been studied. The relative roles of receptor-mediated endocytosis and transcytosis in this region are also unknown. Moreover, other mechanisms for protein and peptide transport have been implicated in alveolar epithelia. Brown *et al.* (1985) reported a saturable Na^+-dependent amino acid cotransport system with 1:1 stoichiometry in suspensions of isolated rat type II alveolar epithelial cells with the neutral amino acid analogue α-methylamino-isobutyric acid (AIB). Sugahara *et al.* (987) extended the findings of brown *et al.* (1985) for a Na^+-dependent amino acid cotransporter for AIB with a 1:1 stoichiometry to monolayer preparations of isolated rat alveolar type II cells in primary culture and also showed that insulin stimulated the rate of AIB uptake. Sodium-dependent lysine flux across excised bullfrog alveolar epithelia has subsequently been reported by Kim and Crandall (1988). In their study, a fourfold asymmetry in unidirectional fluxes of [^{14}C]lysine was detected favoring absorption from the luminal to the pleural surface. Substitution of the choline$^+$ or Li^+ for sodium in Ringer solution abolished the asymmetry in flux, reducing the alveolar to pleural flux to the same rate as the pleural to alveolar flux. The net absorption of [^{14}C]lysine was also significantly reduced by ouabain, and nearly abolished by L-leucine, but only minimally reduced by AIB. Thus, alveolar epithelia may have several regulable processes for the transfer of proteins and peptides out of the alveolus.

6. ROLE OF MACROPHAGE SURVEILLANCE

The role of macrophage surveillance in protein and peptide transport across respiratory epithelia has not been studied. Although an airway macrophage (Brain *et al.*, 1984; Brain, 1986) has been described, its rare presence in the airway lumen implies that it does not have a significant effect on airway macromolecular transport. In contrast, the alveolar macrophage plays a significant role in the removal of particulate debris to the alveolar region and may be in a constant state of activation (Brain, 1986). This cell has been reported to emit toxic superoxide radicals and elastase during phagocytosis and to release proteases and other toxic enzymes upon cell death (Brain *et al.*, 1977; Brain, 1986). These enzymes and O_2 free radicals may catabolize, denature, and/or alter the native protein structure of endogenous and inhaled proteins, interfering with their ability to bind to receptors or other carrier mechanisms for translocation across the epithelia. The macrophage has also been shown to internalize surfactant lipids and proteins (Wright *et al.*, 1987; Wright and Dobbs, 1991). Macrophages bind and internalize SP-A in a mannose-dependent fashion (Wintergerst *et al.*, 1989; Weaver and Whitsett, 1991) which may enhance phagocytosis of opsonized erythrocytes and bacteria. The role this enhanced phagocytosis might play on exogenous peptides and proteins delivered to the alveolar region is unknown.

7. SUMMARY

A variety of endogenous macromolecules are present in both the airway surface liquid (Mentz *et al.*, 1984) and the alveolar epithelial lining fluid (Bignon *et al.*, 1976). The mechanisms responsible for the regulation of the concentrations of macromolecules in these luminal fluids are poorly understood, as are the mechanisms responsible for the macromolecular translocation across respiratory epithelia. The evidence discussed in this review suggests that serum proteins and proteins produced locally within the airway submucosa or alveolar interstitium may leak through the paracellular path and in a few instances be actively secreted into the lumen. This rate of translocation into the ASL or ELF may be increased by inhaled toxins and injury (Holter *et al.*, 1986), increasing the concentrations of these molecules above their basal values. Once present in the ASL or ELF, these macromolecules are actively returned to the interstitium or submucosa by a process involving transcellular endocytosis and/or transcytosis. The effect of macrophage surveillance on this process in the alveolar region is unknown, but may be minimized in the absence of injury by the presence of antioxidants in the ELF (Pacht and Davis, 1988; Cantin *et al.*, 1990).

Similar metabolic fates await peptide and protein drugs that may be delivered to the airways and/or alveoli via inhalation in the future. The ability of the respiratory epithelia to either catabolize or absorb these molecules intact raises important dose–effect and toxicological questions. Moreover, many such molecules may be altered by enzymes located within the ASL or ELF. Hence, understanding the mechanisms regulating all aspects of macromolecular transport across respiratory epithelia will be critical to the development of new protein drugs designed for delivery via inhalation.

REFERENCES

Abrahamson, D. R., and Rodewald, R., 1981, Evidence for the sorting of endocytotic vesicle contents during the receptor-mediated transport of IgG across the newborn rat intestine, *J. Cell Biol.* **91:**270–280.

Alberts, B., Bray, D., Lewis, J., Raff, M., Roberts, K., and Watson, J. D. (eds.), 1989, Cell adhesion, cell junctions and the extracellular matrix, in: *Molecular Biology of the Cell*, Garland, New York, pp. 791–836.

Aungst, B. J., and Rogers, N. J., 1988, Site dependence of absorption-promoting actions of laureth-9, Na salicylate, Na_2EDTA, and aprotinin on rectal, nasal, and buccal insulin delivery, *Pharm. Res.* **5:**305–308.

Bignon, J., Jaurand, M. C., Pinchon, M. C., Sapin, C., and Warnet, J. M., 1976, Immuno-electron microscopic and immunochemical demonstrations of serum proteins in the alveolar lining material of the rat lung, *Am. Rev. Respir. Dis.* **113:**109–120.

Boucher, R. C., 1980, Chemical modulation of airway epithelial permeability, *Environ. Health Perspect.* **35:**3–12.

Boucher, R. C., 1981, Mechanisms of pollutant induced airways toxicity, *Clin. Chest Med.* **2:** 377–392.

Boucher, R. C., Stutts, M. J., Bromberg, P. A., and Gatzy, J. T., 1981a, Regional differences in airway surface liquid composition, *J. Appl. Physiol.* **50:**613–620.

Boucher, R. C., Stutts, M. J., and Gatzy, J. T., 1981b, Regional differences in bioelectric properties and ion flow in excised canine airways, *J. Appl. Physiol.* **51:**706–714.

Bowes, D., Clark, A. E., and Corrin, B., 1981, Ultrastructural localisation of lactoferrin and glycoprotein in human bronchial glands, *Thorax* **36:**108–115.

Brain, J. D., 1986, Toxicological aspects of alterations of pulmonary macrophage function, *Annu. Rev. Pharmacol. Toxicol.* **26:**547–565.

Brain, J. D., Godleski, J. J., and Sorokin, S. P., 1977, Structure, origin, and fate of the macrophage, in: *Lung Biology in Health and Disease*, Volume 5, Part II (J. D. Brain, D. F. Proctor, and L. M. Reid, eds.), Dekker, New York, pp. 849–892.

Brain, J. D., Gehr, P., and Kavet, R. I., 1984, Airway macrophages: The importance of the fixation method, *Am. Rev. Respir. Dis.* **129:**823–826.

Brandli, A. W., Parton, R. G., and Simons, K., 1990, Transcytosis in MDCK cells: Identification of glycoproteins transported bidirectionally between both plasma membrane domains, *J. Cell Biol.* **111:**2909–2921.

Breitfeld, P. P., Casanova, J. E., Simister, N. E., Ross, S. A., McKinnon, W. C., and Mostov, K. E., 1989, Transepithelial transport of immunoglobulins: A model of protein sorting and transcytosis, *Am. J. Respir. Cell Mol. Biol.* **1:**257–262.

Brown, S. E. S., Kim, K. J., Goodman, B. E., Wells, J. R., and Crandall, E. D., 1985, Sodium–

amino acid cotransport by type II alveolar epithelial cells, *J. Appl. Physiol.* **59:**1616–1622.

Buckle, F. G., and Cohen, A. B., 1975, Nasal mucosal hyperpermeability to macromolecules in atopic rhinitis and extrinsic asthma, *J. Allergy Clin. Immunol.* **55:**213–221.

Cantin, A. M., Fells, G. A., Hubbard, R. C., and Crystal, R. G., 1990, Antioxidant macromolecules in the epithelial lining fluid of the normal human lower respiratory tract, *J. Clin. Invest.* **86:**962–971.

Carson, J. L., Collier, A. M., and Boucher, R. C., 1986, Ultrastructure of the respiratory epithelium in the human nose, in: *Pathophysiological Aspects of Allergic and Vasomotor Rhinitis* (N. Mygind, ed.), Munksgaard, Copenhagen, pp. 11–27.

Cheek, J. M., Kim, K. J., and Crandall, E. D., 1989, Active sodium transport by tight monolayers of alveolar epithelia cells, *Am. Rev. Respir. Dis.* **139:**A477.

Davis, C. G., van Driel, I. R., Russell, D. W., Brown, M. S., and Goldstein, J. C., 1987, The low density lipoprotein receptor: Identification of amino acids in cytoplasmic domain required for rapid endocytosis, *J. Biol. Chem.* **262:**4075–4082.

Egan, E. A., 1980, Response of alveolar epithelial solute permeability to changes in lung inflation, *J. Appl. Physiol.* **49:**1032.

Egan, E. A., Olver, R. E., and Strang, L. B., 1975, Changes in non-electrolyte permeability of alveoli and breathing in the lamb, *J. Physiol. (London)* **244:**161–179.

Egan, E. A., Nelson, R. M., and Olver, R. E., 1976, Lung inflation and alveolar permeability to non-electrolytes in adult sheep in vivo, *J. Physiol. (London)* **260:**409.

Gatzy, J. T., 1982, Pathways of hydrophilic solute flow across excised bullfrog lung, *Exp. Lung Res.* **3:**147–161.

Ghitescu, L. A., Fixman, A., Simionescu, M., and Simionescu, N., 1986, Specific binding sites for albumin to plasmalemmal vesicles of continuous capillary endothelium: Receptor-mediated transcytosis, *J. Cell Biol.* **102:**1304–1311.

Gonella, P. A., and Neutra, M. R., 1984, Membrane-bound and fluid-phase macromolecules enter separate prelysosomal compartments in absorptive cells of suckling rat ileum, *J. Cell Biol.* **99:**909–917.

Goodman, M. R., Link, D. W., Brown, W. R., and Nakane, P. K., 1981, Ultrastructural evidence of transport of secretory IGA across bronchial epithelium, *Am. Rev. Respir. Dis.* **123:**115–119.

Gordon, G. S., Moses, A. C., Silver, R. D., Flier, J. S., and Carey, M. C., 1985, Nasal absorption of insulin: Enhancement by hydrophobic bile salts, *Proc. Natl. Acad. Sci. USA* **82:**7419–7423.

Hawgood, S., and Shiffer, K., 1991, Structures and properties of the surfactant-associated proteins, *Annu. Rev. Physiol.* **53:**375–394.

Hirai, S., Ikenaga, T., and Matsuzawa, T., 1977, Nasal absorption of insulin in dogs, *Diabetes* **27:**296–299.

Holter, J. F., Weiland, J. E., Pacht, E. R., Gadek, J. E., and Davis, W. B., 1986, Protein permeability in the adult respiratory distress syndrome, *J. Clin. Invest.* **78:**1513–1522.

Horiuchi, S., Takata, K., and Morino, Y., 1985, Characterization of a membrane-associated receptor from rat sinusoidal liver cells that bind formaldehyde-treated serum albumin, *J. Biol. Chem.* **260:**475–481.

Jacquot, J. R., Benali, R., Sommerhoff, C. P., Finkbeiner, W. E., Goldstein, G., Puchelle, E., and Basbaum, C. P., 1988, Identification of albumin-like protein released by cultured bovine tracheal serous cells, *Am. Rev. Respir. Dis.* **137:**A11.

Johnson, L. G., Cheng, P. W., and Boucher, R. C., 1989, Albumin absorption by canine bronchial epithelium, *J. Appl. Physiol.* **66**:2772–2777.

Kahn, M. J., Lakshminarayanaiah, N., Trotman, B. W., Chun, P., Kaplan, S. A., and Margulies, C., 1982, Calcium binding and bile salt structure, *Hepatology* **2**:732.

Kim, K. J., and Crandall, E. D., 1988, Sodium-dependent lysine flux across bullfrog alveolar epithelium, *J. Appl. Physiol.* **65**:1655–1661.

Kim, K. J., LeBon, T. R., Shinbane, J. S., and Crandall, E. D., 1985, Asymmetric ^{14}C albumin transport across bullfrog alveolar epithelium, *J. Appl. Physiol.* **59**:1290–1297.

Kim, K. J., Cheek, J. M., and Crandall, E. D., 1989, ^{14}C-albumin transport across rat alveolar epithelial monolayers, *Am. Rev. Respir. Dis.* **139**:A479.

Kuroki, Y., Mason, R. J., and Voelker, D. R., 1988, Alveolar type II cells express a high-affinity receptor for pulmonary surfactant protein A, *Proc. Natl. Acad. Sci. USA* **85**: 5566–5570.

McMartin, C., Hutchinson, L. E. F., Hyde, R., and Peters, G. E., 1987, Analysis of structural requirements for the absorption of drugs and macromolecules from the nasal cavity, *J. Pharm. Sci.* **76**:535–540.

Mentz, W. M., Knowles, M. R., Brown, J. B., Gatzy, J. T., and Boucher, R. C., 1984, Measurement of airway surface liquid (ASL) composition in normal human subjects, *Am. Rev. Respir. Dis.* **129**:A315.

Mostov, K. E., and Simister, N. E., 1985, Transcytosis minireview, *Cell* **43**:389–390.

Mostov, K. E., De Bruyn Kops, A., and Deitcher, D. L., 1986, Deletion of the cytoplasmic domain of the polymeric immunoglobulin receptor prevents basolateral localization and endocytosis, *Cell* **47**:359–364.

Neutra, M. R., Phillips, T. L., Mayer, E. L., and Fishkind, D. J., 1987, Transport of membrane-bound macromolecules by M cells in follicle associated epithelium of rabbit Peyer's patch, *Cell Tissue Res.* **247**:537–546.

Normand, I. C. S., Olver, R. E., Reynolds, E. O. R., and Strang, L. B., 1971, Permeability of lung capillaries and alveoli to non-electrolytes in the foetal lamb, *J. Physiol. (London)* **219**:303–330.

Olver, R. E., Schneeberger, E. E., and Walters, D. V., 1981, Epithelial solute permeability, ion transport and tight junction morphology in the developing lung of the fetal lamb, *J. Physiol. (London)* **315**:395–412.

Pacht, E. R., and Davis, W. B., 1988, Role of transferrin and ceruloplasmin in antioxidant activity of lung epithelial lining fluid, *J. Appl. Physiol.* **64**:2092–2099.

Park, C. H., and Maack, T., 1984, Albumin absorption and catabolism by isolated perfused proximal convoluted tubules of the rabbit, *J. Clin. Invest.* **73**:767–777.

Patton, S. E., Gilmore, L. B., Jetten, A. M., Nettesheim, P., and Glook, G. E. R., 1986, Biosynthesis and release of proteins by isolated pulmonary Clara cells, *Exp. Lung Res.* **11**: 277–294.

Penney, D. P., 1988, The ultrastructure of epithelial cells of the distal lung, *Int. Rev. Cytol.* **3**: 231–269.

Pontiroli, A. E., and Pozza, G., 1990, Intranasal administration of peptide hormones: Factors affecting transmucosal absorption, *Diabetic Med.* **7**:770–774.

Pontiroli, A. E., Calderara, A., and Pozza, G., 1989, Intranasal drug delivery: Potential advantages and limitations from a clinical pharmacokinetic perspective, *Clin. Pharmacokinet.* **17**:299–307.

Possmayer, F., 1988, A proposed nomenclature for pulmonary surfactant-associated proteins, *Am. Rev. Respir. Dis.* **138:**990–998.

Price, A. M., Webber, S. E., and Widdicombe, J. C., 1990, Transport of albumin by the rabbit trachea *in vitro*, *J. Appl. Physiol.* **68:**726–730.

Ranga, V., and Kleinerman, J., 1982, The effect of pilocarpine on vesicular uptake and transport of horseradish peroxidase by the guinea pig tracheal epithelium, *Am. Rev. Respir. Dis.* **125:**579–585.

Rice, W. R., Ross, G. T., Singleton, F. M., Dingle, S., and Whitsett, J. A., 1987, Surfactant-associated protein inhibits phospholipid secretion from type II cells, *J. Appl. Physiol.* **63:**692–98.

Rice, W. R., Sarin, V. K., Fox, J. L., Baatz, J., Wert, S., and Whitsett, J. A., 1989, Surfactant peptides stimulate uptake of phosphatidylcholine by isolated cells, *Biochim. Biophys. Acta* **1006:**237–245.

Richardson, J., Bouchard, T., and Ferguson, C. C., 1976, Uptake and transport of exogenous proteins by respiratory epithelium, *Lab. Invest.* **35:**307–314.

Ryan, R. M., Morris, R. E., Rice, W. R., Ciraolo, G., and Whitsett, J. A., 1989, Binding and uptake of pulmonary surfactant (SP-A) by pulmonary type II epithelial cells, *J. Histochem. Cytochem.* **37:**429–440.

Salzman, R., Manson, J. E., Griffing, G. T., Kimmerle, R., Ruderman, N., McCall, A., Stoltz, E. I., Mullin, C., Small, D., Armstrong, J., and Melby, J. C., 1985, Intranasal aerosolized insulin: Mixed-meal studies and long-term use in type I diabetes, *N. Engl. J. Med.* **312:**1078–1084.

Schneeberger, E. E., 1980, Heterogeneity of tight junction morphology in extrapulmonary and intrapulmonary airways of the rat, *Anat. Rec.* **198:**193–208.

Singh, G., Singal, S., Katyal, S. L., Brown, W. E., and Gottron, S. A., 1987, Isolation and amino acid composition of a rat Clara cell specific protein, *Exp. Lung Res.* **13:**299–310.

Solomon, A. K., 1968, Characterization of biological membranes by equivalent pores, *J. Gen. Physiol.* **51:**335s–364s.

Stutts, M. J., Gatzy, J. T., and Boucher, R. C., 1988, Effects of metabolic inhibition on ion transport by dog bronchial epithelium, *J. Appl. Physiol.* **64:**253–258.

Sugahara, K., Voelker, D. R., and Mason, R. J., 1987, Insulin stimulates amino acid transport by alveolar type II epithelial cells in primary culture, *Am. Rev. Respir. Dis.* **135:**617–621.

Svensson, C., Baumgarten, C. R., Pipkorn, U., Alkner, U., and Persson, C. G. A., 1989, Reversibility and reproducibility of histamine induced plasma leakage in nasal airways, *Thorax* **44:**13–18.

Taylor, A. E., and Gaar, K. A., Jr., 1970, Estimation of equivalent pore radii of pulmonary capillary and alveolar membranes, *Am. J. Physiol.* **218:**1133–1140.

Theodore, J., Robin, E. D., Gaudio, R., and Acevedo, J., 1975, Transalveolar transport of large polar solutes (sucrose, inulin, and dextran), *Am. J. Physiol.* **229:**989–996.

Weaver, T. E., and Whitsett, J. A., 1991, Function and regulation of pulmonary surfactant-associated proteins, *Biochem. J.* **273:**249–264.

Webber, S. E., and Widdicombe, J. G., 1989, The transport of albumin across ferret in-vitro trachea, *J. Physiol. (London)* **408:**457–472.

Wintergerst, E., Manz-Keinke, H., Plattner, H., and Schlepper-Schäfer, J., 1989, The interaction of a lung surfactant (SP-A) with macrophages is mannose dependent, *Eur. J. Cell Biol.* **50:**291–298.

Wright, J. R., and Dobbs, L. G., 1991, Regulation of pulmonary surfactant secretion and clearance, *Annu. Rev. Physiol.* **53**:395–414.

Wright, J. R., Wager, R. E., Hawgood, S., Dobbs, L., and Clements, J. A., 1987, Surfactant apoprotein M$_r$ 26,000–36,000 enhances uptake of liposomes by type II cells, *J. Biol. Chem.* **262**:2888–2894.

Wright, J. R., Borchelt, J. D., and Hawgood, S., 1989, Lung surfactant apoprotein SP-A (26–36kDa) binds with high affinity to isolated alveolar type II cells, *Proc. Natl. Acad. Sci. USA* **86**:5410–5414.

Yamamoto, A., Hayakawa, E., and Lee, H. L., 1990, Insulin and proinsulin proteolysis in mucosal homogenates of the albino rabbit: Implications in peptide delivery from nonoral routes, *Life Sci.* **47**:2465–2474.

Yankaskas, J. R., Gatzy, J. T., and Boucher, R. C., 1987, The effects of raised osmolality on canine tracheal epithelial ion transport function, *J. Appl. Physiol.* **62**:2241–2245.

Chapter 8

Dermal Absorption of Peptides and Proteins

Ajay K. Banga and Yie W. Chien

1. INTRODUCTION

Historically, the skin was viewed as an impermeable barrier. However, in recent years, it has been increasingly recognized that intact skin can be used as a port for continuous systemic administration of drugs. The first transdermal drug delivery system which has used this new concept is a scopolamine-releasing transdermal therapeutic system (Transderm-Scop® by Ciba) for motion sickness. This was followed by the marketing of several nitroglycerin-releasing systems. Currently, about ten drugs have been either successfully marketed or are under clinical evaluation for transdermal delivery.

Several biomedical benefits offered by (trans)dermal delivery could also be extended to peptide drugs and, furthermore, there are some advantages unique to peptide drugs. The transdermal route provides a continuous mode of administration, which is highly desirable for peptide drugs since these drugs are extremely short-acting, by virtue of their short biological half-lives. In addition to the avoidance of hepatic "first-pass" elimination, transdermal delivery offers a mechanism to reduce proteolytic degradation encountered by delivery through the oral route (see Chapters 5, 6) and also through most absorptive mucosae (see Chapters 2, 7, 15). However, the

Ajay K. Banga and Yie W. Chien • Controlled Drug-Delivery Research Center, College of Pharmacy, Rutgers—The State University of New Jersey, Piscataway, New Jersey 08855. *Current Address of A.K.B.*: Department of Pharmacal Sciences, School of Pharmacy, Auburn University, Auburn, Alabama 36849.

Biological Barriers to Protein Delivery, edited by Kenneth L. Audus and Thomas J. Raub. Plenum Press, New York, 1993.

dermal route has some major drawbacks which will limit its utilization for peptide delivery. Peptide/protein drugs, because of their hydrophilicity and large molecular dimensions, are not good candidates for dermal delivery. In addition, the enzymatic barrier in the dermal tissue needs to be considered (Section 3).

The limitation of the dermal route for the delivery of peptide/protein drugs perhaps explains the reason why not much work has been done in this area. However, there are a few early publications and, in recent years, there has been a growing interest in this area, as reflected by several publications. As early as 1966, Tregear studied the permeation of macromolecular proteins and polymers through the excised skin from humans and animals. More recently, Menasche et al. (1981) found that percutaneously administered elastin peptides penetrate into the dermis and 30–40% of the administered dose can still be found in the skin 48 hr later. Some proteins and protein hydrolysate found application in cosmetics or toiletries for topical effects. Growth factors for wound healing represent another use for dermal application of peptides/proteins. Several investigations have recently reported the achievement of dermal delivery of peptides/proteins by iontophoresis and these will be discussed in Section 4.1.4.

2. DERMAL TRANSPORT

2.1. Dermal Anatomy

The skin is a complex multilayered organ with a total thickness of about 2–3 mm. Macroscopically, it has two distinct layers: the outer epidermis and the inner dermis. The dermis provides physiological support for the epidermis by supplying it with blood and lymphatic vessels and also with nerve endings (Fig. 1). The epidermis is comprised of several physiologically active epidermal tissues and the physiologically inactive stratum corneum. The physiologically active epidermis contains keratinocytes as the predominant cell type. These cells originate in a layer called the stratum germinativum and undergo continuous differentiation and mitotic activity during the course of migration upward through the layers of spinosum, granulosum, and lucidum. Finally, a layer of dead, flattened keratin-filled cells (corneocytes), which is called the stratum corneum or the horny layer, is produced. The entire epidermis is avascular and is supported by the underlying dermis. The cells migrating from the stratum germinativum layer are slowly dying as they move upward away from their source of oxygen and nourishment. Upon reaching the stratum corneum, these cells are cornified and dead. The time required for the cells to proliferate from the stratum germinativum to the stratum corneum is about 28 days, of which 14 days is spent as corneocytes in the stratum corneum. The corneocytes are then sloughed off from the skin into the environment (about one cell layer per day), a process called desquamation. Many theories have been proposed for the biochemical changes preceding desquamation but the underlying mechanisms are not clearly

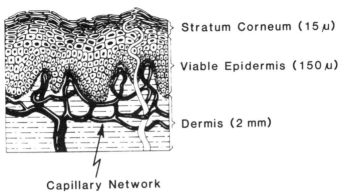

Stratum Corneum (15 μ)

Viable Epidermis (150 μ)

Dermis (2 mm)

Capillary Network

Figure 1. Diagrammatic illustration showing the human skin (upper region) and the capillary network at the dermo-epidermal junction.

known. (Wertz and Downing, 1989). There are several appendages in the skin, which include hair follicles, sebaceous and sweat glands, but these occupy only about 0.1% of the total human skin surface.

Most of the epidermal mass is concentrated in the stratum corneum and this layer forms the principal barrier to the penetration of drugs, i.e., it forms the rate-limiting membrane to transdermal permeation. Elias (1983) puts it succinctly when he says, "Our knowledge about the structure and function of the stratum corneum has unfolded like an ever-changing kaleidoscope over the past three decades." As reviewed by Kligman (1984), this layer has been packed with hexagonal cells, an arrangement which has provided a large surface area with the least mass. Species difference exists, e.g., the cells are stacked in vertical columns in mice but distributed randomly in humans. Each corneocyte is bounded by a thick, proteinaceous envelope with the tough fibrous protein keratin as the main component. Earlier reports based on transmission electron microscopy suggested that the spaces between corneocytes are empty; however, this is now believed to be an artifact of sample preparation. In fact, an intact stratum corneum is a highly ordered structure. As the epithelial cells migrate upward toward the stratum corneum, their plasma membrane seems to thicken due to a deposition of material on its inner and outer surface. This is the process of keratinization during which the polypeptide chains unfold and break down and then resynthesize into keratin, the tough, fibrous protein which forms the main component of the corneocyte. Corneocytes contain a compact arrangement of α-keratin filaments 60–80 Å in diameter and distributed in an amorphous matrix. The intercellular spaces of the stratum corneum are completely filled with broad, multiple lamellae. The lipids constituting these lamellae are unique in that they do not contain phospholipids. They are, instead, mainly composed of ceramides, cholesterol, fatty acids, and cholesteryl esters (Fig. 2). In contrast, the lipid composition in the viable

Ajay K. Banga and Yie W. Chien

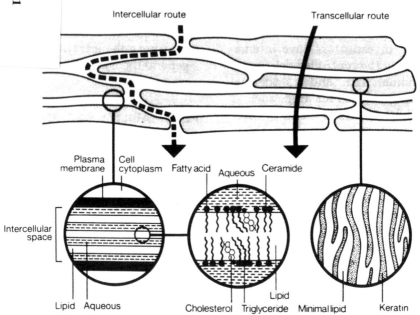

Figure 2. Schematic illustration of human stratum corneum with macroscopic, microscopic, and molecular domains, and suggested routes of drug penetration. (From Barry, 1987.)

epidermis is predominantly composed of phospholipids. These changes in the lipid composition have been demonstrated to occur during the keratinization process of the stratum corneum.

2.2. Transport Mechanisms

It should be emphasized that unlike other epithelial barriers, the outer layer of the dermal barrier, the stratum corneum, is a keratinized tissue. Since the stratum corneum is the principal permeation barrier, the transdermal delivery of small molecules has been considered as a process of interfacial partitioning and molecular diffusion through this barrier. In contrast, the permeation of di- and tripeptides, which are also small molecules, across the gastric mucosa is believed to occur through a carrier-mediated mechanism while proteins, such as IgG and epidermal growth factor, are believed to be absorbed by endocytosis (see Chapters 4, 5). Also, there is indirect evidence for the existence of amino acid carriers in the nasal mucosa of the rat (Lee, 1991).

A mathematical model was developed by Michaels *et al.* (1975) which treats the stratum corneum as a two-phase protein–lipid heterogeneous membrane having the lipid matrix as the continuous phase. Since then, several theoretical skin-permeation models have been proposed which predict the transdermal flux of a drug based on a few physicochemical properties of the drug. These models often make some assumptions about the barrier properties of the skin and predict the transdermal flux of a drug from a saturated aqueous solution, given a knowledge of the water solubility and molecular weight of the drug and its lipid–protein partition coefficient. Most of these theoretical expressions assume a two-compartment model of the stratum corneum based on a heterogeneous two-compartment system of protein-enriched cells embedded in lipid-laden intercellular domains. An analogy of "bricks and mortar" is often given for this model.

Based on this model, drugs can diffuse through the stratum corneum via a transepidermal or a transappendageal route. Figure 2 illustrates two potential routes of transepidermal drug penetration through the stratum corneum. These are between the cells (intercellular) or through the protein-filled cells (transcellular) (Barry, 1987). The relative contribution of these routes depends on the solubility, partition coefficient, and diffusivity of the drug within these protein or lipid phases. The transcellular route will predominate for polar drugs. Elias (1988) provides support for this model by making the observation that differences in lipid content more accurately predict regional variations in skin permeability as compared with stratum corneum thickness. For this reason, facial stratum corneum, whose lipid content is 7–10% by weight, is readily permeable to topical steroids.

The transappendageal route normally contributes to only a very limited extent to the overall kinetic profile of transdermal drug delivery. However, the hair follicles and sweat ducts can act as diffusion shunts for ionic molecules during iontophoretic transport (Section 4.1.4). The electroosmotic effects, which accompany iontophoretic delivery, could also be partly responsible for the observed enhanced flux. Additionally, as electric potential is applied across the skin, the rearrangement of lipid bilayers and/or the flip-flop of the polypeptide helices in the stratum corneum may occur to open up pores, as a result of repulsion between neighboring dipoles (Jung *et al.* 1983), which could create "artificial shunts" leading to enhancement of skin permeability for peptide and protein molecules (Chien, 1991).

3. DERMAL BARRIERS TO PEPTIDE/PROTEIN DELIVERY

3.1. Metabolic/Enzymatic Barrier

In the last several years, growing evidence has accumulated that the skin is not only a barrier for passive diffusion, but also consists of an enzymatic barrier capable of metabolizing drugs. The composition of enzymes and the spectrum of metabolic reactions in the skin are similar to those in the liver. However, the skin has only about 10% of the metabolic activity detected in the liver (Merkle, 1989) and its blood perfusion is rather poor since the total blood flow to the skin is only about 6.25% of that to the liver (Tauber, 1989). Therefore, the skin does not contribute significantly to systemic drug disposition. However, the skin could have a significant biotransformation potential in topical and transcutaneous drug therapy. Also, it should be borne in mind, especially for transcutaneous therapy, that proteolytic enzymes are ubiquitous and traversing the dermal barrier is not the last, but rather the first step. Therefore, peptides and proteins need to be protected from the enzymatic barriers in several anatomical sites before they can exert their pharmacological or therapeutic effects.

The metabolic activities of the skin include a considerable amount of proteolytic activity. Both exo- and endopeptidases are known to be present in the skin. Exopeptidases include those enzymes that cleave peptides and proteins at the N-terminus (e.g., aminopeptidases), at the C-terminus (e.g., carboxypeptidases), or those either having specificity for dipeptides or liberating dipeptides from the N- or C-terminus. Several of these enzymes are found in the skin, with the aminopeptidase (see Chapter 2) being the best known (Hopsu-Havu et al., 1977). On the other hand, endopeptidases cleave peptides and proteins at an internal peptide bond and these include the proteinases. Skin homogenates have been used to study the degradation of peptides by dermal tissues; however, the use of skin homogenates does not allow us to determine the origin of the individual enzymes. The methods available for the histochemical localization of enzymes are rather limited and this prevents a complete understanding of the subcellular compartmentalization of proteolytic enzymes. However, incubation of a peptide with a tissue homogenate is a simple first step in characterizing the proteolytic barrier of any tissue or organ. Lee et al. (1991) have reviewed their investigations on the stability of several peptide/protein candidates against mucosal homogenates of the albino rabbit. The rank order of stability for small peptides was found to be rectal > buccal > nasal > vaginal. For proteins, however, the rank order was nasal > rectal > vaginal > buccal. The stability of insulin in skin homogenates of hairless rat skin was investigated by Banga (1990) and will be discussed in Section 4.1.1.

Some investigations have reported that the metabolic activity of the skin predominantly resides in the viable epidermis, but these studies have not taken into account the mass ratio between the dermis and the epidermis. The activity per unit volume was reported to be higher in the dermis for some enzymes (Tauber, 1989). Also, there could be some significant difference between the species and caution should be observed in extrapolating the results from animal studies to humans. A case in point for the subcutaneous degradation of insulin will be discussed in Section 4.2.1.

3.2. Physical Barrier

As discussed in Section 2.2, the stratum corneum forms the principal barrier to the systemic or dermal delivery of drugs. Over most of the body, the stratum corneum has 15–25 layers of flattened corneocytes with an overall thickness of about 10 μm. The permeation rate of drugs through the stratum corneum has been reported to decrease with an increase in molecular size. A molecular size of about 1000 Da is generally believed to be the molecular size limit for transdermal delivery but there are some literature reports on the permeation of macromolecules (> 50,000 Da) across the stratum corneum by passive diffusion. In addition to molecular size, the lipid–protein partition coefficient of the penetrant is very important. For peptide/protein drugs, both their macromolecular dimensions and hydrophilic nature are not suitable for transdermal delivery. Some of these disadvantages, however, can be overcome by the skin permeability-enhancing techniques, such as iontophoresis, which will be discussed in Section 4.1.

3.3. Microbial Barrier

A wide variety of microorganisms have been isolated from the skin surface. These dermal flora could be resident or transient and will also vary with individual, age, skin site and condition. Collectively, the skin flora possesses a wide range of enzymes, and drug metabolism by skin microorganisms has been reported to take place for drugs such as steroid esters and nitroglycerin (Denyer and McNabb, 1989). Furthermore, microbial growth on the skin would be promoted by the occlusive environment often generated under a transdermal patch. Microbial metabolism could be a potential issue for the dermal delivery of peptide/protein drugs, since these could potentially serve as a food substrate for the skin flora. This could be especially true if the peptide/protein delivery system is going to be in contact with the skin for any prolonged length of time. The use of preservatives may be advantageous in this regard, provided the preservative does not have any skin irritation potential.

4. STRATEGIES FOR DERMAL DELIVERY OF PEPTIDES/PROTEINS

The skin is impermeable to molecules as large as peptides and proteins. The physical and enzymatic barriers discussed above, in combination, create a formidable block against any permeation under normal circumstances. Various techniques have been attempted to surmount these barriers. These include the use of protease inhibitors to suppress enzymatic activity or the use of penetration enhancers to reversibly reduce the barrier resistance of the stratum corneum. Other alternatives include forced delivery under an electric field (iontophoresis) or ultrasonic energy (phonophoresis). These are strategies for delivery through the intact skin. Another approach which has been used for decades is the subcutaneous deposition of drugs via hypodermic injection. In the following sections, these techniques have been grouped as noninvasive and invasive techniques for discussion.

4.1. Noninvasive Delivery through Intact Skin

4.1.1. PROTEASE INHIBITORS

Protease inhibitors provide a viable means to circumvent the enzymatic barrier in achieving the delivery of peptides and proteins. Numerous agents could act as protease inhibitors by a variety of mechanisms, e.g., by tightly binding to or covalent modification of the active sites of protease or by chelating the metal ions essential for proteolytic activity. The selection of an appropriate protease inhibitor could be guided by studying the principal proteases responsible for the degradation of the peptide/protein to be delivered, their subcellular compartmentalization, and the mechanism for the transport of the peptide/protein.

Hori *et al.* (1983) investigated the effect of coadministering peptides and amino acids, as stabilizing agents, on insulin degradation at a subcutaneous injection site in rats. They discovered benzyloxycarbonyl-Gly-Pro-Leu, benzyloxycarbonyl-Gly-

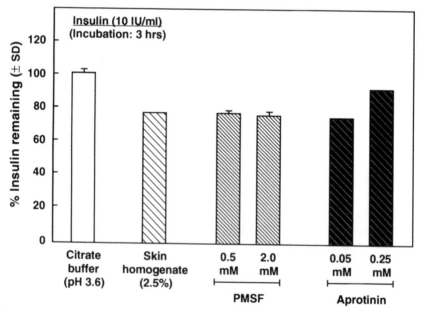

Figure 3. *In vitro* evaluations of protease inhibitors in preventing the enzymatic degradation of insulin by skin homogenates in pH 3.6 citrate buffer. A 10% skin homogenate was centrifuged and filtered to get a clear extract, of which 2.5% (v/v) was added to a solution of porcine insulin (10 IU/ml). After incubation at 37°C for 3 hr, the degradation of insulin was followed by HPLC assay. (Modified from Banga, 1990.)

Pro-Leu-Gly, and dinitrophenyl-Pro-Leu-Gly as the best candidates for protecting insulin from degradation at the subcutaneous site. Several other protease inhibitors that have been evaluated include amastatin, aprotinin, bestatin, boroleucine, p-chloromercuribenzoate, leupeptin, pepstatin, phenylmethylsulfonyl fluoride (PMSF), and puromycin. Sodium glycocholate, in addition to being a penetration enhancer, was also observed to prevent the proteolysis of insulin, though it is not as potent as amastatin and bestatin, the aminopeptidase inhibitors. PMSF has been widely used to prevent the breakdown of enzymes and other proteins during purification and acts by causing the sulfonylation of serine residues. Aprotinin, also known as bovine pancreatic trypsin inhibitor (or trypsin–kallikrein inhibitor), is also a serine protease inhibitor. A polypeptide with a molecular weight of 6512, aprotinin is rather stable and has a broad inhibitory specificity. It has found uses in some radioimmunoassay kits and also for the therapy of acute pancreatitis. While its potential application in preventing the subcutaneous degradation of insulin has been found rather controversial with conflicting reports, the *in vitro* study by Banga (1990) has shown aprotinin's capability of reducing the degradation of insulin by skin homogenates (Fig. 3). It was observed that the concentration of aprotinin required for the protective effect has to be carefully determined for any given experimental (or study) conditions. The total substrate, enzymatic activity, temperature, and time of exposure could be some of the important factors influencing the concentration of aprotinin required. In this study, however, PMSF failed to offer any such protection. It has also been reported that protease inhibitors could be used to potentiate the healing effect of epidermal growth factor in would healing therapy. This is due to the fact that several proteases are known to become activated in the wounded tissues (Okumura *et al.*, 1989).

4.1.2. CHEMICAL MODIFICATION

Chemical modification of the peptide backbone or the modification and/or blocking of the N- and C-terminals could potentially reduce proteolysis. Also, substitution with the unnatural D-amino acids in place of the natural L-amino acids has been used to improve the enzymatic stability of peptides. Another type of modification, which has been reported to decrease the susceptibility of peptides to enzymatic degradation, increase their plasma circulation half-life, as well as decrease their immunogenicity is the covalent attachment of polyethylene glycol (PEG) to the amino groups of a peptide/protein molecule. Chemical modification of peptides has been used to improve the transdermal permeation of peptides. For example, Tyr-D-Ala-Gly-Phe-Leu and its amide, the enzymatically stable analogues of leucine enkephalin (Tyr-Gly-Gly-Phe-Leu), were observed to exhibit significant fluxes in the presence of n-decylmethyl sulfoxide across hairless mouse skin (Choi *et al.*, 1990). The amide showed greater stability to metabolism than did the [D-Ala^2]-leucine enkephalin. Because of the structural complexity of the protein molecules, most chemical modification has been done on peptides and not on proteins.

4.1.3. PENETRATION ENHANCERS

While penetration enhancers have been widely investigated in terms of enhancing the mucosal delivery of peptide/protein drugs, their use for dermal delivery of peptides seems to be somewhat limited. This could be attributed to the fact that the keratinized stratum corneum is much more impermeable than the mostly nonkeratinized mucosal membranes. However, penetration enhancers have been successfully used to enhance dermal delivery of organic-based drug molecules including hydrophilic molecules. Attempts were made to deliver transdermally radiolabeled vasopressin, and treatment of the skin with sodium lauryl sulfate, an anionic surfactant, was found ineffective even at a concentration as high as 20%. However, following the stripping of the skin, the transdermal flux of vasopressin was improved over 70-fold (Banerjee and Ritschel, 1989). On the other hand, the *in vitro* skin permeation fluxes of leucine enkephalin and its synthetic analogues were enhanced by the presence of *n*-decylmethyl sulfoxide, a nonionic surfactant (Choi *et al.*, 1990). A penetration enhancer is usually a small molecule and mechanisms of enhancement include modification of intercellular lipid matrix with increased membrane fluidity. They could also loosen the tight junctions between cells, e.g., increasing the dimensions of the paracellular pathway, in the case of mucosal delivery (Chien *et al.*, 1989). Examples of such enhancers include bile salts, chelating agents, surfactants, and fatty acid derivatives. On the other hand, alkanols, alkanoic acids and their esters, dimethylsulfoxide (DMSO), and 1-dodecylazacycloheptan-2-one (Azone) have been widely investigated as enhancers for transdermal delivery (Chien, 1987).

4.1.4. IONTOPHORESIS

Iontophoresis implies the use of small amounts of physiologically acceptable electric current to drive ionic drugs into the body. In essence, the peptide drug can be made to have a charge by controlling the pH of its solution relative to its isoelectric point and an electrode of like charge can be used to facilitate the movement of peptide molecules through the dermal tissue under the electric potential gradient. Iontophoresis has been used for topical medication for several decades, but its use for systemic medication by transdermal delivery is relatively recent. For comprehensive reviews on the subject, the reader is referred to Tyle (1986) and Banga and Chien (1988).

Iontophoresis-facilitated transdermal delivery has been investigated for several therapeutic peptides and proteins. Several literature references for such studies have been compiled in Table I in the order of increasing molecular weight of the peptides/proteins. Iontophoresis could also provide a potential means to deliver peptide/protein in a pulsatile manner (by turning electric current on and off), to prevent the downregulation of receptors, in addition to continuous administration (Chien *et al.*, 1990).

Table I

Iontophoresis-Facilitated Transdermal Delivery of Peptide/Protein Drugs

Peptide/protein	No. of amino acids	Size	Animal model/ experiment design	Reference
Thyrotropin-releasing hormone (TRH)	3	362	Nude mice/*in vitro*	Burnette and Marrero (1986)
Vasopressin (arginine)	9	1084	Hairless rats/*in vitro*	Lelawongs *et al.* (1989, 1990), Lelawongs (1990), Banga (1990), Chien *et al.* (1990)
LHRH and analogues	9/10	1182	Humans/*in vivo*	Meyer *et al.* (1988)
			Hairless mice/*in vitro*	Miller *et al.* (1990)
Calcitonin (human)	32	3418	Hairless rats/*in vitro*	Banga (1990)
Insulin	51	5808	Pigs/*in vivo*	Stephen *et al.* (1984)
			Diabetic rabbits/*in vivo*	Kari (1986), Chien *et al.* (1988), Meyer *et al.* (1989)
			Hairless rats/*in vitro*	Banga (1990)
			Diabetic hairless rats/*in vitro* and *in vivo*	Siddiqui *et al.* (1987) Chien *et al.* (1988, 1990)
			Diabetic hairless rats/*in vivo*	Liu *et al.* (1988)

Meyer *et al.* (1988) conducted a double-blind cross-over clinical study on the transdermal delivery of leuprolide, an LHRH analogue, for the hormonal treatment of metastatic prostatic carcinoma. It was observed that serum LH concentrations with active patches have been significantly increased relative to placebo patches over the duration of the study. This successful transdermal delivery of therapeutic doses of leuprolide to humans was attributed to electroosmotic effects.

Chien *et al.* (1990) conducted *in vitro* studies investigating the effect of iontophoresis on the skin permeation kinetics of vasopressin. As can be seen from Fig. 4, the skin permeation rate increased almost 190-fold after periodic application of pulsed current, delivered from a constant-current power source (TPIS), with a specific combination of waveform, frequency, on/off ratio, and intensity. The skin permeation profile was found to consist of two phases: an activation phase, during which the pulsed current was applied periodically, and a postactivation phase, during which the current was terminated and the permeability properties of the skin appeared to recover.

One protein that has been extensively investigated for feasibility of using iontophoresis to facilitate the rate of transdermal delivery is insulin. The results from various investigators are somewhat conflicting, which is most likely due to the multitude of factors and variables in operation, and also due to different experimental

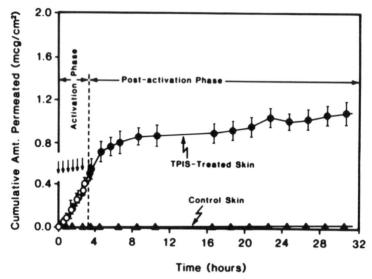

Figure 4. *In vitro* skin permeation profiles of vasopressin, from donor solution at pH 5, across the abdominal skin of hairless rats and the enhancement of skin permeation by TPIS treatment (square waveform; frequency, 2 kHz; on/off ratio, 1:1) at a current intensity of 1.0 mA, which was switched on for 10 min and then off for 30 min repeatedly for six times over a 4-hr period. (From Chien *et al.*, 1990.)

conditions and species. Stephen *et al.* (1984) were successful in delivering a highly ionized monomeric form of insulin through pig skin for systemic effect, but failed to deliver regular (soluble) insulin to human volunteers by iontophoresis. Kari (1986) successfully delivered insulin to rabbits after disruption of the stratum corneum. He also observed that blood glucose levels continued to decrease even after the current was turned off, suggesting the formation of a reservoir in the subcutaneous tissues. Chien *et al.* (1988) developed diabetic models in both hairless rats and New Zealand White rabbits by intraperitoneal injections of streptozotocin and demonstrated the feasibility of transdermal iontophoretic delivery of insulin through intact skin. The results of this and several other studies have been reviewed by Chien (1991). Banga (1990) investigated the formation of an insulin reservoir in the hairless rat skin under *in vitro* conditions (Fig. 5). It was found that the amount retained in the skin (on a unit volume basis) is much higher than the total amount of insulin permeating through the skin over the course of 30 hr at two pH values. At pH 3.6, insulin exists as a positively charged molecule and was delivered under anode; at pH 7.4, insulin is a negatively charged molecule and was delivered under cathode. It is evident that permeation is higher at a pH lower than the isoelectric point of insulin, which could be due to the higher charge density on the molecule and/or due to the higher activity coefficient at the lower pH.

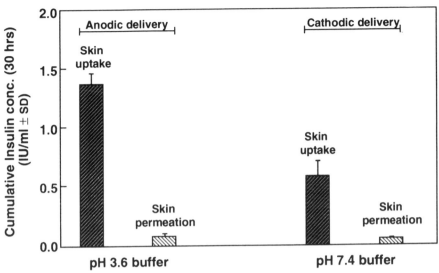

Figure 5. A comparison between the amount of insulin accumulated in the skin, as depot, after 4 hr of iontophoresis and the cumulative amounts of insulin permeated through the hairless rat skin for 30 hr (on a unit volume basis) at pH 3.6 and 7.4 (current intensity, 0.62 mA/cm^2; frequency, 2.0 kHz; on/off ratio, 1:1). (Modified from Banga, 1990.)

4.1.5. PHONOPHORESIS

Phonophoresis implies the transport of drug molecules under the influence of ultrasound. The drug is delivered from a coupling (contact) agent which transfers ultrasonic energy from the ultrasonic device to the skin. The exact mechanism involved is unknown, but enhancement presumably results from thermal, mechanical, and chemical alterations in the skin induced by ultrasonic waves. Although several drugs have been investigated for phonophoretic delivery, the results are not always positive and many of these studies lack adequate controls. Tyle and Agrawala (1989) have conducted a comprehensive review of drug delivery by phonophoresis and listed several drugs, which includes papain and interferon, as peptide/protein candidates being investigated.

4.2. Invasive Delivery through Broken Skin

4.2.1. SUBCUTANEOUS INJECTION/IMPLANT

Subcutaneous injections of insulin have been the practical and primary treatment of diabetes for several decades. Zinc–insulin complexes in amorphous and crystalline forms are prepared to formulate long-acting insulin preparations of varying duration of hypoglycemic activities, which are still popularly used today for the treatment of diabetes. Several long-acting injectables, novel injectors, and infusion pumps developed for insulin have been reviewed by Chien and Banga (1989). It has been reported by several investigators that subcutaneously administered insulin is also metabolized to a significant extent at the injection site. Efforts have also been directed to minimizing this degradation by coadministration of protease inhibitors and this has been discussed in Section 4.1.1. However, the reported magnitudes of degradation appear to be controversial and most studies that reported considerable degradation seem to be those conducted in animals. Kraegen and Chisholm (1985) reported a low degradation of insulin at subcutaneous tissues in most normal and diabetic subjects, which they felt was not of clinical significance for the majority of the patients during subcutaneous insulin infusion therapy. Another approach for subcutaneous delivery of peptides is via the long-acting subdermal polymeric implants. Langer and Folkman (1976) fabricated subdermal implants from hydrox-yethylmethacrylate and (ethylene vinyl acetate) copolymers and demonstrated that macromolecules of up to 2×10^6 Da can be made to release slowly from subdermal implants for periods exceeding 3 months. Brown et al. (1986) implanted an insulin/(ethylene vinyl acetate) copolymer matrix device subcutaneously in diabetic rats and normoglycemia was achieved with a single implant for over 100 days.

A contraceptive implant (Norplant/Wyeth-Ayerst Labs) that releases progestin from a silicone-based subdermal capsule for a period of 5 years was recently approved by the FDA for marketing in the United States (Time, December 24, 1990). The biocompatibility of polymeric implants is an important concern since these implants are intended to reside in the body for extended periods of time. Biodegradable polymers can also be utilized if their degradation products are innocuous and biocompatible. For peptides and proteins, an additional concern is their immunogenicity. The most potent immunogens are those proteins of greater than 100 kDa. Furthermore, chemical complexity also plays a role, e.g., the aromatic amino acid components in the protein molecule contribute more to its immunogenicity than do the nonaromatic amino acid residues.

4.2.2. DISRUPTION OF STRATUM CORNEUM

As discussed earlier, the stratum corneum forms the main barrier to the trans-dermal delivery of drugs, peptides and nonpeptides alike. Removal of this barrier by stripping or shaving produces an increase in the transdermal flux of drugs. The passive permeation of vasopressin was reportedly increased substantially by stripping the skin (Banerjee and Ritschel, 1989), but stripping of the skin was observed to produce no enhancing effect on the skin permeation of vasopressin by iontophoresis, because of the reduction in transmembrane potential (Lelawongs, 1990). On the other hand, Kari (1986) found it necessary to disrupt the stratum corneum of diabetic rabbits to achieve the iontophoretic delivery of insulin. Some peptides would normally be administered only through a broken or damaged skin for a topical effect; a typical example is the delivery of epidermal growth factors in wound healing.

5. DRUG DELIVERY DESIGN

A multitude of factors need to be carefully considered for the proper design of a drug delivery system for the dermal delivery of peptides. These include considerations of the charges on the peptide molecule and the skin in relation to the environmental pH. Presence of proteolytic enzymes in the skin and the binding of peptides to the skin to form a reservoir (or depot) are some of the other important considerations. Preformulation data should be generated for the formulation development of a dosage form or the design of a drug delivery system to achieve optimum stability and maximum bioavailability. Peptide/protein drugs are known to have a strong tendency to adsorb to a variety of surfaces and this must be carefully evaluated as this could lead to misleading interpretations of data. Certain additives, e.g., albumin, can be used to minimize such surface adsorption. Another potential problem is self-aggregation of peptide/protein molecules, especially under stirring. Additives, such as urea, at low concentrations have been reported to minimize both the self-aggregation and adsorption of insulin (Sato *et al.*, 1983).

The situation is even more complicated for iontophoretic delivery. A careful choice of electrode, pH, buffer, and ionic strength is required. For example, it may not be desirable to use a reversible electrode, such as silver–silver chloride, since it may react with peptides to cause precipitation and/or discoloration. Platinum electrode could be used but it will cause pH drifts and the use of buffers is thus indicated. The electrolytes contained in these buffers will compete with the peptides for delivery, thus reducing the efficiency of iontophoretic transport. A compromise is often required to balance several opposing factors. The charge on the peptide molecule can be controlled by adjusting the solution pH and thus delivery can be manipulated for either cathodal or anodal iontophoresis. Insulin was found to have a better permeability as a positively charged molecule delivered under the anode (Fig. 5). The use of hydrogels could provide a pragmatic choice for a delivery system. In actual usage, a hydrogel formulation will be easier to apply and remove from the skin surface and will not require any elaborate devices or membranes to prevent drug leakage as in solution formulations. If a miniaturized programmable iontophoretic device is developed in the future, a hydrogel formulation would be preferable as it could be developed into a unit dose-type dosage form. A hydrogel device might also be helpful in reducing skin hydration during the period of medication and might minimize the convective flows that accompany iontophoretic delivery. A hydrogel formulation will also increase the skin compliance of a transdermal patch. This is because a hydrogel can absorb the secretions of skin glands, i.e., water and other sweat components which, under long-term occlusion, may cause skin hydration and maceration. Three hydrogel systems—polyacrylamide, p-HEMA, and carbopol—were evaluated for the iontophoretic delivery of three model peptides: vasopressin, calcitonin, and insulin (Chien and Banga, 1990). The permeability coefficients were found to follow the rank order vasopressin > calcitonin > insulin (Table II), which are dependent upon molecular weight and other parameters.

6. PROSPECTS OF DERMAL DELIVERY OF PEPTIDES/PROTEINS

In the coming years, therapeutic peptides and proteins are going to gain increasing importance as a result of rapid strides in the biotechnology industry. The therapeutic application and market introduction of this new generation of therapeutic agents will require parallel development of efficient delivery systems by the pharmaceutical industry. The skin, with its accessibility, enormous surface area, and possibility for site targeting, offers a potential means of noninvasive delivery. In addition, the skin has much less proteolytic activity relative to other mucosal barriers. Passive diffusion of macromolecules through the skin is not very likely for most candidates, but the use of enhancers or iontophoresis could hold some promise. Such energy-assisted enhancing mechanisms may cause some proteins to penetrate the stratum corneum, followed by the formation of a depot (or reservoir) in the skin. Thus, in these cases, iontophoresis may turn out to be a means to load the skin tissues with protein drug, from which site the protein will then diffuse slowly into the systemic circulation.

While the skin provides an opportunity for a zero-order delivery analogous to a constant-rate intravenous infusion, such a delivery may not always be advantageous for peptide drugs. Some peptides may require delivery at a pulsatile pattern to avoid downregulation of receptors dictated by their chronopharmacology. Iontophoresis could also potentially provide such a delivery system since the current can be turned on and off periodically to modulate delivery.

The use of protease inhibitors or chemical modification of the peptide should help to minimize the proteolytic degradation in the skin. New agents may be

Table II
Transdermal Iontophoretic Delivery of Peptide/Protein Drugs from Hydrogel Formulations

Peptide/protein	No. of amino acids	Size (Da)	Permeability coefficient (cm/sec)		
			Polyacrylamide	p-HEMA	Carbopol
Vasopressin (arginine)	9	1084	6.16×10^{-7}	1.06×10^{-7}	2.74×10^{-7}
Calcitonin (human)	32	3418	6.20×10^{-8}	1.95×10^{-8}	0.69×10^{-8}
Insulin	51	5808	3.12×10^{-9}	6.54×10^{-9}	—

uncovered or novel techniques may be discovered. For example, Chien and Banga (1990) proposed the coadministration of a protease inhibitor, such as aprotinin, with the iontophoretic delivery of peptide/protein drug to afford maximal protection against proteolysis.

Another concern for the dermal delivery of drugs is the immunological functions of the skin. The skin has Langerhans and other kinds of cells which could be involved in the immunological process. It is possible that small molecules or peptides may serve as antigens or haptens to the immunological apparatus of the skin (Merkle, 1989). Possible implications include delayed hypersensitivity, allergic contact dermatitis, or even loss of biological activity of the drug.

REFERENCES

Banerjee, P. S., and Ritschel, W. A., 1989, Transdermal permeation of vasopressin. I. Influence of pH, concentration, shaving and surfactant on *in-vitro* permeation, *Int. J. Pharm.* **49:** 189–197.

Banga, A. K., 1990, Transdermal iontophoretic delivery of peptide drugs: Delivery mechanisms and hydrogel formulations, Ph.D. thesis, Rutgers—The State University of New Jersey.

Banga, A. K., and Chien, Y. W., 1988, Iontophoretic delivery of drugs: Fundamentals, developments and biomedical applications, *J. Controlled Release* **7:**1–24.

Barry, B. W., 1987, Mode of action of penetration enhancers in human skin, *J. Controlled Release* **6:**85–97.

Brown, L., Munoz, C., Siemer, L., Edelman, E., and Langer, R., 1986, Controlled release of insulin from polymeric matrices, *Diabetes* **35:**692–697.

Burnette, R. R., and Marrero, D., 1986, Comparison between the iontophoretic and passive transport of thyrotropin releasing hormone across excised nude mouse skin, *J. Pharm. Sci.* **75:**738–743.

Chien, Y. W., 1987, *Transdermal Controlled Systemic Medications*, Dekker, New York, Chapter 2.

Chien, Y. W., 1991, Transdermal route of peptide and protein drug delivery, in: *Peptide and Protein Drug Delivery* (V. H. L. Lee, ed.), Dekker, New York, pp. 667–689.

Chien, Y. W., and Banga, A. K., 1989, Potential developments in systemic delivery of insulin, *Drug Dev. Ind. Pharm.* **15:**1601–1634.

Chien, Y. W., and Banga, A. K., 1990, Iontotherapeutic devices, reservoir electrode devices therefore, process and unit dose, U.S. Patent pending.

Chien, Y. W., Siddiqui, O., Sun, Y., Shi, W. M., and Liu, J. C., 1988, Transdermal iontophoretic delivery of therapeutic peptides/proteins: (I) Insulin, *Ann. N.Y. Acad. Sci.* **507:**32–51.

Chien, Y. W., Su, K. S. E., and Chang, S. F., 1989, *Nasal Systemic Drug Delivery*, Dekker, New York.

Chien, Y. W., Lelawongs, P., Siddiqui, O., Sun, Y., and Shi, W. M., 1990, Facilitated transdermal delivery of therapeutic peptides/proteins by iontophoretic delivery devices, *J. Controlled Release* **13:**263–278.

Choi, H. K., Flynn, G. L., and Amidon, G. L., 1990, Transdermal delivery of bioactive peptides: The effect of n-decylmethyl sulfoxide, pH and inhibitors on enkephalin metabolism and transport, *Pharm. Res.* **7**:1099–1106.

Denyer, S. P., and McNabb, C., 1989, Microbial metabolism of topically applied drugs, in: *Transdermal Drug Delivery: Development Issues and Research Initiatives* (J. Hadgraft and R. H. Guy, eds.), Dekker, New York, pp. 113–134.

Elias, P. M., 1983, Epidermal lipids, barrier function and desquamation, *J. Invest. Dermatol.* **80**:44s–49s.

Elias, P. M., 1988, Structure and function of the stratum corneum permeability barrier, *Drug Dev. Res.* **13**:97–105.

Hopsu-Havu, V. K., Fraki, J. E., and Jarvinen, M., 1977, Proteolytic enzymes in the skin, in: *Proteinases in Mammalian Cells and Tissues* (Barrett, ed.), Elsevier/North-Holland, Amsterdam, pp. 547–581.

Hori, R., Komada, F., and Okumura, K., 1983, Pharmaceutical approach to subcutaneous dosage forms of insulin, *J. Pharm. Sci.* **72**:435–439.

Jung, G., Katz, E., Schmitt, H., Voges, K. P., Menestrina, G., and Boheim, G., 1983, Conformational requirements for the potential dependent pore formation of the peptide antibiotics alamethicin, suzukacillin and trichotoxin, in: *Physical Chemistry of Transmembrane Ion Motion* (G. Spach, ed.), Elsevier, Amsterdam.

Kari, B., 1986, Control of blood glucose levels in alloxan-diabetic rabbits by iontophoresis of insulin, *Diabetes* **35**:217–221.

Kligman, A. M., 1984, Skin permeability: Dermatologic aspects of transdermal drug delivery, *Am. Heart J.* **108**:200–206.

Kraegen, E. W., and Chisholm, D. J., 1985, Pharmacokinetics of insulin: Implications for continuous subcutaneous insulin infusion therapy, *Clin. Pharmacokinet.* **10**:303–314.

Langer, R., and Folkman, J., 1976, Polymers for the sustained release of proteins and other macromolecules, *Nature* **263**:797–800.

Lee, V. H. L., 1991, Changing needs in drug delivery in the era of peptide and protein drugs, in: *Peptide and Protein Drug Delivery* (V. H. L. Lee, ed.), Dekker, New York, pp. 1–56.

Lee, V. H. L., Traver, R. D., and Taub, M. E., 1991, Enzymatic barriers to peptide and protein drug delivery, in: *Peptide and Protein Drug Delivery* (V. H. L. Lee, ed.), Dekker, New York, pp. 303–358.

Lelawongs, P., 1990, Transdermal iontophoretic delivery of arginine vasopressin, Ph.D. thesis, Rutgers—The State University of New Jersey.

Lelawongs, P., Liu, J. C., Siddiqui, O., and Chien, Y. W., 1989, Transdermal iontophoretic delivery of arginine vasopressin: (I) Physicochemical considerations, *Int. J. Pharm.* **56**:13–22.

Lelawongs, P., Liu, J. C., Siddiqui, O., and Chien, Y. W., 1990, Transdermal iontophoretic delivery of arginine vasopressin: (II) Evaluation of electrical and operational factors, *Int. J. Pharm.* **61**:179–188.

Liu, J. C., Sun, Y., Siddiqui, O., and Chien, Y. W., 1988, Blood glucose control in diabetic rats by transdermal iontophoretic delivery, *Int. J. Pharm.* **44**:197–204.

Menasche, M., Jacob, M. P., Godeau, G., Robert, A. M., and Robert, L., 1981, Pharmacological studies on elastin peptides (kappa- elastin). Blood clearance, percutaneous penetration and tissue distribution, *Pathol. Biol.* **29**:548–554.

Merkle, H. P., 1989, Transdermal delivery systems, *Meth. Find. Exp. Clin. Pharmacol.* **11:** 135–153.

Meyer, B. R., Kreis, W., Eschbach, J., O'Mara, V., Rosen, S., and Sibalis, D., 1988, Successful transdermal administration of therapeutic doses of polypeptide to normal human volunteers, *Clin. Pharmacol. Ther.* **44:**607–612.

Meyer, B. R., Katzeff, H. L., Eschbach, J. C., Trimmer, J., Zacharias, S. B., Rosen, S., and Sibalis, D., 1989, Transdermal delivery of human insulin to albino rabbits using electrical current, *Am. J. Med. Sci.* **297:**321–325.

Michaels, A. S., Chandrasekaran, S. K., and Shaw, J. E., 1975, Drug permeation through human skin: Theory and *in vitro* experimental measurement, *AIChE J.* **21:**985–996.

Miller, L. L., Kolaskie, C. J., Smith, G. A., and Rivier, J., 1990, Transdermal iontophoresis of gonadotropin releasing hormone (LHRH) and two analogues, *J. Pharm. Sci.* **79:**490–493.

Okumura, K., Kiyohara, Y., Komada, F., Mishima, Y., and Fuwa, T., 1989, Protease inhibitor potentiates the healing effect of EGF (epidermal growth factor) in the wounded or burned skin, 2nd International Symposium on *Disposition and Delivery of Peptide Drugs*, FIP Satellite Symposium, September 1–3, Leiden, The Netherlands.

Sato, S., Ebert, C. D., and Kim, S. W., 1983, Prevention of insulin self-association and surface adsorption, *J. Pharm. Sci.* **72:**228–232.

Siddiqui, O., Sun, Y., Liu, J. C., and Chien, Y. W., 1987, Facilitated transdermal transport of insulin, *J. Pharm. Sci.* **76:**341–345.

Stephen, R. L., Petelenz, T. J., and Jacobsen, S. C., 1984, Potential novel methods for insulin administration: I. Iontophoresis, *Biomed. Biochim. Acta* **43:**553–558.

Tauber, U., 1989, Drug metabolism in the skin: Advantages and disadvantages, in: *Transdermal Drug Delivery: Developmental Issues and Research Initiatives* (J. Hadgraft and R. H. Guy, eds.), Dekker, New York, pp. 99–111.

Tregear, R. T., 1966, The permeability of skin to albumin, dextrans and polyvinylpyrrolidone, *J. Invest. Dermatol.* **46:**24–27.

Tyle, P., 1986, Iontophoretic devices for drug delivery, *Pharm. Res.* **3:**318–326.

Tyle, P., and Agrawala, P., 1989, Drug delivery by phonophoresis, *Pharm. Res.* **6:**355–361.

Wertz, P. W., and Downing, D. T., 1989, Stratum corneum: Biological and biomedical considerations, in: *Transdermal Drug Delivery: Developmental Issues and Research Initiatives* (J. Hadgraft and R. H. Guy, eds.), Dekker, New York, pp. 1–22.

Chapter 9

Rectal and Vaginal Absorption of Peptides and Proteins

*Shozo Muranishi, Akira Yamamoto,
and Hiroaki Okada*

1. INTRODUCTION

Peptide/protein drugs are increasingly becoming a very important class of therapeutic agents as a result of our gaining more understanding of their role in physiology
and pathology as well as the rapid advances in the field of biotechnology/genetic
engineering. These drugs are generally not suitable for oral administration, since they
are poorly absorbed and easily degraded by proteolytic enzymes in the gastrointestinal tract (Lee and Yamamoto, 1990). For systemic delivery of peptide and protein
drugs, parenteral administration is currently required in order to achieve their
therapeutic activities. However, these administration routes are poorly accepted by
patients and may cause an allergic reaction. Thus, alternative routes such as nasal
(Hirai *et al.*, 1981a), buccal (Ishida *et al.*, 1981), rectal (Nishihata *et al.*, 1983),
vaginal (Okada *et al.*, 1982), conjunctival (Yamamoto *et al.*, 1989), and transdermal
(Liu *et al.*, 1988) are being investigated for peptide and protein delivery. Among
these routes, rectal and vaginal are potentially important routes for peptide administration, although they are poorly accepted in several countries. In contrast to the oral
route, the rectal delivery of peptide/protein drugs provides the advantage of greater

Shozo Muranishi and Akira Yamamoto • Department of Biopharmaceutics, Kyoto Pharmaceutical
University, Misasagi, Yamashina-ku, Kyoto, 607, Japan. *Hiroaki Okada* • DDS Research Laboratories, Research Pharmaceutical Division, Takeda Chemical Industries, Ltd., Osaka, 532, Japan.

Biological Barriers to Protein Delivery, edited by Kenneth L. Audus and Thomas J. Raub. Plenum Press,
New York, 1993.

systemic bioavailability, especially with the coadministration of adjuvants. Additional advantage is the avoidance of first-pass elimination (de Boer *et al.*, 1980). On the other hand, in the vaginal membrane, the predominant transport of lipophilic substances takes place by partition and diffusion through the transcellular lipoidal passage.

In this chapter, we introduce the anatomical and physiological aspects of the rectum and vagina, which are related to the characteristics of absorption of drugs and peptides from these routes. We also review the effect of absorption promoters on peptide absorption from the rectum and vagina and their mechanisms of enhancing action. Further, we describe the contribution of the transcellular and paracellular routes and lymphatic transport of drugs via a rectal route and the metabolic property of peptides in the rectal mucosa, which is another important barrier for systemic delivery of these drugs.

2. RECTAL BARRIER

2.1. Anatomy and Physiology of the Rectum

The rectal epithelium is columnar or cuboidal with numerous goblet cells. However, unlike the small intestine, the rectal epithelium does not contain villi. In addition, the mucus-containing goblet cells are interspersed in an organized fashion among absorptive cells in the small intestine, gradually increasing in number toward the large intestine. In the descending colon of humans, goblet cells are numerous, comprising one-eighth of the epithelial cell population (Forstner, 1978). The human rectum has a length of 5 inches and a surface area of only about 200 to 400 cm^2, compared with 2,000,000 cm^2 for the small intestine (Wilson, 1962). Consequently, from a surface area consideration, one would expect absorption to be much less from the rectum than from the upper gastrointestinal tract. Indeed, the rectal bioavailability of peptide drugs is generally very low; e.g., less than 1% for the nonapeptide leuprolide in women, 5–16% for the tetrapeptide tetragastrin in rats, and about 1% for the tripeptide thyrotropin-releasing hormone (TRH) in humans (Hoogdalem *et al.*, 1989). For the polypeptide insulin, a pharmacological availability of 5.3% was observed, using suppositories in rabbits. Therefore, larger polypeptides and proteins require absorption promoters in order to improve the rectal absorption of these macromolecules.

On the other hand, in the vasculature of the rectum, the upper venous drainage system (superior hemorrhoidal vein) is connected to the portal system, whereas the lower venous drainage system (inferior and middle hemorrhoidal veins) is connected directly to the systemic circulation by the ileac vein and vena cava (Banga and Chien, 1988). Thus, an opportunity to reduce the extent of hepatic first-pass elimination exists in the rectum, especially when the drug is administered in the lower region of the rectum (de Boer *et al.*, 1980; Deleede *et al.*, 1983). The rectum also has a large number of lymphatic vessels which offer an opportunity to target drug delivery to the lymphatic circulation (Banga and Chien, 1988).

2.2. Rectal Absorption of Peptides and Proteins

2.2.1. ABSORPTION CHARACTERISTICS OF DRUGS IN THE RECTUM

In general, intestinal absorption of a number of organic compounds could be explained by the pH partition hypothesis (Hogben *et al.*, 1959; Shore *et al.*, 1957). Our previous studies showed that the rectal absorption of sulfonamides, for example, is good agreement with this theory (Kakemi *et al.*, 1965a,b,c). However, we found some discrepancies when the intestinal transport of salicylic acid and barbituric acid was studied. Figure 1 shows pH profiles of the intestinal absorption of salicylic acid in small intestine, colon, and rectum. The absorption amount versus pH profiles of salicylic acid showed the greatest difference in the small intestine, followed by the colon and the rectum, as determined from the curve of the nonionized fraction versus pH (Kakemi *et al.*, 1969). Consequently, it is likely that, of the entire intestine, absorption via the rectal mucous membrane is most consistent with the pH partition theory.

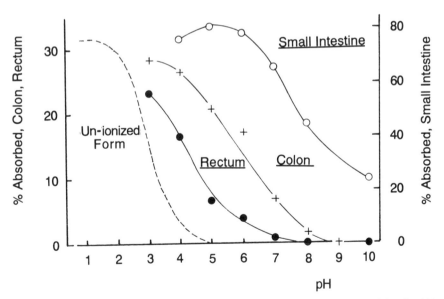

Figure 1. Comparison of absorption rate–pH relationship of salicylic acid in the rectum (●), colon (+), and small intestine (○) of rat.

In addition, colorectal absorption is unlikely to provide a carrier-mediated transport. It was reported that absorption of 5-fluorouracil and nutrient-like substances such as galactose, L-tyrosine, and vitamin B_{12} from the small intestine was a saturable process to transport against a concentration gradient, while rectal and colonic absorption of these drugs did not exhibit a carrier-mediated transport system (Binder, 1970; Cordero and Wilson, 1961; Muranishi *et al.*, 1979).

We therefore speculated from these data that drugs may be transported from the colorectal mucous membrane by a simple diffusion without a carrier-mediated mechanism.

2.2.2. ABSORPTION OF PEPTIDES AND PROTEINS AND EFFECT OF ABSORPTION PROMOTERS

Extensive studies have been conducted regarding the rectal absorption of peptides and proteins, especially insulin. However, in the absence of an absorption-promoting adjuvant, the rectal absorption of these drugs is much less than for i.m., i.v., or s.c. administration. Incomplete absorption is probably due to a combination of poor membrane permeability and metabolism at the absorption site (Lee and Yamamoto, 1990). Thus, a number of absorption promoters have been utilized for improving rectal absorption of larger polypeptides and proteins (Ichikawa *et al.*, 1980; Kamada *et al.*, 1981; Yoshioka *et al.*, 1982; Sithigorngul *et al.*, 1983; Miyake *et al.*, 1984, 1985; Murakami *et al.*, 1988b; Yoshikawa *et al.*, 1986; Hoogdalem *et al.*, 1989). Examples of rectal absorption of peptides and proteins with various absorption promoters are listed in Table I. Of the absorption promoters, the most efficacious rectal promoters that have been studied include salicylate, medium-chain glyceride, enamines, and mixed micelles. The mechanisms by which the absorption of peptides was improved by these promoters, especially mixed micelles, are discussed in detail below.

Among the peptides and proteins summarized in Table I, insulin is probably the most often studied protein with respect to rectal absorption. Nishihata *et al.* (1983) examined the effect of salicylates on rectal absorption of insulin in dogs. They found that sodium salicylate and 5-methoxysalicylate both increased the rectal absorption of insulin.

The absorption-promoting effect of sodium 5-methoxysalicylate was also studied in the rat with respect to rectal delivery of pentagastrin and gastrin. Rectal bioavailability was measured by direct comparison of pharmacological effect with intravenous dose response. As shown in Fig. 2, coadministration of the absorption adjuvant greatly enhanced the rectal bioavailability of the model peptides (Yoshioka *et al.*, 1982). The bioavailability of pentagastrin and gastrin in the absence of absorption promoter was 6 ± 4 and 0, respectively, while the bioavailability of these peptides increased 33 ± 10 and 18 ± 7 with the adjuvant.

Concerning the regional differences of peptide absorption in various nonoral

Table I

Enhancement of Rectal Absorption of Peptides/Proteins by Various Absorption Promoters

Peptides/proteins	Absorption promoters	Animals
Insulin	Various surfactants	Rabbits
	Bile acid	
	Phospholipid	
	Enamine derivatives	Rabbits, rats, dogs
	Sodium salicylate	Dogs
	Sodium 5-methoxysalicylate	
Gastrin	Sodium 5-methoxysalicylate	Rats
Pentagastrin		
Lysozyme	Enamine derivatives	Rabbits
Heparin		
$(Asu^{1,7})$-eel calcitonin	Enamine derivatives	Rats
	Sodium salicylate	
Human epidermal growth factor	Sodium caprate	Rats
	CMC Na	
Interferon (human fibroblast interferon)	Mixed micelle (linoleic acid, HCO60)	Rats
Des-enkephalin-γ-endorphin	Medium-chain glyceride	Rats
	Na_2-EDTA	

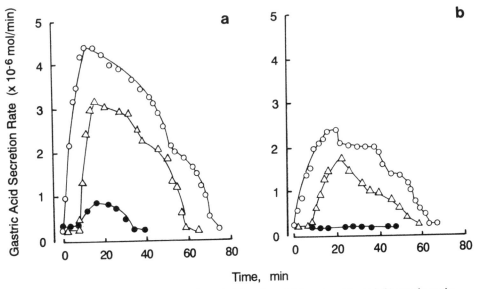

Figure 2. Gastric acid secretion following administration of (a) pentagastrin and (b) gastrin to the anesthetized rat. The solutions were administered: i.v. with 125 μg polypeptide/kg rat in 0.1 ml of saline, pH 8.0 (○), and rectally with 125 μg polypeptide/kg rat in the absence (●) and presence (△) of 25 mg of 5-methoxysalicylate/kg rat in 0.1 ml of saline, pH 8.0.

routes, few studies have determined the rectal absorption of peptides. Recently, Aungst *et al.* (1988) compared insulin absorption from various noninjection sites of administration and found that rectal insulin was more efficacious than nasal, buccal, and sublingual insulin, when administered without absorption-promoting adjuvant (Fig. 3). Similarly, rectal administration of an LH-RH analogue was more efficacious than nasal administration, but less so than i.v. or s.c. injections in rats (Okada *et al.*, 1982). Consequently, it was determined that absorption by the rectal route is better than others.

The mechanisms whereby peptide and protein absorption was improved by absorption promoters were examined from various aspects. These mechanisms involve increase in membrane fluidity, interaction with the ability of calcium ions to maintain the dimension of the intracellular space, solubilization of the mucosal membrane, increase in water flux, reduction of the viscosity of the mucus layer adhering to all mucosal surfaces (Lee, 1986). Further, for peptides and proteins, inhibition of peptidase activity is also an important factor in improving the absorption of these drugs (Hirai *et al.*, 1981b).

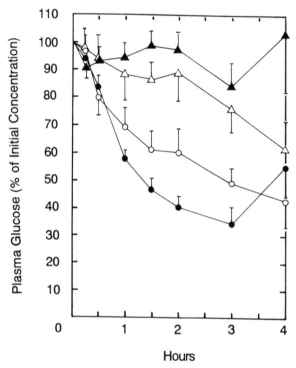

Figure 3.   Plasma glucose concentrations in rats administered 50 U/kg of insulin nasally (○), rectally (●), bucally (△), and sublingually (▲).

We have mainly investigated the effect of mixed micelles on the intestinal absorption of drugs and their mechanisms of enhancing action. Generally, the barrier to movement of drug across the membrane is provided by two major components of the membrane: lipids and proteins. We have studied the influence of the promoters on the lipid bilayer of egg phosphatidylcholine. The incorporation of oleic acid, linoleic acid, or monoolein into the liposomes increased the release rates of procainamide ethobromide, an impermeable compound, while the saturated long-chain fatty acids produced less of an increase (Muranishi et al., 1980). In addition, we have shown by electron spin resonance (ESR) and nuclear magnetic resonance (NMR) that the fatty acids can interact with phosphatidylcholine (Muranishi et al., 1981). The NMR peaks of the phosphate and olefin protons disappeared, and the choline methyl resonance was broadened. These findings suggest that the increase in membrane permeability caused by the fatty acid is associated with the disorder in the interior of the membrane and interaction between the incorporated fatty acid and polar head group of the phospholipid. However, enhanced permeation through the intestinal epithelium cannot be entirely explained by destabilization of the lipid bilayer.

We next studied the contribution of membrane-bound proteins to the enhancement of intestinal permeability using several sulfhydryl (SH)-modifying reagents (Murakami et al., 1988a). Carboxyfluorescein (CF), a poorly absorbable drug, was used as a model drug and the absorption-enhancing effect of oleic acid on the large intestine was estimated by the in vitro permeability of CF using the perfusion technique. The concentration of CF in the serosal effluent was markedly increased about 15 min after administration of the oleic acid micellar solution. N-Ethylmaleimide (NEM), which is one of the SH-modifying agents (SH-modifiers), suppressed the enhancing effect of oleic acid in a concentration-dependent manner. This finding suggested that a SH-related substance is involved in the permeability-enhancing effect of oleic acid. Further, it was found that NEM produced a concentration-dependent decrease in both nonprotein and protein-bound SH level (Fig. 4). On the other hand, diethylmaleate (DEM), an agent known to covalently complex and deplete glutathione, was used in the same experiment of enhanced absorption of oleic acid. DEM did not affect the absorption of CF, and depressed the cellular nonprotein-SH level but not the cellular protein-SH level (Fig. 4). Oleic acid did not affect either the SH level of protein or nonprotein (Fig. 4). The results of the experiment on SH level determination suggest that the SH level of cellular protein is related to the promoting effect of oleic acid.

2.2.3. CONTRIBUTION OF TRANSCELLULAR AND PARACELLULAR ROUTES

Intestinal epithelia have two permeation routes for the absorption of small ions and water: a paracellular route through the junctional complex that binds the cells together and a transcellular route through the lipoidal membrane. Lipid-soluble

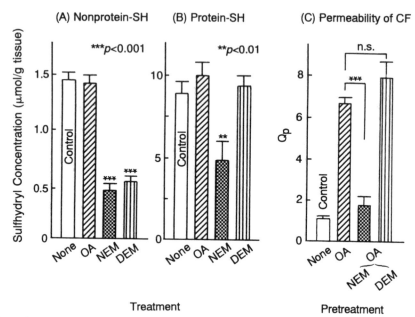

Figure 4. Effect of oleic acid (OA), *N*-ethylmaleimide (NEM), and diethylmaleate (DEM) on tissue sulfhydryl content and permeability of carboxyfluorescein (CF).

organic compounds are believed to pass mostly across the lipoidal barrier (transcellular pathway) of the epithelium. Ionized or polar organic solutes are absorbed with difficulty, probably via both permeation routes. Polar macromolecules cannot be absorbed by either route. To elucidate the transport process of macromolecules, we have used three kinds of colorectal absorption experiments: an *in vitro* everted loop, isolated epithelial cells, and brush border vesicles (Muranishi *et al.*, 1986; Masuda *et al.*, 1986).

The experiment using the perfused everted loop involved permeation through two routes. In contrast, the uptake into the isolated epithelial cells involves only transmembrane movement and includes brush border membrane transport as well as basolateral membrane transport. The uptake into brush border membrane vesicles, prepared by the method of Kessler *et al.* (1978), involves only permeation of the mucosal surface membrane of the cells.

We have evaluated the permeation-enhancing effect on the large intestine of rats using various sizes of fluorescein isothiocyanate (FITC) dextrans as nonabsorbable markers of the transport of macromolecules (Muranishi, 1989). The amounts transferred from the mucosal and serosal side of the everted loop preparation of the large intestine *in vitro* were measured for 1 hr. As a result, the amount appearing on the serosal side decreased with increasing molecular weight from 10 to 70 kDa either with or without an enhancer, linoleic acid–polyoxygenated castor oil (HCO60) mixed micelle (MM). The degree of enhancement was greater for small than for larger dextrans. In studies using isolated epithelial cells of the large intestine, the uptake of smaller dextrans, but not the largest dextran (FD-70), was also much increased by the enhancer. The ratios of FD permeated or taken up with MM, to that without MM, were calculated and compared. As shown in Fig. 5, the uptake into isolated epithelial cells was less than that for the everted loop. There were some differences in the two systems for any given FD; the biggest difference occurred for FD 10 and a slight difference was found even for 70 kDa. In the study of brush border membrane vesicles, the uptake of FDs smaller than 40 kDa was promoted. These results suggested that the brush border membrane of the epithelial cells could become permeable to the macromolecules in the presence of MM. Based on the above permeation study, the participation of paracellular as well as transcellular transport can be considered important in the enhancing effect of MM on the colorectal absorption of macromolecules.

2.2.4. LYMPHATIC TRANSPORT OF PEPTIDES

After passage through the mucosal cells of the intestinal tract, there are both blood and lymph routes by which a compound may be transported before systemic

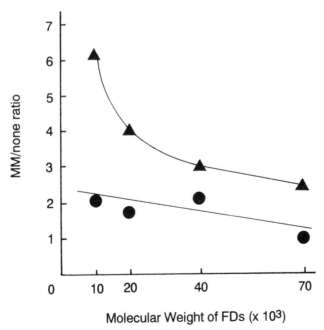

Figure 5. Comparison of enhancing effect by MM between isolated cells and everted loop. ●, isolated cells; ▲ everted loop.

distribution. In general, most of the water-soluble, low-molecular-weight drugs such as sulfonamides, *p*-aminosalicylic acid, and antipyrine show no selective lymph absorption when administered into the small intestine (DeMareo and Levine, 1969; Sieber *et al.*, 1974). It was demonstrated that lymph/plasma (L/P) concentration ratios of these compounds were almost constant over time. However, the lymph pathway is known to play an important role in the absorption of highly lipophilic compounds, e.g., cholesterol, triglyceride, lipid-soluble vitamins, DDT, Sudan blue, and naftidine (Muranishi, 1985). Some efforts have been made to enable a lymphotropic delivery of hydrophilic compounds by designing a lipophilic prodrug, for the purpose of bypassing the liver or preventing lymph metastasis. However, long-chain fatty acid 5-fluorouracil diglycerides designed in our laboratory did not show selective lymphotropic transport. Presently, it is difficult to design a lymphotropic compound for such a hydrophilic drug.

On the other hand, macromolecules such as proteins, dextrans, dextran sulfate, and enzymes, when absorbed intact, are transported selectively by the lymphatic system (Muranishi, 1985). First, FITC-labeled dextrans of various sizes above 10 kDa were chosen as water-soluble macromolecules in order to evaluate the blood–lymph selection mechanism for higher-molecular-weight compounds (Yoshikawa *et al.*, 1984a). The maximal plasma level decreased with increasing molecular weight, whereas the maximal lymph levels remained high irrespective of molecular weight. Therefore, the lymph/plasma ratio was elevated by an increase in molecular weight. We can speculate from these results that the marked increase in lymph/plasma ratios of FD with an increase in their molecular weight is responsible for the difficulty of FD transfer into the blood circulation.

The results obtained in this study indicate that the lymph levels of FD over 17.5 kDa were significant higher than the plasma levels. On the other hand, absorption experiments in the small intestine using the same method showed that the lymph levels of FD were significantly higher than the plasma levels at molecular sizes over 39 kDa. These findings suggest that the threshold of molecular size for permselectivity is between 17.5 and 39 kDa in the small intestine, and between 10.5 and 17.5 kDa in the large intestine.

We also considered the use of an ion-pair complex with dextran sulfate (DS) as the lymphotropic carrier, which is an anionic high-molecular-weight compound. Bleomycin, a cationic glycoprotein that is an anticancer agent, was chosen for complexing with DS, and a lipid–surfactant mixed micelle was used as an effective intestinal absorption promoter for unabsorbable high-molecular-weight compounds (Yoshikawa *et al.*, 1981, 1983). When bleomycin was administered alone, monoolein-taurocholate mixed micelle induced a marked absorption of bleomycin from the intestine; however, its concentrations in the blood and the lymph were almost identical. On the other hand, administering the bleomycin–DS complex together with the mixed micelle selectively produced a very high lymphatic concentration. This lymphotropic selectivity, observed by complexing bleomycin with dextran, was more effective in the large intestine than in the small intestine. Its mechanism may be due to a molecular sieving in the blood–lymph barrier in intestinal tissue. The bleomycin–dextran complex remained stable in the lumen of the large intestine.

Recently, the application of an absorption promoter has been attempted in the delivery of interferon (IFN) via the colorectal route (Yoshikawa et al., 1984b, 1985a,b, 1986). It is reasonable to suggest that little uptake of IFN from the gastrointestinal tract occurs due to its large molecular size. After the administration of human (h) IFN-β–saline solution alone to the large intestine, hIFN-β was not detected in either the serum or the lymph for 5 hr. Neither linoleic acid nor polyoxyethylene derivative of hydrogenated castor oil (HCO60) promoted the absorption of hIFN-β. However, the administration of a mixed micellar system combined with the above components led to a high concentration of hIFN-β in the lymph, but minimal in the serum (Table II). The ratio of hIFN in the lymph to that in serum ranged from 17 to 48 for 5 hr after administration. These ratios were much higher than those of dextrans (D_{20} and D_{40}) having a molecular weight close to that of hIFN-β. The extremely high lymphotropic property of hIFN-β might be attributed to such factors as the shape and configuration of the molecule, in addition to its molecular weight.

2.2.5. METABOLIC PROPERTIES OF RECTAL EPITHELIUM

The far-from-complete bioavailability of peptides and proteins from various mucosal routes, even in the presence of absorption promoters, suggests that there is at least one more reason limiting peptide and protein absorption, namely that these peptides are still hydrolyzed by the peptidases in the respective mucosae (Lee and Yamamoto, 1990).

In the intestinal brush border membrane, the most abundant peptidase is aminopeptidase, accounting for up to 3.5–8% of its total protein content. This enzyme has been shown to be primarily responsible for completing the hydrolysis of orally administered peptides and proteins. However, there have been few reports on the activity of this peptidase in the various absorptive mucosae, or how the activity in the rectum differs from the aminopeptidase activity in the other absorptive

Table II
Cumulative Amount of Human IFN-β at 6 hr
after Administration in Lumen of Rat Large Intestine

Adjuvant	Cumulative amount (IU)
MM	1487.3 ± 233.2[a]
Saline control	74.8 ± 7.1
HCO60	97.0 ± 24.7
Linoleic acid	96.4 ± 23.4

[a]Values are mean ± S.E.

mucosae. Stratford and Lee (1986) have recently developed a method to determine the type and activity of aminopeptidases in homogenates of various absorptive mucosae, so as to define the aminopeptidase barrier to peptide absorption from the nonoral routes. 4-Methoxy-2-naphthylamides of leucine, alanine, arginine, and glutamic acid were used as substrates. Based on the pattern of substrate hydrolysis and the effect of activators and inhibitors on the rate of substrate hydrolysis, four to five aminopeptidases were estimated to be present in the mucosal homogenates studied. As a result, aminopeptidase activity in the rectal mucosa did not differ substantially from that in various nonoral mucosae.

In addition, Dodda-Kashi and Lee (1986) determined the hydrolytic rate of methionine enkephalin (YGGFM), leucine enkephalin (YGGFL), and (D-Ala^2)-Met-enkephalinamide (YAGFM) in homogenates of various absorptive mucosae. These enkephalins were most rapidly hydrolyzed in the rectal and buccal homogenates, followed by the nasal and then the vaginal homogenate, but the differences in the rates were small. These findings suggest that the same enzymatic barrier to enkephalin absorption possibly exists in both the rectal and other nonoral mucosae.

We have examined the rate of hydrolysis of insulin and proinsulin in homogenates of various nonoral absorptive mucosae (Yamamoto *et al.*, 1990). As summarized in Table III, proteolytic activity for insulin was highest in the nasal and rectal homogenates, followed by the ileal, vaginal, conjunctival, and buccal homogenates in that order. Therefore, in the rectum, insulin and proinsulin are expected to be degraded more rapidly than are conjunctival and buccal homogenates. There was no marked difference in the rate of hydrolysis of these peptides between rectal and another intestinal mucosa (ileum). From these results, it was suggested that peptides and proteins were relatively degraded in the rectal mucosa as well as the intestinal and other nonoral mucosae, although the rectal route was believed to have low protease activity.

Table III

Half-lives for the Hydrolysis of Insulin and Proinsulin in
Homogenates of Various Absorptive Mucosae of the Albino Rabbit

Mucosa	Half-life (min)	
	Insulin	Proinsulin
Nasal	44.8 ± 4.9[a]	70.5 ± 4.2
Buccal	268.3 ± 21.4	395.8 ± 112.0
Ileal	98.1 ± 6.4	55.7 ± 7.0
Rectal	71.6 ± 11.6	122.7 ± 18.3
Vaginal	106.0 ± 16.2	163.2 ± 24.3
Conjunctival	234.5 ± 16.0	186.7 ± 8.9

[a]Values are mean ± S.D.

2.3. Conclusion

Although rectal drug formulations do not have the level of patient acceptability in most countries, this route offers some advantages for systemic delivery of peptide and protein. First of all, it is possible to avoid hepatic first-pass elimination by rectal administration of peptides, since the lower part of the rectal venous drainage is directly connected to the general circulation. Second, rectal absorption of peptides is typically very low due to poor membrane permeability characteristics and extensive hydrolysis in the rectal mucosa. However, enhancement by absorption promoters was more predominant in the large intestine than in the small intestine. Therefore, this apparent sensitivity of the colorectal mucous membrane to the adjuvants is an attractive characteristic for rectal dosage form design. Third, the selective transfer of macromolecules and drug–macromolecule conjugate into the lymphatic system was achieved after intestinal administration. This lymphotropic selectivity of the conjugate was more effective in the large intestine than in the small intestine. In summary, it was suggested that the rectal route may be applicable for the systemic delivery of peptides and proteins.

3. VAGINAL BARRIER

3.1. Anatomy and Physiology of the Vagina

The vaginal wall membrane in humans consists of an epithelial layer (epithelial lamina and lamina propria), muscular layer, and tunica adventitia, and is regulated by cyclic alteration of the reproductive system, which is directly controlled by hormones such as estrogens, progesterone, LH, and FSH (Fig. 6) (Ganong, 1983).

The epithelium before puberty is very thin but following puberty it increases in thickness with estrogen activity (Walz *et al.*, 1978). In the adult stage, the vaginal surface during the follicular phase appears homogeneous with large superficial polygonal cells with a high degree of proliferation caused by estrogen stimulation and the presence of cornification. The narrow intercellular edges are of a dense structure, corresponding to the cellular cluster of a nonkeratinized squamous epithelium. This proliferation of cells concomitantly increases the epithelial thickness and number of layers (Fig. 6) (Burgos and Roig de Bargas-Linares, 1978). The vaginal epithelium is composed of five different cell layers: basal (single row), parabasal (two rows), intermediate (about ten rows), transitional (about ten rows), and superficial (about ten rows). The cyclical variations of the vaginal epithelium generally involve proliferation, differentiation, and desquamation. The intermediate, transitional, and superficial layers are strongly affected by the cycle and become the thickest layers at

[Menstrual cycle]

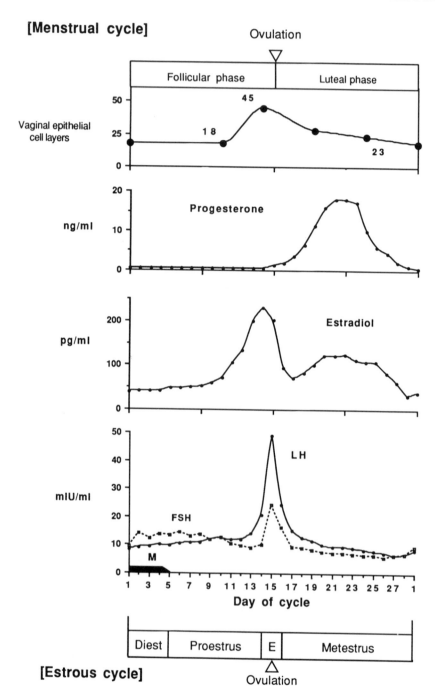

[Estrous cycle]

ovulation. During the luteal phase, desquamation occurs on the superficial epithelial layer extending as far as the intermediate cells. This cyclic desquamation is preceded by loosening of intercellular grooves and a pore-like widening of intercellular crevices following ovulation. The intercellular channels are narrow during the early follicular phase, but become widely opened at ovulation and during the luteal phase. During the luteal phase the protein pool of the lamina propria, i.e., albumin and globulins, transudes through the dilated intercellular channels.

The lamina propria specialized supporting structure for the epithelial cells contains a blood supply, a lymphatic drainage system, and a network of nerve fibers. It is composed of a connective tissue, consisting of collagen fibers, ground substance, and cells. The ground substance represents a pool of proteins for nutrition of the vaginal mucosa. The cells are fibroblasts, macrophages, mast cells, lymphocytes, plasma cells, neutrophils, and eosinophils. Each cell plays an important role in the physiology of the vagina. Leukocytes penetrate the intercellular channels and separate the vaginal epithelial cells, opening desmosomes and dilating the intracellular space, and mechanically detach the surface epithelium in the desquamation process during the luteal phase.

In postmenopausal women the vaginal epithelium becomes extremely thin, cell boundaries in the surface are less distinct, the microridges are reduced, and the vagina is often invaded by leukocytes. During pregnancy the most marked change in the vagina is the increased vascularity and venous stasis. The epithelial layer is greatly thickened. On the extended vaginal surface, distended and densely convoluted intercellular microridges and tender cellular borders have been seen. Following delivery, the vagina requires several weeks to reestablish its prepregnancy appearance.

The vagina epithelium is aglandular but is usually covered with a surface film of moisture (Wagner and Levin, 1978). The pH of the vaginal lumen is controlled mainly by the lactic acid produced from the cellular glycogen by the action of the normal microflora, Doderlein's bacilli. The pH is neutral until the onset of puberty but following the latter, falls and varies between 4.0 and 5.0 depending on the ovarian cycle and is the highest during menstruation. The lowest values are found near the anterior fornix and the highest ones near the vestibule. This acidity plays a clinically important role in preventing the proliferation of pathogenic bacteria. During pregnancy the pH varies between 3.8 and 4.4 owing to the increase in cellular glycogen content, whereas in the postmenopausal state the decrease in cellular glycogen causes elevation of the pH to 7.0–7.4.

The arterial blood supply in the vagina is derived from the visceral branches of the internal iliac artery, and venous drainage occurs mainly via the uterine vein to the internal iliac vein (Platzer et al., 1978). Fine-meshed networks of lymph capillaries

←———

Figure 6.   Plasma hormone concentrations during a normal 28-day human menstrual cycle and timing of the rodent estrous cycle (4–5 days). M, menstruation; E, estrus.

are located in the lamina propria and tunica muscularis of the vagina, and lymph drains to the iliac, sacral, gluteal, rectal, and inguinal lymph nodes.

Among nonhuman primates, monkeys and baboons exhibit an ovarian cycle and reproductive system similar to that of the human female. The rat, mouse, and guinea pig have short estrous cycles which are completed in 4 to 5 days (Fig. 6) (Turner and Bagnara, 1976). The rat vaginal mucosa during diestrus is thin, and leukocytes migrate through it. During proestrus the vaginal wall is the thickest and consists of fresh cells. During estrus many mitoses occur in the vaginal mucosa as a result of estrogenic action, and the superficial layers become squamous and cornified. During metestrus many leukocytes appear in the vaginal lumen along with a few cornified cells. Ovariectomy causes a marked involution of the vagina; the vaginal mucus becomes thin and mitotic division is seldom encountered. Adult nonpregnant domestic rabbits are in a constant state of estrus, and ovulation is induced by coitus or some comparative cervical stimulation. The ovaries of wild rabbits, however, are inactive in the winter but enlarge in the spring. Cats and ferrets exhibit a similar type of estrous cycle. Ovulation in dogs is spontaneous, and there are generally two estrous periods per year. Proestrus lasts for about 10 days and is followed by a 6- to 10-day estrus. Loss of blood occurs through the vagina during proestrus. Each estrus is followed by a functional luteal phase lasting approximately 60 days.

3.2. Vaginal Absorption of Peptides and Proteins

3.2.1. ABSORPTION OF PEPTIDES AND PROTEINS

The first evidence of the absorption of peptides through the vaginal epithelium as a clinical approach to drug delivery was reported using insulin in depancreatized dogs (Fisher, 1923) and cats (Robinson, 1927). It is now known that several peptide hormones, antigenic proteins, and penicillin are absorbed intact through the epithelial membrane, and the bioavailability is greater than that by the oral route due to high intercellular permeability and reduced first-pass effect (Aref *et al.*, 1978; Benziger and Edelson, 1983; Okada, 1991). TSH, peanut protein, and bovine milk immunoglobulins have been reported to be absorbed in an intact form through the vaginal membrane in the human or baboon. The vaginal absorption of bacterial antigens, viral DNA particles, and the sperm head caused local secretion of the respective specific antibodies and strong local immunological responses. In these systems, the vaginal intercellular channels play an essential role in the passage of different immunogens from the vaginal lumen to the lamina propria and the blood or lymphatic vessels. Absorption of bacterial and viral antigens through the epithelium is important in the local production of antibodies to prevent infection of the genital organs. It was recently discovered that antibodies against the sperm head prevent sperm from penetrating the cervical mucus to cause immunological infertility (Wang *et al.*,

1985). However, in the development of drug delivery systems it must be noted that antibodies induced following chronic administration possibly neutralize peptides and proteins or prevent their penetration as explicitly demonstrated in the gastrointestinal tract (Walker and Isselbacher, 1977) and nasal mucosa (Spit *et al.*, 1989).

Most of the vaginal preparations on the market are used for a topical effect on the vaginal membrane, such as antibacterial, antifungal, or antiviral agents. To obtain systemic effects the vaginal application of insulin and LH-RH analogues was recently investigated. Synthetic LH-RH (Humphrey *et al.*, 1973), (D-Ala[6], des-Gly[10]) and (D-Leu[6], des-Gly[10]) LH-RH ethylamide (leuprolide) (Nishi *et al.*, 1975; De La Cruz *et al.*, 1975) were absorbed from the vaginal lumen in ovariectomized steroid-blocked rats and intact immature female rats. The vaginal application of these peptides induced a greater elevation in serum LH and FSH levels than did oral administration. Application of a tablet containing 2 mg of leuprolide in the posterior fornix of the human vagina during the early or midfollicular phase elevated the plasma levels of gonadotropins and estrogen, corresponding to about 0.6% bioavailability (Saito *et al.*, 1977). When insulin in Cetomacrogol 1000/PEG 400 was administered vaginally to rats with streptozotocin-induced diabetes, the blood glucose was reduced to 50% of the initial level 4 hr later; this effect was less than achieved by the rectal route (Touitou *et al.*, 1978). Rapid and potent hypoglycemic effects were produced in rats and rabbits with alloxan diabetes by the vaginal administration of insulin suspended in a polyacrylic acid jelly (Morimoto *et al.*, 1982).

A series of studies on the vaginal absorption of leuprolide (leuprorelin) has been carried out in rats to establish a rational dosage form for treating hormone-dependent mammary tumors (Okada, 1983). Figure 7 shows the pharmacological effect in diestrous rats after leuprolide was administered by seven different routes (Okada

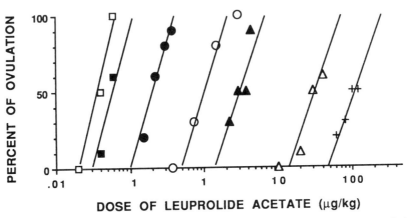

Figure 7. Ovulation induction by leuprolide after administration intravenously (□), subcutaneously (■), vaginally (○), vaginally with 10% citric acid (●), rectally (▲), nasally (△), and orally (+) in diestrous rats. (From Okada *et al.*, 1982.)

et al., 1982). The absolute bioavailability was 0.05% by the oral route (in a mixed micellar solution), 1.2% by the rectal route, and 0.11% and 1.8–3% by the nasal route (without and with an absorption promoter; 1% sodium glycocholate, surfactin, or laureth-9). The bioavailability by the vaginal route was relatively large, 3.8%, not using an absorption promoter. The nasal absorption in this experiment was likely underestimated as a consequence of drainage of the test solution; the bioavailability estimated in the other experiments was 18.7% in rats with the orifice and outlet of the nasal cavity closed and 2.9% in men by a nasal aqueous drop without a promoter (unpublished data). With mucosal application of drugs, retention of the drug at the site of absorption is very important for achieving precise and sufficient absorption (Harris *et al.*, 1988; Illum *et al.*, 1990).

The barrier properties of the nasal, rectal, and vaginal membranes in the rabbit were investigated *in vitro* (Corbo *et al.*, 1990). Progesterone (a lipophilic marker) penetrated through the vaginal membrane more rapidly than mannitol (a hydrophilic marker), and mannitol more easily penetrated through the vaginal than the rectal membrane. The nasal mucosa allowed the greatest transport of both model compounds.

Regarding the enzymatic barrier, few enzymes have been found in the vaginal epithelium and most avoid the vaginal blood drainage by introduction to the portal vein and liver. Reduced first-pass effects after vaginal application of estrogens (Rigg *et al.*, 1978) and propranolol (Patel *et al.*, 1983) have been reported. Recently, comparative studies on the peptidase activity against enkephalins, substance P, insulin, and proinsulin in the absorptive mucous membranes in the rabbit have demonstrated that the supernatants of homogenates of the vaginal, nasal, buccal, rectal, and ileal mucous membranes exhibit similar proteolytic activity (Lee, 1988; Lee and Yamamoto, 1990). For small peptides, the rectal route is the most active, followed by the buccal, nasal, and vaginal routes. For proteins, the nasal route is more active than the rectal, vaginal, and buccal routes. However, to avoid the enzymatic barrier the morphological organization of the enzymes during the penetration process should be considered as well as the gross activity; e.g., pancreatic proteases localized on the glycocalyx mucus in the intestine, membranous or cytosolic proteolysis (80% cytosolic in the vaginal mucosa), and transcellular or intercellular transport.

3.2.2. EFFECTS OF CHANGES IN THE REPRODUCTIVE CYCLE

The effects of the ovarian cyclic changes on the vaginal absorption of peptides were clearly demonstrated using rats (Okada *et al.*, 1983b, 1984). After vaginal insertion of an insulin suppository in rats, a slight decrease in the glucose level was observed during proestrus, whereas a remarkable decrease was observed during metestrus and diestrus. The urinary excretion (percentage of dose) of phenol red, a water-soluble maker, within 6 hr after vaginal administration was 2.5% during proestrus, 5.5% during estrus, 37.5% during metestrus, and 31.4% during diestrus.

The vaginal absorption of the undissociated salicylic acid from a low-pH buffer was rapid and almost the same in proestrus and diestrus. The absorption of the ionized acid was different: 66% of the dose in 1 hr during diestrus and 29% during proestrus. These results indicate that the permeability for hydrophobic compounds is less affected by the ovarian cycle since the compound is transported mainly trans-cellularly, while hydrophilic compounds are transported mainly through intercellular channels. The vaginal absorption of leuprolide in rats was also remarkably influenced by the estrous cycle (Fig. 8). This effect can be explained by the changes in the thickness and pore-like pathway of the epithelium; the apparent porosity during diestrus is presumed to be more than ten times that during proestrus and estrus. In therapy using leuprolide, it is fortunate that continuous treatment halts the cycle at diestrus in rats, corresponding to the luteal and early follicular phases in humans. Therefore, during treatment the vaginal mucosa is thin, similar to that in diestrous or ovariectomized rats, and the absorption is enhanced with less variation (Okada *et al.*, 1984).

The effects of the reproductive cycle on the vaginal absorption of penicillin in humans have been reported (Rock *et al.*, 1947). After insertion of a vaginal suppository, high blood levels of penicillin, sufficient to be therapeutic, were found near the end of the menstrual cycle and during menopause, but absorption was somewhat diminished during the ovulation phase and late pregnancy. The vaginal membrane permeability for vidarabine, a hydrophilic antiviral agent, was found to be affected by the estrous cycle in the mouse (Hsu *et al.*, 1983) and guinea pig (Durrani

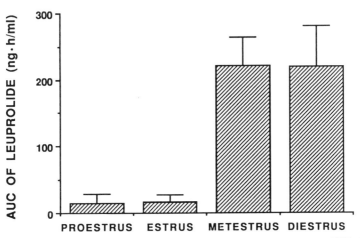

Figure 8. Effects of the estrous cycle on vaginal absorption of leuprolide in rats. AUC of serum leuprolide concentrations for 6 hr after vaginal administration at a dose of 500 μg/kg/0.2 ml of 5% citric acid solution (pH 3.5) (mean ± SE, $n=5$). (From Okada *et al.*, 1984.)

et al., 1985) in *in vitro* studies. The permeability coefficients were 10–100 times higher during early diestrus or diestrus than during estrus. These observations are well explained by the morphological changes in the vaginal epithelium during the ovarian cycle and pregnancy.

Estrogen therapy and steroidal contraceptives influence the vaginal fluid and epithelial thickness and vascularity, resulting in a change of the vaginal absorption of drugs (Sjoberg *et al.*, 1988; Pschera *et al.*, 1989). The influence of the ovarian cycle on the protease activity in the vagina was also demonstrated in rats (Havran and Oster, 1977) and women (Fishman and Mitchell, 1959). The trypsin-like activity in rat vaginal smears was found to be maximal at proestrus.

3.2.3. ABSORPTION ENHANCEMENT

Although the vagina is permeable to many peptides and proteins, the bio-availability may be insufficient for systemic therapy in most cases. Therefore, absorption enhancement is generally required to elicit more reliable therapy by the vaginal route.

The screening of promoters for the vaginal absorption of leuprolide has been carried out in diestrous rats (Okada *et al.*, 1982). The absorption was facilitated markedly by polybasic carboxylic acids (citric, succinic, tartaric, and glycocholic acids). The absolute bioavailability increased from 3.8% to about 20% when poly-basic carboxylic acids were added. Interestingly, the absorption was only poorly enhanced by surfactants such as sodium glycocholate, sodium oleate and laureth-9, which are known to enhance rectal and nasal absorption of hydrophilic drugs. Vaginal absorption of synthetic LH-RH and insulin in diestrous rats was enhanced 30- and 5-fold, respectively, by addition of citric acid (Okada *et al.*, 1983a). The absolute bioavailability of insulin after vaginal administration was elevated to 18%. The mechanism of the absorption enhancement caused by organic acids was assumed to result from their acidifying and chelating ability. The vaginal absorption of leuprolide increased as the pH of the solution decreased to 2.0. The enhancing potency of organic acids correlated well with their chelating ability. Distinct staining of the vaginal epithelial membrane by Evan's blue injected intravenously was elicited by treatment with carboxylic acids but not with their calcium salts (Okada, 1983). This observation indicates that the blood–vaginal epithelial barrier is loosened by organic acids, ascribed to the uptake of Ca^{2+} due to their chelating ability. The deep staining was observed 30 min and 1 hr after treatment with a 10% citric acid solution, but the stain gradually faded. The change in the vaginal epithelium produced by the acids reversed rapidly, indicating that the change in the epithelium was moderate and transient.

The tight junctions (zonula occludens, ZO) forming continuous zonular structures (strand network) circumscribing the cell apex provide a barrier to substances which cross the epithelial membrane through the paracellular space (Fig. 9). They

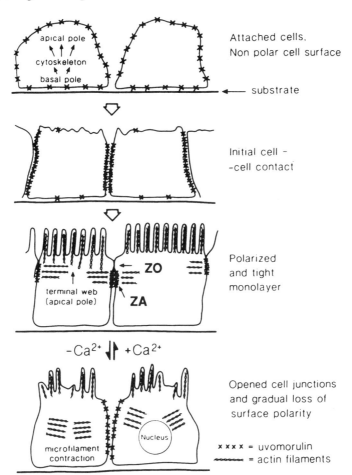

Figure 9.   Model for assembly and maintenance of epithelial tight junctions. ZO, zonula occludens (tight junction); ZA, zonula adherens. Opening of tight junctions by removal of Ca^{2+} results from Ca^{2+}-dependent loss of uvomorulin-mediated adhesion, causing dissociation of ZA. (From Gumbiner, 1987.)

can be very dynamic structures and can be influenced by the junctional complex elements including the zonula adherens (ZA), Ca^{2+}-dependent adhesion molecules (cadherins), desmosomes (spotlike junction), gap junctions, and actin filaments of the cytoskeleton (Gumbiner, 1987). Cadherins, N-, P-, and E-cadherin (uvomorulin), are found to be particularly important in the initial recognition between cells (differentiation) and in the cell adhesion of various tissues leading to the formation of more rigid intercellular structures such as tight and gap junctions (Hirano *et al.*, 1987). Uvomorulin provides the Ca^{2+}-dependent adhesive contact that forms the ZA.

Successively, under the influence of the actin filaments, the ZA becomes positioned in the apical region and provide the adhesive strength and localization necessary for the assembly of an organized ZO from individual tight junction elements (Fig. 9) (Gumbiner and Simons, 1986). The functional tight junctions can form very rapidly in the absence of protein synthesis; tight junction precursor molecules are readily available, perhaps in the epithelial plasma membrane, and can be rapidly recruited to form a complete ZO (Martinez-Palomo *et al.*, 1980). The integrity of the tight junctions is reversibly altered by extracellular Ca^{2+} concentration resulting indirectly from Ca^{2+} effects on other junctional elements like cadherins, rather than from direct effects on the ZO. The channel permeability of gap junctions, cell-to-cell channels that provide pathways for the direct flow of hydrophilic substances of up to 1–2 kDa, is also reversibly regulated by the intracellular Ca^{2+} and H^+ concentrations (Loewenstein, 1984).

Schuchner *et al.* (1974) demonstrated by ultrastructural observations that in the human vaginal epithelium the desmosomes open and the size of the intercellular spaces increases after daily treatment for 6 days with a tampon impregnated with a solution of 0.2 M EDTA (pH 3.5); these changes were reversible. Cho *et al.* (1989) demonstrated by monitoring the transepithelial electrical resistance that citrate may serve as a Ca^{2+} chelator thereby reversibly opening tight junctions in the cultured cell monolayer of canine kidney epithelium.

After vaginal application of a cotton ball soaked with a leuprolide solution containing 5% citric acid (pH 3.5), high and long-lasting serum levels of the peptide were observed in rats, the absolute bioavailability for 6 hr being 25.8% (Fig. 10) (Okada *et al.*, 1984). We demonstrated in rats that repeated daily vaginal administration of leuprolide (500 μg/kg, jelly containing 5% citric acid causes a strong desensitization of the pituitary (Okada *et al.*, 1983c) and significant regression of hormone-dependent mammary tumors induced by DMBA (Okada *et al.*, 1983d).

In preliminary human studies, persistent serum concentrations of leuprolide and saturated gonadotropin releasing response in the dosage range of 5–25 mg have been provided after vaginal insertion during day 8–10 of the cycle using the jelly described above; the maximal serum level was observed about 6 hr after insertion (unpublished data). Although therapeutic levels were provided at the 10 mg dose, the bioavailability compared with the subcutaneous route was about 1–4% (26 subjects) and fluctuated greatly: 10–90% in 6 subjects. Thus, it should be noted that the epithelial barriers in the human and experimental animals may be different; the number of cell layers of the vaginal epithelium varies from 18 to 45 in humans (Fig. 6) but is only 2 in ovariectomized rats (Richardson *et al.*, 1989).

Richardson *et al.* (1989) demonstrated marked enhancement of the vaginal absorption of gentamicin using lysophosphatidylcholine (LPC), palmitoylcarnitine (PLC), laureth-9, and citric acid in ovariectomized rats. Severe desquamation of the epithelium was observed with laureth-9 and LPC; this might be ascribed to the solubilization of lipid molecules in the tight junction by detergents (Helenius and Simons, 1975). PLC was the most effective absorption enhancer and showed only

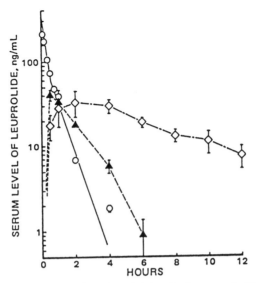

Figure 10. Serum leuprolide concentrations after administration intravenously (○), subcutaneously (▲), and vaginally (◇) in diestrous rats. Dose: intravenous and subcutaneous, 100 μg/kg; vaginal, 500 μg/kg/0.2 ml of 5% citric acid solution (pH 3.5) (mean ± SE, $n= 5$). (From Okada *et al.*, 1984.)

moderate epithelial damage. Citric acid enhanced the absorption fivefold, while causing only minor epithelial damage, similar to the results in our studies. We found that α-cyclodextrin facilitates vaginal absorption of leuprolide in rats by about six-fold as well as the nasal absorption of leuprolide and insulin in rats and dogs (Hirai *et al.*, 1987). This effect is interestingly elicited by the removal of palmitic and oleic acid, which are minor membrane components. More research is required to find an absorption promoter possessing stronger promoting ability with only transient and minor tissue alteration; however, these properties might appear to be conflicting.

3.3. Conclusions

The vagina is a complex organ with multiple functions. The vaginal epithelium consists of noncornified, stratified squamous cells providing a tight barrier against the external environment. It is drastically influenced by estrogen throughout a woman's life from birth to menarche and menopause and during the menstrual cycle in adults. During the luteal and early follicular phases in the human, during metestrus and diestrus in the mouse and rat, and in menopausal women and ovariectomized rats, the vaginal epithelium becomes explicitly thin and porous; this is attributed to loosened intercellular channels and desquamation of the superficial cell layer. This elevated porosity results in greater permeation by peptides and proteins. First-pass effects by topical and liver proteases are almost completely avoided using the vaginal route of administration, and this route may be suitable for peptides and proteins. However, the bioavailability is somewhat insufficient and fluctuates greatly; an absorption promoter is required. The polybasic carboxylic acids, which have a chelating action in an acidic milieu, exert a potent absorption-enhancing effect with only moderate tissue damage. To break through the epithelial barriers, further studies to find absorption promoters with less adverse reactions will be required; their utilization will depend on the balance of practical merit and demerit.

REFERENCES

Aref, I., El-Sheikha, Z., and Hafez, E. S. E., 1978, Absorption of drugs and hormones in the vagina, in: *The Human Vagina* (E. S. E. Hafez and T. N. Evans, eds.), Elsevier/North-Holland, Amsterdam, pp. 179–191.

Aungst, B. J., Rogers, N. J., and Shefter, E., 1988, Comparison of nasal, rectal, buccal, sublingual and intramuscular insulin efficacy and the effects of a bile salt absorption promoter, *J. Pharmacol. Exp. Ther.* **244:**23–27.

Banga, A. K., and Chien, Y. W., 1988, Systemic delivery of therapeutic peptides and proteins, *Int. J. Pharm.* **48:**15–50.

Benziger, D. P., and Edelson, J., 1983, Absorption from the vagina, *Drug Metab. Rev.* **14;**137–168.

Binder, H. J., 1970, Amino acid absorption in the mammalian colon, *Biochim. Biophys. Acta* **219:**503–506.

Burgos, M. H., and Roig de Bargas-Linares, C. E., 1978, Ultrastructure of the vaginal mucosa, in: *The Human Vagina* (E. S. E. Hafez and T. N. Evans, eds.), Elsevier/North-Holland, Amsterdam, pp. 63–93.

Cho, M. J., Scieszka, J. F., and Burton, P. S., 1989, Citric acid as an adjuvant for transepithelial transport, *Int. J. Pharm.* **52:**79–81.

Corbo, D. C., Liu, J.-C., and Chien, Y. W., 1990, Characterization of the barrier properties of mucosal membranes, *J. Pharm. Sci.* **79:**202–206.

Cordero, N., and Wilson, T. H., 1961, Comparison of transport capacity of small and large intestine, *Gastroenterology* **41:**500–504.

De Boer, A. G., Breimer, D. D., Pronk, J., and Gubbens-Stibbe, J. M., 1980, Rectal bioavailability of lidocaine in rats: Absence of significant first-pass elimination, *J. Pharm. Sci.* **69:**804–807.

De La Cruz, A., De La Cruz, K. G., Arimura, A., Coy, D. H., Vilchez-Martinez, J. A., Coy, E. J., and Schally, A. V., 1975, Gonadotrophin-releasing activity of two highly active and long-acting analogs of luteinizing hormone-releasing hormone after subcutaneous, intra-vaginal and oral administration, *Fertil. Steril.* **26:**894–900.

Deleede, L. C. J., De Boer, A. G., Roozen, C. P. J. M., and Breimer, D. D., 1983, Avoidance of first-pass elimination of rectally administered lidocaine in relation to the site of absorption in rats, *J. Pharmacol. Exp. Ther.* **225:**181–185.

DeMareo, T. J., and Levine, R. R., 1969, Role of the lymphatics in the absorption and distribution of drugs, *J. Pharmacol. Exp. Ther.* **169:**142–151.

Dodda-Kashi, S., and Lee, V. H. L., 1986, Enkephalin hydrolysis in homogenates of various absorptive mucosae of the albino rabbit: Similarities in rats and involvement of amino-peptidases, *Life Sci.* **38**:2019–2028.

Durrani, M. J., Kusai, A., Ho, N. F. H., Fox, J. L., and Higuchi, W. I., 1985, Topical vaginal drug delivery in the guinea pig. I. Effect of estrous cycle on the vaginal membrane permeability of vidarabine, *Int. J. Pharm.* **24**:209–218.

Fisher, N. F., 1923, The absorption of insulin from the intestine, vagina and scrotal sac, *Am. J. Physiol.* **67**:65–71.

Fishman, W. H., and Mitchell, G. W., 1959, Studies on vaginal enzymology, *Ann. N.Y. Acad. Sci.* **83**:105–121.

Forstner, J. F., 1978, Intestinal mucins in health and disease, *Digestion* **17**:234–263.

Ganong, W. F., 1983, *Review of Medical Physiology*, 11th ed., Lange Medical Publications/Maruzen Asia, Singapore.

Gumbiner, B., 1987, Structure, biochemistry, and assembly of epithelial tight junctions, *Am. J. Physiol.* **253**:C749–C758.

Gumbiner, B., and Simons, K., 1986, A functional assay for proteins involved in establishing an epithelial occluding barrier: Identification of a uvomorulin-like polypeptide, *J. Cell Biol.* **102**:457–468.

Harris, A. S., Svensson, E., Wagner, Z. G., Lethagen, S., and Nilsson, I. M., 1988, Effect of viscosity on particle size, deposition, and clearance of nasal delivery systems containing desmopressin, *J. Pharm. Sci.* **77**:405–408.

Havran, R. T., and Oster, G., 1977, Trypsin-like activity in the vaginal epithelial cells of the rat, *J. Histochem. Cytochem.* **25**:1178–1186.

Helenius, A., and Simons, K., 1975, Solubilization of membranes by detergents, *Biochim. Biophys. Acta* **415**:29–79.

Hirai, S., Yashiki, T., and Mima, H., 1981a, Effect of surfactants on the nasal absorption of insulin in rats, *Int. J. Pharm.* **9**:165–172.

Hirai, S., Yashiki, T., and Mima, H., 1981b, Mechanisms for the enhancement of the nasal absorption of insulin by surfactants, *Int. J. Pharm.* **9**:173–184.

Hirai, S., Okada, H., and Yashiki, T., 1987, Pharmaceutical composition and its nasal or vaginal use, *U.S. Patent* 4, 659,696.

Hirano, S., Nose, A., Hatta, K., Kawakami, A., and Takeichi, M., 1987, Calcium-dependent cell–cell adhesion molecules (cadherins): Subclass specificities and possible involvement of actin bundles, *J. Cell Biol.* **105**:2501–2510.

Hogben, C. A. M., Tocco, D. J., Brodie, B. B., and Schanker, L. S., 1959, On the mechanism of intestinal absorption of drugs, *J. Pharmacol. Exp. Ther.* **125**:275–282.

Hoogdalem, E. J., Heijligers-Feijen, C. D., De Boer, A. G., Verhoef, J. C., and Breimer, D. D., 1989, Rectal absorption enhancement of des-enkephalin-γ-endorphin (DERE) by medium-chain glyceride and EDTA in conscious rats, *Pharm. Res.* **6**:91–95.

Hsu, C. C., Park, J. Y., Ho, N. F. H., Higuchi, W. I., and Fox, J. L., 1983, Topical vaginal drug delivery. I. Effect of the estrous cycle on vaginal membrane permeability and diffusivity of vidarabine in mice, *J. Pharm. Sci.* **72**:674–680.

Humphrey, R. R., Dermody, W. C., Brink, H. O., Bousley, F. G., Schottin, N. H., Sakowski, R., Vaitkus, J. W., Veloso, H. T. and Reel, J. R., 1973, Induction of luteinizing hormone (LH) release and ovulation in rats, hamsters and rabbits by synthetic luteinizing hormone-releasing factor (LRF), *Endocrinology* **92**:1515–1525.

Ichikawa, K., Ohata, I., Mitomi, M., Kawamura, S., Maeno, H., and Kawata, H., 1980, Rectal absorption of insulin suppositories in rabbits, *J. Pharm. Pharmacol.* **32:**314–318.

Illum, L., Farraj, N. F., Davis, S. S., Johansen, B. R., and O'Hagan, D. T., 1990, Investigation of the nasal absorption of biosynthetic human growth hormone in sheep—Use of a bioadhesive microsphere delivery system, *Int. J. Pharm.* **63:**207–211.

Ishida, M., Machida, Y., Nambu, N., and Nagai, T., 1981, New mucosal dosage form of insulin, *Chem. Pharm. Bull.* **29:**810–816.

Kakemi, K., Arita,T., and Muranishi, S., 1965a, Absorption and excretion of drugs. XXV. On the mechanism of rectal absorption of sulfonamides, *Chem. Pharm. Bull.* **13:**861–869.

Kakemi, K., Arita, T., and Muranishi, S., 1965b, Absorption and excretion of drugs. XXVI. Effect of water soluble bases on rectal absorption of sulfonamide, *Chem. Pharm. Bull.* **13:** 969–975.

Kakemi, K., Arita, T., and Muranishi, S., 1965c, Absorption and excretion of drugs. XXVII. Effect of nonionic surface-active agents on rectal absorption of sulfonamides, *Chem. Pharm. Bull.* **13:**976–985.

Kakemi, K., Arita, T., Hori, R., Konishi, R., and Nishimura, K., 1969, Absorption and excretion of drugs. XXXIV. An aspect of mechanism of drug absorption from the intestinal tract in rats, *Chem. Pharm. Bull.* **17:**255–261.

Kamada, A., Nishihata, T., Kim, S., Yamamoto, M., and Yata, N., 1981, Study of enamine derivatives of phenylglycine as adjuvants for the rectal absorption of insulin, *Chem. Pharm. Bull.* **29:**2012–2019.

Kessler, M., Acutor, O., Storelli, C., Murer, H., Muller, M., and Semenza, G., 1978, A modified procedure for the rapid preparation of efficiently transporting vesicles from small intestinal brush border membranes, *Biochim. Biophys. Acta* **506:**136–154.

Lee, V. H. L., 1986, Enzymatic barriers to peptide and protein absorption and the use of penetration enhancers to modify absorption, in: *Delivery Systems for Peptide Drugs* (S. S. Davis, L. Illum, and E. Tomlinson, eds.), Plenum Press, New York, pp. 87–104.

Lee, V. H. L., 1988, Enzymatic barriers to peptide and protein absorption, *CRC Crit. Rev. Ther. Drug Carrier Syst.* **5:**69–97.

Lee, V. H. L., and Yamamoto, A., 1990, Presentation and enzymatic barriers to peptide and protein absorption, *Adv. Drug Deliv. Rev.* **4:**171–207.

Liu, J.-C., Sun, Y., Siddiqui, O., Chien, Y. W., Shi, W., and Li, J., 1988, Blood glucose control in diabetic rats by transdermal iontophoretic delivery of insulin, *Int. J. Pharm.* **44:** 197–204.

Loewenstein, W. R., 1984, Channels in the junctions between cells, *Curr. Top. Membr. Transp.* **21:**221–252.

Martinez-Palomo, A., Meza, I., Beaty, G., and Cereijido, M., 1980, Experimental modulation of occluding junctions in a cultured transporting epithelium, *J. Cell Biol.* **87:**736–745.

Masuda, Y., Yoshikawa, H., Takada, K., and Muranishi, S., 1986, The mode of enhanced enteral absorption of macromolecules by lipid-surfactant mixed micelles I, *J. Pharmacobio-Dyn.* **9:**793–798.

Miyake, M., Nishihata, T., Wada, N., Takeshima, E., and Kamada, A., 1984, Rectal absorption of lysozyme and heparin in rabbits in the presence of non-surfactant adjuvants, *Chem. Pharm. Bull.* **32:**2020–2025.

Miyake, M., Nishihata, T., Nagano, A., Kyobashi, Y., and Kamada, A., 1985, Rectal absorption of [Asu^{1,7}]-eel calcitonin in rats, *Chem. Pharm. Bull.* **33:**740–745.

Morimoto, K., Takeeda, T., Nakamoto, Y., and Morisaka, K., 1982, Effective vaginal absorption of insulin in diabetic rats and rabbits using polyacrylic acid aqueous gel bases, *Int. J. Pharm.* **12**:107–111.

Murakami, M., Takada, K., Fujii, T., and Muranishi, S., 1988a, Intestinal absorption enhanced by unsaturated fatty acid: Inhibitory effect of sulfhydryl modifiers, *Biochim. Biophys. Acta* **939**:238–246.

Murakami, T., Kawakita, H., Kishimoto, M., Higashi, Y., Amagase, H., Hayashi, T., Nojima, N., Fuwa, T, and Yata, N., 1988b, Intravenous and subcutaneous pharmacokinetics and rectal bioavailability of human epidermal growth factor in the presence of absorption promoter in rats, *Int. J. Pharm.* **46**:9–17.

Muranishi, S., 1985, Modification of intestinal absorption of drugs by lipoidal adjuvants, *Pharm. Res.* **2**:108–118.

Muranishi, S., 1989, Absorption enhancers: Mechanisms and application, in: *Novel Drug Delivery and Its Therapeutic Application* (L. F. Prescott, and W. S. Nimmo, eds.), Wiley, New York, pp. 69–77.

Muranishi, S., Yoshikawa, H., and Sezaki, H., 1979, Absorption of 5-fluorouracil from various regions of gastrointestinal tract in rats. Effect of mixed micelles, *J. Pharmacobio-Dyn.* **2**: 286–294.

Muranishi, S., Takada, K., Yoshikawa, H., and Murakami, M., 1986, Enhanced absorption and lymphatic transport of macromolecules via the rectal route, in: *Delivery Systems for Peptide Drugs* (S. S. Davis, L. Illum, and E. Tomlinson, eds.), Plenum Press, New York, pp. 177–189.

Muranushi, N., Nakajima, Y., Kinugawa, M., Muranishi, S., and Sezaki, H., 1980, Mechanism for the inducement of the intestinal absorption of poorly absorbed drugs by mixed micelles. II. Effect of incorporation of various lipids on the permeability of liposomal membranes, *Int. J. Pharm.* **4**:280–291.

Muranushi, N., Takagi, N., Muranishi, S., and Sezaki, H., 1981, Effect of fatty acid and monoglycerides on permeability of lipid bilayer, *Chem. Phys. Lipids* **28**:269–279.

Nishi, N., Arimura, A., Coy, D. H., Vilchez-Martinez, J. A., and Schally, A. V., 1975, The effect of oral and vaginal administration of synthetic LH-RH and [D-Ala[6],des-Gly[10]-NH[2]]-LH-RH ethylamide on serum LH levels in ovariectomized, steroid-blocked rats, *Proc. Soc. Exp. Biol. Med.* **148**:1009–1012.

Nishihata, T., Rytting, J. H., Kamada, A., Higuchi, T., Routh, M., and Caldwell, L., 1983, Enhancement of rectal absorption of insulin using salicylates in dogs, *J. Pharm. Pharmacol.* **35**:148–151.

Okada, H., 1983, Vaginal administration of a potent luteinizing hormone-releasing hormone analog (leuprolide), *J. Takeda Res. Lab.* **42**:150–208.

Okada, H., 1991, Vaginal route of peptide and protein drug delivery, in: *Peptide and Protein Drug Delivery* (V. H. L. Lee, ed.), Dekker, New York, pp. 633–666.

Okada, H., Yamazaki, I., Ogawa, Y., Hirai, S., Yashiki, T., and Mima, H., 1982, Vaginal absorption of a potent luteinizing hormone-releasing hormone analog (leuprolide) in rats. I. Absorption by various routes and absorption enhancement, *J. Pharm. Sci.* **71**:1367–1371.

Okada, H., Yamazaki, I., Yashiki, T., and Mima, H., 1983a, Vaginal absorption of a potent luteinizing hormone-releasing hormone analogue (leuprolide) in rats. II. Mechanism of absorption enhancement with organic acids, *J. Pharm. Sci.* **72**:75–78.

Okada, H., Yashiki, T., and Mima, H., 1983b, Vaginal absorption of a potent luteinizing hormone-releasing hormone analogue (leuprolide) in rats. III. Effect of estrous cycle on vaginal absorption of hydrophilic model compounds, *J. Pharm. Sci.* **72**:173–176.

Okada, H., Yamazaki, I., Sakura, Y., Yashiki, T., Shimamoto, T., and Mima, H., 1983c, Desensitization of gonadotropin-releasing response following vaginal consecutive administration of leuprolide in rats, *J. Pharmacobio-Dyn.* **6**:512–522.

Okada, H., Sakura, Y., Kawaji, H., Yashiki, T,. and Mima, H., 1983d, Regression of rat mammary tumors by a potent luteinizing hormone-releasing hormone analogue (leuprolide) administered vaginally, *Cancer Res.* **43**:1869–1874.

Okada, H., Yamazaki, I., Yashiki, T., Shimamoto, T., and Mima, H., 1984, Vaginal absorption of a potent luteinizing hormone-releasing hormone analogue (leuprolide) in rats. IV. Evaluation of the vaginal absorption and gonadotropin responses by radioimmunoassay, *J. Pharm. Sci.* **73**:298–302.

Patel, L. G., Warrington, S. J., and Pearson, R. M., 1983, Propranolol concentrations in plasma after insertion into the vagina, *Br. Med. J.* **287**:1247–1248.

Platzer, W., Poisel, S., and Hafez, E. S. E., 1978, Functional anatomy of the human vagina, in: *The Human Vagina* (E. S. E. Hafez and T. N. Evans, eds.), Elsevier/North Holland, Amsterdam, pp. 39–53.

Pschera, H., Hjerpe, A., and Carlstrom, K., 1989, Influence of the maturity of the vaginal epithelium upon the absorption of vaginally administered estradiol-17-beta and progesterone in postmenopausal women, *Gynecol. Obstet. Invest.* **27**:204–207.

Richardson, J. L., Minhas, P. S., Thomas, N. W., and Illum, L., 1989, Vaginal administration of gentamicin to rats. Pharmaceutical and morphological studies using absorption enhancers, *Int. J. Pharm.* **56**:29–35.

Rigg, L. A., Hermann, H., and Yen, S. S. C., 1978, Absorption of estrogens from vaginal creams, *N. Engl. J. Med.* **298**:195–197.

Robinson, G. D., 1927, Absorption from the vagina, *J. Pharmacol. Exp. Ther.* **32**:81–88.

Rock, J., Barker, R. H., and Bacon, W. B., 1947, Vaginal absorption of penicillin, *Science* **105**:13.

Saito, M., Kumasaki, T., Yaoi, Y., Nishi, N., Arimura, A., Coy, D. H., and Schally, A. V., 1977, Stimulation of luteinizing hormone (LH) and follicle-stimulating hormone (FSH) by [D-leu^6,des-Gly^{10}-NH$_2$]-LH-releasing hormone ethylamide after subcutaneous, intravaginal and intrarectal administration to women, *Fertil. Steril.* **28**:240–244.

Schuchner, E. B., Foix, A., Borenstein, C. A., and Marchese, C., 1974, Electron microscopy of human vaginal epithelium under normal and experimental conditions, *J. Reprod. Fertil.* **36**:231–233.

Shore, P. A., Brodie, B. B., and Hogben, C. A. M., 1957, The gastric secretion of drugs: A pH-partition hypothesis, *J. Pharmacol. Exp. Ther.* **119**:361–369.

Sieber, S. M., Cohn, V. H., and Wynn, W. T., 1974, The entry of foreign compounds into the thoracic duct lymph of the rat, *Xenobiotica* **4**:265–284.

Sithigorngul, P., Burton, P., Nishihata, T., and Caldwell, L., 1983, Effects of sodium salicylate on epithelial cells of the rectal mucosa of the rats: A light and electron microscopic study, *Life Sci.* **33**:1025–1032.

Sjoberg, I., Cajander, S., and Rylander, E., 1988, Morphometric characteristics of the vaginal epithelium during the menstrual cycle, *Gynecol. Obstet. Invest.* **26**:136–144.

Spit, B. J., Hendriksen, E. G. J., Bruijntjes, J. P., and Kuper, C. F., 1989, Nasal lymphoid tissue in the rat, *Cell Tissue Res.* **255**:193–198.

Stratford, R. E. and Lee, V. H. L., 1986, Aminopeptidase activity in homogenates of various absorptive mucosae in the albino rabbit: Implications in peptide delivery, *Int. J. Pharm.* **30**:73–82.

Touitou, E., Donbrow, M., and Azaz, E., 1978, New hydrophilic vehicle enabling rectal and vaginal absorption of insulin, heparin, phenol red and gentamicin, *J. Pharm. Pharmacol.* **30**:662–663.

Turner, C. D., and Bagnara, J. T., 1976, *General Endocrinology*, 6th ed., Saunders, Philadelphia.

Wagner, G., and Levin, R. J., 1978, Vaginal fluid, in: *The Human Vagina* (E. S. E. Hafez and T. N. Evans, eds.), Elsevier/North-Holland, Amsterdam, pp. 121–137.

Walker, W. A., and Isselbacher, K. J., 1977, Intestinal antibodies, *N. Engl. J. Med.* **297**: 767–773.

Walz, K. A., Metzger, H., and Ludwig, H., 1978, Surface ultrastructure of the vagina, in: *The Human Vagina* (E. S. E. Hafez and T. N. Evans, eds.), Elsevier/North-Holland, Amsterdam, pp. 55–61.

Wang, C., Baker, H. W. G., Jennings, M. G., Burger, H. G., and Lutjen, P., 1985, Interaction between human cervical mucus and sperm surface antibodies, *Fertil. Steril.* **44**:484–488.

Wilson, T. H., 1962, *Intestinal Absorption*, Saunders, Philadelphia.

Yamamoto, A., Luo, A. M., Dodda-Kashi, S., and Lee, V. H. L., 1989, The ocular route for systemic insulin delivery in the albino rabbit, *J. Pharmacol. Exp. Ther.* **249**:249–255.

Yamamoto, A., Hayakawa, E., and Lee, V. H. L., 1990, Insulin and proinsulin proteolysis in mucosal homogenates of the albino rabbits: Implications in peptide delivery from nonoral routes, *Life Sci.* **47**:2465–2474.

Yoshikawa, H., Muranishi, S., Kato, C., and Sezaki, H., 1981, Bifunctional delivery system for selective transfer of bleomycin into lymphatics via enteral route, *Int. J. Pharm.* **8:** 291–302.

Yoshikawa, H., Muranishi, S., Sugihara, N., and Sezaki, H., 1983, Mechanism of transfer of bleomycin into lymphatics by bifunctional delivery system via lumen of small intestine, *Chem. Pharm. Bull.* **31**:1726–1732.

Yoshikawa, H., Takada, K., and Muranishi, S., 1984a, Molecular weight dependence of permselectivity to rat small intestinal blood–lymph barrier for exogenous macromolecules from lumen, *J. Pharmacobio-Dyn.* **7**:1–6.

Yoshikawa, H., Takada, K., Muranishi, S., Satoh, Y., and Naruse, N., 1984b, A method to potential enteral absorption of interferon and selective delivery into lymphatics, *J. Pharmacobio-Dyn.* **7**:59–62.

Yoshikawa, H., Satoh, Y., Naruse, N., Takada, K., and Muranishi, S., 1985a, Comparison of disappearance from blood and lymphatic delivery of human fibroplast interferon in rats by different administration routes, *J. Pharmacobio-Dyn.* **8**:206–210.

Yoshikawa, H., Takada, K., Satoh, Y., Naruse, N., and Muranishi, S., 1985b, Potentiation of enteral absorption of human interferon alpha and selective transfer into lymphatics in rats, *Pharm. Res.* **2**:249–250.

Yoshikawa, H., Takada, K., Satoh, Y., Naruse, N., and Muranishi, S., 1986, Development of interferon suppositories. I. Enhanced rectal absorption of human fibroblast interferon by fusogenic lipid via lymphotropic delivery in rats, *Pharm. Res.* **3**:116–117.

Yoshioka, S., Caldwell, L., and Higuchi, T., 1982, Enhanced rectal bioavailability of polypeptides using sodium 5-methoxysalicylate as an absorption promoter, *J. Pharm. Sci.* **71**: 593–594.

III

Vascular Barriers

Chapter 10

Vascular Endothelial Barrier Function and Its Regulation

Asrar B. Malik and Alma Siflinger-Birnboim

1. INTRODUCTION

The vascular endothelium has a variety of functions [hemostasis, defense reaction (inflammatory response), angiogenesis], among which the control of the exchange of substances between blood and tissues is of prime importance. Capillary permeability to plasma proteins is a critical factor in regulating tissue–fluid balance. The endothelial cell monolayer lining the vessel wall is a porous (semipermeable) membrane through which fluid and solutes are transported. The substances transported include water, respiratory gases and other small lipid-soluble molecules, ions, small lipid-insoluble organic molecules, large hydrophilic and lipophilic proteins. The transport of molecules from the plasma to the vascular endothelium is governed by several factors (Renkin, 1977; Simionescu and Simionescu, 1984): (1) vascular driving forces (i.e., hydrostatic and oncotic pressure gradients), (2) physiochemical properties of the permeant molecule, and (3) the surface properties of the endothelial membrane (Table I). Transport to tissues also depends to a significant extent on the physiochemical properties of the components underlying the endothelium (basement membrane, extracellular matrix, interstitial fluid). This review discusses some critical and novel aspects of transport of solutes and water across the endothelial barrier. Although we have discussed much current literature in the field, it is highly likely that there have been oversights.

Asrar B. Malik and Alma Siflinger-Birnboim • Department of Physiology and Cell Biology, The Albany Medical College of Union University, Albany, New York 12208.

Biological Barriers to Protein Delivery, edited by Kenneth L. Audus and Thomas J. Raub. Plenum Press, New York, 1993.

Table I

Factors Governing the Transport of Plasma Molecules
across Vascular Endothelium

Plasma and hemodynamic forces
  Hydrostatic pressure gradient
  Oncotic pressure gradient
Properties of the permeant molecules
  Molecular size
  Molecular shape
  Molecular charge
  Molecular chemistry (binding of molecules to cell surface receptors)
  Generation of transendothelial gradients by the concentrations of the permeating molecule
Properties of endothelial cells
  Endothelial surface charge
  Structure of the endothelial cell surface
  Location in vasculature (site specificity)

---

2. ROUTES OF ENDOTHELIAL TRANSPORT

The morphology of the vascular endothelium varies from organ to organ and from one location to another within the vasculature of an organ (Simionescu and Simionescu, 1984). Based on variations in the continuity of the endothelium and its basement membrane, three general categories of endothelium have been determined: continuous, fenestrated, and discontinuous endothelium (Simionescu and Simionescu, 1984).

The continuous capillary endothelium occurs in large vessels, arterioles, most venules of skeletal muscle, heart, smooth muscles, lung, skin, subcutaneous tissue, sinus, and mucous membranes. The fenestrated endothelial cells are found in intestinal mucosa, endocrine and exocrine glands, and glomerular and peritubular renal capillaries. The discontinuous endothelial cells are found predominantly in the liver, spleen, and bone marrow.

The complexity of the endothelial barrier and the different hypothesized routes of transport in a "typical" continuous endothelial cell monolayer are shown in Fig. 1 and summarized in Table II. The transport of solutes in general occurs through paracellular and transcellular routes (Renkin, 1977). Whether a solute is transported via a transcellular or a paracellular route depends to a considerable degree on properties of the solute such as its size, charge, and existence of cell surface receptors (e.g., as in the case of transferrin). Water, small lipid-soluble substances, and respiratory gases are transported across the cell membrane pathway. Vesicles are responsible for transcytosis of solutes and, to some extent, water across the membrane (Renkin, 1977). In some cases, plasmalemmal vesicles can fuse to form large channels (Renkin, 1977). The interendothelial clefts comprising pathways of varying

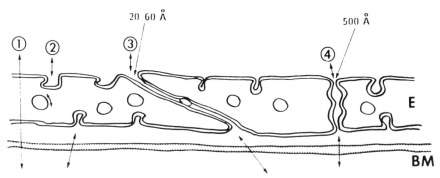

Figure 1. Schematics of hypothetical pathways of transport in continuous endothelium. 1, transcellular pathway; 2, vesicular pathway; 3, small and large pore pathways; 4, fused plasmalemmal channel pathway. E, endothelial cell; BM, basement membrane. See Table II for detailed explanation.

diameters are the chief paracellular routes of solute and water transport (Renkin, 1988).

3. STARLING'S EQUATION

Several general principles have been developed to describe transport of molecules across porous membranes. Starling (1896) first quantified the physical forces responsible for the transcapillary exchange of water as the difference between

Table II
Routes of Endothelial Transport

---

1. Transcellular (cell membrane) pathway
   Consists of three barriers in series (plasma membrane, cytoplasm, plasma membrane)
2. Vesicular pathway
   Equilibrates luminal and abluminal fluids
   Transport may be dependent on concentration gradient and rate of vesicular turnover
3. Small- and large-"pore" pathways
   Continuous route from the luminal to abluminal sides
   Flux of water and solutes across these pathways can be diffusive and convective
   Diffusive and convective transport may be coupled
   Molecular sieving (i.e., restriction of solute permeation as molecular size approaches the pore dimension) is a characteristic of these pathways
4. Fused plasmalemmal channels
   Transient fusion of two or more vesicles
   Exhibit the same characteristics as junctional large "pores"

---

capillary hydrostatic and colloid osmotic pressures (see Chapter 15). Kedem and Katchalsky (1958) refined Starling's concept by subdividing solute flux into "diffusive" and "convective" components.

The rate of water transport J_v is described by:

$$J_v = L_pS[(P_c - P_i) - \sigma(\pi_c - \pi_i)] \tag{1}$$

where L_p is hydraulic conductivity, S is capillary surface area, P_c and P_i are capillary and interstitial hydrostatic pressures, respectively, π_c and π_i are capillary and interstitial colloid osmotic pressures, and σ is protein reflection coefficient of the vessel wall ($\sigma = 0$ if the membrane is freely permeable to the molecule crossing the membrane and $\sigma = 1$ if the membrane "rejects" the molecule making it impermeable). The movement of solutes across the porous endothelial barrier, J_s, is defined by:

$$J_s = J_v(1 - \alpha)\overline{C}_s + PS(\Delta C) \tag{2}$$

where PS is permeability coefficient–surface area product, ΔC is solute concentration difference across the endothelial monolayer, and $\overline{C}_s$ is mean concentration of the solute within the endothelial "pores" (or interendothelial junctional clefts).

4. ENDOTHELIAL CELL MONOLAYERS IN CULTURE

Cultured endothelial cell monolayer preparations provide a means of studying the permeability characteristics of the endothelial barrier. The use of confluent endothelial monolayers surmounts some of the complexities associated with the study of the intact vascular barrier due to its complex structure consisting of several barriers with different restrictive properties. The use of cell cultures offers advantages such as homogeneity of cell type, ease of handling, ability to study transport under controlled conditions, and manipulation of parameters influencing the transport (Table I). Over the last several years, approaches have been developed to measure the barrier properties of the cultured endothelial monolayers in which critical variables such as hydrostatic and oncotic pressures and surface area can be regulated or varied and in which shape change of endothelial cells (a primary determinant of endothelial permeability) can be monitored (Garcia et al., 1986; Siflinger-Birnboim et al., 1987). Monolayers of endothelial cells from different sites have been grown by some investigators on porous microcarrier beads (Boiadjeiva et al., 1984; Bottaro et al., 1986; Killackey et al., 1986) and most often on porous filters with different porosities and thicknesses (Shasby et al., 1985; Bizios et al., 1986; Garcia et al., 1986; Siflinger-Birnboim et al., 1986, 1987; Cooper et al., 1987; Del Vecchio et al., 1987; Albelda et al., 1988; Casnocha et al., 1989; Lynch et al., 1990; Oliver, 1990). The most commonly used system (illustrated in Fig. 2) is constructed by gluing a 13-mm-diameter gelatin- and fibronectin-treated polycarbonate filter (pore size

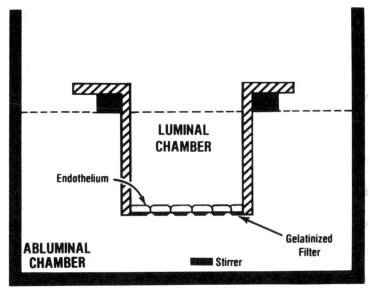

Figure 2. *In vitro* system for the measurement of molecular flux across the cultured endothelial monolayer. (From Siflinger-Birnboim *et al.*, 1987.)

0.8 μm; 10 μm thickness) to a polystyrene cylinder (13 mm outside diameter, 9 mm inner diameter) to form the luminal (upper) chamber (Cooper *et al.*, 1987; Siflinger-Birnboim *et al.*, 1987). The endothelial cells are seeded on the filter inside the luminal chamber at a density of 2×10^5 cells/ml in Dulbecco's modified Eagle's medium, 20% fetal bovine serum, 0.01 mM nonessential amino acids, and gentamicin sulfate (50 μg/ml). The luminal chambers are incubated at 37°C in a humidified CO_2 incubator (5% CO_2, 95% room air) for 3–4 days, allowing the monolayers to become confluent.

Permeability of the endothelial barrier to molecules of various sizes has been studied (Del Vecchio *et al.*, 1987; Siflinger-Birnboim *et al.*, 1987) by equipping the luminal chamber containing 0.7 ml of media with a Styrofoam flotation collar and allowing it to float in a 25-ml volume of fluid contained in a beaker (the abluminal chamber). The floating luminal chamber permits the sequential sampling of fluid form the abluminal chamber without the development of a hydrostatic pressure gradient. The transendothelial macromolecular diffusive flux was determined by measuring the net transfer of tracer amounts of radiolabeled molecules across the monolayer (Siflinger-Birnboim *et al.*, 1987). The clearance rate of a tracer molecule, which is a measure of endothelial permeability, was calculated from the slope of the line produced by plotting volume of fluid cleared (μl) per unit time (min). The system allows the measurement of clearance rate for the endothelium on the filter and for the filter without the endothelium. The endothelial permeability of the tracer molecule is calculated by the ratio of clearance rate-to-surface area of the filter as previously described (Cooper *et al.*, 1987; Siflinger-Birnboim *et al.*, 1987). This system also allows calculation of endothelial permeability of most tracers (except lipophilic molecules) independent of the effects of unstirred layers (Cooper *et al.*, 1987; Siflinger-Birnboim *et al.*, 1987; Barry and Diamond, 1984; Pedley, 1983).

The endothelial cells grown on the filter become confluent within 3–4 days postseeding with no intercellular gaps obvious in the cell layer as examined by acridine orange staining of the monolayers on the filter (Phillips *et al.*, 1988) (Fig. 3). The endothelial monolayer begins to demonstrate restricted diffusion to macromolecules within 3–4 days postseeding. Cross sections of endothelial cells revealed that adjacent cells were fused at several discrete points. The fusion at these sties appears to be a cell-specific phenomenon since junctional membranes were more complex and well-developed in bovine pulmonary microvessel endothelial cell monolayers than in the mainstem pulmonary artery endothelial cells (Table III) (Fig. 4). This difference was associated with a greater restrictiveness of the cultured pulmonary microvessel endothelial cells to solutes of varying molecular size (see Fig. 9).

The development of an organized extracellular matrix with increased duration of seeding, may allow the cells to form a more restrictive barrier (i.e., more complex interendothelial junctions) (Madri and Williams, 1983). This may be achieved in

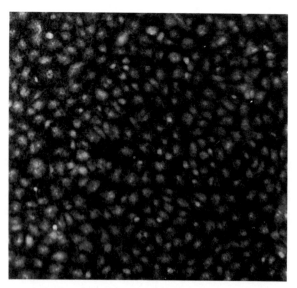

Figure 3. Confluent endothelial monolayers grown on gelatin- and fibronectin-coated polycarbonate filters and visualized by fluorescence microscopy after acridine orange staining.

Table III
Morphometric Characterization of Pulmonary Endothelial Cells (EC) in Culture

	No. of vesicles/μm[a]	No. of complexes[b]/ junction	Junction path length (μm)
Pulmonary microvessel EC	26.7 ± 11.3	4.8 ± 3.0[c]	0.75 ± 0.57
Pulmonary artery EC	5.7 ± 4.6	0.6 ± 0.8	0.61 ± 0.29
Ratio	4.7	8.4	1.22

[a]Approximately 50 μm of apical endothelial membrane was examined (determined in the nonnuclear region of the cell).
[b]Junctional complex stricture sites defined as junctional elements with close apposition of membranes surrounded by electron-dense matrix fibrillar material; 40 cell-to-cell junctions were examined.
[c]Values represent means ± SD.

BPAEC

MV

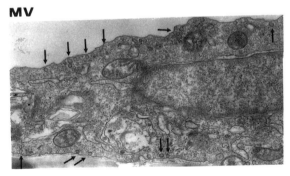

Figure 4. Transmission electron micrographs of representative bovine pulmonary artery (BPAEC) and bovine pulmonary endothelial cell monolayers (MV) grown to confluency on gelatin- and fibronectin-coated polycarbonate filters. 20,000×. The microvessel cells show significantly more vesicles on both luminal and abluminal sides of the plasma membrane than do BPAEC (arrows).

culture by varying the time of seeding before monolayer use from 3 to 14 days at which time the cells have reached a confluent state and have developed extracellular matrix constituents (i.e., fibronectin, laminin, heparin sulfate, collagens I and IV) (Madri and Williams, 1983).

The basis of these differences between endothelial cells from the same vascular bed is not clear nor is it clear whether these differences reflect any particular *in situ* characteristics of cells from these two sites in the pulmonary circulation. These cultured endothelial monolayer findings indicate that the cultured endothelial cells need to be used with caution because of the widely different barrier function characteristics and because, in culture, studies are made in the absence of flow conditions and of tissues underlining the endothelium *in vivo* which can contribute to solute and water transport.

5. EFFECT OF MOLECULAR SIZE ON ENDOTHELIAL PERMEABILITY: THE "PORE" THEORY

The "pore" theory (Landis and Pappenheimer, 1963; Crone and Levitt, 1948) describes the transport of lipid-insoluble molecules through cylindrical or long slit-shaped water-filled channels between the cells. The theory describes transport according to the molecular size of the permeant molecules, but it does not distinguish by which pathways these molecules are transported (Table II). The theory is based on the premise that selected areas form a system of "pores" whose combined total area represents less than 0.1% of the total endothelial surface area (Simionescu and Simionescu, 1984). The size-dependent selectivity of the endothelial monolayer to plasma proteins is determined by the ratio of molecular radius-to-pore radius (Pappenheimer *et al.*, 1951; Renkin, 1985).

In vivo studies, in fact, indicate the presence of "pores" of heterogeneous radii in vascular endothelial cells (Taylor and Granger, 1984). For example, the "pore" sizes in the pulmonary vessel endothelium have been estimated to be 50 Å and 200 Å in a model based on lung lymph clearances (Taylor and Gaar, 1970). In most analyses the values ranged from 5 to 280 Å (Lassen and Trap-Jensen, 1970; O'Donnell and Vargas, 1986). The ratio of small to large "pores" is estimated to be from 30,000:1 based on dextran transport (Grotte, 1956), to 438:1 based on lymph protein fluxes (Taylor and Granger, 1984).

The critical shortcomings of the "pore" theory data are that they fail to account for the role of electrostatic charge of the surface of the endothelial barrier and the molecular charge of the molecule being transported, and do not consider the likelihood that transport of some molecules (e.g., albumin) can occur by transcytotic mechanisms dependent on binding of albumin to cell surface receptors (Ghitescu *et al.*, 1986; Milici *et al.*, 1987; Schnitzer *et al.*, 1988a,b, 1990; Siflinger-Birnboim *et al.*, 1988b; Siflinger-Birnboim and Malik, 1989; Siflinger-Birnboim *et al.*, 1992).

The use of lymph to make inferences about the "pore" sizes of the endothelial barrier is also fraught with problems. Lymph solutes become concentrated during their passage through the lymphatic circulation (Taylor and Granger, 1984; Taylor *et al.*, 1985); therefore, lymph does not necessarily reflect the interstitial fluid or events occurring at the level of the endothelial cell monolayer.

In an attempt to validate the "pore" theory, the selectivity of the vascular endothelial cell barrier (i.e., the monolayer without its extracellular matrix components) to different-sized molecules (varying from 0.18 to 340 kDa) (listed in Table IV) was determined using cultured endothelial monolayers (Siflinger-Birnboim *et al.*, 1987). The results indicated in a clear-cut fashion that the permeability of the endothelial monolayer to tracer molecules decreased with increasing molecular weight (Fig. 5). The relationship shown in Fig. 5 held when the permeability of the endothelial cells was corrected for effects of diffusion by dividing permeability values by diffusion coefficients (Fig. 6A). The gelatin-coated filters without endothelial cells exhibited no such selectivity (Fig. 6B). The transendothelial flux of protein fractions in a 20% fetal calf serum solution confirmed the selective nature of the cultured endothelium in contrast to lack of selectivity observed using gelatin-coated microporous filters without the endothelial monolayer (Fig. 7). When modeled according to the "pore" theory, these data were better represented by a two-pore model (Siflinger-Birnboim *et al.*, 1987) (Fig. 8). From these studies, the calculated "pore" radii were 65 and 304 Å for small- and large-"pore" pathways, respectively, and a small-to-large "pore" ratio of 160:1 (Siflinger-Birnboim *et al.*, 1987). Neutral dextrans (6 kDa to 500 kDa) also diffused across the endothelium in a similar manner (Bizios *et al.*, 1986). The endothelial monolayer, therefore, behaved as a heteroporous system discriminating between solutes according to their molecular size. This

Table IV

Physical Characteristics of Tracer Molecules

Molecule	Size (Da)	$r_e{}^a$	$D_{37}{}^b$
Mannitol	182	4.4	0.9
Sucrose	342	5.2	0.721
Inulin	5,500	11–15	0.296
Cytochrome C	12,000	16.5	0.13[c]
α-Thrombin	36,600	28	0.08[d]
Ovalbumin	43,000	27.6	0.11
Albumin	69,000	36.1	0.093
Plasminogen	82,000	45.1	0.043
Fibrinogen	340,000	106	0.033

[a]r_e, Stokes–Einstein radius (A).
[b]Diffusion coefficient in water at 37°C = $D_{37} \times 10^{-5}$ cm²/ sec.
[c]Free diffusion coefficient in water at 20°C.
[d]Free diffusion coefficient in water at 27°C.

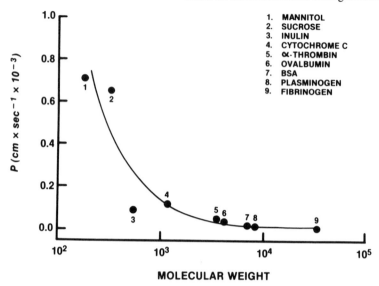

Figure 5.   Permeability of the endothelial monolayer to molecules of differing molecular weight at 37°C (Table IV). (From Del Vecchio *et al.*, 1987.)

selectivity is consistent with a diffusional pathway through "pores" of heterogenous radii. The "pore" theory, however, does not account for molecular–pore interactions due to frictional and electrostatic effects within these hypothetical "pores" (Curry, 1980; Crone and Levitt, 1984) and the possibility of vesicular transport of molecules such as albumin (Milici *et al.*, 1987; Schnitzer *et al.*, 1988a).

6. EFFECTS OF CELLULAR AND MOLECULAR CHARGE ON ENDOTHELIAL PERMEABILITY

It is known that the intact endothelial cell membrane is nonthrombogenic (Danon and Skutelsky, 1976). This is, in part, due to the fact that the circulating blood cells and the endothelial surface are negatively charged (Danon and Skutelsky, 1976; Skutelsky and Danon, 1976; Polikan *et al.*, 1979). The endothelial cell surface has a complex molecular composition consisting of sialyl residues which are responsible for its negative charge (N. Simionescu *et al.*, 1981; M. Simionescu *et al.*, 1982). The endothelial cell membrane and the vesicles and "channel" structures contain micro-domains of anionic sites due to specific distribution of glycosaminoglycans, sialo-conjugates, and monosaccharide residues (Simionescu *et al.*, 1981; Simionescu and Simionescu, 1984; Ghinea and Simionescu, 1985). The existence of these micro-domains has been demonstrated using gold-labeled albumin which was shown to decorate sites on the endothelial surface (Ghitescu *et al.*, 1986; Milici *et al.*, 1987).

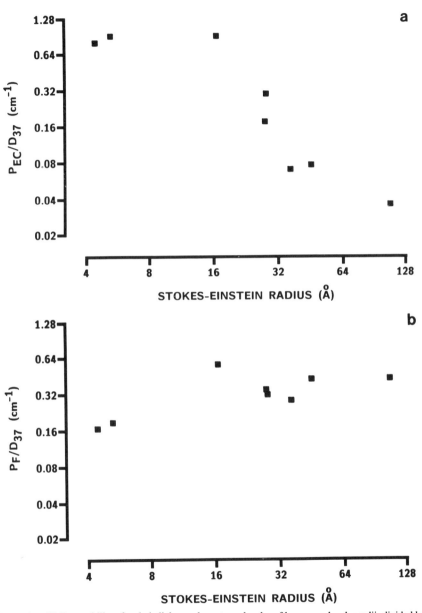

Figure 6. (A) Permeability of endothelial monolayer to molecules of known molecular radii, divided by free diffusion coefficient (P_{EC}/D_{37}) × 10^{-2} as a function of molecular radius. (From Siflinger-Birnboim *et al.*, 1987.) (B) Permeability of filter without the endothelium to molecules of known molecular radii, divided by free diffusion coefficient (P_F/D_{37}) × 10^{-2} as a function of hydrated radius. (From Siflinger-Birnboim *et al.*, 1987.)

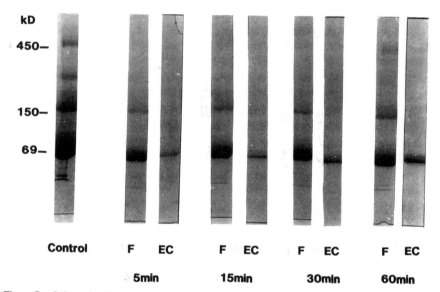

Figure 7. Polyacrylamide gradient gel (5–15%) electrophoresis in 0.1% SDS on 20% fetal calf serum in DMEM obtained from luminal (control) and abluminal chambers [filter alone (F), endothelial monolayer on the filter (EC). Initially (time zero), the luminal chamber contained the calf serum in DMEM alone. Aliquots were obtained from the abluminal chamber at the time points shown. (From Siflinger-Birnboim *et al.*, 1987.)

The permselectivity of the endothelial cell to plasma proteins such as albumin may be related to the anionic charge of the albumin molecule as well as the surface charge on the endothelial cell. For example, the flux of albumin across fenestrated glomerular capillaries is critically dependent on the negative charge of the glomerular capillary endothelial cells as well as the extracellular matrix proteins comprising the basement membrane (Michel, 1984). Disruption of the negatively charged sites results in "leakage" of albumin across the capillary–matrix complex (Michel, 1984).

The endothelial cell membrane surface charge (Simionescu *et al.*, 1981; Ghinea and Simionescu, 1985) influences the transport of albumin (isoelectric point 4.1) since the distribution of charge sites provides a means of preferentially "gating" albumin across the endothelial monolayer. Neutralization of negative charges on the endothelial cell with cationic ferritin increased the flux of [^{125}I]albumin (Table V), indicating that the cell surface negative charge contributes to albumin transport. Another charge-related effect on albumin transport is due to the negative charge distribution of interstitial macromolecules, e.g., heparin sulfate, chondroitin sulfate, and other complex proteoglycans (Perry *et al.*, 1983; Lanken *et al.*, 1985; Taylor *et al.*, 1985).

Because of the negative charge of the albumin molecule, the permeability to albumin is greater than the permeability predicted from its molecular size. For

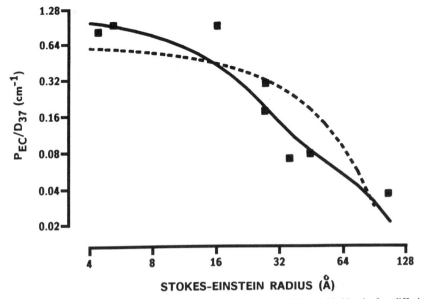

Figure 8. One- and two-pore modeling analysis. Endothelial permeability divided by the free diffusion coefficient $(P_{EC}/D_{37}) \times 10^{-2}$ is plotted on a log scale against the log of molecular radius. The dashed line represents the one-pore model fit and the solid line represents the two-pore model fit. (From Siflinger-Birnboim *et al.*, 1987.)

example, dextran sulfate (negatively charged dextran, 500 kDa) was found to be threefold more permeable than neutral dextran of the same molecular size (Table VI) (Bizios *et al.*, 1986). Sheep lung lymph data also indicate that plasma-to-lymph transport of negative dextrans across the pulmonary vascular endothelial barrier is greater than transport of neutral dextrans (Lanken *et al.*, 1985). Therefore, the

Table V

Effect of Cationic Ferritin
on [125I]-Albumin Clearance Rate[a]

Cationic ferritin (mg/ml)	[125I]-Albumin clearance rate (μl/min)	
	Study #1	Study #2
0 (control)	0.108 ± 0.021[b]	0.226 ± 0.020
1.0	0.136 ± 0.015*	0.287 ± 0.023*

[a]The pulmonary microvessel endothelial cell monolayer was pre-incubated with cationic ferritin for 15 min.
[b]Values are mean ± SEM (*$p < 0.05$ from control).

Table VI
Effect of Charge of the Permeant Molecule of Permeability
of Cultured Endothelial Monolayer to Dextrans[a]

	Abluminal concentration (pg/ml)	
	Filter	Endothelium
Neutral	36.7 ± 2.4[b]	3.1 ± 0.3
Sulfate (negative)	44.1 ± 6.1	9.4 ± 1.2
Neutral	33.4	10.9 ± 1.3

[a]Dextran was added to the luminal chamber. The abluminal chamber was sampled at 45 min and the concentration of dextran transported through the endothelium was measured using spectrophotometrical evaluation.
[b]Values are mean ± SEM.

negatively charged molecules appear to be transported preferentially across the endothelial barrier because of "gating" phenomenon even though the endothelial cell membrane has a net negative charge (Lanken *et al.*, 1985; Bizios *et al.*, 1986).

7. REGIONAL VASCULAR DIFFERENCES IN ENDOTHELIUM ON ENDOTHELIAL PERMEABILITY

Endothelial cells from different sites in the vasculature have many common features, but specific properties of the cells distinguish endothelial cells from large versus small vessels within an organ and between organs (e.g., between the blood–brain barrier and pulmonary vascular endothelial cells) (Zetter, 1981; Simionescu and Simionescu, 1984; Gerritsen, 1987; Belloni and Tressler, 1990; Siflinger-Birnboim *et al.*, 1991a). The [125I]albumin permeability value in cultured pulmonary microvessel endothelial cells was about two to five times lower than the values of similarly cultured cells from the mainstem pulmonary artery (Fig. 9). Interestingly, the microvessel cells were significantly more restrictive with respect to sucrose and inulin permeability (Fig. 9) indicative of less transport occurring via paracellular pathways.

The greater restrictiveness to sucrose and inulin versus albumin suggests that other routes of albumin transport (possible vesicles) are involved in albumin flux across pulmonary microvessel endothelial cells relative to pulmonary artery endothelial cells. Micrographs of the monolayers revealed that a greater number of vesicles were present on the apical surface of the microvessel cells relative to the pulmonary artery endothelial cells (Table III). This is clearly evident on examination of the apical and basolateral membranes of the two cell monolayers grown on microporous filters (Fig. 4). In addition, there were more membrane strictures observed within inter-

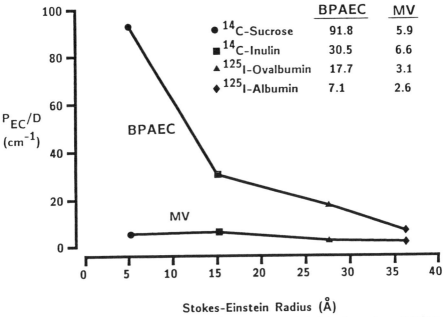

Figure 9. Comparison of selectivity of bovine pulmonary artery endothelial cell monolayers (BPAEC) and bovine pulmonary microvessel endothelial cell monolayers (MV) to molecules of various sizes. Actual endothelial permeability–diffusion coefficient ratios (P_{EC}/D) (37°C) are shown for the two cell lines.

cellular junctions in microvessel cells (Table III). Thus, the greater complexity of the intercellular junctions could account for the significantly lower inulin and sucrose permeabilities in the pulmonary microvessel endothelial cells. The transport path length is an unlikely explanation for the relatively low microvessel endothelial permeability values since the path length of the junctions is about 25% greater in microvessel endothelial cell monolayers than in pulmonary artery endothelial mono-layers (Table III).

Studies of endothelial cells from different sites in the vasculature have demon-strated organ-specific antigens on capillary endothelial cells which may be respon-sible for the different permeability in the regional vascular beds (Auerbach *et al.*, 1985). For example, there are phenotypic differences in the lectin-binding domains of different vascular endothelial cells (Del Vecchio *et al.*, 1993). Specific lectins [*Ricinus communis* agglutinin (RCA) and peanut agglutinin (PNA)] bind to pulmon-ary microvessel endothelial glycoproteins (galactose-containing glycoproteins of 220–160, 60, and 30–40 kDa). Belloni and Nicolson (1988) have also recently described differences in cell surface glycoproteins of endothelial cells in different vascular beds which may be related to the permeability characteristics of these vascular endothelial cells *in situ*.

8. ACTIVE TRANSPORT OF ALBUMIN

A study by Shasby (Shasby and Shasby, 1985) using cultured endothelial monolayers from porcine pulmonary arteries indicated that the transendothelial albumin flux is asymmetric; i.e., the transport albumin from the abluminal to the luminal side of the endothelial monolayer was ~ tenfold greater than transport from the luminal to the abluminal side. Siflinger-Birnboim *et al.* (1986) using a similar system were unable to confirm this observation using bovine or ovine pulmonary artery endothelial cell monolayers. In the latter study, transendothelial albumin flux was measured in the absence of hydrostatic and oncotic pressure differences by adding [^{125}I] albumin tracer either to the luminal or to the abluminal side of the endothelium or by simultaneously added [^{125}I]albumin to the luminal side of the endothelium and [^{131}I]albumin to the abluminal side of the endothelium (Siflinger-Birnboim *et al.*, 1986). The results indicated that the flux of albumin in either direction was symmetric. It is possible that the observed differences in the bidirectional albumin transport may be due to factors such as differences in cell types, presence of hydrostatic pressure favoring transport in one direction, problems with increased free tracer [^{125}I]albumin on the abluminal side, and error due to the effect of unstirred layers. Shasby's results were particularly intriguing because metabolic poisons such as cyanide inhibited the asymmetric albumin transport (Shasby and Shasby, 1985), and hence they remain all the more inexplicable. The nature of albumin transport across the endothelial barrier would need to be revised if this mechanism of albumin transport is confirmed by other studies.

9. RECEPTOR-MEDIATED ALBUMIN TRANSCYTOSIS

The endothelium does not only act as a semipermeable membrane that enables the transport of these macromolecules according to molecular size and charge. Recent studies indicate that transendothelial flux of proteins such as insulin, transferrin, and albumin may involve recognition by receptors located on the luminal side of the endothelial cell (Jefferies *et al.*, 1984; King and Johnson, 1984; Ghitescu *et al.*, 1986; Milici *et al.*, 1987; Schnitzer *et al.*, 1988a; Siflinger-Birnboim *et al.*, 1991b).

Albumin has been shown to bind to the endothelial glycocalyx components predominantly within plasmalemmal vesicles on the luminal side of the endothelium (Yokoto, 1983; Jefferies *et al.*, 1984; King and Johnson, 1984; Ghitescu *et al.*, 1986; Milici *et al.*, 1987; Schnitzer *et al.*, 1988a). Once bound, it apparently crosses the endothelial cell by shuttling of vesicles from the luminal to the abluminal side of the cell (Ghitescu *et al.*, 1986; Milici *et al.*, 1987). Albumin was shown to bind to cultured microvascular rat endothelium in a specific, saturable, and reversible manner (Schnitzer *et al.*, 1988a). Several "albumin receptors" such as a 60-kDa glycoprotein (gp60) (Schnitzer *et al.*, 1988b) and 18 and 31-kDa cell surface proteins (Ghinea *et al.*, 1988, 1989) have recently been described.

Albumin binding to endothelial cells and its potential role in transendothelial flux of albumin have been explored using cultured BPAEC monolayers (Siflinger-Birnboim *et al.*, 1991b). In the presence of unlabeled albumin (0 to 60 mg/ml), binding of tracer [^{125}I]albumin to BPAEC monolayers grown in fibronectin-coated 24-well plates was saturated at 5 mg/ml unlabeled albumin (Fig. 10), indicating that albumin competed with [^{125}I]albumin for specific sites on the BPAEC surface. Binding of [^{125}I]albumin to BPAEC monolayers was reduced at 4°C compared to 37°C and [^{125}I]albumin binding was reversible. The binding of [^{125}I]albumin was independent of the concentration of other unlabeled protein (e..g, gelatin) (Fig. 10). The apparent equilibrium binding affinity constant (K_d) for albumin binding to BPAEC is 6×10^{-7} M with a maximum number of 1.45×10^6 BSA molecules per cell. However, albumin binds with a higher affinity in pulmonary microvessel endothelial cells (Schnitzer *et al.*, 1988a), which is consistent with a greater number of vesicles in these cells (Table III).

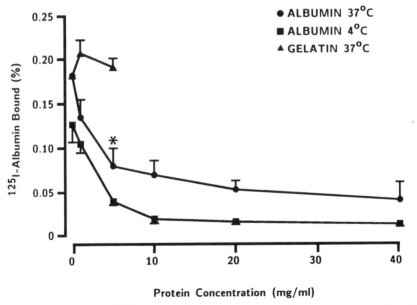

Figure 10.   Competition for [^{125}I]albumin binding sites on the endothelial monolayer in the presence of increased concentrations of unlabeled albumin and gelatin. The curve was displaced downwards at 4°C compared to 37°C ($p < 0.05$) in the presence of unlabeled albumin. Saturation of [^{125}I]albumin binding occurred at concentrations greater than 5 mg/ml (asterisk). The effect was not observed in the presence of increasing concentrations of gelatin. Values are mean ± SEM (Siflinger-Birnboim *et al.*, 1991b).

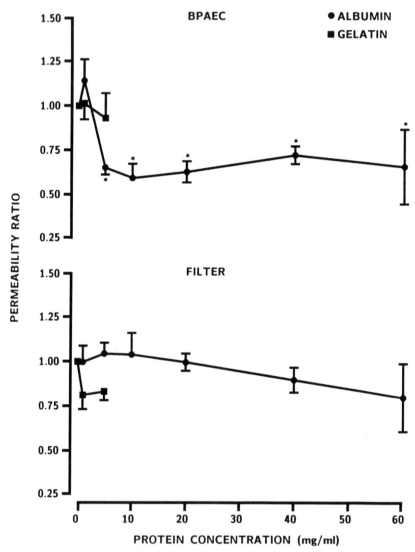

Figure 11. Permeability to [125I]albumin for the bovine pulmonary artery endothelial cell monolayers (BPAEC) grown on the filter and for the filter without the endothelial cells as a function of increasing unlabeled albumin and gelatin concentration. The permeability ratio is the [125I]albumin permeability in the presence of unlabeled protein divided by the [125I]albumin permeability in the absence of unlabeled protein. Values are mean ± SEM. Asterisks indicate values different from one (Siflinger-Birnboim *et al.*, 1991b).

The effect of albumin binding to endothelial cells on endothelial permeability to [125I]albumin was studied by adding unlabeled bovine serum albumin (0 to 60 mg/ml) and a constant amount of [125I]albumin tracer to BPAEC monolayers grown on microporous filters as described above. The transendothelial permeability to [125I]-albumin decreased by about 40% at unlabeled albumin concentrations of 5 mg/ml and remained at this level at higher albumin concentrations (Fig. 11). Addition of other unlabeled proteins (e.g., gelatin) to the medium did not alter the transendothelial permeability to [125I]albumin (Fig. 11). This effect is not observed for the filters without the endothelial cells (Fig. 11).

The lectin *Ricinus communis* agglutinin (RCA) binds to the 60-kDa albumin-binding endothelial surface glycoprotein (gp60) in rat epididymal fat pad endothelial cells (Schnitzer *et al.*, 1988b). RCA also precipitates a 60-kDa glycoprotein on BPAEC plasmalemmal membrane, indicating that gp60 was also present in these cells (Siflinger-Birnboim *et al.*, 1991b). The addition of RCA produced a 60% decrease in [125I]albumin binding to BPAEC, which did not occur when RCA was complexed with its cognate hapten monosaccharide β-D-galactose; therefore, RCA appears to interact with a galactose-containing component of gp60. Addition of RCA to the BPAEC monolayer reduced [125I]albumin permeability by 40% (Fig. 12), which was similar to the decrease in albumin permeability observed with excess unlabeled albumin (Fig. 11). Several other lectins such as *Ulex europaeus* agglutinin (UEA) and soybean agglutinin (SBA) (which did not bind to gp60) had no effect on [125I]albumin

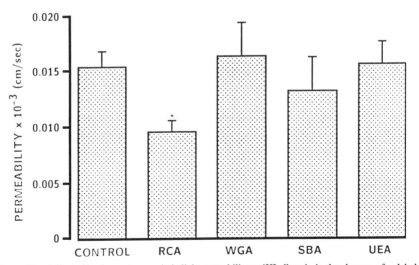

Figure 12.   Effect of lectins on transendothelial permeability to 125I albumin in the absence of unlabeled albumin. Maximum effect was observed at lectin concentrations of 50 μg/ml. RCA, *Ricinus communis* agglutinin; WGA, wheat germ agglutinin; SBA, soybean agglutinin; UEA, *Ulex europaeus* agglutinin. Values are mean ± SEM. Asterisk indicates different from control (Siflinger-Birnboim *et al.*, 1991b).

permeability (Fig. 12). These results are the first to demonstrate a role of albumin binding to endothelial cell membrane glycoprotein in the transendothelial albumin flux. Although the proportion of albumin transport in BPAEC dependent on binding of albumin is 40%, there are relatively few vesicles in BPAEC compared to microvessel cells (Table III). Therefore, it is conceivable that pulmonary microvessel cells have a greater proportion of albumin transport dependent on albumin binding.

A specific receptor-linked transport of albumin may be physiologically important in the transport of lipids such as albumin-linked long-chain free fatty acids and hormones across the vessel wall (Peters, 1975). The upregulation of gp60 may be a factor in the inflammatory response characterized by increased endothelial permeability to albumin. The signals regulating synthesis and expression of endothelial cell surface gp60 are not known. Upregulation of gp60 in vascular endothelial cells such as the blood–brain barrier (which is normally impermeable to albumin) may allow the transport of solutes by a receptor-linked mechanism.

10. MECHANISMS OF INCREASED ENDOTHELIAL PERMEABILITY

It is becoming clear that endothelial permeability increases secondary to the activation of second messenger pathways. In the following sections, some of the important mechanisms contributing to the increase in permeability are discussed.

10.1. Characteristics of Increased Endothelial Permeability

The effects of thrombin, a prototypic inflammatory agent mediating increases in endothelial permeability, have been examined in several recent studies (Galdal et al., 1984; Garcia et al., 1986; Siflinger-Birnboim et al., 1988a; Minnear et al., 1989; Aschner et al., 1990). Another well-studied mediator, histamine, also increased endothelial monolayer permeability in vitro probably by activation of common second messenger pathways (Majno and Palade, 1961; Rotrosen and Gallin, 1986; Bottaro et al., 1986; Killackey et al., 1986). Events occurring on the endothelial cell membrane are critical for the initiation of the cascade of events leading to the increase in permeability. The permeability-increasing effects of histamine are mediated by binding to histamine receptors (N. Simionescu et al., 1982). In contrast, the permeability increase induced by thrombin is the result of more complex interactions between thrombin and the cell membrane involving binding to the membrane as well as proteolytic cleavage of membrane phospholipids (Aschner et al., 1990). Thrombin binding alone was insufficient in explaining the thrombin-mediated increase in endothelial permeability (Aubrey et al., 1979; Lollan and Owen, 1980). Thrombin's active catalytic site is a critical requirement for the response, indicating that proteolysis of cell membrane components and activation of second messenger pathways are responsible for the increase in permeability (Garcia *et al.*, 1986; Aschner *et al.*, 1990).

The effect of thrombin on endothelial permeability is rapid (within 5 min after its addition to the endothelial monolayer) (Lum *et al.*, 1989) and the response is reversible within 30 min after cell wash (Garcia *et al.*, 1986; Phillips *et al.*, 1989). The increase in endothelial permeability is associated with the formation of inter-cellular gaps and with alterations in the cytoskeletal elements described below.

10.2. Cytoskeletal Alterations

The increase in permeability is characteristically associated with loss of the peripheral actin filaments and the centralization of cytosolic actin filaments (Phillips *et al.*, 1989). This response is evident within the time course of the permeability increase. The reversal of increased permeability by removing the mediator is also correlated with a reexpression of peripheral actin filaments (Garcia *et al.*, 1986; Phillips *et al.*, 1989). The increased endothelial permeability involves changes in cell shape secondary to alterations in the endothelial actin filaments resulting in the formation of intercellular gaps (Garcia *et al.*, 1986).

The role of intercellular gaps induced by the change in cell shape has been confirmed by a recent observation involving osmotic shrinkage of endothelial cells by the addition of hypertonic solution to the cell bathing medium (Shepard *et al.*, 1987). This resulted in formation of intercellular gaps and an increase in endothelial permeability to [^{125}I]albumin, an effect which was reversed by rehydrating the cells (Shepard *et al.*, 1987).

A study by Phillips *et al.* (1989) points to the role of peripheral actin filaments in junctional stability and the regulation of endothelial permeability under certain stimuli. Pretreatment of endothelial cells with 0.3 μM 7-nitrobenz-2-oxa-1,3-diazole (NBD)-phallacidin, a specific actin-stabilizing agent, prevented the changes in actin filament distribution and markedly attenuated the increase in albumin permeability induced by α-thrombin. Disruption of the endothelial F-actin with cytochaslasins B and D resulted in an increase in macromolecular permeability of vessels in intact lungs and cultured endothelial cell monolayers (Shasby *et al.*, 1982), further support-ing a role of F-actin in the response. These findings indicate that F-actin filaments, particularly the peripheral bands, contribute to the maintenance of endothelial barrier function and the shift in distribution in F-actin mediates the increase in endothelial permeability.

F-actin may be an important cytoskeletal protein for regulation of cell shape for several reasons (Rotrosen and Gallin, 1986; Savion *et al.*, 1982; Shasby *et al.*, 1982; Phillips *et al.*, 1989). F-actin is a globular monomer that assembles reversibly to form long fibers (Stossel, 1984). Changes in the state of assembly in different parts of the cell account for differences in cytoplasmic consistency, and thereby can cause a change in cell shape (Stossel, 1984). F-actin fibers, if sufficiently stiff or organized into bundles, maintain the endothelial cells in a particular configuration (Stossel, 1984). F-actin acting in conjunction with myosin and other actin-binding proteins, vinculin and myosin light chain, can cause contracture of cells (Stossel, 1984). Therefore, the properties of F-actin point to its dynamic role in changing endothelial cell shape in response to inflammatory mediators.

The formation of intercellular gaps by the shift in F-actin does not imply that this mechanism is solely responsible for the regulation of transport of albumin. Stelzner *et al.* (1989) demonstrated that cholera toxin-induced cAMP resulted in decreased permeability with loss of peripheral bands, which suggests that peripheral actin band alterations may not be a critical determinant of the permeability alterations. The role of F-actin distribution may depend on the specific stimulus. It has also been shown that transendothelial albumin flux increased, following thrombin challenge of pulmonary endothelial cell monolayers, but IgG (160 kDa) flux was greater than albumin (69 kDa) flux (Siflinger-Birnboim *et al.*, 1988b). Had intercellular gaps been solely responsible for the effect, the IgG flux should have increased to the same extent as did the albumin flux. The greater increase in albumin transport implies that formation of endothelial vesicles also contributes to the increased albumin permeability.

10.3. Intracellular Ca^{2+} Shifts

Recent evidence indicates that endothelial permeability is critically dependent on a rise in intracellular Ca^{2+} concentration. The thrombin-induced increase in transendothelial [^{125}I]albumin clearance rate was inhibited by decreasing the availability of cytosolic Ca^{2+} (Lum *et al.*, 1989). The direct application of the Ca^{2+} ionophore, A23187, also increased transendothelial albumin permeability (Selden and Pollard, 1983), decreased transendothelial (Olesen, 1987) and transepithelial (Palant *et al.*, 1983) resistances, and increased hydraulic conductivity of intact microvessels (He *et al.*, 1990). The increase in [Ca^{2+}]$_i$ may signal the increase in endothelial permeability because Ca^{2+} is a known regulator of cytoskeletal assembly, structure, and contractility (Bennett and Weeds, 1986; Mooseker *et al.*, 1986). Reorganization of the actin cytoskeletal network involves a sequence of polymerization and depolymerization steps of actin as well as interactions of F-actin with other cytoskeletal proteins such as intermediate filament vimentin (Bershalsky *et al.*, 1990) and microbubules (Bershalsky *et al.*, 1990; Bhalla *et al.*, 1990). Both intermediate filaments (Bennett and Weeds, 1986) and microtubules (Marcum *et al.*, 1978; Bennett and Weeds, 1986) are regulated by changes in [Ca^{2+}]$_i$. Several actin-binding proteins are known to affect the polymerization state, cross-linking, and bundling activity of

F-actin in response to changes of $[Ca^{2+}]_i$ (Stossel *et al.*, 1985; Bennett and Weeds, 1986).

Ca^{2+} is also required for the activation of the phosphorylating enzymes Ca^{2+}–calmodulin-dependent-kinase and the phospholipid-dependent protein kinase (PKC) (England, 1986). Although the functional significance of phosphorylation of cytoskeletal proteins in mediating the increase in endothelial permeability remains to be determined, these kinases phosphorylate cytoskeletal proteins such as vinculin (Werth *et al.*, 1983), α-actinin (Stossel *et al.*, 1985), myosin light chain (Stossel *et al.*, 1985; England, 1986; Olesen, 1987), vimentin (Huang *et al.*, 1984), and microtubule-associated proteins (Selden and Pollard, 1983). The phosphorylated myosin light chain determines actin organization and cell shape change in fibroblasts (Bayley and Rees, 1986; Lamb *et al.*, 1988) and cell retraction in endothelial cells (Wysolmerski and Lagunoff, 1985).

Activation of PKC with the tumor promoter phorbol 13-myristate 12-acetate decreases the transepithelial resistances (Ojakian, 1981; Mullins and O'Brien, 1986), produces reorganization of actin and vinculin in several cell types (Schliwa *et al.*, 1984; Keller *et al.*, 1989), and increases in endothelial permeability (Lynch *et al.*, 1990). Cytosolic Ca^{2+} is required for activation of these kinases, and thus may be involved in the phosphorylation of cytoskeletal proteins. The intracellular events hypothesized to mediate the increase in endothelial permeability are summarized in Fig. 13.

Thrombin and other permeability-increasing agents cause a rapid initial rise in cytosolic $[Ca^{2+}]$ which is followed by a second phase of slow decay (Jaffe *et al.*, 1987; Ryan *et al.*, 1988; Lum *et al.*, 1989). Typical of the Ca^{2+} transient behavior in many cell types, the initial Ca^{2+} rise in the endothelial cells is due to Ca^{2+} mobilized from intracellular stores, whereas the second phase is caused by Ca^{2+} influx (Ryan *et al.*, 1988; Lum *et al.*, 1989). The initial Ca^{2+} rise likely occurs in response to increased inositol polyphosphate generation, particularly inositol 1,4,5-trisphospate, derived from phospholipase C-activated hydrolysis of phosphoinositides (Jaffe *et al.*, 1987; Putney *et al.*, 1989). The functional significance of the initial Ca^{2+} rise in regulating the permeability increase may involve the activation of Ca^{2+}-dependent enzymes such as phospholipase A_2 and PKC (Fig. 13).

Although the regulation of the second phase of the Ca^{2+} response is less well understood, this response is probably the critical signal mediating the increase in permeability (Fig. 14). Goligorsky *et al.* (1989) have shown that inhibition of the lipoxygenase pathway of arachidonate metabolism abolishes this long-lived Ca^{2+} rise and changes in F-actin cytoskeletal reorganization, suggesting that lipoxygenase products open Ca^{2+} channels and that the resultant prolonged rise in $[Ca^{2+}]_i$ provides the signal for the cytoskeletal reorganization and the increase in permeability. The lipoxygenase products, the monohydroxyeicosatetraenoic acids (HETEs), have been reported to directly increase vascular permeability (Burhop *et al.*, 1988), and may do it by opening Ca^{2+} channels as described above. The HETEs, therefore, may be the key intracellular messengers responsible for this second critical phase of the

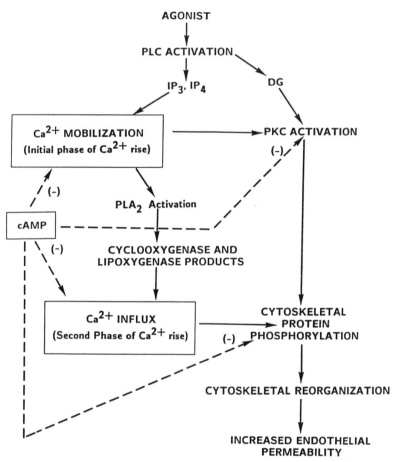

Figure 13. Schematics of hypothetical pathways by which Ca^{2+} and cAMP may regulate increases in endothelial permeability. PKC, protein kinase C; PLC, phospholipase C; IP_3, inositol 1,3,4- and 1,4,5-trisphosphate; IP_4, inositol 1,3,4,5,-tetrakisphosphate; DG, diacylglycerol; PLA_2, phospholipase A_2.

Ca^{2+} increase. The sequence of events by which the rise in Ca^{2+} "triggers" the increase in permeability are summarized in Fig. 13.

10.4. Protein Kinase C Activation

Activation of PKC is needed to phosphorylate cytoskeletal proteins, and thus realign F-actin filaments. Activation of PKC, which can occur as a result of the generation of 1,2-diacylglycerol, decreases transepithelial resistance (Gainer, 1985),

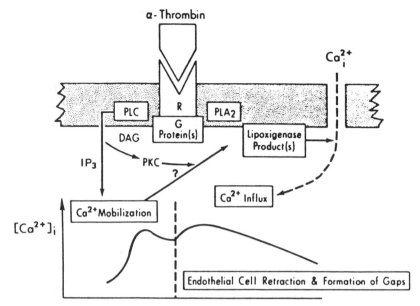

Figure 14. The hypothetical pathways of endothelial cell activation by thrombin leading to cytoskeletal reorganization, gap formation between endothelial cells and increases in endothelial permeability. PLC, phospholipase C; PLA_2, phospholipase A_2; DAG, diacylglycerol; PKC, protein kinase C. (From Goligorsky *et al.*, 1989.)

suggesting that the regulation of epithelial barrier transport occurs via a PKC-dependent pathway. The increased epithelial permeability was also associated with the phosphorylation of cytoskeletal proteins and decreased cell–cell contact (Ojakian, 1981; Gainer, 1985; Mullins and O'Brien, 1986).

Exposure of confluent BPAEC monolayers in culture to PMA or 1-oleoyl 2-acetyl glycerol (OAG) increased PKC activity in a concentration-dependent manner and also increased the transendothelial flux of [^{125}I]albumin (Lynch *et al.*, 1990). Neither 4α-phorbol 12,13-didecanoate nor 1-mono-oleoyl glycerol, which did not activate PKC, altered endothelial permeability. The increase in [^{125}I]albumin permeability induced by PMA was inhibited by the isoquinolinylsulfonamide derivative H7 (Lynch *et al.*, 1990), a strong PKC inhibitor (Hidaka *et al.*, 1984), but not by the control compound HA1004 (Hidaka *et al.*, 1984). After 16 hr of exposure to PMA, cytosolic PKC activity was significantly reduced, and the [^{125}I]albumin permeability returned to baseline. Further challenge with PMA at this time resulted in no increase in PKC activity, indicating a downregulation or depletion of the enzyme; interestingly, the subsequent PMA challenge did not increase endothelial permeability. Exposure of endothelial monolayers to PLC or α-thrombin (both of which increased membrane phosphatidylinositide turnover) also induced concentration-dependent activation of PKC and increases in [^{125}I]albumin endothelial permeability. Both the thrombin- and PLC-induced permeability increases were inhibited by H7. These results indicate that PKC activation is an important signal transduction pathway by which extracellular mediators increase transendothelial molecular flux.

Studies have not examined the relationship between the increase in [Ca^{2+}]$_i$ and PKC activation in mediating the increase in endothelial permeability. It is tempting to propose that increased [Ca^{2+}]$_i$ is linked to PKC activation, which in turn phosphorylates cytoskeletal proteins and causes cytoskeletal reorganization resulting in a "rounding up" of endothelial cells and/or increased vesicular transport. This may be the common pathway by which a variety of inflammatory mediators increase endothelial permeability.

10.5. Basement Membrane and Matrix Components

The behavior of cells is different when the cells are grown and maintained on extracellular matrix versus nonbiological surfaces (Phillips and Tsan, 1988; Madri *et al.*, 1988). Moreover, cell behavior can be modulated depending on the composition and organization of the matrix components or tissue used (Phillips and Tsan, 1988; Madri *et al.*, 1988). An organized basement membrane and extracellular matrix surrounding the endothelium may control transendothelial solute flux and this may be dependent on the particular extracellular matrix [e.g., glomerular matrix is known to restrict albumin transport because of the negative charge of its glycosaminoglycan constituents (Taylor and Granger, 1984)]. *In vivo* studies indicate that the interstitial matrix is capable of 14-fold reduction in diffusive transport of albumin (Fox and Wayland, 1979). Collagen gels used to examine the barrier function of the matrix components were selective to molecules ranging from 39 to 110 kDa (Shaw and Schy, 1979). Application of extracellular matrix consisting primarily of type I collagen to microporous filters produced a 10-fold reduction in the transport of [^{125}I]albumin. Albumin restriction was not increased by coating the filters with fibronectin, indicating that only certain matrix components are required to impose a restrictive barrier. The core matrix proteins (such as collagens I and IV) may be key determinants of transendothelial protein flux. Other matrix proteins such as laminin, because of its abundance in the matrix, may also determine cell–substratum adhesion and cell–cell contact (Michel, 1984), and thereby may regulate endothelial cell shape in response to inflammatory mediators. Layering endothelial cells on matrix composed of collagens I and IV and laminin may enable these cells to form a more restrictive barrier through the development of complex intercellular junctions.

The interrelationships among the matrix components and how they regulate endothelial macromolecular permeability are poorly understood. Extracellular matrix can be remodeled by the endothelial cell proteases, causing endothelial cells layered on this matrix to become more permeable (Partridge *et al.*, 1991). If the matrix causes endothelial cells to become more permeable, it is equally possible that other matrix alterations can cause the endothelial cells to become more restrictive.

11. DECREASES IN ENDOTHELIAL PERMEABILITY INDUCED BY CYCLIC NUCLEOTIDES

There is good evidence indicating that increases in $[cAMP]_i$ produced by agents such as cholera toxin, forskolin, and isoproterenol, decrease endothelial permeability (Casnocha et al., 1989; Stelzner et al.,1989; Siflinger-Birnboim and Malik, 1990; Oliver, 1990) and inhibit the permeability-increasing effects of several mediators including thrombin (Casnocha et al., 1989; Minnear et al., 1989) and histamine (Killackey et al., 1986; Carson et al., 1989). This effect of cAMP is associated with an increase in cytoskeletal F-actin (Stelzner et al., 1989) and inhibition of the F-actin reorganization caused by permeability-increasing agents such as thrombin (Minnear et al., 1989).

It has been shown in microvessel endothelial monolayers grown on gelatin- and fibronectin-coated filters as described above, that endothelial permeability to sucrose, inulin, ovalbumin, and albumin was reduced in the presence of isoproterenol $(2 \times 10^{-6}$ M) (which produced a threefold increase in intracellular cAMP) compared with control (untreated) endothelial cells (Fig. 15). The decrease in permeability of the small molecules (sucrose and inulin) was greater than that of the large molecules (ovalbumin and albumin), indicating that increased cAMP concentration primarily reduced the transport through paracellular pathways. The cytoskeletal and cell shape changes responsible for this effect are likely to be opposite to those occurring during increased permeability.

However, the mechanism of action of cAMP remains unknown. The permeability decrease may involve a cAMP-mediated activation of cAMP-dependent kinases which phosphorylate proteins such as myosin light chain kinase (Fig. 14). Their phosphorylation can inhibit the kinase activity, and thus may inhibit myosin light chain phosphorylation and its interaction with F-actin (Bayley and Rees, 1986; Lamb et al., 1988). Another possibility is that increases in cellular cAMP may inhibit increases in $[Ca^{2+}]_i$ and PKC activation (Lanza et al., 1987; Takuwa et al., 1988; McCann et al., 1989).

12. ENDOTHELIAL WATER PERMEABILITY

The capillary endothelium provides the primary resistance to transvascular flow of water (Johnson, 1966; Curry, 1980; Taylor and Granger, 1984), although the relative distribution of transcapillary water flow through the paracellular and transcellular pathways remains controversial (Curry, 1980; Michel, 1984). Albumin is a

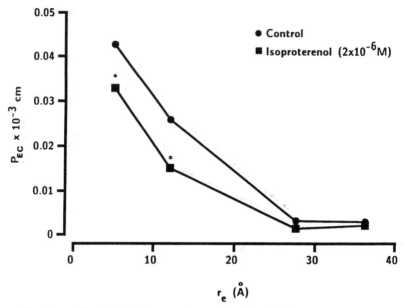

Figure 15. Permeability (P_{EC}) of pulmonary microvessel endothelial cell monolayers in the presence of isoproterenol (2×10^{-6} M). Selectivity of the monolayers to molecules of various molecular radius is maintained in the presence of isoproterenol.

major determinant of endothelial water permeability since albumin serves to regulate the vessel wall hydraulic conductivity (Curry, 1980; Curry and Michel, 1980; Michel *et al.*, 1985; Huxley and Curry, 1987). This observation is the cornerstone of the "fiber matrix" hypothesis in that an interaction between albumin and the glycocalyx and interendothelial molecules (hyaluronic acid and sulfated proteoglycans) is responsible for regulating water flow across the endothelial barrier (Curry, 1980; Curry and Michel, 1980).

Using a modification of the system described above for the measurement of endothelial permeability to albumin and other solutes, the hydraulic conductivity has been measured in endothelial cell monolayers grown on polycarbonate filters (Fig. 16) (Powers *et al.*, 1989; McCandless *et al.*, 1991). Hydraulic conductivity across the filters alone was $3.2 \pm 0.3 \times 10^{-3}$ cm/sec per cm H_2O, whereas hydraulic conductivity across the endothelial monolayer on the filter was $17.4 \pm 2.7 \times 10^{-5}$ cm/sec per cm H_2O [these values are 10- to 100-fold greater than those reported for intact vessel walls (Michel *et al.*, 1985), which reflects the lack of series resistance due to absence of basement membrane and adjacent interstitium (Fox and Wayland, 1979)]. Exposure of monolayers to albumin (0.5 to 2.5 mg/ml) decreased the hydraulic conductivity in a concentration- dependent manner (Powers *et al.*, 1989). This did not appear to be due to mechanical "plugging" of interendothelial clefts or other

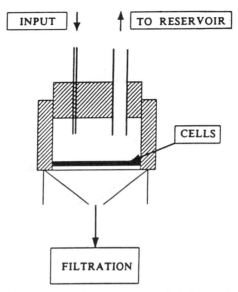

Figure 16. Schematic of the apparatus used to measure transendothelial protein and water flux at various hydrostatic pressures. (From Powers *et al.*, 1989.)

pathways since the decrease in hydraulic conductivity could be reversed by subsequent removal of the albumin from the media and since the hydraulic conductivity did not decrease further when the medium albumin concentration was increased from 2.5 mg/ml to 10 mg/ml (Powers *et al.*, 1989). The effect of albumin was the result of physiochemical interaction of albumin with endothelial cells since the response did not occur when albumin was replaced with 70-kDa dextran (McCandless *et al.*, 1991). Michel *et al.* (1985) using frog mesenteric vessels suggested that the interaction of albumin with the capillary endothelium occurred by the association of arginyl residues on the albumin molecule with negative charge on the endothelium. This interaction has been confirmed in mammalian vascular endothelial cells in which the arginyl residues of albumin are required for the response, possibly as a result of a charge interaction of albumin with the endothelial cell mediated via the arginyl residues (Powers *et al.*, 1989).

13. CONCLUSIONS

The endothelial barrier allows the free exchange of water but is restrictive, to varying degrees in different microvascular beds, to the transport of solutes. For example, in the brain microvessels, the endothelial barrier restricts the transport of albumin, whereas the endothelial barrier is semipermeable in the fenestrated and continuous endothelial cells. The endothelial cell monolayer demonstrates selectivity, i.e., the permeation of the transported molecules is inversely related to their molecular size. The "pore" theory describes to a degree the transendothelial flux of solutes but fails to take into account flux via transcellular routes such as vesicles and does not take into account the role of charge in transport. In addition to molecular size, transport of solutes is dependent on the charge of solutes and endothelial cell membrane, as well as the ability of the molecules to bind to receptors and to be internalized by endothelial cells. Receptor-mediated transcytosis of albumin may contribute to about 50% of albumin transport with the remainder of the transport via the paracellular routes.

Increased endothelial permeability in inflammatory states is dependent on the shape and configuration of endothelial cells as determined by alterations in F-actin and the interaction of endothelial cells with the substratum matrix proteins. The increase in permeability is governed by activation of intracellular second messenger pathways.

ACKNOWLEDGMENT. Supported by NIH grants HL32418, HL27016, and HL45638.

REFERENCES

Albelda, S. M., Sampson, P. M., Haselton, F. R., McNiff, J. M., Mueller, S. N., Williams, S. K., Fishman, A. P., and Levine, E. M., 1988, Permeability characteristics of cultured endothelial cell monolayers, *J. Appl. Physiol.* **64**:308–322.

Aschner, J. L., Lennon, J. M., Fenton, J. W., II, Aschner, M., and Malik, A. B., 1990, Enzymatic activity is necessary for thrombin-mediated increase in endothelial permeability, *Am. J. Physiol.* **259**:L270–L275.

Aubrey, B. J., Hoak, J. C., and Owen, W. G., 1979, Binding of human thrombin to cultured human endothelial cells, *J. Biol. Chem.* **254**;4092–4095.

Auerbach, R., Alby, L., Morrissey, L. W., Tu, M., and Joseph, J., 1985, Expression of organ-specific antigens on capillary endothelial cells, *Microvasc. Res.* **29**:401–411.

Barry, P. H., and Diamond, J. M., 1984, Effects of unstirred layers on membrane phenomena, *Physiol. Rev.* **64**:763–872.

Bayley, S. A., and Rees, D. A., 1986, Myosin light chain phosphorylation in fibroblast shape change, detachment and patching, *Eur. J. Cell Biol.* **42**:10–16.

Belloni, P. N., and Nicolson, G. L., 1988, Differential expression of cell surface glycoproteins on various organ-derived microvascular endothelial and endothelial cell cultures, *J. Cell. Physiol.* **136**:389–398.

Belloni, P. N., and Tressler, R. J., 1990, Microvascular endothelial cell heterogeneity: Interaction with leukocytes and tumor cells, *Cancer Metastasis Rev.* **8**:353–389.

Bennett, J., and Weeds, A., 1986, Calcium and the cytoskeleton, *Br. Med. Bull.* **42**:385–390.

Bershalsky, A. D., Ivanova, O. Y., Lyass, L. A., Pletyushkina, O. Y., Vasilev, J. M., and Gelfand, I. M., 1990, Cytoskeletal reorganization responsible for the phorbol ester-induced formation of cytoplasmic processes: Possible involvement of intermediate filaments, *Proc. Natl. Acad. Sci. USA* **87**:1884–1888.

Bhalla, D. K., Rasmussen, R. E., and Tjen, S., 1990, Interactive effects of O_2, cytochalasin D, and vinblastine on transendothelial transport and cytoskeleton in rat airways, *Am. J. Respir. Mol. Biol.* **3**:119–129.

Bizios, R., Blumenstock, F. A., Del Vecchio, P. J., and Malik, A. B., 1986, Permselectivity of cultured endothelial monolayers, *J. Cell Biol.* **106**:192a.

Boiadjeiva, S., Hallberg, C., Hogstrom, M., and Busch, C., 1984, Exclusion of trypan blue from microcaries by endothelial cells; An *in vitro* barrier function test, *Lab. Invest.* **50**: 239–246.

Bottaro, D., Shepro, D., Peterson, S., and Hechtman, H. B., 1986, Serotonin, norepinephrine and histamine mediation of endothelial cell barrier function *in vitro*, *J. Cell. Physiol.* **128**: 189–194.

Burhop, K. E,. Selig, W. M., and Malik, A. B., 1988, Monohydroxyeicosatetraenoic acids (5-HETE and 15-HETE) induce pulmonary vasoconstriction and edema, *Circ. Res.* **62**: 687–698.

Carson, M. R., Shasby, S., and Shasby, D. M., 1989, Histamine and inositol phosphate accumulation in endothelium: cAMP and a G protein, *Am. J. Physiol.* **257**:L259–L264.

Casnocha, C. A., Eskin, S. G., Hall, E. R., and McIntire, L. V., 1989, Permeability of human endothelial monolayers: Effect of vasoactive agonists and cAMP, *J. Appl. Physiol.* **62**: 1997–1005.

Cooper, J. A., Del Vecchio, P. J., Minnear, F. L., Burhop, K. E., Selig, W. M., and Malik, A. B., 1987, Measurement of albumin permeability across endothelial monolayers *in vitro*, *J. Appl. Physiol.* **62**:1076–1083.

Crone, C., and Levitt, D., 1984, Capillary permeability to small solutes, in: *Handbook of Physiology: The Cardiovascular System*, Volume IV (E. M. Renkin and C. C. Michel, eds.), American Physiological Society, Bethesda, pp. 411–466.

Curry, F. E., 1980, Mechanism and thermodynamics of transcapillary exchange, in: *Handbook of Physiology*, Section 2, The Cardiovascular System, Volume IV, Microcirculation, Part 1 (E. M. Renkin and C. C. Michel, eds.), American Physiological Society, Bethesda, pp. 309–374.

Curry, F. E., and Michel, C. C., 1980, A fiber matrix model of capillary permeability, *Microvasc. Res.* **20**:96–99.

Danon, D., and Skutelsky, E., 1976, Endothelial surface charge and its possible relationship to thrombogenesis, *Ann. N. Y. Acad. Sci.* **275**:47–63.

Del Vecchio, P. J., Siflinger-Birnboim, A., Shepard, J. M., and Malik, A. B., 1987, Endothelial monolayer permeability to macromolecules, *Fed. Proc.* **46**:2511–2515.

Del Vecchio, P. J., Belloni, P. N., Holleran, L. A., Lum, H., Siflinger-Birnboim, A., and Malik, A. B., 1993, Pulmonary microvascular endothelial cells: Cell surface glycoproteins and barrier function *In Vitro Cell. Dev. Biol.* (in press).

England, P. J., 1986, Intracellular calcium receptor mechanisms, *Br. Med. Bull.* **42**:375–383.

Fox, J. R., and Wayland, H., 1979, Interstitial diffusion of macromolecules in the rat mesentery, *Microvasc. Res.* **18**:255–276.

Gainer, M., 1985, Diacylglycerol inhibits gap junction communication in cultured epithelial cells: Evidence for a role of protein kinase C, *Biochem. Biophys. Res. Commun.* **126**: 1109–1113.

Galdal, K. S., Evensen, S. A., Hoglund, S., and Nilsen, E., 1984, Actin pools and actin microfilament organization in cultured human endothelial cells after exposure to thrombin, *Br. J. Haematol.* **58:**617–625.

Garcia, J. G. N., Siflinger-Birnboim, A., Bizios, R., Del Vecchio, P. J., Fenton, J. W., II, and Malik, A. B., 1986, Thrombin-induced increases in albumin transport across cultured endothelial monolayers, *J. Cell. Physiol.* **128:**96–104.

Gerritsen, M. E., 1987, Functional heterogeneity of vascular endothelial cells, *Biochem. Pharmacol.* **36:**2701–2711.

Ghinea, N., and Simionescu, N., 1985, Anionized and cationized hemeundecapeptides as probes for cell surface charge and permeability studies: Differential labelling of endothelial plasmalemmal vesicles, *J. Cell Biol.* **100:**606–612.

Ghinea, N., Fixman, A., Alexandru, D., Popov, D., Hasu, M., Ghitescu, L., Eskenazy, M., Simionescu, M., and Simionescu, N., 1988, Identification of albumin binding proteins in capillary endothelial cells, *J. Cell Biol.* **107:**231–239.

Ghinea, N., Eskenazy, M., Simionescu, M., and Simionescu, N., 1989, Endothelial albumin binding proteins are membrane-associated components exposed on the endothelial cell surface, *J. Biol. Chem.* **264:**4755–4758.

Ghitescu, L., Fixman, A., Simionescu, M., and Simionescu, N., 1986, Specific binding sites for albumin restricted to plasmalemmal vesicles of continuous capillary endothelium: Receptor-mediated transcytosis, *J. Cell Biol.* **102:**1304–1311.

Goligorsky, M. S., Menton, D. N., Laszlo, A., and Lum, H., 1989, Nature of thrombin-induced sustained increase in cytosolic calcium concentration in cultured endothelial cells, *J. Biol. Chem.* **264:**16771–16775.

Grotte, G., 1956, Passage of dextran molecules cross the blood–lymph barrier, *Acta Chir. Scand.* **211:**1–84.

He, P., Pagakis, N., and Curry, F. E., 1990, Measurement of cytoplasmic calcium in single microvessels with increased permeability, *Am. J. Physiol.* **258:**H1366–H1374.

Hidaka, H., Inagaki, M., Kawamoto, S., and Sasaki, Y., 1984, Isoquinolinosulfonamides, novel, and potent inhibitors of cyclic nucleotide-dependent protein kinase and protein kinase C, *Biochemistry* **23:**5036–5041.

Huang, C.-K., Hill, J. M., Borman, B.-J., Jr., Mackin, W. M., and Becker, E. L., 1984, Chemotactic factors induced vimentin phosphorylation in rabbit peritoneal neutrophil, *J. Biol. Chem.* **259:**1386–1389.

Huxley, V. H., and Curry, F. E., 1987, Effect of superfusate albumin on single capillary hydraulic conductivity, *Am. J. Physiol.* **252:**H395–H401.

Jaffe, E. A., Grulich, J. Weksler, B. B., Hampel, G., and Watanabe, K., 1987, Correlation between thrombin-induced prostacyclin production and inositol triphosphate and cytosolic free calcium levels in cultured human endothelial cells, *J. Biol. Chem.* **262:**8557–8565.

Jefferies, W. A., Brandon, M. R., Hunt, S. V., Williams, A. F., Gatter, K. C., and Mason, D. Y., 1984, Transferrin receptor on endothelium of brain capillaries, *Nature* **312:**162–163.

Johnson, J. A., 1966, Capillary permeability, extracellular space estimation and lymph flow, *Am. J. Physiol.* **211:**1261–1263.

Kedem, O., and Katchalsky, A., 1958, Thermodynamic analysis of the permeability of biological membranes to non-electrolytes, *Biochim. Biophys. Acta* **27:**229–246.

Keller, H. V., Niggli, V., and Zimmerman, A., 1989, Diacylglycerols and PMA induce actin polymerization and distinct shape changes in lymphocytes: Relation to fluid pinocytosis and locomotion, *J. Cell Sci.* **93:**457–465.

Killackey, J. J. F., Johnson, M. G., and Movat, H. Z., 1986, Increased permeability of microcarrier-cultured endothelial monolayers in response to histamine and thrombin. A model for the *in vitro* study of increased vasopermeability, *Am. J. Pathol.* **122**:50–61.

King, G. L., and Johnson, S. M., 1984, Receptor-mediated transport of insulin across endothelial cells, *Science* **227**:1583–1586.

Lamb, N. J. C., Fernandez, A., Conti, M. A., Adelstein, R., Glass, D. B., Welcj, W. J., and Feramisco, J. R., 1988, Regulation of actin microfilament integrity in living non-muscle cells by the cAMP-dependent protein kinase and the myosin light chain kinase, *J. Cell Biol.* **106**:1955–1971.

Landis, E. M., and Pappenheimer, J. R., 1963, Exchange of substances through the capillary walls, in: *Handbook of Physiology: Circulation*, Section 2, Volume II (W. F. Hamilton and P. Dow, eds.), American Physiological Society, Washington, D.C., pp. 961–1034.

Lanken, P. N., Hansen-Flaschen, J. H., Sampson, P. M., Pietra, G. G., Haselton, F. R., and Fishman, A. P., 1985, Passage of unchanged dextrans from blood to lymph in awake sheep, *J. Appl. Physiol.* **59**:580–591.

Lanza, F., Beretz, A., Stierle, A., Corre, G., and Cazenave, J. P., 1987, Cyclic nucleotide phosphodiesterase inhibitors prevent aggregation of human platelets by rising cAMP and reducing cytoplasmic free calcium mobilization, *Thromb. Res.* **45**:477–484.

Lassen, N. A., and Trap-Jensen, J., 1970, Estimation of the fraction of the interendothelial slit which must be open in order to account for hydrophilic molecules in skeletal muscle in man, in: *Capillary Permeability* (C. Crone and N. A. Lassen, eds.), New York Academy of Sciences, pp. 647–653.

Lollan, P., and Owen, W. G., 1980, Clearance of thrombin from circulation in rabbits by high-affinity biding sites on endothelium, *J. Clin. Invest.* **66**:1222–1230.

Lum, H., Del Vecchio, P. J., Schneider, A. S., Goligorsky, M. S., and Malik, A. B., 1989, Calcium dependence of the thrombin-induced increase in endothelial albumin permeability, *J. Appl. Physiol.* **66**:1471–1476.

Lynch, J. J., Ferro, T. J., Blumenstock, F. A., Brockenauer, A. M., and Malik, A. B., 1990, Increased endothelial albumin permeability mediated by protein kinase C activation, *J. Clin. Invest.* **85**:1991–1998.

McCandless, B. K., Powers, M. R., Cooper, J. A., and Malik, A. B., 1991, Hydraulic conductivity of endothelial monolayers: Effect of albumin, *Am. J. Physiol.* **260**:L571–L576.

McCann, J. D., Bhalla, R. C., and Welsch, M. J., 1989, Release of intracellular calcium by two different second messengers in airway epithelium, *Am. J. Physiol.* **257**:L116–L124.

Madri, J. A., and Williams, S. K., 1983, Capillary endothelial cell cultures: Phenotypic modulation by matrix components, *J. Cell Biol.* **97**:153–165.

Madri, J. A., Pratt, B. M., and Yannariello-Brown, J., 1988, Endothelial cell–extracellular matrix interactions: Matrix as a modulator of cell function, in: *Endothelial Cell Biology* (N. Simionescu and M. Simionescu, eds.), Plenum Press, New York, pp. 167–188.

Majno, G., and Palade, G. E., 1961, Studies on inflammation. I. The effect of histamine and serotonin on vascular permeability: An electron microscopic study, *J. Biophys. Biochem. Cytol.* **11**:571–605.

Marcum, J. M., Dedman, J. R., Brinkley, B. R., and Means, A. R., 1978, Control of microtubule assembly–disassembly by calcium-dependent regulator proteins, *Proc. Natl. Acad. Sci. USA* **75**:3771–3775.

Michel, C. C., 1984, The fluid movement through capillary walls, in: *Handbook of Physiology:*

The Cardiovascular System, Section 2, Volume IV, Part I (E. M. Renkin and C. C. Michel, eds.), American Physiological Society, Bethesda, pp. 142–156.

Michel, C. C., Phillips, M. E., and Turner, M. R., 1985, The effects of native and modified bovine serum albumin on the frog mesenteric capillaries, *J. Physiol. (London)* **360:**333–346.

Milici, A. J., Watrous, N. E., Stukenbrok, H., and Palade, G. E., 1987, Transcytosis of albumin in capillary endothelium, *J. Cell Biol.* **105:**2603–2612.

Minnear, F. L., DeMichele, M. A. A., Moon, D. G., Rieder, C. L., and Fenton, II, J. W., 1989, Isoproterenol reduces thrombin-induced pulmonary endothelial permeability *in vitro*, *Am. J. Physiol.* **257:**H1613–H1623.

Mooseker, M. S., Coleman, T. R., and Conzelman, K. A., 1986, *Ciba Found. Symp.* **122:** 232–249.

Mullins, J. M., and O'Brien, T. G., 1986, Effects of tumor promoter on LLC-Pk, renal epithelial tight junctions and transepithelial fluxes, *Am. J. Physiol.* **251:**C597–C602.

O'Donnell, M., and Vargas, F. F., 1986, Electrical conductivity and its use in estimating an equivalent pore size for arterial endothelium, *Am. J. Physiol.* **250:**H16–H21.

Ojakian, G. K., 1981, Tumor promoter-induced changes in the permeability of epithelial cell tight junctions, *Cell* **23:**95–103.

Olesen, S. P., 1987, Regulation of ion permeability in frog brain venules. Significance of calcium, cyclic nucleotides and protein kinase C, *J. Physiol. (London)* **387:**59–68.

Oliver, J. A., 1990, Adenylate cyclase and protein kinase C mediates opposite actions on endothelial junctions, *J. Cell. Physiol.* **145:**536–542.

Palant, C. E., Duffey, B. K., Mookerjee, B. K., Ho, S., and Bentzel, C. J., 1983, Ca^{2+} regulation of tight-junction permeability and structure in Necturus gallbladder, *Am. J. Physiol.* **245:**C203–C212.

Pappenheimer, J. R., Renkin, E. M., and Borrero, L. M., 1951, Filtration, diffusion and molecular sieving through peripheral capillary membranes, *Am. J. Physiol.* **162:**13–46.

Partridge, C. A., Horvath, C. J., and Malik, A. B., 1991, Tumor necrosis factor increases endothelial monolayer permeability through alterations of extracellular matrix, *FASEB J.* **5:**656.

Pedley, J. T., 1983, Calculation of unstirred layer thickness in membrane transport experiments: A survey, *Q. Rev. Biophys.* **16:**115–150.

Perry, M. A., Berroit, J. N., Kovietys, P. R., and Granger, D. N., 1983, Restricted transport of cationic macromolecules across interstitial capillaries, *Am. J. Physiol.* **245:**G568–G572.

Peters, T. J., 1975, Serum albumin, in: *The Plasma Proteins: Structure, Function, and Genetic Control*, Volume I (F. W. Putman, ed.), Academic Press, New York, pp. 133–181.

Phillips, P. G., and Tsan, M.-F., 1988, Direct staining and visualization of endothelial monolayers cultured on synthetic polycarbonate filters, *J. Histochem. Cytochem.* **36:**551–554.

Phillips, P. G., Lum, H., Malik, A. B., and Tsan, M.-F., 1989, Phallacidin prevents thrombin-induced increases in endothelial permeability to albumin, *Am. J. Physiol.* **257:**C562–C567.

Polikan, P., Gimbrone, M. A., Jr., and Contran, R. S., 1979, Distribution and movement of anionic cell surface sites in cultured human vascular endothelial cells, *Atherosclerosis* **32:** 69–80.

Powers, M. R., Blumenstock, F. A., Cooper, J. A., and Malik, A. B., 1989, Role of albumin arginyl sties in albumin-induced reduction of endothelial hydraulic conductivity, *J. Cell. Physiol.* **141:**558–564.

Putney, J. W., Jr., Takemura, H., Hughes, A. R., Horstman, D. A., and Thastrup, O., 1989, How do inositol phosphates regulate calcium signalling? *FASEB J.* **3**:1899–1905.

Renkin, E. M., 1977, Multiple pathways of capillary permeability, *Circ. Res.* **41**:735–743.

Renkin, E. M., 1985, Capillary transport of macromolecules: Pores and other endothelial pathways, *J. Appl. Physiol.* **58**:315–325.

Renkin, E. M., 1988, Transport pathways and processes, in: *Endothelial Cell Biology* (N. Simionescu and M. Simionescu, eds.), Plenum Press, New York, pp. 51–68.

Rotrosen, D., and Gallin, J. I., 1986, Histamine type I receptor occupancy increases endothelial cytosolic calcium, reduces F-actin, and promotes albumin diffusion across cultured endothelial monolayers, *J. Cell Biol.* **103**:2379–2387.

Ryan, U. S., Avdonin, V., Posin, Y. E., Popov, E. G., Danilov, S. M., and Tkachuk, V. A., 1988, Influence of vasoactive agents on cytoplasmic free calcium in vascular endothelial cells, *J. Appl. Physiol.* **65**:2221–2227.

Savion, N., Vlodavsky, I., Greenburg, G., and Gospodarowicz, D., 1982, Synthesis and distribution of cytoskeletal elements in endothelial cells as a function of cell growth and organization, *J. Cell. Physiol.* **110**:129–141.

Schliwa, M., Nakamura, T., Porter, K. R., and Euteneuer, U., 1984, A tumor promoter induces rapid and coordinated reorganization of actin and vinculin in cultured cells, *J. Cell Biol.* **99**:1045–1059.

Schnitzer, J. E., and Carley, W. W., and Palade, G. E., 1988a, Specific albumin binding to microvascular endothelium in culture, *Am. J. Physiol.* **254**:H425–H437.

Schnitzer, J. E., Carley, W. W., and Palade, G. E., 1988b, Albumin interacts specifically with a 60 kDa microvascular endothelial glycoprotein, *Proc. Natl. Acad. Sci. USA* **85**:6773–6777.

Schnitzer, J. E., Shen, C.-P., and Palade, G. E., 1990, Lectin analysis of common glycoproteins detected on the surface of continuous microvascular endothelium *in situ* and in culture: Identification of sialoglycoproteins, *Eur. J. Cell Biol.* **52**:241–251.

Selden, S. C., and Pollard, T. D., 1983, Phosphorylation of microtubule-associated proteins regulates their interaction with actin filaments, *J. Biol. Chem.* **258**:7064–7071.

Shasby, M. D., and Shasby, S. S., 1985, Active transendothelial transport of albumin: Interstitium to lumen, *Circ. Res.* **57**:903–908.

Shasby, D. M., and Shasby, S. S., Sullivan, S. M., and Peach, M. J., 1982, Role of endothelial cell cytoskeleton in the control of endothelial permeability, *Circ. Res.* **51**:657–661.

Shasby, D. M., Lind, S. E., Shasby, S. S., Goldsmith, J. C., and Hunninghake, G. W., 1985, Reversible oxidant-induced increases in albumin transfer across cultured endothelium: Alterations in cell shape and calcium homeostasis, *Blood* **65**:605–614.

Shaw, M., and Schy, A., 1979, Molecular distribution within a 1% collagen gel column, *J. Chromatogr.* **170**:449–452.

Shepard, J. M., Goderie, S. K., Malik, A. B., and Kimelberg, H. K., 1987, Effects of alterations in endothelial cell volume on albumin permeability, *J. Cell. Physiol.* **133**:389–394.

Siflinger-Birnboim, A., and Malik, A. B., 1989, Effect of specific lectins on transendothelial albumin transport, *FASEB J.* **3**:A1139.

Siflinger-Birnboim, A., and Malik, A. B., 1990, Effects of cAMP on neutrophil (PMN)-mediated increase in endothelial permeability, *FASEB J.* **4**:A838.

Siflinger-Birnboim, A., Del Vecchio, P. J., Cooper, J. A., and Malik, A. B., 1986, Transendothelial albumin flux: Evidence against asymmetric albumin transport, *J. Appl. Physiol.* **61:** 2035–2039.

Siflinger-Birnboim, A., Del Vecchio, P. J., Cooper, J. A., Blumenstock, F. A., Shepard, J. M., and Malik, A. B., 1987, Molecular sieving characteristics of the cultured endothelial monolayer, *J. Cell. Physiol.* **132:**111–117.

Siflinger-Birnboim, A., Cooper, J. A., Del Vecchio, P. J., Lum, H., and Malik, A. B., 1988a, Selectivity of the endothelial monolayer: Effects of increased permeability, *Microvasc. Res.* **36:**216–227.

Siflinger-Birnboim, A., Lum, H., Blumenstock, F. A., and Malik, A. B., 1988b, Binding of albumin to the endothelial monolayer, *Microvasc. Res.* **2:**A1879.

Siflinger-Birnboim, A., Schnitzer, J. E., Del Vecchio, P. J., and Malik, A. B., 1991a, Differences in glycosylation, barrier function and structure in cultured pulmonary microvascular and artery endothelial cells, *FASEB J.* **5:**A754, 2227.

Siflinger-Birnboim, A., Schnitzer, J. E., Lum, H., Blumenstock, F. A., Shen, J.C.-P., Del Vecchio, P. J., and Malik, A. B., 1991b, Lectin binding to gp60 decreases specific albumin binding and transport in pulmonary arterial endothelial monolayers, *J. Cell Physiol.* **149:**575–584.

Simionescu, M., and Simionescu, N., 1984, Ultrastructure of the microvascular wall: Functional correlations, in: *Handbook of Physiology: The Cardiovascular System*, Section 2, Volume IV (E. M. Renkin and C. C. Michel, eds.), American Physiological Society, Bethesda, pp. 41–101.

Simionescu, N., Simionescu, M., and Palade, G. E., 1981, Differentiated microdomains on the luminal surface of the capillary endothelium. I. Preferential distribution of anionic sites, *J. Cell Biol.* **90:**605–613.

Simionescu, M., Simionescu, N., and Palade, G. E., 1982, Differentiated microdomains on the luminal surface of the capillary endothelium: Distribution of lectin receptors, *J. Cell Biol.* **94:**406–413.

Simionescu, N., Heltinu, C., Antohe, F., and Simionescu, M., 1982, Endothelial cell receptors for histamine, *Ann. N.Y. Acad. Sci.* **401:**132–148.

Skutelsky, E., and Danon, D., 1976, Redistribution of surface anionic sites on the luminal front of blood vessel endothelium after interaction with polycationic ligand, *J. Cell Biol.* **71:** 232–241.

Starling, E. H., 1896, On the absorption of fluids from the connective tissue spaces, *J. Physiol. (London)* **19:**312–326.

Stelzner, T. J., Weil, J. V., and O'Brien, R. F., 1989, Role of cyclic adenosine monophosphate in the induction of endothelial barrier properties, *J. Cell. Physiol.* **139:**157–166.

Stossel, T. P., 1984, Contribution of actin to the structure of the cytoplasmic matrix, *J. Cell Biol.* **99:**15S–21S.

Stossel, T. P., Chapponier, C., Ezzell, R. M., Hartwig, J. H., Janney, P. A., Kwiatlowsky, D. J., Lind, S. E., Southwick, D. B., Yin, H. L., and Zaner, K. S., 1985, Nonmuscle actin-binding proteins, *Annu. Rev. Cell Biol.* **1:**353–402.

Takuwa, Y., Takuwa, N., and Rasmussen, H., 1988, The effects of isoproterenol on intracellular calcium concentration, *J. Biol. Chem.* **263:**762–768.

Taylor, A. E., and Gaar, K. A., 1970, Estimation of equivalent pore radii of pulmonary artery and alveolar membranes, *Am. J. Physiol.* **218:**1133–1140.

Taylor, A. E. and Granger, D. N., 1984, Exchange of macromolecules across the microcirculation, in: *Handbook of Physiology: The Cardiovascular System*, Section 2, Volume IV (E. M. Renkin and C. C. Michel, eds.), American Physiological Society, Bethesda, pp. 467–520.

Taylor, A. E., Townsley, M. I., and Korthuis, R. J., 1985, Macromolecular transport across microvessel walls, *Exp. Lung Res.* **8:**97–123.

Werth, D. K., Niedal, J. E., and Pastan, I., 1983, Vinculin, a cytoskeletal substrate of protein kinase C, *J. Biol. Chem.* **258:**11423–11426.

Wysolmerski, R. B., and Lagunoff, D., 1985, The effect of ethchlorvynol on cultured endothelial cells. A model for the study of the mechanisms of increased vascular permeability, *Am. J. Pathol.* **119:**505–512.

Yokoto, A., 1983, Immunocytochemical evidence for transendothelial transport of albumin and fibrinogen in rat heart and diaphragm, *Biomed. Res.* **4:**577–586.

Zetter, B., 1981, The endothelial cells of large and small blood vessels, *Diabetes* **30**(Suppl. 2): 24–28.

Chapter 11

Transcytosis of Macromolecules through the Blood–Brain Fluid Barriers *in Vivo*

Richard D. Broadwell

1. INTRODUCTION

The term *transcytosis* refers to the combined, sequential events of intracellular internalization or endocytosis of an extracellular, non-lipid-soluble micro-/macro-molecule, transport of that molecule through the cell, and secretion or exocytosis of the molecule from the cell opposite the side of entry. The process has been described for a variety of molecules and a host of cell types; however, no cell type in which transcytosis has been proposed is more controversial than the endothelium of the blood–brain barrier (BBB). A plethora of publications on morphology from the 1970s through the present promotes a transendothelial transfer of the protein tracer horse-radish peroxidase (HRP) from blood to brain as a consequence of central nervous system (CNS) exposure to pathophysiological, physical, or chemical insults. Other publications focusing on the BBB under normal conditions advocate the transcytosis of peroxidase from blood to brain through segments of specific arterioles and from brain to blood through capillaries and arterioles (for an extensive list of references, see Broadwell, 1989). The conclusions derived from these studies are based largely on repeated attempts to interpret dynamic cellular events from two-dimensional, static electron micrographs (Fig. 1). At best, these data are suggestive, do not warrant the stated conclusions, and, therefore, are open to criticism (Broadwell, 1989). More recently, the BBB literature has emphasized the potential for receptor-mediated trans-

Richard D. Broadwell • Division of Neurological Surgery, Department of Surgery, and Department of Pathology, University of Maryland School of Medicine, Baltimore, Maryland 21201.

Biological Barriers to Protein Delivery, edited by Kenneth L. Audus and Thomas J. Raub. Plenum Press, New York, 1993.

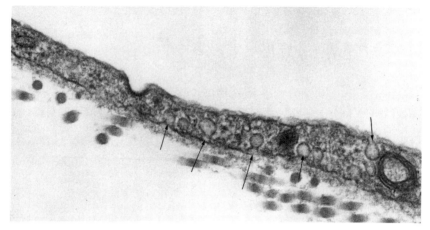

Figure 1. Many nonfenestrated endothelial cells, under normal conditions and within or outside the CNS, exhibit a plethora of vesicular profiles. Some of these profiles do indeed represent bona fide 40- to 70-nm-wide vesicles and are involved in shuttling membrane and macromolecules (e.g., enzymes) among constituents of the endomembrane system of organelles (e.g., endoplasmic reticulum, Golgi complex, endosomes, lysosomes, plasmalemma). Other profiles have the appearance of vesicles but are not vesicles. For example, some "vesicular profiles" may be portions of organelles (e.g., lysosomal tubules) cut in cross section. Additional "vesicular profiles" in proximity to the luminal and abluminal plasma membranes are *static* invaginations or caveolae (arrows) in the plasmalemma; these invaginations, appreciated best in serial thin sections, look like vesicles in random thin sections and have been interpreted as such engaged in transendothelial transport. A critical appraisal of this interpretation is provided in the text (see also Fig. 3 and Broadwell, 1989).

cytosis of peptides (Pardridge, 1986, 1991) and other blood-borne molecules that bind to the luminal surface of the normal, intact cerebral endothelium (Broadwell, 1989).

Circumvention of the BBB has obvious clinical implications for the delivery of chemotherapeutic substances to combat CNS infections, tumors, enzyme and neurotransmitter deficiencies, and toxins associated with brain disease or dysfunction. The nonfenestrated cerebral endothelium, which represents one of the cellular components of the mammalian BBB, is not the only blood–brain fluid barrier within the CNS. Additional barriers include the blood–CSF barrier associated with epithelia of the choroid plexus and other circumventricular organs (e.g., median eminence, area postrema), the nose–brain barrier associated with epithelia of the nasal mucosa, and the arachnoid mater–CSF barrier. Each of these barriers is believed to be, in part, intercellular tight junctional complexes that preclude the extracellular movement of non-lipid-soluble micro-/macromolecules bidirectionally between the external environment (i.e., air, blood) and the CNS milieu.

Despite reports advocating experimental manipulation to open the blood–brain fluid barriers transiently, most notably the BBB (Brightman *et al.*, 1973; Rapoport, 1985, 1988), the barriers under normal conditions are not absolute. Each is circumvented in a noninvasive fashion by endogenous and exogenous blood-borne proteins/ peptides moving through patent extracellular routes and/or traversing intracellular pathways related to adsorptive and receptor-mediated endocytic processes. Before transcytosis through the blood–brain fluid barriers can be interpreted correctly, the potentially viable intracellular and extracellular avenues must be considered. Additionally and just as importantly, the intracellular fate of external macromolecules associated with internalized cell surface membrane must be defined with regard to the processes of fluid-phase, adsorptive, and receptor-mediated endocytoses. The discussion to follow will address these topics. The cellular secretory process (Palade, 1975) and membrane behavior from a cell biological perspective will be emphasized with regard to fission and fusion of cell membranes among constituents of the endomembrane system of organelles (e.g., Golgi complex, endosomes, lysosomes, plasmalemma).

2. ENDOCYTIC PROCESSES DEFINED

The three endocytic processes from the least to the most specific are fluid- or bulk-phase endocytosis, adsorptive endocytosis, and receptor-mediated endocytosis (Table I; see also Chapters 3 and 4). Each of the three processes involves the internalization of external molecules with cell surface membrane. Cells comprising the individual blood–brain fluid barriers all internalize or retrieve their cell surface membrane, a normal cell biological event for exchanging or recycling old, worn-out plasmalemma for newly synthesized plasma membrane. This cell function is associated with fluid- or bulk-phase endocytosis and permits extracellular macromolecules not binding to the plasmalemma to enter cells nonselectively and indiscriminately within 40- to 70-nm-wide endocytic vesicles of plasmalemmal origin; the endocytic vesicles are directed to endosomes (a prelysosomal compartment) and/or to dense body lysosomes (Fig. 2).

Native HRP, arguably the most well-known and utilized fluid-phase tracer, has no difficulty in gaining entry to cells in general by fluid-phase endocytosis. Perox-

Table I
Endocytic Processes

	Example	Specificity	Organelles
Fluid phase	HRP, ferritin	Nonspecific	Vesicles, endosomes, lysosomes
Adsorptive phase	WGA, ricin, cationized probes	Specific oligosaccharides	Vesicles, endosomes, lysosomes, Golgi complex
Receptor-mediated	Insulin, transferrin	Specific receptors	Vesicles, endosomes, lysosomes, Golgi complex (?)

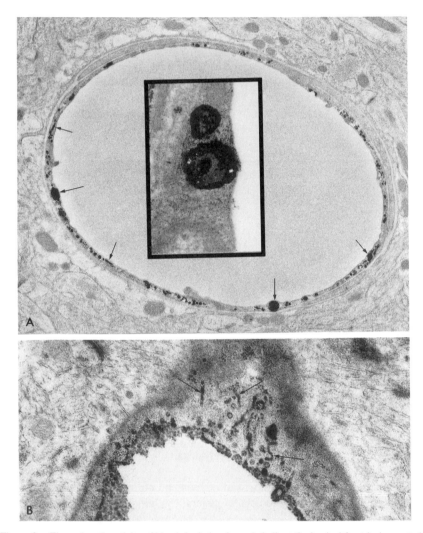

Figure 2. The endocytic activity of blood–brain barrier endothelia at the luminal front is demonstrable upon exposure to the fluid-phase tracer horseradish peroxidase administered intravenously. Within 5 min postinjection, blood–brain barrier endothelia exhibit peroxidase-labeled endocytic vesicles, dense bodies, and tubular profiles (A, B; arrows); labeled dense bodies are comparable morphologically to endothelial, acid phosphatase-positive, secondary lysosomes (A, inset). Tubular profiles (B, arrows) harboring HRP reaction product have not been identified to establish parajunctional channels by way of membrane continuities with the luminal and abluminal plasma membranes (Balin *et al.*, 1987); however, some tubules are confluent with peroxidase-labeled dense bodies and also stain positively for acid hydrolase activity (Balin *et al.*, 1987; Broadwell and Salcman, 1981), indicating they too are secondary lysosomes. Peroxidase reaction product is not evident on the abluminal surface or within the perivascular clefts of cerebral endothelia deep in the CNS removed from sites not possessing a blood–brain barrier (e.g., pial surface, circumventricular organs) in animals injected intravenously with the tracer.

idase is a 40-kDa glycoprotein that does not bind to membranes. The fate of macromolecules entering most mammalian cells by fluid-phase endocytosis (e.g., HRP, serum proteins, uncharged molecules) is degradation within acid hydrolase-containing secondary lysosomes (Balin et al., 1986, 1987; Balin and Broadwell, 1988; Broadwell, 1989; Broadwell and Salcman, 1981). Transcellular transport or transcytosis of fluid-phase macromolecules does not occur through the blood–brain fluid barriers (for a discussion, see Balin and Broadwell, 1988; Broadwell, 1989). The potential for native HRP to undergo transcellular transport or transcytosis may occur in specific cell types, e.g., in the epithelium of the seminal vesicle (Mata and David-Ferriera, 1973) and in somatotrophs of the anterior pituitary gland (Broadwell and Oliver, 1983). HRP introduced extracellularly to these two cell types in vivo becomes sequestered within the innermost Golgi saccule of the cells for subsequent packaging, export, and possible exocytosis. For both cell types, native HRP may behave as a membrane-bound marker, comparable to a lectin or ligand, rather than as a soluble or fluid-phase marker. The binding may involve the carbohydrate moieties of the peroxidase molecule. Available data suggest that HRP (Straus, 1981) may bind to mannose 6-phosphate receptors on the cell surface (Sly et al., 1981).

Adsorptive endocytosis concerns molecules like lectins [e.g., wheat germ agglutinin (WGA), ricin] that bind to carbohydrate moieties on the cell surface and positively charged (cationized) molecules that bind to negatively charged cell surface components. Conversely, receptor-mediated endocytosis is identified with the binding of a ligand (e..g, insulin, transferrin) to a cell surface receptor specific for that ligand; the binding may trigger the internalization of the receptor–ligand complex (Dautry-Varsat and Lodish, 1984). The intracellular fate and trafficking of macromolecules associated with cell surface membrane entering cells by adsorptive endocytosis and that by receptor-mediated endocytosis may be mutually exclusive; similarities do exist, however, among fluid-phase, adsorptive, and receptor-mediated endocytic processes (Table I).

3. CHARACTERISTICS OF THE BLOOD–BRAIN FLUID BARRIERS

The BBB and the blood–CSF barrier represent the two most intensively investigated blood–brain fluid barriers in mammals. Circumferential belts of tight junctional complexes among nonfenestrated cerebral endothelial (Brightman, 1977; Reese and Karnovsky, 1967) and among choroid plexus epithelia (Balin and Broadwell, 1988; Brightman, 1968) are the distinguishing characteristic of these two cellular barriers. Similar junctional complexes exist among cells of the arachnoid mater (Balin et al., 1986; Nabeshima et al., 1975), but those among epithelial cells of the nasal mucosa (Balin et al., 1986), the median eminence (Broadwell et al., 1983a, 1987a), and perhaps other circumventricular organs (see below) are believed not to be circumferentially tight to the extracellular passage of many molecules. Interendothelial tight junctions within the CNS may be initiated developmentally by astrocytes (Arthur *et al.*, 1987; Janzer and Raff, 1987; Senjo *et al.*, 1986; Tao-Cheng *et al.*, 1987).

The "enzymatic" barrier provided by intracellular secondary lysosomes, to which internalized macromolecules and associated cell surface membrane are directed in all three endocytic processes (see Chapter 3), is enhanced by additional, nonlysosomal hydrolytic enzymes such as monoamine oxidase; this particular enzyme in BBB endothelial catabolizes dopamine derived by intraendothelial decarboxylation of blood-borne L-dopa, thus denying entry of dopamine to the CNS through the BBB (Bertler *et al.*, 1963, 1966).

Although the recognized blood–brain fluid barrier cells are not categorized functionally as phagocytes *per se*, other cell types located behind and intimately associated with the blood–brain fluid barriers may be phagocytic in function. These additional cell types likely serve as an auxiliary line of defense once the initial cell barriers are breached (Baker and Broadwell, 1992; Balin *et al.*, 1986; Broadwell, 1989; Broadwell and Brightman, 1976; Broadwell and Salcman, 1981). Potential phagocytic cells associated with the blood–brain fluid barriers include perivascular pericytes, microglia, and macrophages, arachnoid and subarachnoid macrophages, and macrophages lying on the surfaces of ependymal and choroid plexus epithelia. The subarachnoid and perivascular phagocytes throughout the CNS in rodents and primates are labeled with the blood-borne, fluid-phase tracer HRP, suggesting that this probe molecule is successful in circumventing the blood–brain fluid barriers by extracellular routes (see below).

The internalization of cell surface membrane and endocytosis are demonstrable circumferentially in choroid plexus and median eminence epithelia as evidenced by exposure of these epithelia to blood-borne and CSF-borne peroxidase; sequestration of HRP reaction product is evident within epithelial vesicles, endosomes, tubular profiles, multivesicular bodies, and dense body lysosomes (Balin and Broadwell, 1988; Broadwell *et al.*, 1987a). Conversely, BBB endothelia do not exhibit a demonstrable endocytic activity circumferentially. Organelles in BBB endothelia identical to those in choroid epithelia are labeled with *blood-borne* peroxidase but fail to be labeled when the abluminal plasmalemma of BBB endothelia is bathed for 5 min through 24 hr in peroxidase delivered into the CNS by intraventricular injection (Balin *et al.*, 1986, 1987; Broadwell, 1989; Broadwell *et al.*, 1983a). Pits, invaginations, or caveolae in the abluminal plasmalemma of cerebral endothelia readily fill with CSF-borne peroxidase (Fig. 3). Abluminal surface pits have been misinterpreted as endothelial vesicles engaged in the transcytosis of protein from blood to brain and from brain to blood under normal and experimental conditions (see Broadwell, 1989 for a critical, in-depth discussion and references; see also Fig. 1). A comparison of membrane behavior at the luminal versus abluminal front of the BBB suggests that, unlike choroid epithelia of the blood–CSF barrier, endothelia of the BBB are polarized with regard to the internalization or recycling of cell surface membrane and endocytosis of macromolecules (Broadwell, 1989; Broadwell *et al.*, 1983a; Villegas and Broadwell, 1993). This polarity in the BBB suggests further that transcytosis of macromolecules through nonfenestrated cerebral endothelia, if indeed the process occurs significantly, is vectorial, from blood to brain but not from brain to blood. Selected characteristics of the blood–brain fluid barriers are summarized in Table II.

4. EXTRACELLULAR PATHWAYS CIRCUMVENTING THE BLOOD–BRAIN FLUID BARRIERS

Extracellular pathways circumventing the blood–brain fluid barriers (Table III) are comparable in the CNS of rodents and subhuman primates (Balin *et al.*, 1986; Broadwell, 1989; Broadwell *et al.*, 1987a,b, 1992a; Broadwell and Sofroniew, 1993). The most highly documented extracellular route is through the circumventricular organs (e.g., median eminence, organum vasculosum of the lamina terminalis, subfornical organ, and area postrema), all of which contain fenestrated capillaries and, therefore, lie outside the BBB (Fig. 4). Blood-borne macromolecules, specifically fluid-phase molecules, escaping fenestrated vessels supplying the circumventricular organs move extracellularly into adjacent brain areas located behind the BBB (for references see Broadwell and Brightman, 1976; Broadwell *et al.*, 1987a; Gross, 1987; Gross and Weindl, 1987).

Additional extracellular avenues into the CNS are associated with sites believed to possess patent intercellular junctional complexes: the nasal mucosa, epithelial linings of the median eminence and area postrema, and possibly endothelia of large vessels on the pial surface and/or occupying the Virchow–Robin spaces. The degeneration–regeneration of cells in the nasal mucosa allows the intercellular clefts of this epithelium to be patent to fluid-phase macromolecules instilled in the nares (i.e., airborne or applied topically) or delivered to the epithelium in the blood through leaky capillaries. The extracellular route continues along the olfactory nerve into the subarachnoid space at the level of the olfactory bulb (Balin *et al.*, 1986). The absence of a nose–brain barrier is of importance when the nose is considered a site for delivery of drugs, viruses, and environmental toxins associated with neurological disease (Balin *et al.*, 1986; Barthold, 1988; Langston, 1985; Talamo *et al.*, 1989).

Junctional complexes among ependymal cells lining the area postrema and median eminence are suspected of being discontinuous and not circumferentially tight, unlike the tight junctional complexes among BBB endothelia and epithelia of the choroid plexus; the apparent absence of a blood–CSF barrier at these two circumventricular organs heralds the bidirectional exchange of micro-/macromolecules between the blood and CSF (Broadwell *et al.*, 1983a, 1987a; Gotow and Hashimoto, 1979, 1981; Richards, 1978). Even if the junctional complexes among ependymal cells in the median eminence and area postrema were circumferentially tight, blood-borne, fluid-phase substances leaking from vessels in the circumventricular organs can move extracellularly around the tight junctions and enter the CSF through gap junctions among ependymal cells adjacent to the media eminence and area postrema.

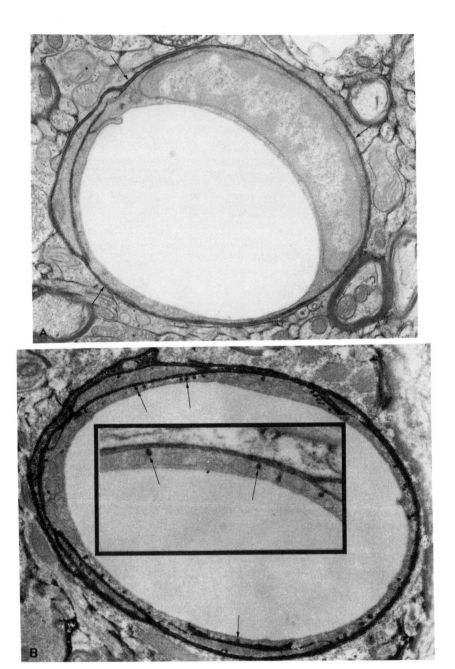

Table II
Characteristics of the Mammalian Blood–Brain Fluid Barriers

Cell types
    Blood–brain barrier: nonfenestrated endothelia
    • Circumferential belts of tight junctional complexes
    • Secondary lysosomes containing acid hydrolases
    • Nonlysosomal hydrolytic enzymes (e.g., monoamine oxidase)
    Blood–cerebrospinal fluid barrier
    Choroid plexus epithelium
    • Circumferential belts of tight junctional complexes
    • Secondary lysosomes containing acid hydrolases
    Arachnoid mater
    • Circumferential belts of tight junctional complexes
    Phagocytes
    • Circumventricular organ/subarachnoid macrophages
    • Perivascular cells: pericytes, microglia, macrophages
    • Supraependymal macrophages, Kolmer cells
Polarity of the blood–brain barrier
    The endothelium is polarized with regard to demonstrable internalization or recycling of its cell surface membrane and endocytosis of non-lipid-soluble macromolecules. These events occur from the blood side but not from the brain side of the endothelium.
Blood–brain barrier versus brain–blood barrier
    Blood–brain barrier is not absolute, whereas its counterpart, the brain–blood barrier, may be. Potential adsorptive and receptor-mediated transcytoses of macromolecules through the barrier is from blood to brain but not from brain to blood; hence, transcytosis through the endothelium appears to be vectorial.
Blood–cerebrospinal fluid barrier is not polarized
    Internalization of cell surface membrane associated with fluid-phase and adsorptive endocytoses is circumferential in epithelial cells of the choroid plexus and median eminence. Adsorptive transcytosis through these epithelial cells is bidirectional.

←

Figure 3. Exposure of the abluminal surface (A, arrows) of cerebral endothelia for 5 min–12 hr to peroxidase delivered by ventriculocisternal perfusion yields no concentration of peroxidase-labeled organelles comparable to that observed in the same endothelia exposed to blood-borne HRP (see Fig. 1). This observation suggests that the endocytic activity and internalization of abluminal surface membrane in BBB endothelia are insignificant or minor at best. Invaginations or pits in the abluminal plasmalemma (B and inset, arrows) fill with peroxidase delivered intraventricularly and have been misinterpreted in the literature as endocytic and/or transporting vesicles. Abluminal surface pits also can fill with blood-borne HRP that has circumvented the blood–brain barrier extracellularly under normal conditions (see Fig. 4) or that has moved through the experimentally manipulated barrier; abluminal pits labeled in this fashion have been misinterpreted as vesicles engaged in transendothelial transport and exocytosis of the tracer from blood to brain (for a discussion, see Broadwell, 1989; Broadwell *et al.*, 1983a; Broadwell and Sofroniew, 1993).

Table III

Circumvention of the Blood–Brain Fluid Barriers: Extracellular Routes

---

Permeable blood vessels
- Circumventricular organs (e.g., median eminence, area postrema)
- Pial surface

These sites lie outside the barrier and, therefore, are not immunologically privileged sites within the CNS.

Patent intercellular junctional complexes
- Nasal epithelium; absence of a nose–brain barrier
- Ependymal lining of median eminence and area postrema

Intracerebral transplants
- Blood vessels supplying solid tissue grafts of peripheral origin are leaky and are indigenous to the grafted tissue; the grafts are deficient in blood–brain and brain–blood barriers.
- Blood vessels supplying cell suspension grafts of peripheral origin are leaky and are of host origin; the vessels have forfeit their blood–brain and brain–blood barrier characteristics.

Absence of a BBB to blood-borne macromolecules at the pial surface of the brain was reported initially in mice injected intravenously with the fluid-phase tracer HRP (Balin *et al.*, 1986) and now is confirmed in rat and monkey (Broadwell, 1989; Broadwell *et al.*, 1992a; Broadwell and Sofroniew, 1993). We suspect the "leak" lies at the level of larger vessels on the pial surface and/or within the Virchow–Robin spaces. Blood-borne peroxidase entering the subarachnoid space by this route easily moves extracellularly through the pia mater and glial limitans into the extracellular clefts of the underlying neuropil and through the Virchow–Robin spaces for widespread distribution within the perivascular clefts (Fig. 5). The circumventricular organs also contribute to the dissemination of blood-borne HRP through the peri-

---

→

Figure 4. Seven areas within the mammalian CNS posses blood vessels permeable to peptides and many proteins; therefore, they lie outside the blood–brain barrier. These seven sites, collectively termed the circumventricular organs, are outlined in black in panel A and include the organum vasculosum of the lamina terminalis (OVLT) at the rostroventral tip of the third ventricle, the median eminence (ME) and its ventral extension from the floor of the third ventricle, the neurohypophysis (NH) or posterior pituitary gland, the subfornical organ (SFO), the choroid plexus (CP) in the lateral and fourth ventricles, the pineal gland (PL) situated dorsally, and the area postrema (AP) most caudally; the subcommissural organ (SCO) does not contain leaky blood vessels and is not considered a circumventricular organ by definition (Broadwell and Brightman, 1976). Vessels supplying each of the circumventricular organs are leaky to blood-borne peroxidase (B, median eminence), which *in vivo* moves extracellularly into adjacent brain parenchyma (B, arrow) supplied with blood–brain barrier endothelia. The leaky vessels are outlined (C, arrows; median eminence) in sections from immersion-fixed brains following incubation of the sections to reveal the endogenous peroxidase activity of red cells trapped within the vascular tree (Broadwell *et al.*, 1987b). Ultrastructurally, the leaky vessels appear highly fenestrated (D and inset, arrows; median eminence) and have open interendothelial junctions, thus permitting unobstructed, bidirectional movement of non-lipid-soluble micro-/macromolecules between the parenchyma of the circumventricular organ and the blood. (A from Weindl, 1973.)

vascular clefts. Subarachnoid macrophages followed by perivascular phagocytes lying superficially and deep within the brain parenchyma are labeled with blood-borne HRP in that order. Fluid-phase markers may be propelled through the perivascular clefts *in vivo* by the pulsatile activity of arterioles (Broadwell and Sofroniew, 1993). The patent extracellular pathways serve to explain how perivascular phagocytes throughout the rodent and monkey CNS become exposed to and are labeled with blood-borne peroxidase in less than 1 hr postinjection (Balin *et al.*, 1986; Broadwell, 1989; Broadwell and Sofroniew, 1993).

The above-referenced CNS sites leaky to fluid-phase molecules circumventing the BBB may not be insignificant; they allow the ingress of blood-borne, fluid-phase,

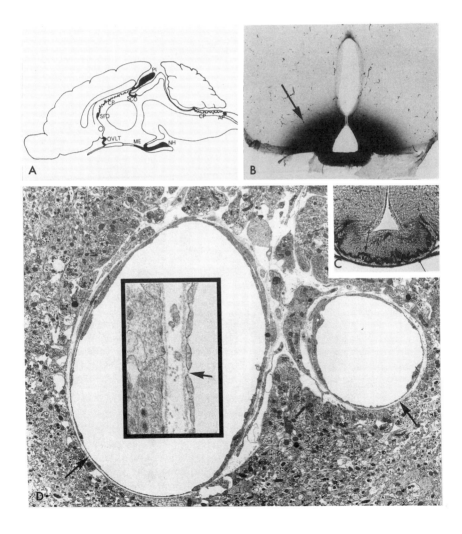

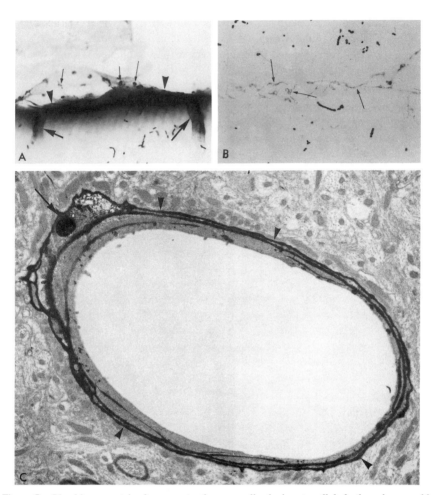

Figure 5. Blood-borne protein also can enter the mammalian brain extracellularly through permeable vessels supplying the pial surface. Once having gained access to the subarachnoid space and the pial surface (A, arrowheads), blood-borne peroxidase passes easily within the Virchow–Robin spaces (A, large arrows) and farther along the perivascular clefts (C, arrowheads) into the CNS; peroxidase is endocytosed by phagocytic cells on the pial surface (A, small arrows) and throughout the CNS perivascular spaces (B, C, arrows).

endogenous molecules the size of IgG (165 kDa) and IgM (500 kDa) (Broadwell and Sofroniew, 1993) and, therefore, represent nonimmunologically privileged sites in the CNS (Broadwell, 1989; Broadwell *et al.*, 1992a; Broadwell and Sofroniew, 1993; Santos and Valdimarsson, 1982). Absence of a BBB in the circumventricular organs and subarachnoid space is compensated for partially by populations of microglia, macrophages, and class II cells of the major histocompatibility complex (MHC) occupying these sites (Baker and Broadwell, 1992). WGA-HRP as an adsorptive-phase tracer is suspected not to follow the extracellular avenues into brain from the blood due to avid binding of the WGA molecule to the plasma membrane of cells (Balin and Broadwell, 1988; Broadwell *et al.*, 1988; Villegas and Broadwell, 1993). Whether or not blood-borne, adsorptive, and receptor-mediated phase molecules (e.g., cationized probes, ferrotransferrin, insulin) that do not bind to the luminal surface of BBB endothelia gain access to extracellular routes circumventing the BBB remains to be determined.

5. CIRCUMVENTING THE BLOOD–BRAIN FLUID BARRIERS THROUGH INTRACELLULAR PATHWAYS AND TRANSCYTOSIS

Non-lipid-soluble micro-/macromolecules from the periphery are capable of circumventing the blood–brain fluid barriers by specific intracellular routes, each of which is related to one of the three different endocytic processes discussed above. Suspected intracellular pathways circumventing the blood–brain fluid barriers are considered below and are listed in Table IV.

5.1. Fluid-Phase Endocytosis

When administered intravenously, the fluid-phase tracer HRP is endocytosed by axon terminals supplying circumventricular organs and peripheral tissues (i.e., ganglia, muscle) possessing permeable blood vessels. Peroxidase taken into the axon terminals first appears in vesicles the size of synaptic vesicles and subsequently undergoes retrograde axoplasmic transport in vesicles, vacuoles, and tubular profiles for sequestration in dendrites and perikarya. Neuronal systems so labeled with blood-borne peroxidase include cranial and spinal cord motor and preganglionic autonomic neurons and hypothalamic neurosecretory cells afferent to the neurohypophysis and

Table IV
Circumvention of the Blood–Brain Fluid Barriers: Intracellular Routes

---

Cerebral endothelium: adsorptive and receptor-mediated transcytoses of specific blood-borne micro-/macromolecules

Choroid plexus epithelium: bidirectional, adsorptive transcytosis of protein between the blood and CSF

Primary olfactory neurons: anterograde axoplasmic transport and transsynaptic transfer (adsorptive transcytosis) of extracellular protein

Neurosecretory cells, motor and autonomic neurons: retrograde axoplasmic transport and transsynaptic transfer (adsorptive transcytosis) of blood-borne protein and virus

---

perhaps to other circumventricular organs (Broadwell and Brightman, 1976, 1979, 1983). Retrogradely transported HRP is not secreted from the parent perikarya and dendrites; peroxidase-labeled organelles undergoing retrograde transport eventually fuse with perikaryal secondary lysosomes or become secondary lysosomes after fusing with primary lysosomes derived from the inner saccule of the Golgi complex (Broadwell, 1980; Broadwell and Brightman, 1979; Broadwell *et al.*, 1980). The retrograde labeling of well-defined neuronal perikarya positioned behind the BBB suggests toxins, neurovirulent viruses, and other substances can enter the same neuronal groups, as does blood-borne peroxidase, from cerebral and extracerebral blood.

First-order olfactory neurons in the nasal mucosa exposed to blood-borne HRP or to HRP instilled directly into the nares likewise will endocytose the tracer protein, which then undergoes anterograde axoplasmic transport in lysosomes to axon terminals innervating the glomeruli of the olfactory bulb (Broadwell and Balin, 1985); organelles transporting peroxidase into the olfactory terminals do not secrete their contents.

The transcytosis of fluid-phase markers is not documented conclusively for cells of the blood–brain fluid barriers (Balin and Broadwell, 1988; Broadwell, 1989); however, a retrograde transneuronal transfer of herpesvirus (Ugolini *et al.*, 1989) and the C fragment of tetanus toxin (Fishman and Carrigan, 1988) are reported from the periphery through lower motoneurons into the cortex and brain stem. The transcytosis of these substances may be a consequence of adsorptive or receptor-mediated endocytosis.

5.2. Adsorptive Transcytosis

Potential transcytosis of lectins and cationized molecules is not specific for cells of the blood–brain fluid barriers; the potential exists for all cells exposed to the molecules, those located within the CNS as well as those situated peripherally. To date, WGA (36 kDa) conjugated to HRP is the only molecule documented morphologically to be associated with adsorptive transcytosis through the blood–brain fluid barriers, specifically BBB endothelia and choroid epithelia (Balin and Broadwell, 1988; Broadwell, 1989; Broadwell *et al.*, 1988; Villegas and Broadwell, 1993). Biochemical data advocate cationized serum proteins as additional molecules for adsorptive transcytosis through BBB endothelia (Pardridge *et al.*, 1990; Triguero *et al.*, 1989). Because lectins and cationized molecules are cytotoxic, they may not represent the best vehicles for brain delivery of molecules normally excluded entry by the blood–brain fluid barriers.

Subsequent to the binding of WGA to *N*-acetylglucosamine and sialic acid moieties on the plasmalemma of cells (Gonatas and Avrameas, 1973), the intracellular pathway WGA-HRP follows through cells of the blood–brain fluid barriers and neurons projecting outside the BBB is initially similar to that of fluid-phase molecules. Organelles sequestering WGA-HRP early on include endocytic 40- to 70-nm vesicles, tubular profiles, vacuoles that may represent endosomes or a prelysosomal compartment, and dense bodies comparable morphologically to secondary lysosomes. With the passage of time (e.g., 1–3 hr) and unlike with fluid-phase molecules, the transmost saccule of the Golgi complex also is labeled with reaction product for WGA-HRP. Labeling of the inner Golgi saccule precedes the apparent transcytosis of the lectin conjugate from blood to brain through BBB endothelia (Broadwell *et al.*, 1988; Villegas and Broadwell, 1993), and the transsynaptic transfer of WGA-HRP from the periphery in primary olfactory neurons afferent to the olfactory bulb (Broadwell and Balin, 1985) and in the retrograde direction through hypothalamic neurosecretory neurons (Balin and Broadwell, 1987; Villegas and Broadwell, 1989).

Support for transcytosis of WGA-HRP through BBB endothelia includes: (1) sequestration of reaction product for blood-borne WGA-HRP within the transmost Golgi saccule of cerebral endothelia; (2) reaction product filling the perivascular clefts; (3) labeling of perivascular phagocytes throughout the CNS, suggesting that adsorptive transcytosis through the BBB is global; (4) WGA-HRP reaction product in the transmost Golgi saccule of perivascular phagocytes; and (5) WGA-HRP occupying extracellular clefts and processes in the neuropil beyond the basal lamina surrounding the perivascular phagocytes and endothelia (Fig. 6) (Broadwell *et al.*, 1988; Villegas and Broadwell, 1993).

Adsorptive transcytosis of WGA-HRP figures prominently with the Golgi complex and, therefore, with the Palade (1975) scheme of the cellular secretory process. The transmost Golgi saccule, which in most mammalian cell types exhibits acid phosphatase enzyme activity cytochemically, gives rise to primary lysosomes and additional vesicles/vacuoles involved in delivering membrane and macromolecules such as enzymes to other organelles; this Golgi saccule also is charged with packaging molecules for export and exocytosis at the cell surface (for references see Balin and Broadwell, 1988; Broadwell and Balin, 1985; Broadwell and Oliver, 1981, 1983). WGA-HRP labeling of the Golgi complex may be a consequence of saturating the endosomal compartment by the internalization of cell surface membrane tagged with the lectin conjugate.

The endosome compartment is the common denominator among fluid-phase, adsorptive, and receptor-mediated endocytic processes; it represents a sorting center and first intracellular stop in the endocytic pathway for internalized cell surface membrane associated with lectin, the ligand–receptor complex, and fluid-phase molecules (Dautry-Varsat and Lodish, 1984; Gonatas *et al.*, 1984; Helenius *et al.*, 1983; Steinman *et al.*, 1983); some internalized membrane directed to the endosome compartment may recycle to the cell surface after leaving the lectin/ligand within the endosome. In fluid-phase endocytosis, internalized cell surface membrane likewise recycles from endosomes to the plasmalemma after endocytic vesicles have deposited their contents in endosomes. We have speculated that when individual endosomes are saturated with endocytic membrane associated with WGA-HRP, the "normal" intracellular endocytic pathway may be perturbed. As a consequence, internalized

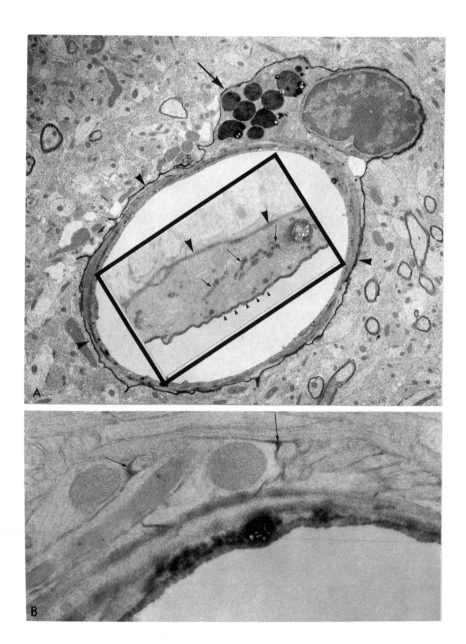

membrane with attached lectin may be diverted to the transmost Golgi saccule wherein reside specific enzymes that contribute to the processing of membrane macromolecules, such as the addition or replacement of sialic acid (Bennett and O'Shaughnessy, 1981; Bennett *et al.*, 1981) to which WGA binds on the cell surface. Membrane and WGA-HRP recycled through the transmost Golgi saccule would be packaged for export to other organelles (e.g., endosomes, secondary lysosomes, plasmalemma) as well as for exocytosis. Because a population of sialic acid-rich glycoproteins is routed through the Golgi complex constitutively, the possibility cannot be excluded that particular subpopulations of cell surface glycoproteins and lipids are recycled directly through the transmost Golgi saccule.

Although we have yet to identify the transcytosis of macromolecules from brain to blood through BBB endothelia (see below and Broadwell, 1989; Broadwell *et al.*, 1983a; Villegas and Broadwell, 1993), transcytosis of WGA-HRP does occur bidirectionally through choroid plexus epithelia and perhaps without involvement of the Golgi complex. WGA-HRP delivered into the lateral cerebral ventricle is endocytosed avidly by choroid epithelia at the microvillus face of the cells and is transcytosed through the epithelia within 10 min for binding to fenestrated endothelia located at the opposite pole of the epithelia; this transcytosis appears to occur in advance of WGA-HRP labeling of the Golgi complex and is speculated to utilize the endosome compartment as an intermediary in the transcytotic pathway (Balin and Broadwell, 1988). If our speculation is correct, the transcytotic pathway through choroid epithelia is similar to that reported for IgG through intestinal epithelia (Rodewald and Kraehenbuhl, 1984). We also find that the transcytosis of blood-borne WGA-HRP occurs through choroid epithelia but with difficulty. The event requires 18–24 hr to be identified ultrastructurally and is minor at best, with the signal for transcytosis of the lectin conjugate represented by WGA-HRP labeling of the microvillus border and of phagocytic (Kolmer) cells residing on the surface of the microvilli (Villegas and Broadwell, 1993). The binding of blood-borne WGA-HRP to luminal and abluminal surfaces of fenestrated endothelia supplying the choroid plexus is so prominent that extracellular availability of the lectin for adsorptive endocytosis by choroid epithelia may be compromised. Suspected membrane trafficking within choroid epithelia is diagrammed in Fig. 7.

←
_____

Figure 6. The lectin wheat germ agglutinin conjugated to horseradish peroxidase (WGA-HRP) and administered intravenously to rodents undergoes adsorptive transcytosis through the blood–brain barrier. The lectin conjugate in time labels perivascular phagocytes (A, arrow) throughout the CNS. Prior to 3 hr postinjection, WGA-HRP reaction product is identified ultrastructurally on the luminal surface membrane (A inset, small arrowheads) and within the inner saccule of the Golgi complex (A inset, arrows), vesicles, tubules, and dense bodies. Reaction product occupies the perivascular clefts (A and inset, large arrowheads) and extracellular spaces (B, arrows) in the neuropil beyond the endothelial basal lamina at postinjection times of 3 hr and longer (Broadwell, 1989; Broadwell *et al.*, 1983a; Villegas and Broadwell, 1993).

CSF

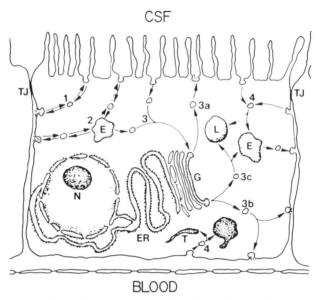

BLOOD

Figure 7. Three potential intracellular pathways through epithelia of the choroid plexus are proposed for the transcytosis of CSF-borne and blood-borne proteins and peptides that bind to the plasmalemma. Such macromolecules would enter the choroid epithelium by adsorptive endocytosis (e.g., lectins) and receptor-mediated endocytosis (e.g., ligands). Potential transcytotic pathways include direct vesicular transport (1) bidirectionally between the basolateral surface and apical or microvillus surface, indirect vesicular transport (2) by way of endosomes (E), or vesicular transport with recruitment of the Golgi complex (G) either directly from the cell surface or indirectly from endosomes (3). The transmost or inner Golgi saccule gives rise to a host of transporting vesicles; some of the vesicles are destined to engage in exocytosis at the apical surface (3a) or basolateral surface (3b) of the choroid epithelium, while others represent primary lysosomes (3c) that ferry acid hydrolytic enzymes to endosomes (prelysosomes) and secondary lysosomes (L). Blood-borne or CSF-borne macromolecules entering the choroid epithelium in vesicles derived from the plasmalemma by fluid-phase endocytosis (e.g., native HRP, ferritin), adsorptive endocytosis (WGA, cationized probes), or receptor-mediated endocytosis (e.g., ligands) are most likely directed to endosomes or secondary lysosomes (4); internalized plasmalemma possessing receptors freed from ligand/lectin deposited in endosomes may recycle to the cell surface. The endoplasmic reticulum (ER) and tubular profiles (T) do not pariticipate in the transepithelial transport and exocytosis of extracellular macromolecules (Balin and Broadwell, 1987; Cataldo and Broadwell, 1986). Because the choroid epithelium is not polarized with regard to its endocytic activity and internalization of cell surface membranes, a bidirectional transcytosis of extracellular macromolecules through the epithelium between blood and CSF is conceivable. N, nucleus; TJ, tight junctional complex.

5.3. Receptor-Mediated Transcytosis

Data advocating the receptor-mediated transfer across BBB endothelia abound for ligands, specifically a host of peptides (e.g., insulin, transferrin, vasopressin); however, the data are more biochemical (Banks and Kastin, 1985, 1990; Banks *et al.*,

1987; Barrera *et al.*, 1989; Duffy and Pardridge, 1987; Fishman *et al.*, 1987; Pardridge, 1986) than morphological (Broadwell, 1989; Broadwell *et al.*, 1992b) and are not without controversy (Meisenberg and Simmons, 1983; Ermisch *et al.*, 1985). One laboratory is of the belief that selected peptides may be involved in a bidirectional transport across the BBB (Banks *et al.*, 1988, 1989).

The potential pathways for receptor-mediated transcytosis through and membrane trafficking within BBB endothelia are diagrammed in Fig. 8; the pathways are: (1) direct vesicular transport from the luminal to the abluminal side; (2) indirect vesicular transport utilizing the endosome compartment as an intermediary; and (3) vesicular transport by way of the Golgi complex. Similar pathways apply to receptor-mediated transcytosis through epithelia of the blood–CSF barrier (Fig. 7).

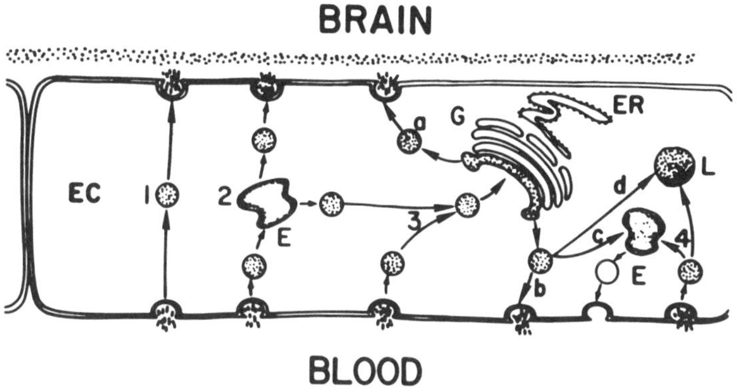

Figure 8. Potential transcytotic routes through endothelial cells (EC) of the blood–brain barrier and membrane events associated with fluid-phase, adsorptive, and receptor-mediated endocytic processes are represented in pathways 1–4. A direct transendothelial vesicular transfer (1) or an indirect route (2) through the endosome compartment (E) may be utilized for the receptor-mediated transcytosis of blood-borne ligands (e.g., transferrin, insulin); available data suggest the indirect route is the more plausible of the two (see text). Macromolecules entering the endothelium by adsorptive endocytosis are channeled to endosomes (2) and either directly or indirectly to the transmost saccule of the Golgi complex (3); this Golgi saccule is responsible for packaging macromolecules for exocytosis at the abluminal (a) and luminal (b) faces of the endothelium. Alternatively, some Golgi-derived vesicles represent primary lysosomes charged with delivering acid hydrolases to endosomes (E; pathway c) and secondary lysosomes (L; pathway d). Fluid-phase macromolecules and some macromolecules taken into the endothelium by receptor-mediated and adsorptive endocytic processes are directed to endosomes and secondary lysosomes (4) for degradation. In pathway 4, the internalized cell surface membrane may recycle to the luminal plasmalemma as vesicles from the endosome compartment. Reports that the endoplasmic reticulum (ER) of the cerebral endothelium may be involved in the transcytosis of blood-borne macromolecules (Mollgard and Saunders, 1975, 1977) are not confirmed (Broadwell *et al.*, 1983b; Cataldo and Broadwell, 1986). Endocytosis and retrieval of cell surface membrane at the abluminal front are not conspicuous in comparison to the same events at the luminal plasmalemma (see text for discussion).

Our data of ferrotransferrin (f-TRF; 75 kDa) conjugated to peroxidase for receptor-mediated transcytosis through the BBB are, at best, suggestive and raise important questions regarding the process in BBB endothelia (Broadwell, 1989; Broadwell *et al.*, 1992b; Tangoren *et al.*, 1988). Immunohistochemistry has demonstrated that endothelia possessing the receptor for f-TRF are restricted to cerebral vessels (Jeffries *et al.*, 1984). Receptor recognition of f-TRF may promote the transport and delivery of transferrin and iron from plasma into brain. The f-TRF-HRP we have used is run on gel electrophoresis and chromatographed to ensure that the conjugate administered into the carotid artery of rats (6 mg in 0.5–1.0 ml saline) is not contaminated with free or unbound HRP. At 1 hr postinjection, reaction product for f-TRF-HRP is observed within endothelial endocytic vesicles, tubules, spherical endosomes, and dense bodies, and within the perivascular clefts and perivascular phagocytes. Subarachnoid macrophages, the circumventricular organs, and Golgi saccules in BBB endothelia appear free of reaction product. [To date, a limited number of *in vitro* studies suggest that the transferrin receptor in some cell types may recycle through the Golgi complex; see Fishman and Fine (1987), Snider and Rogers (1985), and Woods *et al.* (1986).] Infrequently, reaction product fills a presumptive "exocytic vesicle" or pit positioned at the abluminal plasmalemma. The data suggest, although not without argument, that the f-TRF-HRP conjugate may undergo receptor-mediated transcytosis through the BBB. Because of the specificity of the endothelial TRF receptor in brain, blood-borne iron-bound TRF may be an ideal vehicle for ferrying substances across the barrier.

If receptor-mediated transcytosis of f-TRF-HRP does occur through cerebral endothelia, the treatment of f-TRF by this endothelium differs from that in other cell types. In the hepatocyte, for example, the f-TRF associated with the cell surface f-TRF receptor is internalized and directed to endosomes wherein iron is dissociated from transferrin and transferred to the iron-storing protein ferritin; the iron-free apotransferrin remains bound to its membrane receptor and is recycled with it to the cell surface (Dautry-Varsat and Lodish, 1983). The iron-free apotransferrin is released to bind additional iron when the receptor–apotransferrin complex encounters the neutral pH of the extracellular medium.

A possible hypothesis is that the fate of f-TRF in the endosome of BBB endothelia is dissociation of the ligand from its receptor, which is recycled to the luminal surface. f-TRF is anticipated to be transferred to a vesicle of endosome origin for export and exocytosis at the abluminal front. Iron would dissociate from the TRF once exocytosis into the perivascular space occurs. Perivascular phagocytes are in position to endocytose the free TRF. This scheme excludes consideration of a direct transendothelial vesicular transport comparable to that diagrammed in Fig. 8. To date, direct transendothelial vesicular transport is only a suggestion in the BBB literature without confirmation. Immunocytochemistry utilizing a monoclonal antibody against a BBB protein (Sternberger and Sternberger, 1987) fails to support movement of luminal surface membrane to abluminal surface; the antibody recognizes a luminal membrane antigen that is internalized but is not directed to the abluminal surface (N. Sternberger and C. Shear, personal communication).

6. THE ENIGMA OF TRANSCYTOSIS THROUGH THE BLOOD–BRAIN BARRIER

The event of transcytosis through the BBB implies that the membrane of exporting vesicles engaged in exocytosis at the abluminal front of the endothelium fuses with the plasmalemma. In so doing, exocytic vesicle membrane is added necessarily to abluminal surface membrane. The Palade (1975) scheme of the cellular secretory process states that exocytosis and endocytosis are complementary events, events well recognized in bona fide secretory cells such as the neuron and those of endocrine and exocrine glands. With the addition of secretory vesicle or granule membrane to the plasmalemma for exocytosis to commence, the endocytic process signals the retrieval of cell surface membrane as a compensatory response to exocytosis, thereby ensuring that the overall surface area of cell membrane remains static with each exocytic event. The endocytic activity at the luminal front of the BBB is demonstrable upon even brief exposure to blood-borne tracer (Fig. 2) but not so at the abluminal front exposed for 5 min through 24 hr to native peroxidase or WGA-HRP filling the perivascular clefts following ventriculocisternal perfusion of the proteins (Balin *et al.*, 1986; Broadwell, 1989; Broadwell *et al.*, 1983a; Villegas and Broadwell, 1993); in the latter preparation, peroxidase-labeled endosomes and dense bodies are exceedingly rare within the endothelia, and labeled tubular profiles and Golgi saccules are nonexistent (Fig. 3). Presumptive exocytic and endocytic vesicles at the abluminal face are impossible to discern morphologically and unequivocally from static pits studding the abluminal plasmalemma. Abluminal surface pits are appreciated readily in preparations incubated to reveal the alkaline phosphatase activity inherent to the plasma membranes and fill with peroxidase that has gained access to the perivascular clefts from the blood through patent extracellular routes (considered below), ventriculocisternal perfusion or from experimental manipulation of the BBB (Broadwell, 1989).

The enigma of a *significant* transcytosis of blood-borne proteins and peptides through the BBB concerns how this endothelial barrier compensates for the absence of a demonstrable endocytic activity subsequent to exocytosis at its abluminal plasma membrane. Such a discrepancy in the cellular secretory process applied to the BBB requires clarification for defining transcytosis through the BBB. The problem is not encountered with the bidirectional transcytosis through the blood–choroid epithelial barrier, because the choroid epithelium demonstrates endocytic activity circumferentially. The apparent absence of macromolecular transcytosis through the cerebral endothelium from brain to blood suggests the CNS possesses a *brain–blood barrier* that indeed may be absolute, whereas its counterpart, the blood–brain barrier, is not (Broadwell, 1993).

7. INTRACEREBRAL TRANSPLANTS AND THE BLOOD–BRAIN BARRIER

Immunohistochemistry with antibodies directed against MHC class I antigen on the luminal surface of endothelia has demonstrated that blood vessels inherent to intracerebrally placed grafts of CNS or peripheral origin are sustained and anastomose with host cerebral vessels (Broadwell et al., 1990, 1991, 1992a). Consequently, solid brain tissue grafted intracerebrally presents a BBB to blood-borne macromolecules (Broadwell, 1988; Broadwell et al., 1987b, 1989, 1990, 1991, 1992a). Cell suspensions of astrocytes or neurons injected intracerebrally are supplied with host CNS vessels and a BBB to circulating protein (Broadwell et al., 1991). Solid tissue of the anterior pituitary gland (Broadwell et al., 1987b, 1991), adrenal medullary gland (Rosenstein, 1987), muscle, skin, and superior cervical ganglia (Rosenstein and Brightman, 1986; Wakai et al., 1986) grafted intracerebrally do not exhibit a BBB nor do intracerebral cell suspensions of peripheral origin (e.g., fibroblasts, PC12 cells); the latter are supplied exclusively with host BBB vessels that lose their BBB properties and become fenestrated and/or have patent interendothelial junctional complexes (Broadwell et al., 1990, 1991, 1992a). Not only will peripheral tissue/cell suspension grafts create a "window" in the BBB for passive entry of blood-borne substances, such as chemotherapeutics, to the brain, but the absence of a BBB in these grafts necessarily means the absence of a brain–blood barrier as well. This fact introduces a significant problem to the intracerebral application of peripheral tissue/ cell suspension grafts for production and release of a neurotransmitter or peptide in clinical treatment for neurodegenerative disorders (e.g., Parkinson's disease). Any neurotransmitter or peptide anticipated to be secreted from intracerebral, peripheral tissue/cell suspension grafts likely would fail to enter the surrounding host neuropil in large enough concentration to effect a significant clinical improvement; rather, the secreted product most likely would enter the non-BBB vessels supplying the graft and be removed within the general circulation. Our investigations in mammals (Broadwell et al., 1987, 1989, 1990, 1991, 1992b) demonstrate that the presence versus absence of blood–brain and brain–blood barriers within intracerebrally positioned grafts is dictated by the grafted tissue/cells and not by the surrounding host tissue contributing vessels to the graft. The same extracellular routes circumventing the BBB in the normal brain would apply to the host CNS and grafted CNS tissue/cells as well. For this reason, criteria for assessing the potential for transcytosis through the BBB under normal conditions likewise would apply to BBB vessels supplying intracerebrally grafted solid/cell suspension CNS grafts.

8. SUMMARY AND CONCLUSIONS

Potential intracellular and extracellular pathways that blood-borne substances may follow for circumventing the blood–brain fluid barriers and entry to the CNS are multiple. The extracellular avenues, patent to blood-borne protein the size of IgG and IgM, and movement of blood-borne macromolecules through perivascular clefts deep into the CNS complicate the interpretation and identification of bona fide transcytosis through the BBB. The often-stated belief in literature reviews of the BBB that nonfenestrated cerebral endothelia fail to engage in demonstrable endocytosis and possess few vesicles under normal conditions is invalid. Endocytic vesicle formation and vesicular traffic among constituents of the endomembrane system are no different in BBB endothelia than in other cell types. Available biochemical and morphological data advocate the transcytosis of selected blood-borne protein and peptides through nonfenestrated cerebral endothelia; however, absence of demonstrable endocytic activity at the abluminal front compared to a very prominent endocytic activity at the luminal surface of BBB endothelia argues against bidirectional membrane trafficking through the BBB and supports the concept of a *brain–blood barrier*. The latter is no less significant functionally than the BBB and may be more so in deterring transendothelial transfer of peptides and proteins bidirectionally through the non-fenestrated cerebral endothelium. The difficulty in interpreting transcytosis through BBB endothelia is not encountered for epithelia of the blood–CSF barrier at the level of the choroid plexus. Choroid epithelia engage in endocytosis circumferentially; hence, the potential for transcytosis and circumvention of the blood–CSF barrier through an intraepithelial route exists bidirectionally in the choroid plexus.

REFERENCES

Arthur, F. E., Shivers, R. R., and Bowman, P. D., 1987, Astrocyte-mediated induction of tight junctions in brain capillary endothelium: An efficient in vitro model, *Dev. Brain Res.* **36:** 155–159.

Baker, B. J., and Broadwell, R. D., 1992, Cellular line of defense upon breachment of the blood–brain barrier, *Abstr. Soc. Neurosci.* **18:**1281.

Balin, F. J., and Broadwell, R. D., 1987, Lectin-labeled membrane is transferred to the Golgi complex in mouse pituitary cells in vivo, *J. Histochem. Cytochem.* **35:**489–498.

Balin, B. J., and Broadwell, R. D., 1988, Transcytosis of protein through the mammalian cerebral epithelium and endothelium. I. Choroid plexus and the blood–cerebrospinal fluid barrier, *J. Neurocytol.* **17:**809–826.

Balin, B. J., Broadwell, R. D., Salcman, M., and El-Kalliny, M., 1986, Avenues for entry of peripherally administered protein to the CNS in mouse, rat and monkey, *J. Comp. Neurol.* **251:**260–280.

Balin, B. J., Broadwell, R. D., and Salcman, M., 1987, Evidence against tubular profiles contributing to the formation of transendothelial channels through the blood–brain barrier, *J. Neurocytol.* **16**:721–728.

Banks, W. A., and Kastin, A. J., 1985, Permeability of the blood–brain barrier to neuro-peptides: The case for penetration, *Psychoneuroendocrinology* **10**:385–399.

Banks, W. A., and Kastin, A. J., 1990, Peptide transport systems for opiates across the blood–brain barrier, *Am. J. Physiol.* **259**:E1–E10.

Banks, W. A., Kastin, A. J., Horvath, A., and Michals, E. A., 1987, Carrier-mediated transport of vasopressin across the blood–brain barrier of the mouse, *J. Neurosci. Res.* **18**:326–332.

Banks, W. A., Kastin, A. J., Fasold, M. B., Barrera, C. M., and Augereau, G., 1988, Studies of the slow bidirectional transport of iron and transferrin across the blood–brain barrier, *Brain Res. Bull.* **21**:881–885.

Banks, W. A., Kastin, A. J., and Durham, D. A., 1989, Bidirectional transport of interleukin-1 alpha across the blood–brain barrier, *Brain Res. Bull.* **23**:433–437.

Barrera, C. M., Banks, W. A., and Kastin, A. J., 1989, Passage of Tyr-MIF-1 from blood to brain, *Brain Res. Bull.* **23**:439–442.

Barthold, S. W., 1988, Olfactory neural pathway in mouse hepatitis virus nasoencephalitis, *Acta Neuropathol.* **76**:502–506.

Bennett, G., and O'Shaughnessy, D., 1981, The site of incorporation of sialic acid residues into glycoproteins and the subsequent fates of these molecules in various rat and mouse cell types by radioautography after injection of [^{3}H] N-acetylmannosamine. I. Observations in hepatocytes, *J. Cell Biol.* **88**:1–15.

Bennett, G., Kan, F. W. K., and O'Shaughnessy, D., 1981, The site of incorporation of sialic acid residues into glycoproteins and the subsequent fate of these molecules in various rat and mouse cell types as shown by radio-autography after injection of [^{3}H] N-acetyl-mannosamine. II. Observations in tissues other than liver, *J. Cell Biol.* **88**:16–31.

Bertler, A., Falck, B., and Rosengren, E., 1963, The direct demonstration of a barrier mechanism in brain capillaries, *Acta Pharmacol. Toxicol.* **20**:317–321.

Bertler, A., Falck, B., Owman, C. H., and Rosengren, E., 1966, Localization of mono-aminergic blood–brain barrier mechanism, *Pharmacol. Rev.* **18**:369–385.

Brightman, M. W., 1968, The intracerebral movement of proteins injected into blood and cerebrospinal fluid of mice, *Prog. Brain Res.* **29**:19–31.

Brightman, M. W., 1977, Morphology of blood–brain interfaces, *Exp. Eye Res.* **15**:1–25.

Brightman, M. W., Hori, M., and Rapport, S. I., 1973, Osmotic opening of tight junctions in cerebral endothelium. *J. Comp. Neurol.* **152**:317–326.

Broadwell, R. D., 1980, Cytochemical localization of acid hydrolases in neurons of the mammalian central nervous system, *J. Histochem. Cytochem.* **28**:87–89.

Broadwell, R. D., 1988, Addressing the absence of a blood–brain barrier within transplanted brain tissue, *Science* **24**:473.

Broadwell, R. D., 1989, Transcytosis of macromolecules through the blood–brain barrier. A critical appraisal and cell biological perspective, *Acta Neuropathol.* **79**:117–128.

Broadwell, R. D., 1993, Endothelial cell biology and the enigma of transcytosis through the blood–brain barrier, *J. Cereb. Blood Flow Metab.* (in press).

Broadwell, R. D., and Balin, B. J., 1985, Endocytic and exocytic pathways of the neuronal secretory process and trans-synaptic transfer of wheat germ agglutinin–horseradish peroxidase in vivo, *J. Comp. Neurol.* **242**:632–650.

Broadwell, R. D., and Brightman, M. W., 1976, Entry of peroxidase to neurons of the central and peripheral nervous system from extra-cerebral and cerebral blood, *J. Comp. Neurol.* **166:**257–284.

Broadwell, R. D., and Brightman, M. W., 1979, Cytochemistry of undamaged neurons transporting exogenous protein in vivo, *J. Comp. Neurol.* **185:**31–74.

Broadwell, R. D., and Brightman, M. W., 1983, Horseradish peroxidase: A tool for study of the neuroendocrine cell and other peptide secreting cells, *Methods Enzymol.* **103:**187–218.

Broadwell, R. D., and Oliver, C., 1981, The Golgi apparatus, GERL, and secretory granule formation within the hypothalamo-neurohypophysial system of control and hyper-osmotically stressed mice, *J. Cell Biol.* **90:**474–484.

Broadwell, R. D., and Oliver, C., 1983, An enzyme cytochemical study of the endocytic pathways in anterior pituitary cells of the mouse in vivo, *J. Histochem. Cytochem.* **31:** 325–335.

Broadwell, R. D., and Salcman, M., 1981, Expanding the definition of the blood–brain barrier to protein, *Proc. Natl. Acad. Sci. USA* **78:**7820–7824.

Broadwell, R. D., and Sofroniew, M., 1993, Immunohistochemical identification of serum proteins circumventing the blood–brain barrier *Exp. Neurol.* (in press).

Broadwell, R. D., Brightman, M. W., and Oliver, C., 1980, Neuronal transport of acid hydrolase and peroxidase within the lysosomal system of organelles: Involvement of agranular reticulum-like cisterns, *J. Comp. Neurol.* **190:**519–532.

Broadwell, R. D., Balin, B., Salcman, M., and Kaplan, R. S., 1983a, A blood–brain barrier? Yes and no, *Proc. Natl. Acad. Sci. USA* **80:**7352–7356.

Broadwell, R. D., Cataldo, A. M., and Salcman, M., 1983b, Cytochemical localization of glucose 6-phosphatase activity in cerebral endothelial cells, *J. Histochem. Cytochem.* **31:** 818–822.

Broadwell, R. D., Balin, B. J., and Cataldo, A. M., 1987a, Fine structure and cytochemistry of the mammalian median eminence, in: *Circumventricular Organs and Body Fluids* (P. M. Gross, ed.), CRC Press, Boca Raton, Fla., pp. 61–85.

Broadwell, R. D., Charlton, H. M., Balin, B., and Salcman, M., 1987b, Angioarchitecture of the CNS, pituitary gland and intracerebral grafts revealed with peroxidase cytochemistry, *J. Comp. Neurol.* **260:**47–62.

Broadwell, R. D., Balin, B. J., and Salcman, M., 1988, Transcytosis of blood-borne protein through the blood–brain barrier, *Proc. Natl. Acad. Sci. USA* **85:**632–636.

Broadwell, R. D., Charlton, H. M., Ganong, W. F., Salcman, M., and Sofroniew, M., 1989, Allografts of CNS tissue possess a blood–brain barrier. I. Grafts of medial preoptic area in hypogonadal mice, *Exp. Neurol.* **105:**135–151.

Broadwell, R. D., Charlton, H. M., Ebert, P., Hickey, W. F., Villegas, J. C., and Wolf, A. L., 1990, Angiogenesis and the blood–brain barrier in solid and dissociated cell grafts within the CNS, *Prog. Brain Res.* **82:**95–101.

Broadwell, R. D., Charlton, H. M., Ebert, P., Hickey, W. F., Shirazi, Y., Villegas, J., and Wolf, A. L., 1991, Allografts of CNS tissue possess a blood–brain barrier. II. Angiogenesis in solid tissue and cell suspension grafts, *Exp. Neurol.* **112:**1–28.

Broadwell, R. D., Baker, B. J., Ebert, P., Hickey, W. F., and Villegas, J., 1992a, Intracerebral grafting of solid tissues and cell suspensions: The blood–brain barrier and host immune response, *Prog. Brain Res.* **91:**95–102.

Broadwell, R. D., Baker, B. J., and Friden, B., 1992b, Receptor-mediated transcytosis through the blood–brain barrier: Ferro-transferrin and its receptor, *Abstr. Soc. Neurosci.* **18:**1130.

Cataldo, A. M., and Broadwell, R. D., 1986, Cytochemical identification of cerebral glycogen and glucose 6-phosphatase activity under normal and experimental conditions. I. Choroid plexus and ependymal epithelia, endothelia, and pericytes, *J. Neurocytol.* **15**:511–524.

Dautry-Varsat, A., and Lodish, H. F., 1983, The Golgi complex and the sorting of membrane and secreted protein, *Trends Neurosci.* **6**:484–490.

Dautry-Varsat, A., and Lodish, H. F., 1984, How receptors bring proteins and particles into cells, *Sci. Am.* **250**:52–58.

Ermisch, A., Rhule, H. J., Landgraf, R., and Hess, J., 1985, Blood–brain barrier and peptides, *J. Cerebral Blood Flow Metab.* **5**:350–357.

Fishman, J. B., and Fine, R. E., 1987, A trans Golgi-derived exocytic coated vesicle can contain both newly synthesized cholinesterase and internalized transferrin, *Cell* **48**:157–164.

Fishman, J. B., Rubin, J. B., Handrahan, J. V., Connor, J. R., and Fine, R. E., 1987, Receptor-mediated transcytosis of transferrin across the blood–brain barrier, *J. Neurosci. Res.* **18**:299–304.

Fishman, P. S., and Carrigan, D. R., 1988, Motoneuron uptake from the circulation of the binding fragment of tetanus toxin, *Arch. Neurol.* **45**:558–561.

Gonatas, J. K., and Avrameas, S., 1973, Detection of plasma membrane carbohydrates with lectin peroxidase conjugates, *J. Cell Biol.* **59**:436–445.

Gonatas, J. K., Steiber, A., Hickey, W. F., Herbert, S. H., and Gonatas, J. O., 1984, Endosomes and Golgi vesicles in adsorptive and fluid phase endocytosis, *J. Cell Biol.* **99**:1379–1390.

Gotow, T., and Hashimoto, P. H., 1979, Fine structure of the ependyma and intercellular junctions in the area postrema of the rat, *Cell Tissue Res.* **201**:207–225.

Gotow, T., and Hashimoto, P. H., 1981, Graded differences in tightness of ependymal intercellular junctions within and in the vicinity of the rat median eminence, *J. Ultrastruct. Res.* **76**:292–311.

Gross, P. M., 1987, *Circumventricular Organs and Body Fluids*, Volumes I–III, CRC Press, Boca Raton, Fla.

Gross, P. M., and Weindl, A., 1987, Peering through the windows of the brain, *J. Cerebral Blood Flow Metab.* **7**:663–672.

Helenius, A., Mellman, I., Wall, D., and Hubbard, A., 1983, Endosomes, *Trends Biochem. Sci.* **8**:245–249.

Janzer, R. C., and Raff, M. C., 1987, Astrocytes induce blood–brain barrier properties in endothelial cells, *Nature* **325**:253–257.

Jeffries, W. A., Brandon, M. R., Hunt, S. V., Williams, A. F., Gatter, K. C., and Mason, D. Y., 1984, Transferrin receptor on endothelium of brain capillaries, *Nature* **312**:162–163.

Langston, J. W., 1985, MPTP and Parkinson's disease, *Trends Neurosci.* **8**:79–83.

Mata, L. R., and David-Ferriera, J. F., 1973, Transport of exogenous peroxidase to Golgi cisternae in the hamster seminal vesicle, *J. Microscop.* **17**:103–110.

Meisenberg, G., and Simmons, W. H., 1983, Peptides and the blood–brain barrier, *Life Sci.* **32**:2611–2623.

Mollgard, K., and Saunders, N. R., 1975, Complex tight junctions of epithelial and endothelial cells in early foetal brain, *J. Neurocytol.* **4**:453–468.

Mollgard, K., and Saunders, N. R., 1977, A possible transepithelial pathway via endoplasmic reticulum in foetal sheep choroid plexus, *Proc. R. Soc. London Ser. B* **199**:321–326.

Nabeshima,S., Reese, T. S., Landis, D. M. D., and Brightman, M. W., 1975, Junctions in the meninges and marginal glia, *J. Comp. Neurol.* **164**:127–169.

Palade, G., 1975, Intracellular aspects of the process of protein synthesis, *Science* **189**: 347–358.

Pardridge, W. M., 1986, Receptor-mediated peptide transport through the blood–brain barrier, *Endocrine Rev.* **7**:314–330.

Pardridge, W. M., 1991, *Peptide Drug Delivery to the Brain*, Raven Press, New York.

Pardridge, W. M., Triguero, D., and Buciak, J. L., 1990, β-Endorphin chimeric peptides: Transport through the blood–brain barrier in vivo and cleavage of disulfide linkage by brain, *Endocrinology* **126**:977–984.

Rapoport, S. I., 1985, Tight junctional modification as compared to increased pinocytosis as the basis of osmotically-induced opening of the blood–brain barrier. Further evidence of the tight junctional mechanism and against pinocytosis, *Acta Neurol. Scand.* **72**:107.

Rapoport, S. I., 1988, Osmotic opening of the blood–brain barrier, *Ann. Neurol.* **24**:677–680.

Reese, T. S., and Karnovsky, M. J., 1967, Fine structural localization of a blood–brain barrier to exogenous peroxidase, *J. Cell Biol.* **34**:207–217.

Richards, J. G., 1978, Permeability of intercellular junctions in brain epithelia and endothelia to exogenous amine: Cytochemical localization of extracellular 5-hydroxydopamine, *J. Neurocytol.* **7**:61–70.

Rodewald, R., and Kraehenbuhl, J. P., 1984, Receptor-mediated transport of IgG, *J. Cell Biol.* **99**:159s–164s.

Rosenstein, J. M., 1987, Adrenal medulla grafts produce blood–brain barrier dysfunction, *Brain Res.* **414**:192–196.

Rosenstein, J. M., and Brightman, M. W., 1986, Alternations of the blood–brain barrier after transplantation of autonomic ganglia into the mammalian central nervous system, *J. Comp. Neurol.* **250**:339–351.

Santos, T. Q., and Valdimarsson, H., 1982, T-dependent antigens are more immunogenic in the subarachnoid space than in other sites, *J. Neuroimmunol.* **2**:215–222.

Senjo, M. T., Ishibashi, T., Terashiman, T., and Inoue, Y., 1986, Correlation between astrogliogenesis and blood–brain barrier formation; immunocytochemical demonstration by using astroglia-specific enzyme glutathione S-transferase, *Neurosci. Lett.* **66**:39–42.

Sly, W. S., Fischer, H. D., and Gonzales-Horiega, A., 1981, Role of 6-phosphomannolsyl-enzyme receptor in intracellular transport of adsorptive pinocytosis of lysosomal enzymes, *Methods Cell Biol.* **23**:191–197.

Snider, M. D., and Rogers, O. C., 1985, Intracellular movement of cell surface receptors after endocytosis: Resialylation of asialotransferrin receptor in human erythroleukemia cells, *J. Cell Biol.* **100**:826–834.

Steinman, R. M., Mellman, I. S., Muller, W. A., and Cohn, Z. A., 1983, Endocytosis and the recycling of plasma membrane, *J. Cell Biol.* **96**:1–27.

Sternberger, N., and Sternberger, L., 1987, Blood–brain barrier protein recognized by monoclonal antibody, *Proc. Natl. Acad. Sci. USA* **84**:8169–8173.

Straus, W., 1981, Cytochemical detection of mannose-specific receptors for glycoproteins with horseradish peroxidase as a ligand, *Histochemistry* **73**:39–45.

Talamo, B. R., Rudel, R. A., Kosik, K. S., Lee, V. M. Y., Neff, S., Adelman, L., and Kauer, J. S., 1989, Pathological changes in olfactory neurons in patients with Alzheimer's disease, *Nature* **337**:736–739.

Tangoren, M., Broadwell, R. D., Moriyama, E., Oliver, C., and Wolf, A., 1988, How significant is the blood–brain barrier? *Soc. Neurosci. Abstr.* **14**:617.

Tao-Cheng, J. H., Nagy, Z., and Brightman, M. W., 1987, Tight junctions of brain endothelium in vitro are enhanced by astroglia, *J. Neurosci.* **7:**3293–3299.

Triguero, D., Buciak, J. B., Yang, J., and Pardridge, W. M., 1989, Blood–brain barrier transport of cationized immunoglobulin G: Enhanced delivery compared to native protein, *Proc. Natl. Acad. Sci. USA* **86:**4761–4765.

Ugolini, G., Kuypers, H. G., and Strick, P. L., 1989, Transneuronal transfer of herpes virus from peripheral nerves to cortex and brainstem, *Science* **243:**89–91.

Villegas, S., and Broadwell, R., 1989, Retrograde trans-synaptic transfer of a blood-borne protein, *Soc. Neurosci. Abstr.* **15:**821.

Villegas, J., and Broadwell, R. D., 1993, Transcytosis of protein through the mammalian cerebral epithelium and endothelium. II. Adsorptive transcytosis of WGA-HRP and the blood–brain and brain–blood barriers, *J. Neurocytol.* **22** (in press).

Wakai, S., Meiselman, S. E., and Brightman, M. W., 1986, Focal circumvention of blood–brain barrier with grafts of muscle, skin and autonomic ganglia, *Brain Res.* **386:**209–222.

Weindl, A., 1973, Neuroendocrinal aspects of circumventricular organs, in: *Frontier of Neuroendocrinology* (W. F. Ganong and L. Martini, eds.), Oxford University Press, New York, pp. 3–31.

Woods, J. W., Dorizaux, M., and Farquhar, M. G., 1986, Transferrin receptors recycle to cis and middle as well as trans Golgi cisternae in Ig-secreting myeloma cells, *J. Cell Biol.* **103:**277–286.

IV

Elimination Barriers

Chapter 12

Renal Uptake and Disposal of Proteins and Peptides

Ralph Rabkin and David C. Dahl

1. INTRODUCTION

While the normal kidney provides an effective barrier to the elimination of large proteins from the systemic circulation, it serves as a major route for the elimination of most small proteins (Maack *et al.*, 1985; Rabkin and Kitaji, 1983). In health the glomerular capillary wall almost completely restricts the passage of molecules of molecular radius greater than ~ 42 Å (~ 50 kDa) while increasingly favoring the passage of proteins as their size decreases (Brenner *et al.*, 1978). Thus, serum proteins the size of albumin (36 Å; 60 kDa) or greater are severely restricted by the renal vascular barrier and the normal kidney plays a negligible role in their metabolism (Waldmann *et al.*, 1972). In contrast, the renal elimination of small proteins that circulate as monomers with radii < 20 Å (< 25 kDa) is highly efficient. These small proteins escape from the renal circulation through the glomerular capillaries and, to a variable though usually minor extent, through the peritubular capillaries (Baylis, 1982). The contribution of the kidney to the metabolism of small proteins has been particularly well studied and, depending on the contribution of extrarenal sites, the kidney may account for up to 80% of the total metabolism of certain proteins.

Whether the kidney serves as a barrier to or a conduit for the removal of proteins

Ralph Rabkin • Department of Medicine, Nephrology Section, Stanford University School of Medicine, Stanford, and Palo Alto Department of Veterans Affairs Medical Center, Palo Alto, California 94304. *David C. Dahl* • Department of Medicine, University of Minnesota, and Regional Kidney Disease Program, Hennepin County Medical Center, Minneapolis, Minnesota 55415.

Biological Barriers to Protein Delivery, edited by Kenneth L. Audus and Thomas J. Raub. Plenum Press, New York, 1993.

is determined largely by the permeability properties of the glomerular filtration barrier and the size, conformation, and charge of the protein (Brenner *et al.*, 1978; Bohrer *et al.*, 1979). Glomerular filtration, with the delivery of protein into the proximal tubular lumen, is the major route for renal protein elimination. There are highly efficient mechanisms for the disposal of filtered protein. Complex polypeptides and proteins are rapidly internalized by endocytosis and then hydrolyzed within the cell to peptide fragments and constituent amino acids (Maack *et al.*, 1985; Wall and Maack, 1985; see Chapter 3). The amino acids are transported out of the cell into the interstitial compartment and are returned to the systemic circulation. In contrast, small linear peptides such as angiotensin and bradykinin are hydrolyzed by the peptide-rich luminal brush border membrane of the proximal tubules and the products, predominantly amino acids, are transported into and across the cell and returned to the circulation (Carone and Peterson, 1980). These proximal tubular events are highly efficient and for most proteins and peptides, only small amounts of the filtered material, often less than 1%, escape into the urine (see Chapter 2). Trapping of protein within cellular components of the glomeruli with local destruction does occur but the glomerulus as a site of protein metabolism is trivial (Rabkin and Kitaji, 1983).

In addition to the glomerular filtration–tubular degradative pathway, the kidney also eliminates proteins from the postglomerular peritubular capillary network (Rabkin and Kitaji, 1983; Rabkin and Petersen, 1984) (Fig. 1). Significant elimination via this latter route occurs only for small bioactive proteins, such as insulin,

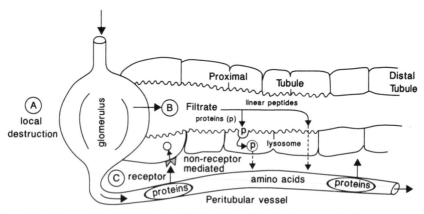

Figure 1. Pathways of protein elimination. (A) Local degradation by the glomerulus represents a minor site of elimination for small peptide hormones such as angiotensin, bradykinin, atrial natriuretic peptide, and possibly calcitonin. (B) Glomerular filtration and tubular absorption is the major route for proteins and peptides. Proteins and complex peptides require internalization prior to degradation. Small linear peptides are hydrolyzed by the luminal membrane. (C) Postglomerular peritubular elimination constitutes an important though lesser route of elimination for some protein hormones. Their basolateral uptake is largely receptor mediated and is followed by intracellular degradation.

vasopressin, and parathyroid hormone, which have hormone-specific receptors located in the basolateral tubular cell membrane. For insulin, this pathway accounts for 40% of its renal elimination (Rabkin *et al.*, 1984). However, for most protein hormones the major importance of this pathway is delivery of hormone to target cells.

A feature of the renal (whole organ) elimination of low molecular-weight proteins is that it is extremely difficult or impossible to saturate. This presumably reflects glomerular filtration as the major route of clearance. Consequently, as exemplified by studies of protein hormones over a wide range of plasma levels, the kidney eliminates low-molecular-weight proteins from a constant volume of plasma per unit time (Emmanouel *et al.*, 1978; Hruska *et al.*, 1975; Rabkin and Colwell, 1969; Rabkin *et al.*, 1970). This serves to help maintain plasma protein hormone concentrations at basal levels, for when the plasma hormone concentration rises the absolute amount removed increases proportionately (Fig. 2). This does not, however, represent a true feedback regulatory mechanism.

Aside from the role of the kidney in eliminating and metabolizing proteins, many proteins in turn may affect renal function. These are comprised largely of protein and peptide hormones with specific receptors in the kidney through which the hormone initiates its signals (Rabkin and Mahoney, 1987). Examples include parathyroid hormone, vasopressin, angiotensin, and atrial natriuretic peptide (ANP), hormones with major renotropic actions. In addition to the physiologic action of

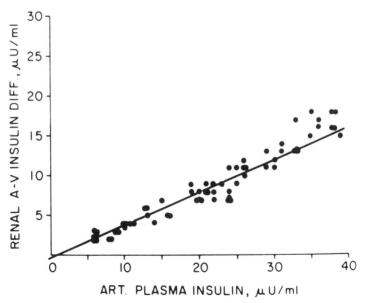

Figure 2. Relationship between plasma arterial insulin concentration and renal arteriovenous insulin difference measured in humans. Over the range of insulin levels studied, the kidney eliminates a constant percentage. (From Rabkin *et al.*, 1970, with permission.)

protein hormones, there are proteins that may have toxic effects on renal structure and function when present in high concentration. These include myeloma proteins which can produce renal failure (Weiss *et al.*, 1981; Sanders *et al.*, 1988) and the enzyme lysozyme which can alter cellular transport function (Cojocel *et al.*, 1982; Mason *et al.*, 1975). Circulating immunoglobulins with antigenic sites in the kidney may also adversely affect renal structure and function and can result in permanent renal damage (Kashtan *et al.*, 1990; Saxena *et al.*, 1989).

The key role played by the kidney is readily apparent in renal failure where profound alterations in protein elimination may occur (Rabkin *et al.*, 1983; Schardijn and van Eps, 1987). This may not only be due to the loss of the kidney as a site of elimination, but in addition extrarenal sites of metabolism may be depressed when uremia intervenes (Rabkin *et al.*, 1979a). In some instances, the rate of hormone secretion may also be altered in the uremic state (Emmanouel *et al.*, 1980). Thus, if a reduction in protein secretion does not accompany reduced elimination, the protein will accumulate in the extracellular compartment. If bioactive, these proteins may then exert adverse biologic effects.

2. METHODS OF STUDY

Understanding of the renal elimination and metabolism of protein has been greatly facilitated by the use of radiolabeled proteins and the development of sensitive radioimmunoassays. Problems associated with the use of radiolabeled proteins include loss of bioactivity and altered renal handling (Shade and Share, 1976; Martin *et al.*, 1987). These problems have been minimized by the use of gentle methods of iodination such as the lactoperoxidase or the Iodogen method and by lowering the average number of radioactive iodine atoms to less than one per protein molecule. Nevertheless, where practical, the use of endogenous or radioisotope-free exogenous native or recombinant protein is preferable. The study of endogenous proteins can, on occasion, be complicated by the presence of circulating heterogeneous forms that are indistinguishable by radioimmunoassay but differ in their bioactivity and half-lives. This may be, as will be discussed later, a special problem in renal failure where precursor forms of proteins or their immunoreactive fragments may accumulate (Emmanouel *et al.*, 1980; Kuku *et al.*, 1976; Lee *et al.*, 1977; Arnaud *et al.*, 1974). To avoid these problems, the use of site-specific antibodies, gel electrophoresis, gel filtration chromatography, or high-performance liquid chromatography (HPLC) has proved helpful. For both endogenous and exogenous proteins, the presence of circulating binding proteins may further complicate their study and steps to distinguish between bound and free protein are required. Good examples are the insulinlike growth factors (Powell *et al.*, 1986; Sara and Hall, 1990).

Considerable information has been obtained from studies performed *in vivo* in which the concentration of hormone in arterial and renal vein blood has been measured simultaneously with that of renal blood flow (Hruska *et al.*, 1975; Katz and Rubenstein, 1973; Rabkin *et al.*, 1970). The renal (organ) clearance of the protein of interest is derived from these values and allows an estimation of the contribution of the kidney to the total (whole body) metabolic clearance rate of the protein. To estimate the contribution of the glomerular filtration pathway to the renal (organ) clearance of protein, it is necessary to measure the glomerular filtration rate and then to correct for the sieving coefficient of the protein. The sieving coefficient is a measure of a molecule's freedom of passage across the filtration barrier and its determination for purposes of studying the glomerular elimination barrier is described later (Section 3.1). For proteins, however, apart from a few exceptions, the true sieving coefficient is not known and to obtain this value directly requires micropuncture measurement in animals. Such studies are few. Indirect approaches have been more commonly employed (Maack *et al.*, 1979; Rabkin and Kitaji, 1983). When peritubular removal is minimal, the sieving coefficient has been derived from the ratio of renal (organ) clearance to glomerular filtration rate (GFR). Otherwise, the sieving coefficient may be derived from the ratio of the urinary clearance of a protein to the clearance of a glomerular filtration marker such as inulin obtained in the presence of an inhibitor of tubular protein absorption. Since the action of inhibitors is not usually limited to tubular function but may have more general effects, inhibition studies are best performed in an isolated perfused kidney preparation (Maack *et al.*, 1979; Rabkin and Kitabchi, 1978). Some estimates of protein sieving coefficients are given in Table I. With small peptides (radius < 10 Å), it is reasonable to assume that there is negligible restriction to passage through the glomerular filtration barrier. With some protein hormones such as insulin, the renal (organ) clearance exceeds the GFR, indicating that elimination is not restricted to glomerular filtration and that clearance from the postglomerular peritubular circulation is

Table I

Glomerular Sieving Coefficients of Small Proteins[a]

Protein	Size (Da)	Stokes–Einstein radius (Å)	Glomerular sieving coefficient
Insulin	6,000	~15.0	0.89
Lysozyme	14,600	19.0	0.75
Myoglobin	16,900	18.8	0.75
Bovine PTH	9,000	~21.4	0.69
Rat GH	20,000	20.4	0.65
Horseradish peroxidases	40,000	31.8	0.007 (anionic)
		29.8	0.06 (neutral)
		30.0	0.34 (cationic)
Bence–Jones (λ-L chain)	44,000	27.7	0.085

[a]From Maack *et al.* (1979).

occurring (Chamberlain and Stimmler, 1967; Rabkin *et al.*, 1970). With other proteins, the presence of peritubular clearance is uncovered only when the GFR is reduced experimentally (Rabkin *et al.*, 1981; Shade and Share, 1977).

A useful approach to studying both the peritubular and glomerular elimination pathway *in vivo* has been developed by Silverman and colleagues (Silverman *et al.*, 1984; Whiteside *et al.*, 1988). This multiple indicator dilution technique requires cannulation of both renal artery and vein and the injection of reference tracers into the renal artery. This is followed by timed serial collection of venous blood and urine. Radiolabeled dextrans serve as molecular size markers, [^{125}I]albumin as a plasma component marker, [^{14}C]inulin as a glomerular filtration reference, and a radiolabeled protein, e.g., insulin (Whiteside *et al.*, 1988), is used as the study protein. With use of appropriate internal standards, the appearance of tracers in the renal vein and urine are plotted against time to produce renal vein and urine outflow curves. As dextrans and inulin are not actively transported or metabolized in the body, they are appropriate markers for the study of transcapillary exchange. The renal outflow curves are interpreted by means of mathematical models as described by Silverman *et al.* (1984) and Whiteside *et al.* (1988).

Because of the limitations of *in vivo* studies, approaches using the isolated perfused kidney, isolated nephron segments, cultured kidney cells, and subcellular organelles have been used to further our understanding of the renal elimination pathways. Particularly useful has been the isolated perfused rat kidney (Maack *et al.*, 1979; Maack, 1986; Rabkin *et al.*, 1979b). This preparation allows for the use of proteins and also inhibitors in concentrations impractical *in vivo*. Furthermore, because of the high perfusion flow and low filtration fraction (GFR/glomerular blood flow), peritubular events are exaggerated in this preparation and are more readily studied. When necessary, peritubular elimination can be studied in the absence of glomerular protein elimination by the use of a nonfiltering isolated kidney preparation. This is achieved by perfusing ureter-ligated preparations with hyperoncotic albumin at a lowered perfusion pressure (Maack, 1986).

Luminal events are commonly studied by measurement of the urinary clearance of a protein. When assaying proteins or polypeptides in urine, it is important to recognize that even though they may be stable in plasma and other body fluids they may disintegrate in urine. This is known to occur with β_2-microglobulin and neurotensin (Schardijn and van Eps, 1987; Bjerke *et al.*, 1989). For β_2-microglobulin, raising the urine pH above 6.0 stabilizes the protein. Since the urinary clearance of a protein may be affected by changes in the GFR and hence filtered load, it is useful to express urinary clearance as a fraction of the simultaneously determined GFR. Unfortunately, this does not allow for changes in the permeability properties of the filtration barrier and it may not always be possible to distinguish between changes in tubular absorption of a protein and changes in filtered load. To study luminal events directly *in vivo*, micropuncture techniques have been successfully employed in animals. This requires anesthesia, surgical exposure of the kidney, and the insertion of a micropipette into a superficial nephron segment on the kidney surface (Carone and Peterson, 1980; Seikaly *et al.*, 1990). Direct study of luminal and contraluminal events has been achieved with isolated microperfused nephron segments (Bourdeau *et al.*, 1973; Nielsen *et al.*, 1986) and more recently with cultured kidney cell monolayers grown on filters suspended in culture wells (Maratos-Flier *et al.*, 1987; Rabkin *et al.*, 1989) (Fig. 3). Current advances in cell culture techniques (Audus *et al.*, 1990), methods of subcellular organelle preparation, and electron immunocytochemistry are all serving to advance our knowledge of the renal elimination pathways. Finally, the impact of molecular biology techniques, which allow for the genetic engineering of proteins, will have a profound effect on our understanding of the cellular processing of proteins. It should, of course, be kept in mind that caution must be exercised when considering information derived from these powerful *in vitro* approaches since the data may not be representative of the situation *in vivo*.

3. VASCULAR ELIMINATION BARRIERS

3.1. Glomerular Vascular Barrier

Quantitatively, the most important route for the renal elimination of protein is via glomerular filtration with subsequent catabolism by the proximal tubule or excretion in the urine. The properties of the glomerular filtration barrier determine which proteins will be cleared by this route (Anderson *et al.*, 1991). The clearance of a particular protein is influenced by factors such as its size, charge, and geometric configuration. Intraglomerular hemodynamic factors also influence the filtration of proteins. These factors will be considered in turn, but first the glomerular capillary wall will be briefly described.

The glomerular capillary wall is a three-layered structure consisting of (from capillary lumen to urinary space): an endothelial cell layer, the glomerular basement membrane, and an epithelial cell layer (Tisher and Brenner, 1989) (Fig. 4). Frequent fenestrations exist in the endothelium. The epithelial cells, or podocytes, send out interdigitating cytoplasmic projections termed foot processes. Gaps exist between the foot processes and these are covered by the so-called slit diaphragms. The glomerular basement membrane consists of three layers of varying electron density, the relatively lucent lamina rara interna and externa, and the more electron-dense lamina densa. The biochemical nature of the glomerular barrier has been well studied and extensively reviewed (Abrahamson, 1987). Both endothelial and epithelial plasma membranes contain a 140-kDa sialoprotein termed podocalyxin (Kerjaschki *et al.*, 1984). This, and other, polyanions likely contribute to the charge selectivity of the glomerular barrier. The basement membrane is composed of a scaffolding of type IV collagen ($\sim$ 550–600 kDa). Enmeshed within this framework are a number of other glycoproteins. The polyanionic proteoglycans, most importantly heparan sulfate, are

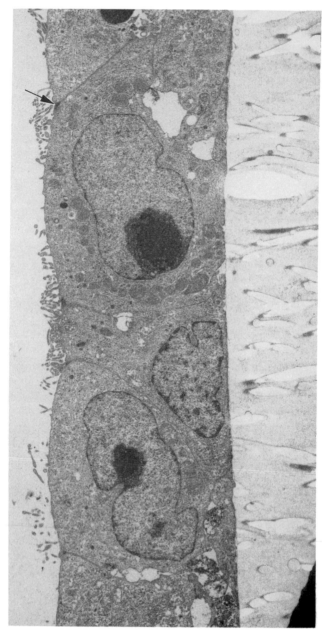

Figure 3. Electron micrograph of a confluent monolayer of opossum kidney cells grown on a polycarbonate filter suspended in a culture well. Note presence of tight junctions (arrow) and microvilli on apical surface of the cells, two features of polarized epithelium. This confluent monolayer effectively divides the culture well into two compartments. Addition of protein of interest to one compartment allows the separate study of apical or basal events. (From Rabkin *et al.*, 1989, with permission.)

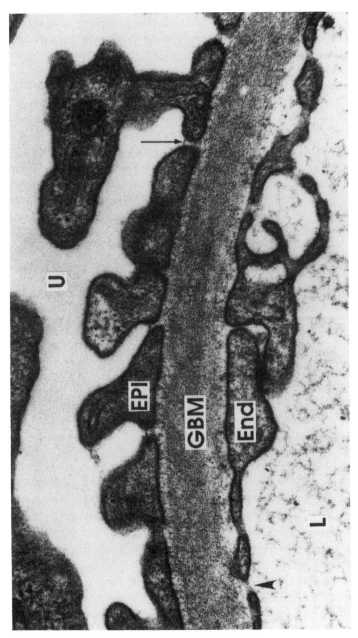

Figure 4. Electron micrograph of human glomerular capillary wall. The capillary lumen (L) is lined by endothelium (End) which is penetrated by fenestrae (arrowhead). The glomerular basement membrane (GBM) is located between the endothelium and the foot processes of the epithelial cells (EPI). Between the foot processes are filtration slits (arrow) covered by slit diaphragms. Beyond the epithelial cells is the urinary space (U). × 96,000. (Micrograph courtesy of Dr. W. Robert Anderson.)

believed to impart to the glomerular basement membrane its charge-selective properties. Laminin and fibronectin help anchor the basement membrane to adjacent cells.

If the concentration of a particular substance is measured simultaneously in the
glomerular ultrafiltrate in Bowman's space and in the glomerular capillary plasma,
then the extent to which that substance's filtration is hindered by the glomerular
barrier can be determined. The ratio of these two concentrations ($C_{\text{urinary space}}/C_{\text{plasma}}$)
is termed the sieving coefficient. For a substance which is freely filtered such as
water, electrolytes, and small uncharged solutes with an effective Stokes–Einstein
radius of less than 18 Å, this value equals one (Fig. 5). Although the determination of
sieving coefficients for various proteins seems an attractive method for studying
glomerular filtration, in fact, this value has been directly determined for very few
proteins, and then only in animals by means of micropuncture studies (Galaske *et al.*,
1979). Protein concentrations in the final urine cannot be used for the purposes of
determining the sieving coefficient because, once filtered, proteins are extensively
reabsorbed and catabolized by the renal tubules.

These problems have been circumvented in part by using dextrans as test
substances to explore the properties of the glomerular capillary barrier. These
nonprotein polymers are not reabsorbed, catabolized, or secreted by the renal tubule.
The urinary clearance of dextran is compared with the clearance of inulin, which is
regarded as an ideal reference marker for GFR since it is freely filtered and is also not

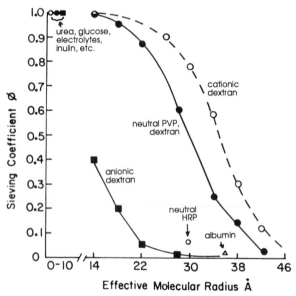

Figure 5. Sieving curves of several macromolecules. The different sieving coefficients reflect the
influence of size, charge, shape, and rigidity of the molecules. (From Arendshorst and Navar, 1988, with
permission.)

reabsorbed, catabolized, or secreted by the renal tubule. Thus, the ratio of the dextran clearance to the inulin clearance is equal to the sieving coefficient. When dextran size is plotted against fractional dextran clearance, it is apparent that the glomerular barrier is size selective and that glomerular filtration is progressively restricted as molecular radius increases (Brenner et al., 1978; Arendshorst and Navar, 1988) (Fig. 5). The most useful model of the glomerular barrier based on analysis of such data describes the glomerular capillary wall as a heteroporous barrier (Deen et al., 1985). In this model, most of the glomerular wall functions as a membrane with pores of equal size. However, the glomerular wall also contains a shunt pathway through which a small portion of the glomerular filtrate passes. The shunt pathway is nonselective in that it does not restrict proteins on the basis of size. The increased proteinuria of certain glomerular diseases is accounted for by a larger fraction of the glomerular filtrate (up to 2%) traversing the shunt and not by an increase in the size of the glomerular pores.

The influence of molecular charge on glomerular permselectivity can be studied by comparing the fractional clearances of anionic dextrans (e.g., dextran sulfate) or cationic dextrans (e.g., diethylaminoethyl dextran) with those of neutral dextran (Brenner et al., 1978; Bohrer et al., 1978). In such experiments, the clearance of anionic molecules is impaired relative to that of neutral substances and, conversely, the clearance of cationic molecules is enhanced (Fig. 5). Thus, the glomerular barrier is charge selective as well as size selective. The influence of charge on glomerular filtration is especially prominent for molecules with radii greater than 20 Å (Maack et al., 1985). The role of molecular configuration in determining the glomerular filtration of proteins has not been as well studied as that of molecular size and charge. Bohrer et al. (1979) compared the fractional clearances of Ficoll (Pharmacia), a spherical molecule, with those of dextran, a flexible coil. At similar molecular radii, the clearance of dextran was significantly greater than that of Ficoll, indicating that molecular shape does influence filtration to some extent.

In addition to the molecule properties of charge, size, and shape, glomerular filtration of proteins is also influenced by the local hemodynamic conditions that obtain in the glomerular capillary. For example, certain pathophysiologic conditions are associated with an absolute or relative increase in the fraction of plasma water filtered (filtration fraction), which leads to an increase in the concentration of proteins within the glomerular capillaries. It has been suggested that such an increased protein concentration favors the passage of proteins across the glomerular capillary wall by (1) creating a more favorable gradient for diffusion and also by (2) increasing the concentration of protein in fluid traversing the glomerulus (Bohrer et al., 1977). Also, increasing glomerular capillary hydraulic pressure may enhance proteinuria by inducing a size-selective defect in the glomerular capillary wall (Yoshioka et al., 1986).

The actual clearance of some relatively small proteins from the glomerular circulation may be reduced dramatically by binding to large plasma proteins. This is exemplified by growth hormone and the insulin-like growth factors. For growth hormone, a specific high-affinity binding protein homologous with the extracellular domain of the growth hormone receptor is present in human plasma (Herington *et al.*, 1986; Leung *et al.*, 1987). Growth hormone complexed to this binding protein is degraded at only one-tenth the rate of the free hormone (Baumann *et al.*, 1989). Since the kidney is a major site of growth hormone elimination and this occurs largely by glomerular filtration (Johnson and Maack, 1977; Rabkin *et al.*, 1981), the reduced total body clearance of the hormone–binding protein complex relative to free hormone may be due in part to impaired filtration of the hormone–binding protein complex (Baumann *et al.*, 1987).

Specific binding proteins also exist in human plasma for IGF-I and IGF-II (Sara and Hall, 1990; Rutanen and Pekonen, 1990; Powell *et al.*, 1986). Circulating IGF is bound to these binding proteins, primarily to the 150-kDa form. The half-life of bound IGF is significantly prolonged relative to that of free IGF in a number of experimental models (Drakenberg *et al.*, 1990; Cascieri *et al.*, 1988). For example, when IGF-I or IGF-II is injected into hypophysectomized rats, which do not possess the 150-kDa binding protein, the half-lives are about 20 to 30 min, versus $\sim$ 4 hr in control rats (Zapf *et al.*, 1986). Restricted glomerular filtration and peritubular extraction of the IGF–binding protein complex could contribute to the prolonged half-life of bound IGF, although experimental proof of this hypothesis is lacking.

3.2. Peritubular Vascular Barrier

The renal tubules are surrounded by a network of capillaries which is derived from the efferent arteriole of the glomerulus (Fig. 6). The endothelium lining the capillaries has numerous fenestrae which are bridged by thin diaphragms. Separating the peritubular capillaries from the tubules is a narrow interstitium (Lemley and Kriz, 1991). Proteins leaving the peritubular vessels enter the interstitium either to be returned to the circulation by lymphatic flow (Wolgast, 1985; Atkins *et al.*, 1988; Pinter, 1988) or to be trapped by the tubular cells. Large proteins diffuse across the capillary walls, predominantly through the endothelial fenestrae, and enter the renal interstitium where they tend to accumulate because of the relatively low rate of lymphatic clearance (Venkatachalam and Karnovsky, 1972). The protein content of the interstitium reaches a concentration of 1–2 g/dl with an albumin/globulin ratio of 2 (in plasma it is unity) and contributes to the pressure gradient between the interstitium and tubular cells. This favors solute and water reabsorption (Wolgast, 1985).

With respect to the permeability of the postglomerular vessels to protein, it should be noted that in contrast to the glomerulus where there is outward movement of fluid, the convective component of fluid flow in the peritubular vessels is directed inwards as fluid transported across the tubular cells is reabsorbed (Silverman *et al.*, 1984). Hence, passage of protein out of the peritubular vessels into interstitium

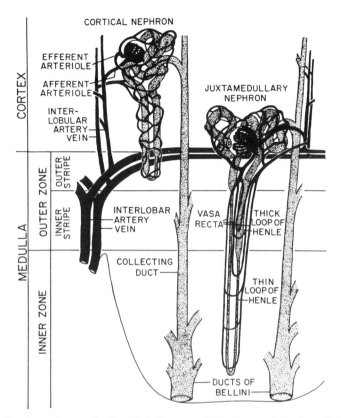

Figure 6. Renal vascular supply. Simplified diagram outlining intrarenal blood supply. Peritubular capillaries arise from the efferent arterioles. (From Pitts, 1974, with permission.)

occurs only by diffusion. Thus, movement of protein out of the peritubular capillaries is reduced relative to glomerular capillaries (Silverman *et al.*, 1984). As in glomerular capillaries, the passage of proteins out of the peritubular vessels is also restricted by size, shape, and charge (Baylis, 1982; Deen *et al.*, 1976; Wolgast, 1985). Because of the electrostatic barrier, negatively charged macromolecules permeate the vessel wall more slowly than do neutral molecules of the same size. In addition to the passage of proteins between endothelial cells or through their fenestrae or pores (Rippe and Haraldsson, 1987), it is likely that transcytosis may also occur (Simionescu *et al.*, 1987). Transcytosis, which is discussed in detail in Chapter 10, is a process whereby plasma proteins are endocytosed by the vascular endothelial cells and transported within vesicles in intact form across the cell to be exocytosed on the extravascular side. This process is well described in endothelial cells from a variety of large and small vessels (Bar *et al.*, 1988; Simionescu *et al.*, 1987). Because of technical difficulties, transcytosis has not been studied in the peritubular vessels. While the role of transcytosis in the passage of large proteins such as albumin remains controversial (Rippe and Haraldsson, 1987), this process is important in the delivery of small protein hormones such as insulin and IGF-I (Bar *et al.*, 1988). Transcytosis of protein hormones is mediated by hormone-specific receptors on the cell surface and is characterized by a relative lack of degrading activity within the transport vesicles. In contrast to this process, there is indirect evidence that the postglomerular vessels may participate directly in the degradation of some small peptide hormones such as angiotensin and ANP (Seikaly *et al.*, 1990; Berg *et al.*, 1988; Murthy *et al.*, 1986). Proteins that move out of the peritubular vessels and enter the interstitial space either are returned to the circulation by lymphatic drainage or are taken up at the contraluminal pole of the tubular cells. Their subsequent processing by the cell is discussed in Section 4.2.

This peritubular pathway serves two important roles. First, it routes several protein hormones to their site of action. Indeed, for most renotropic hormones such as parathyroid hormone, calcitonin, vasopressin, and insulin, their receptors are located only or predominantly on the contraluminal aspect of the tubular cells (Rabkin and Mahoney, 1987). Second, the peritubular pathway serves to eliminate certain proteins. The magnitude of this elimination process varies according to the protein, from negligible for β_2-microglobulin (Schardijn and van Eps, 1987) to 40% for the protein hormone insulin (Rabkin *et al.*, 1970; Chamberlain and Stimmler, 1967). Other protein hormones with a major peritubular elimination pathway include calcitonin (Simmons *et al.*, 1988), parathyroid hormone (Martin *et al.*, 1987), and angiotensin II (Seikaly *et al.*, 1990; Reams *et al.*, 1990). Apart from circulating hormones, the peritubular pathway may also be important with respect to the removal of intrarenally generated hormone. Recently, Seikaly *et al.* (1990) noted that the concentration of angiotensin in early postglomerular vessels was 1000-fold greater than in the systemic circulation. Since renal venous angiotensin concentrations are of a similar order of magnitude to that in systemic plasma, they concluded that substantial elimination of angiotensin occurs downstream in the postglomerular circulation.

In summary, the glomerular filtration, tubular absorption pathway is the major route for the renal elimination of circulating proteins with the postglomerular peritubular pathway playing a significant role for some protein hormones. The major importance of the peritubular pathway is delivery of protein hormones to their site of action.

4. EXTRAVASCULAR PATHWAYS IN THE KIDNEY

4.1. Apical Tubular Cell Uptake and Urinary Excretion

The fate of the filtered peptide or protein after it enters the lumen of the proximal tubule is determined by its size, shape, molecular structure, and charge (Maack *et al.*, 1985; Carone and Peterson, 1980). Large complex polypeptides and proteins are endocytosed by the proximal tubular epithelium while small linear peptides undergo hydrolysis after contact with the peptidases located in the brush border membrane (Stephenson and Kenny, 1987). Absorption of proteins and constituent amino acids and oligopeptides (Ganapathy and Leibach, 1986; Silbernagl, 1988) derived from surface hydrolysis is near complete and only a small fraction of the filtered material is excreted intact in the urine. For most proteins or peptides, this represents less than 1–2% of the filtered load (Maack *et al.*, 1985; Rabkin and Kitaji, 1983).

Descriptions of the proximal tubular processing of small linear peptides have been reviewed by Carone and Peterson (1980) and a brief overview and update follows. In general, much of our information has been derived from studies in which radiolabeled peptides were microinjected into surface tubules of the rat kidney *in vivo* or microinfused into isolated rabbit nephron segments. When the linear peptides bradykinin (Carone *et al.*, 1976), angiotensin I (Peterson *et al.*, 1979), and angiotensin II (AII) (Pullman *et al.*, 1978) were studied, analysis of metabolites revealed that after luminal surface exposure, these peptides were typically hydrolyzed to amino acids and transported across the tubular cell. This luminal process is extremely efficient and only small amounts of intact hormone or metabolites are excreted in the urine. Indeed, these events are reminiscent of the fate of ingested peptides coming into contact with the intestinal brush border membrane (Ugolev and De Laey, 1973). Later in this chapter a more complete description of the renal handling of a typical linear peptide hormone, AII, is provided. Brush border peptidases are described in detail in Chapter 2.

Small peptides with more complex structures may be handled somewhat differently than linear peptides. Oxytocin, which possesses a ring configuration completed by a disulfide bond, is not hydrolyzed on the luminal surface and its rate of absorption is relatively slow (Peterson *et al.*, 1977). Luteinizing hormone-releasing hormone (LH-RH), a linear peptide with a C-terminal amide and an N-terminal pyroglutamyl residue, is resistant to the action of some proteases and is incompletely degraded by the luminal membrane (Stetler-Stevenson *et al.*, 1981). The resultant peptide fragments, and perhaps residual intact LH-RH, are absorbed and further processing occurs within the cell. Constituent amino acids and some peptide fragments are then released into the interstitial compartment. Stepwise cleavage of atrial natriuretic peptide, a small peptide hormone with a ring structure, has also been observed in studies with isolated luminal membranes (Berg *et al.*, 1988). The effects of structure on the handling of filtered polypeptides are further exemplified by insulin and glucagon. Comparing the tubular handling of these two hormones, Peterson *et al.* (1982) noted that hydrolysis of [^{125}I]glucagon with absorption of metabolites followed microinfusion into proximal, but not distal, surface nephron of rats. In contrast, there was no luminal hydrolysis of microinfused [^{125}I]insulin which was absorbed intact in the proximal tubules. This latter finding is consistent with electron microscopic autoradiographic studies which indicate that insulin is internalized by endocytosis (Bourdeau *et al.*, 1973; Nielsen *et al.*, 1987). This difference in handling of the two peptides is presumably related to the more complex structure of insulin, which includes two disulfide bonds, while glucagon is a relatively simple linear peptide. Although it is established that insulin is internalized by endocytosis, degradation by isolated luminal membranes does occur (Rabkin *et al.*, 1982; Talor *et al.*, 1982), but at a rate severalfold slower than for glucagon (Peterson *et al.*, 1982). It is likely, as stressed by Peterson *et al.* (1982), that the two mechanisms for the tubular handling of peptides are not mutually exclusive and some overlap may occur. Thus, while complex peptides resistant to membrane hydrolysis are internalized by endocytosis and simpler peptides susceptible to brush border peptidases are internalized as metabolic products, other peptides depending on their resistance to hydrolysis may involve both pathways to a variable degree. A typical example is arginine vaso-pressin, a nonapeptide with a disulfide cyclic structure, which has two pathways for its metabolism: a major mechanism involving luminal peptidases and a minor one involving endocytosis and lysosomal hydrolysis (Carone *et al.*, 1987).

After entering the lumen of the proximal tubules, large complex polypeptides and proteins are absorbed by means of endocytosis (Figs. 7 and 8). Absorption,

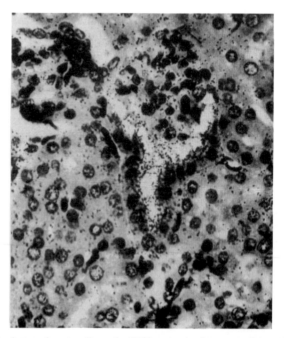

Figure 7. Light microscopic autoradiograph of kidney cortex from a mouse sacrificed 1 min after injection of ^{125}I-labeled growth hormone. Autoradiographic grains are located mainly over luminal half of cells. ×500. (From Rabkin *et al.*, 1973, with permission.)

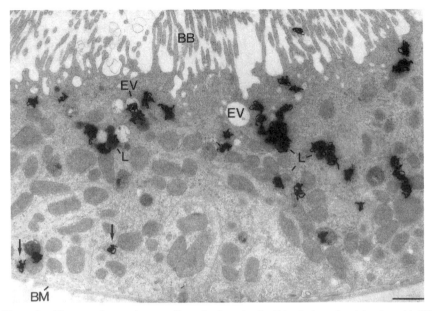

Figure 8. Electron microscopic autoradiograph of proximal rabbit tubule perfused *in vitro* with ^{125}I-labeled lysozyme for 20 min and then chased for 40 min. Autoradiographic grains are located over endocytotic vacuoles (EV) and lysosomes (L). A few grains lie over the basal aspect of the cell. BB, brush border; BM, basement membrane. ×13,500. (From Nielsen *et al.*, 1986, with permission.)

which is influenced by axial luminal fluid flow rate and protein concentration (Nielsen *et al.*, 1987; Nielsen and Nielsen, 1988), is initiated by binding of filtered protein to luminal membrane. In many instances binding is charge related. Evidence for this is provided by Just and Habermann (1973) who reported a close correlation between the net positive charge of peptides and their binding to isolated luminal membranes. They proposed that a basic component of the peptide binds to the negatively charged luminal membranes. Indeed, when they treated isolated membranes with sialidase to remove sialic acid, the binding of basic peptides was markedly decreased. These investigators (Just and Habermann, 1977; Just *et al.*, 1977) noted that binding of polybasic aminonucleosides to isolated luminal membrane is proportional to the number of free amino groups. Further support for charge-related binding is provided by Beyer *et al.* (1983). They noted that binding of lysozyme (basic) to isolated luminal membrane is competitively inhibited by basic amino acids such as lysine, ornithine, or arginine, but not by neutral amino acids, and the authors were able to reproduce these effects in microperfused rat proximal tubules (Cojocel *et al.*, 1981). Studying humans, Mogensen and Sølling (1977) observed that intravenous infusions of lysine and arginine inhibited tubular protein absorption. Since high concentrations of lysine may damage proximal tubular cells (Sumpio and Maack, 1982), the measured increase in protein excretion may, in part, reflect a toxic effect of basic amino acids.

Reflecting a charge-related interaction between protein and luminal cell membrane is the presence of competition between proteins of similar charge. For example, in microperfusion experiments Cojocel et al. (1981) noted that cytochrome c (pI 10.6) inhibited reabsorption of lysozyme (pI 11.0) in the rat proximal tubule. Similarly, Sumpio and Maack (1982) described a dose-dependent inhibition of cytochrome c tubular absorption by lysozyme with the isolated perfused rat kidney. On the other hand, competition between proteins of net opposite charge does not occur (Maack et al., 1979; Sumpio and Maack, 1982). Similarly, in vivo studies in which anionic or basic proteins were infused into dogs and rabbits failed to increase urinary excretion of proteins of opposite net charge (Foulkes, 1982; Harrison and Barnes, 1970). Finally, it should be noted that specific binding sites for a small number of protein hormones including insulin (Rabkin et al., 1982; Talor et al., 1982; Meezan et al., 1988) and calcitonin (Marx et al., 1973) have been described on the luminal membrane. While binding to the apical insulin receptor initiates endocytosis (Yagil et al., 1988), the relationship to hormone action is unclear.

Since absorption of proteins appears to depend on a nonselective charge-related interaction between the exposed positively charged amino acids and the positively charged luminal membrane, it might be expected that highly basic protein would readily compete for uptake with anionic proteins which have fewer negatively charged exposed amino acids. Since this does not occur, Sumpio and Maack (1982) have proposed a selective constraint model to reconcile these findings (Fig. 9). In this model it is proposed that access of proteins to the site of endocytosis, which is localized at the base of the microvilli, is influenced by the charge of the microvilli. The closely juxtaposed, negatively charged microvilli may facilitate the access of basic proteins while impeding the access of anionic proteins. When levels of basic proteins are high, they will occupy a sufficient number of charged binding sites on the microvilli so as to electrically impede the passage of basic, but not anionic, proteins to the site of endocytosis. This model includes a role for factors such as molecular size, shape, and rigidity and urine flow in determining delivery of protein to the site of endocytosis and, hence, protein absorption.

Although the above studies indicate that most filtered proteins are absorbed by a nonselective absorptive process that is, in part, charge related, competition among proteins is not commonly seen under physiologic conditions. This is a consequence of the large capacity of the endocytotic system relative to the filtered load. Consequently, under normal circumstances (Sumpio and Maack, 1982), a constant fraction of filtered protein is absorbed until the tubular maximum is exceeded at which time fractional excretion of the protein increases. This occurs in situations when plasma levels are elevated or when the glomerular filtration barrier is defective. Despite the increase in fractional excretion of the protein presented in excess to the tubular cells, the fractional excretion of other proteins does not increase unless there is associated

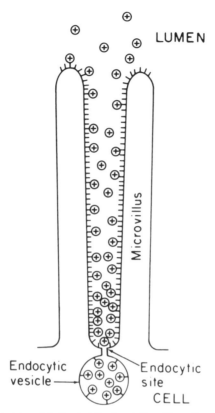

Figure 9. Selective constraint model. See text for description. (From Sumpio and Maack, 1982, with permission.)

tubular cell damage (Sumpio and Hayslett, 1985). This differs from the experimental situation where proteins were infused in high concentration and may be related to the greater filtered load presented to the renal tubules in the experimental model.

The intracellular routing of internalized proteins in renal tubular epithelial cells is similar to that described in Chapters 4, 5, and 7 for epithelial cells in general. What follows is a brief description of events observed in the kidney. After interacting with the plasma membrane at the base of the microvilli, the protein is internalized in clathrin-coated vesicles (van Deurs and Christensen, 1984; Wall and Maack, 1985). As they traverse the cell, these early endosomes lose their clathrin coat and acquire an ATP-driven proton pump, which lowers the intravesicular pH (see Chapter 3), causing the protein to dissociate from its binding site. Sorting of vesicular traffic follows, with routing of endosomes to lysosomes, to the original (retroendocytosis) or opposite pole (transcytosis) of the cell. The vesicles that recycle contain the binding site which is reinserted into the plasma membrane. Since uncoupling of protein from its binding site may be incomplete, a small amount of the internalized protein may also recycle to the cell surface where it is released into the extracellular fluid. In the kidney, this process of retroendocytosis has been best described for insulin (Dahl *et al.*, 1989) and will be discussed in detail later. The most common destination for endocytosed protein is the lysosome. After endosomes and lysosomes fuse, hydrolysis of protein follows with efflux of constituent amino acids into the peritubular space. The rate at which proteins are broken down varies according to their resistance to lysosomal enzymes (Clapp *et al.*, 1988). Until recently, it was generally believed that the lysosomes were the sole site of intracellular protein hydrolysis in the kidney. However, as will be discussed later in the section on insulin, hydrolysis may commence in endosomes (Fawcett and Rabkin, 1990).

Evidence for transcytosis of protein from the apical to the basolateral cell surface has been provided but this appears to be a trivial process. Nielsen *et al.* (1986, 1987) microperfused isolated rabbit proximal tubules with radiolabeled insulin and lysozyme and found trace amounts released on the basolateral side. Similarly, after exposing the apical surface of cultured proximal dog tubule cells to albumin, Goligorsky and Hruska (1986) noted basolateral release of albumin. However, most or all of the proteins transported across the cell can be accounted for by fluid-phase endocytosis. Hence, apical-to-basolateral transcytosis represents a trivial pathway.

Finally, it is worth considering the potential benefits to the body of proximal tubular cell uptake and metabolism of filtered proteins and peptides. This is not readily apparent since the nutritional gain of absorbing amino acids derived from their metabolism is trivial (Maack *et al.*, 1985) and whether proteins and peptides are metabolized or excreted in the urine they are essentially lost to the body. One likely benefit is that filtered bioactive materials are deactivated. If they were not, then they might exert unwanted effects on more distal nephron sites, especially as their concentrations rise as water is reclaimed from tubular fluid. Indeed, it is possible to deliberately inhibit proximal hydrolysis of specific peptides in order to enhance their distal nephron actions. This could be of therapeutic value and is being explored for ANP (Section 5.2).

4.2. Basolateral Tubular Cell Protein Uptake

Whereas the apical pole of the proximal tubular cells is rich in peptidases and has a prominent endocytic system which is easily detected by electron microscopy, this is not true for the basolateral pole. Thus, while proximal internalization of filtered protein is a highly efficient process and easy to monitor, basolateral internalization of large proteins is a minor process, and is detected with difficulty (Bourdeau and

Carone, 1974; Nielsen and Christensen, 1985). Similarly, compared with the apical membranes, basolateral membranes have relatively low or absent degrading activity (Talor *et al.*, 1982, 1983; Berg *et al.*, 1988). Ultrastructural studies utilizing horseradish peroxidase, ferritin, or catalase as tracers have revealed that these proteins are internalized by small endocytotic invaginations in the basolateral membrane (Feria-Velasco, 1974; Venkatachalam and Karnovsky, 1972; Nielsen and Christensen, 1985). Protein-containing endosomes ultimately fuse with lysosomes which often have a multivesicular appearance unlike that of the lysosomes to which protein taken up from the apical pole localizes. This prompted Nielsen and Christensen (1985) to suggest that a different population of lysosomes may serve as the final destination of proteins absorbed from the basal side. In an attempt to quantitate the basolateral uptake of albumin, Bourdeau and Carone (1974) exposed microperfused rabbit tubule segments, isolated from the proximal tubule thick ascending loop and cortical collecting tubules, to [^{125}I]albumin. Basolateral uptake was negligible compared with luminal uptake. Taken together with clearance studies in the intact animal, these reports indicate that elimination of large proteins by basolateral uptake is quantitatively inconsequential.

On the other hand, there is strong evidence indicating that clearance from the peritubular circulation is important for the elimination of some protein hormones. For example, it has been estimated that peritubular clearance accounts for approximately 40% of the insulin eliminated by the human kidney, based on direct measurements *in vivo* (Chamberlain and Stimmler, 1967; Rabkin *et al.*, 1970). Other protein hormones with a major peritubular pathway include calcitonin (Simmons *et al.*, 1988) and AII (Seikaly *et al.*, 1990; Reams *et al.*, 1990). The presence of a significant peritubular pathway for parathyroid hormone is controversial (Martin *et al.*, 1987; Daugaard *et al.*, 1988).

Since basolateral internalization of large proteins is negligble and direct study of basolateral events in the intact kidney is impractical, it has been debated for several years whether protein hormones cleared from the peritubular circulation are degraded on the basolateral cell surface or within the cell after being internalized (Rabkin and Kitaji, 1983). As discussed earlier, the low or absent degrading activity associated with basolateral plasma membranes makes intracellular degradation the more likely event and this is supported by two different *in vitro* studies. In a cell culture model, Rabkin *et al.* (1989) observed that insulin exposed to the basal cell surface underwent receptor-mediated internalization followed by degradation. Utilizing isolated perfused rabbit proximal tubules and electron microscopic autoradiography, Nielsen *et al.* (1987) noted that [^{125}I]insulin added to the tubule outer surface was endocytosed and delivered to lysosomes. These investigators also noted basal-to-luminal transport of small amounts of [^{125}I]insulin in the form of TCA-precipitable radioactivity. Whether this represents transcytosis of intact insulin or release of large insulin metabolites requires further study. Albeit a quantitatively minor process, unidirectional transcytosis of epidermal growth factor from the basolateral to the apical surface has been described in the distal nephron-like cultured Madin–Darby canine kidney cell line (Maratos-Flier *et al.*, 1987). While possessing epidermal growth factor receptors restricted to the basolateral surface, these cells are essentially devoid of insulin receptors and do not transport insulin in either direction.

In summary, it appears that basolateral internalization and degradation represents a significant cellular pathway for the processing of some protein hormones with receptors located on the basolateral cell surface. While transcytosis may occur, its contribution to renal protein metabolism is trivial. For nonhormonal proteins, basolateral uptake plays essentially no role in their elimination.

5. RENAL HANDLING OF BIOACTIVE PROTEINS AND PEPTIDES: REPRESENTATIVE EXAMPLES

5.1. Insulin

The renal handling of insulin is described in some detail as it exemplifies the complex elimination pathways which protein hormones may follow. Endogenous insulin secreted by the pancreas enters the portal circulation and during its first passage through the liver ~ 50% is extracted. Proinsulin also undergoes significant hepatic extraction, while extraction of C-peptide is negligible. On reaching the systemic circulation, insulin is removed by several tissues including kidney, muscle, liver, and fat (Rabkin *et al.*, 1984). The renal contribution to the systemic clearance of insulin, proinsulin, and C-peptide averages 30, 50, and 70%, respectively (Katz and Rubenstein, 1973). Insulin (6 kDa) and its related peptides are extracted from the renal circulation mainly by glomerular filtration with minor restriction to its passage through the glomerular capillary wall (sieving coefficient ~ 0.85). The filtered insulin is absorbed in the proximal tubule and less than 1% appears in the urine. In addition to filtration, a large amount is extracted from the peritubular circulation (Rabkin and Kitabchi, 1978). In humans, this accounts for ~ 40% of the total renal clearance (Chamberlain and Stimmler, 1967; Rabkin *et al.*, 1970). Insulin removed by this pathway undergoes either partial or complete degradation (Petersen *et al.*, 1982; Duckworth *et al.*, 1988). The partially modified insulin molecule has a similar molecular weight and size as insulin but lacks immunoreactivity.

Insulin-specific binding sites are present in isolated luminal and contraluminal membranes, suggesting that tubular uptake is, at least in part, receptor mediated (Talor *et al.*, 1982; Rabkin *et al.*, 1982, 1986). In addition to true insulin receptors, Meezan *et al.* (1988) have described luminal insulin recognition sites that are broadly specific and of low affinity. The exact contribution of true receptors and recognition sites to luminal insulin internalization has yet to be clarified. The contraluminal receptors which are severalfold more plentiful than the luminal receptors are not confined to the proximal tubule but are distributed throughout the nephron (Nakamura *et al.*, 1986). Binding to contraluminal, but not luminal, membrane receptors is followed by receptor phosphorylation (Hammerman, 1985). Taken together, it appears that contraluminal receptors participate in hormone action while luminal receptors probably serve a role in the absorption of filtered insulin. Insulin-degrading activity is present throughout the length of the nephron but is maximal in the proximal tubule which is the site of filtered insulin absorption (Nakamura *et al.*, 1986). Subcellular sites of proximal tubular insulin-degrading activity include lysosomes, mitochondria, and cytosol (Hjelle *et al.*, 1984). Degrading activity has also been identified in luminal membranes and to a lesser extent in contraluminal membranes (Talor *et al.*, 1982; Rabkin *et al.*, 1982). Glomeruli possess insulin receptors, but exhibit little degrading activity (Meezan and Freychet, 1982).

While electron microscopic autoradiographic studies have shown that filtered insulin absorbed in the proximal tubule localizes to lysosomes (Nielsen *et al.*, 1987; Bourdeau *et al.*, 1973), there is mounting evidence that insulin degradation is not confined to lysosomes. Suggesting extralysosomal degradation is the observation that the major insulin-degrading enzymes, insulin-protease and glutathione-insulin transhydrogenase, are located in extralysosomal sites, especially cytosol, and are not found in lysosomes (reviewed in Rabkin *et al.*, 1984). Furthermore, intermediate products of insulin degradation released from intact kidneys and cultured kidney tubular cells have an HPLC pattern similar to that observed when insulin is exposed to purified insulin protease, a nonlysosomal enzyme (Duckworth *et al.*, 1988, 1989) (Fig. 10). Whether any of these products retain bioactivity is unknown. Also, Herrman *et al.* (1988), studying the subcellular distribution of [^{125}I]insulin and [^{14}C]cytochrome *c* in rat kidneys, noted that insulin is distributed in a pattern different from that of cytochrome *c* (12.3 kDa), which localizes to lysosomes. This suggested that insulin is degraded in a compartment different from the classic compartment of protein degradation. The possibility that endosomes may form this compartment is suggested from studies with cultured kidney cells in which retroendocytosis of insulin taken up from the apical cell surface was demonstrated. In these studies, both intact insulin and large intermediates were released from recycling endosomes (Dahl *et al.*, 1989, 1990). Finally, a recent study from the author's laboratory (Fawcett and Rabkin, 1990) provides direct evidence for endosomal processing of insulin (Fig. 11). In this study, isolated rat renal cortical endosomes loaded with [^{125}I]insulin *in vivo* degraded insulin in a pH- and ATP-dependent manner. Taken together, it appears that insulin is internalized by means of endocytosis with degradation commencing in endosomes and reaching completion in lysosomes. Regarding insulin cleared from the peritubular circulation, it appears, as discussed earlier (Section 4.2), that the hormone is internalized by tubular cells via basolateral receptor-mediated endocytosis. As occurs with luminal uptake, this is followed by intracellular processing with stepwise degradation to constituent amino acids (Duckworth *et al.*, 1989). That the kidney produces large intermediates during its stepwise degradation of insulin is of great interest, for the possibility arises that one or more of these intermediate products may possess bioactivity. The concept that an early metabolic product might

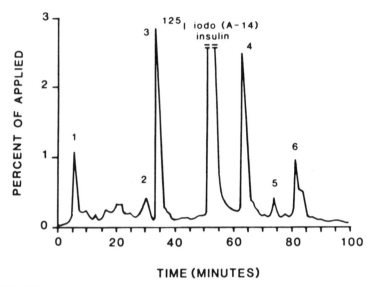

Figure 10. HPLC elution pattern of insulin-size radiolabeled material extracted from the recycling perfusate of an isolated filtering rat kidney. The kidney was perfused for 70 min with ^{125}I-labeled [A14] insulin and perfusate was then chromatographed on a Sephadex G-50 column. Insulin-size radioactivity was concentrated, dissolved in HPLC buffer, and injected on a reverse-phase HPLC. ^{125}I-labeled insulin elutes at 53 min and six other distinct peaks are present. (From Duckworth *et al.*, 1989, with permission.)

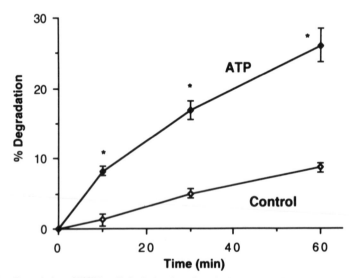

Figure 11. Degradation of [^{125}I] insulin by isolated renal cortical endosomes. Rats injected intravenously with [^{125}I] insulin were sacrificed 5 min later. [^{125}I] insulin-containing endosomes were isolated from cortical homogenates by differential and Percoll gradient centrifugation. Endosomes were incubated in an intracellular buffer, pH 7.4, at 37°C in the absence (controls) or presence of 10 mM ATP. Degradation of [^{125}I] insulin was determined by solubility in 10% trichloroacetic acid. (From Fawcett and Rabkin, 1990.)

exhibit bioactivity is important not only because of the possibility that products may serve as mediators of hormone action, but it is conceivable that proteins, including engineered proteins, may have metabolic intermediates with unwanted properties including immunogenicity.

5.2. Atrial Natriuretic Peptide

Although produced in many extracardiac sites, including the CNS, lung, adrenal and pituitary glands, arch of aorta, cardiac ventricles, and kidney medullary collecting tubules, the vast majority of ANP is produced in the cardiac atria (Inagami, 1989). ANP is rapidly cleared from the plasma (Gerbes and Vollmar, 1990) with a half-life of about 30 sec in rats (Katsube *et al.*, 1986) and 60 sec in uninephrectomized dogs (Woods, 1988). The kidneys extract significant amounts of ANP from the circulation. Woods (1988) found the renal extraction ratio of endogenous ANP to be 35% in the dog and this value increased significantly to 69% during ANP infusion. Other investigators (Hollister *et al.*, 1989) have documented similar extraction ratios of endogenous hormone (42% in the dog and 35% in humans). Since the filtration fraction in humans is approximately 20%, an extraction ratio of 35% suggests a major extrafiltration pathway in the kidney. Nevertheless, the kidneys' contribution to total ANP clearance is modest and in humans accounts for about 20% of total ANP clearance (Vierhapper *et al.*, 1990). In uninephrectomized dogs the renal clearance rate of the hormone represents only 14% of the whole body metabolic clearance rate (Woods, 1988). The half-life of exogenous ANP approximately doubles in rats after bilateral nephrectomy, but even after this maneuver the half-life is only 64 sec. (Katsube *et al.*, 1986). In humans with end-stage renal failure, the metabolic clearance rate of exogenous ANP is approximately 1.0 liter/min and the half-life 4.6 min, compared with 2.6 liters/min and 3.5 min, respectively, in normals (Tonolo *et al.*, 1988). Plasma ANP levels are increased about sixfold in hemodialysis patients but this may be due in large part to volume expansion stimulating release of ANP (Hasegawa *et al.*, 1986). In normal subjects, all segments of the nephron have the capacity to degrade ANP (Fig. 12), but degrading activity is greatest in the proximal tubule and relatively insignificant in the glomerulus (Berg *et al.*, 1988). In subcellular fractions of homogenized kidney cortex, the greatest ANP-degrading activity resides in isolated luminal membranes with negligible activity in basolateral membranes (Fig. 13). This suggests that ANP, like other small peptides, is hydrolyzed primarily by brush border peptidases in the proximal tubule with neutral endopeptidase-24.11 initiating breakdown (Kenny and Stephenson, 1988). However, the fact that ANP degradation is only mildly impaired in the nonfiltering isolated perfused kidney compared with the filtering kidney suggests that other intrarenal pathways exist for the elimination of this peptide hormone (Maack, 1986). As Berg *et al.* (1988) have suggested, the renal vasculature itself may play a role in the degradation of ANP. Nonrenal vasculature is known to have a substantial ANP-degrading capability (Murthy *et al.*, 1986).

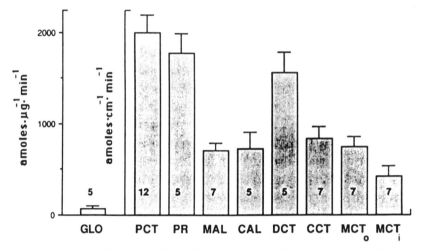

Figure 12. ANP-degrading activity of isolated nephron segments. Segments were permeabilized so as to expose ^{125}I-labeled rat α-ANP to intracellular enzymes. Degrading activity was highest in proximal convoluted tubules, pars recta, and distal convoluted tubules but is present throughout the nephron. Glomeruli degrade ANP poorly. (From Berg *et al.*, 1988, with permission.)

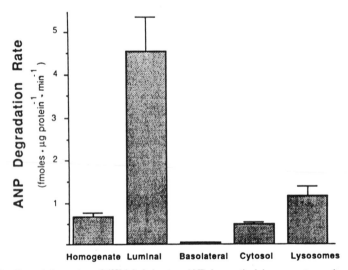

Figure 13. Degradation rates of ^{125}I-labeled rat α-ANP by cortical homogenates and various cell fractions. (From Berg *et al.*, 1988, with permission.)

All of the ANP extracted by the kidney may not be immediately degraded, but a significant portion may be bound by clearance receptors (Maack *et al.*, 1987; Fuller *et al.*, 1988) which are biologically inactive. These receptors avidly bind ANP but do not mediate any of the hormone's known physiologic actions. Up to 99% of the ANP receptors of the kidney may be of this biologically silent type (Maack *et al.*, 1987). While Maack and colleagues found these clearance receptors to comprise the vast majority of ANP receptors in the cortex, medulla, and papilla, others (Brown *et al.*, 1990) have found evidence for clearance receptors only in the glomeruli and renal vasculature up to the proximal interlobular arteries.

Maack *et al.* (1987) suggested that clearance receptors may modulate plasma hormone levels by acting as storage sites for ANP. Furthermore, clearance receptors appear to be important for the degradation of ANP (Maack *et al.*, 1988). In comparison with biologically active ANP receptors, structural requirements for binding to clearance receptors are less strict and clearance receptors bind ANP analogues in addition to ANP. When ANP analogues which bind clearance receptors but not biologically active receptors are administered to experimental animals, the analogues displace native ANP from clearance receptors, reduce the metabolic clearance rate and hydrolysis of the hormone, increase ANP levels in plasma, and lead to enhanced physiologic effects of native ANP (Maack *et al.*, 1988). Accordingly, Maack and colleagues have suggested that these differential binding requirements might be exploited therapeutically. Another pharmacologic approach to modifying ANP metabolism to therapeutic advantage is the use of enzyme inhibitors. Administration of inhibitors of neutral endopeptidase 24.11 is associated with suppressed ANP degradation and enhanced solute and water excretion in response to ANP (Margulies *et al.*, 1990; Richards *et al.*, 1990; Gerbes and Vollmar, 1990). While the endopeptidase plays a role in the peripheral degradation of ANP, its predominance in the proximal renal brush border membrane suggests that by inhibiting proximal tubular hydrolytic activity the hormone is spared to act on distal nephron segments. Another pharmacologic approach is development of analogues resistant to neutral endopeptidase activity (Kenny and Stephenson, 1988). Thus, pharmacologic modification of physiologic processes may permit strategies whereby the action of ANP can be enhanced for the treatment of hypertension and fluid-retaining states.

5.3. Angiotensin

Removal of circulating AII by the kidney is substantial and extraction ratios for this hormone have been reported as 64 to 75% in the dog (Bailie *et al.*, 1971; Oparil and Bailie, 1973; Reams *et al.*, 1990). Although AII is sensitive to proteolytic enzymes in plasma, the metabolism of AII in the renal circulation is too rapid to be accounted for by this process alone (Oparil and Bailie, 1973). Furthermore, circulating AII extracted by the kidney is metabolized and is not sequestered in the kidney

(Oparil and Bailie, 1973). As a decapeptide and octapeptide, respectively, AI and AII are presumably freely filtered at the glomerulus. A number of experiments have demonstrated that filtered AII is catabolized at the proximal tubule brush border and the constituent amino acids then reabsorbed. Pullman *et al.* (1975) microinfused [^{14}C]-AII into the proximal and distal surface nephrons of anesthetized rats. Recovery of radiolabel from urine was only 11% when proximal tubules were infused and most of the recovered radiolabel was in the form of AII metabolites. In contrast, when distal tubules were infused, 95% of the infused radioactivity was recovered in urine, almost all in the form of intact AII. Because of the rapidity of metabolite formation, these findings are most consistent with proximal tubule brush border hydrolysis of AII followed by absorption of metabolites. These results are supported by experiments in which proximal straight tubule segments from the rabbit were perfused *in vitro* with ^{3}H- or ^{14}C-labeled AI (Peterson *et al.*, 1979) or AII (Peterson *et al.*, 1977). There was marked reabsorption of radiolabel from these tubule segments without sequestration in the segments themselves. The collection fluid was characterized by electrophoresis and found to contain radiolabeled amino acids indicating that the angiotensins had been hydrolyzed. Also supporting the concept of angiotensin hydrolysis at the proximal tubule brush border is the finding that isolated renal cortical brush border preparations are enriched in angiotensinase (Ward *et al.*, 1976) and rapidly hydrolyze AI and AII (Peterson *et al.*, 1979). Isolated glomeruli also contain angiotensinase, but at less than one-tenth the concentration found in brush border preparations (Ward *et al.*, 1977). Recently, angiotensin-specific receptors which serve to transduce the modulatory actions of AII on tubular cell solute transport have been identified in brush border membranes (Douglas, 1987).

The kidney itself generates AII (Dzau, 1987; Rosivall *et al.*, 1987; Reams *et al.*, 1990; Seikaly *et al.*, 1990) and this fact complicates studies of renal angiotensin clearance. Seikaly *et al.* (1990) noted that the concentration of AII in early post-glomerular vessels is 1000-fold higher than in the systemic circulation. Since renal venous angiotensin concentrations are of a similar order of magnitude to that in systemic plasma, they concluded that substantial elimination of angiotensin occurs downstream in the peritubular circulation.

6. RENAL HANDLING OF GLYCOPROTEINS

The renal metabolism of glycoproteins has not been extensively studied. The kidneys are, however, an important site of disposal for glycoprotein hormones and a brief overview follows. Studies in rats after acute renal ablation suggest that renal clearance accounts for 94, 78, and 32% of the total metabolic clearance rate of luteinizing hormone, follicle-stimulating hormone, and erythropoietin, respectively (Emmanouel *et al.*, 1984a,b). Renal metabolism of glycoproteins is known to differ in some respects from that of proteins. First, renal clearance of glycoproteins is relatively slow compared to that of most low-molecular-weight proteins. Presumably, this is a result of restricted glomerular filtration due to their size and physiochemical properties. Also, the existence of a significant peritubular pathway is unlikely, again because of their size and also because of a lack of basolateral tubular membrane receptors. Despite this, the kidney's role in the elimination of glycoprotein hormones is important because their total metabolic clearance rates are also slow. Second, renal glycoprotein metabolism differs from protein metabolism in that tubular absorption is far less efficient. Substantial amounts of filtered glycoproteins appear in the final urine. For follicle-stimulating hormone, erythropoietin, and luteinizing hormone, urinary clearance accounts for 35, 7, and 4%, respectively, of their total metabolic clearance rates (Emmanouel et al., 1984a,b). Thus, the contribution of urinary clearance to the overall metabolic clearance of these hormones is relatively high compared with that of protein.

A good example of the renal metabolism of a glycoprotein is human chorionic gonadotropin (hCG). Its renal metabolism is of particular interest because this hormone and its metabolites appear in the serum and urine of pregnant women and individuals with certain malignancies and is thus of diagnostic importance. The urinary clearance of hCG is 22% (Wehmann and Nisula, 1981). When radiolabeled hCH is injected into rats, kidney, liver, and ovaries take up most of the radioactivity (Markkanen et al., 1979). Quantitatively, the kidney is the most important site of accumulation, taking up approximately five times more radioactivity than the liver or ovaries. Most of this radioactivity in the kidney represents degradation products and is localized to proximal tubules (Markkanen et al., 1979). Autoradiographic studies suggest that hCG is filtered at the glomerulus, absorbed by endocytosis at the proximal tubular cell, and degraded in lysosomes (Markkanen and Rajaniemi, 1979). Despite desialylated hCG being smaller than intact hCG, its urinary clearance is substantially less (Amr et al., 1985). As it is likely that the desialylated molecule is filtered to at least the same extent as the intact molecule, this suggests that the carbohydrate portion of the hCG molecule influences its uptake at the proximal tubular cell apical membrane. When renal homogenates of rats infused with hCG are analyzed, they mostly contain a metabolite of the β subunit, termed the β-core fragment (Lefort et al., 1986). The degradation rate of this fragment is slow. β-core fragment is found in the urine of pregnant women (Kato and Braunstein, 1988) where it appears to be derived from the metabolism of hCG, likely intrarenally (Wehmann et al., 1989).

7. EFFECT OF RENAL FAILURE ON ELIMINATION OF PROTEINS

In kidney failure, the plasma levels of low-molecular-weight proteins which are largely dependent on renal elimination are often elevated. In early renal failure, this is due to decreased renal blood flow and thus reduced delivery of the protein to the kidney, for usually the proportion of delivered hormone extracted by the kidney remains constant. In more advanced renal failure, the proportion of hormone extracted also falls (Rabkin et al., 1970). As a consequence of primary tubular disease or the secondary tubular damage which accompanies progressive renal failure, absorption of filtered proteins becomes impaired and fractional urinary clearance of filtered proteins increases (Kaysen et al., 1986; Waldmann et al., 1972). In diseases that affect the glomerulus, the permselective properties of the glomerular wall become altered and an increase in the passage of proteins normally restricted by the filtration barrier occurs (Anderson et al., 1991). This also leads to an increase in urinary losses. Low-molecular-weight proteins may also appear in increased quantities in the urine of individuals with normal kidneys (Kaysen et al., 1986). This occurs when the production of a protein that is relatively freely filtered at the glomerulus is increased to such an extent that the resorptive capacity of the proximal tubule is overwhelmed. Examples of such an overload proteinuria include urinary immunoglobulin light chains in multiple myeloma (Wochner et al., 1967) and lysozyme in certain leukemias (Muggia et al., 1969). When abnormally high concentrations of proteins are presented to the renal tubule, they may be toxic as illustrated by renal failure induced by myeloma proteins. Taken together, it is clear that quantitation of urinary protein excretion as a measure of protein production is fraught with pitfalls.

There are mechanisms in addition to reduced renal clearance which contribute to the elevated plasma protein levels found in renal failure. In some instances, the uremic state may impair the extrarenal clearance of proteins (Rabkin et al., 1979a; Hruska et al., 1981) and in other instances enhanced secretion of protein contributes to the elevated concentrations found in chronic renal failure. Enhanced secretion occurs with parathyroid hormone (Slatopolsky et al., 1990), prolactin (Sievertsen et al., 1980), and perhaps growth hormone (Mehls et al., 1990). Also, some protein hormones circulate in heterogeneous forms in renal failure, all of which may bind an antibody probe, but only some of which may possess biologic activity. These disparate forms are often made up of low-molecular-weight components which are likely protein fragments; normal-weight proteins corresponding to the active hormone or usual protein; and high-molecular-weight substances which may represent aggregates or prohormones (Fig. 14). Examples include parathyroid hormone (Arnaud et al., 1974), calcitonin (Lee et al., 1977), insulin (Jaspan et al., 1977), and glucagon (Kuku et al., 1976). The reduced clearance of proteins normally metabolized by the kidneys often accounts for this heterogeneity. Finally, there may be substances in uremic serum which interfere with the assay of proteins (Powell et al., 1986).

The accumulation of proteins and protein hormones in uremic serum often has deleterious consequences for the patient with renal failure. In renal insufficiency, serum levels of β_2-microglobulin (11.8 kDa) are increased and in hemodialysis patients may rise 50-fold (Vincent et al., 1978). Over time, polymerized β_2-microglobulin may deposit in the bones and joints of dialysis patients in the form of

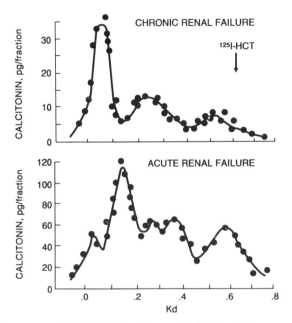

Figure 14. Gel chromatography (Bio-Gel P-10) elution profile of calcitonin immunoreactivity in plasma from two patients with renal failure. Radiolabeled monomer elutes in position indicated by arrow (From Lee *et al.*, 1977, with permission.)

amyloid fibrils and may cause bone disease (Gorevic *et al.*, 1985). Bone disease in patients with renal failure may also result from the elevated parathyroid hormone levels. Other manifestations of protein hormone accumulation, such as insulin-induced hypoglycemia, have been reviewed by Emmanouel *et al.* (1980). From this account of the disturbances of protein homeostasis that occur in renal failure, it is readily apparent that the kidney plays an important role in the elimination of proteins.

ACKNOWLEDGMENTS.   Studies from the author's laboratory described in this chapter were supported by NIHDDK Grant AM 32342 and the Department of Veterans Affairs.

REFERENCES

Abrahamson, D. R., 1987, Structure and development of the glomerular capillary wall and basement membrane, *Am. J. Physiol.* **253**:F783–F794.

Amr, S., Rosa, C., Birken, S., Canfield, R., and Nisula, B., 1985, Carboxyterminal peptide fragments of the beta subunit are urinary products of the metabolism of desialylated human choriogonadotropin, *J. Clin. Invest.* **76**:350–356.

Anderson, S., Garcia, D. L., and Brenner, B. M., 1991, Renal and systemic manifestations of glomerular disease, in: *The Kidney* (B. M. Brenner and F. C. Rector, Jr., eds.), 4th ed., Saunders, Philadelphia, pp. 1831–1870.

Arendshorst, W. J., and Navar, L. G., 1988, Renal circulation and glomerular hemodynamics, in: *Diseases of the Kidney* (R. W. Schrier and C. W. Gottschalk, eds.), 4th ed., Little, Brown, Boston, pp. 65–117.

Arnaud, C. D., Goldsmith, R. S., Bordier, P. J., and Sizemore, G. W., 1974, Influence of immunoheterogeneity of circulating parathyroid hormone on results of radioimmuno-assays of serum in man, *Am. J. Med.* **56**:785–793.

Atkins, J. L., O'Morchoe, C. C. C., and Pinter, G. G., 1988, Total lymphatic clearance of protein from the renal interstitium, *Contrib. Nephrol.* **68**:238–244.

Audus, K. L., Bartel, R. L., Hidalgo, I. J., and Borchardt, R. T., 1990, The use of cultured epithelial and endothelial cells for drug transport and metabolism studies, *Pharm. Res.* **7**: 435–451.

Bailie, M. D., Rector, F. C., Jr., and Seldin, D. W., 1971, Angiotensin II in arterial and renal venous plasma and renal lymph in the dog, *J. Clin. Invest.* **50**:119–126.

Bar, R. S., Boes, M., Dake, B. L., Booth, B. A., Henley, S. A., and Sandra, A., 1988, Insulin, insulin-like growth factors, and vascular endothelium, *Am. J. Med.* **85**(Suppl. 5A):59–70.

Baumann, G., Amburn, K. D., and Buchanan, T. A., 1987, The effect of circulating growth hormone-binding protein on metabolic clearance, distribution, and degradation of human growth hormone, *J. Clin. Endocrinol. Metab.* **64**:657–660.

Baumann, G., Shaw, M. A., and Buchanan, T. A., 1989, In vivo kinetics of a covalent growth hormone-binding protein complex, *Metabolism* **38**:330–333.

Baylis, C., 1982, Transport of molecules across the glomerular peritubular renal capillaries, *Physiologist* **25**:377–383.

Berg, J. A., Hayashi, M., Fujii, Y., and Katz, A. I., 1988, Renal metabolism of atrial natriuretic peptide in the rat, *Am. J. Physiol.* **255**:F466–F473.

Beyer, G., Bode, F., and Baumann, K., 1983, Binding of lysozyme to brush border membranes of rat kidney, *Biochim. Biophys. Acta* **732**:372–376.

Bjerke, T., Christensen, E. I., and Boye, N., 1989, Tubular handling of neurotensin in the rat kidney as studied by micropuncture and HPLC, *Am. J. Physiol.* **256**:F100–F106.

Bohrer, M. P., Deen, W. M., Robertson, C. R., and Brenner, B. M., 1977, Mechanism of angiotensin II-induced proteinuria in the rat, *Am. J. Physiol.* **233**:F13–F21.

Bohrer, M. P., Baylis, C., Humes, H. D., Glassock, R. J., Robertson, C. R., and Brenner, B. M., 1978, Permselectivity of the glomerular capillary wall: Facilitated filtration of circulating polycations, *J. Clin. Invest.* **61**:72–78.

Bohrer, M. P., Deen, W. M., Robertson, C. R., Troy, J. L., and Brenner, B. M., 1979, Influence of molecular configuration on the passage of macromolecules across the glomerular capillary wall, *J. Gen. Physiol.* **74**:583–593.

Bourdeau, J. E., and Carone, F. A., 1974, Protein handling by the renal tubule, *Nephron* **13**:22–34.

Bourdeau, J. E., Chen, E. R. Y., and Carone, F. A., 1973, Insulin uptake in the renal proximal tubule, *Am. J. Physiol.* **225**:1399–1404.

Brenner, B. M., Hostetter, T. H., and Humes, H. D., 1978, Glomerular permselectivity: Barrier function based on discrimination of molecular size and charge, *Am. J. Physiol.* **234**:F455–F460.

Brown, J., Salas, S. P., Singleton, A., Polak, J. M., and Dollery, C. T., 1990, Autoradiographic localization of atrial natriuretic peptide receptor subtypes in rat kidney, *Am. J. Physiol.* **259:**F26–F39.

Carone, F. A., and Peterson, D. R., 1980, Hydrolysis and transport of small peptides by the proximal tubule, *Am. J. Physiol.* **238:**F151–F158.

Carone, F. A., Pullman, T. N., Oparil, S., and Nakamura, S., 1976, Micropuncture evidence of rapid hydrolysis of bradykinin by rat proximal tubule, *Am. J. Physiol.* **230:**1420–1424.

Carone, F. A., Christensen, E. I., and Flouret, G., 1987, Degradation and transport of AVP by proximal tubule, *Am. J. Physiol.* **253:**F1120–F1128.

Cascieri, M. A., Saperstein, R., Hayes, N. S., Green, B. G., Chicchi, G. G., Applebaum, J., and Bayne, M. L., 1988, Serum half-life and biological activity of mutants of human insulin-like growth factor I which do not bind to serum binding proteins, *Endocrinology* **123:**373-381.

Chamberlain, M. J., and Stimmler, L., 1967, The renal handling of insulin, *J. Clin. Invest.* **46:** 911–919.

Clapp, W. L., Park, C. H., Madsen, K. M., and Tisher, C. C., 1988, Axial heterogeneity in the handling of albumin by the rabbit proximal tubule, *Lab. Invest.* **58:**549–558.

Cojocel, C., Franzen-Sieveking, M., Beckmann, G., and Baumann, K., 1981, Inhibition of renal accumulation of lysozyme (basic low molecular weight protein) by basic proteins and other basic substances, *Pfluegers Arch.* **390:**211–215.

Cojocel, C., Dociu, N., and Baumann, K., 1982, Early nephrotoxicity at high plasma concentrations of lysozyme in the rat, *Lab. Invest.* **46:**149–157.

Dahl, D. C., Tsao, T., Duckworth, W. C., Mahoney, M. J., and Rabkin, R., 1989, Retroendocytosis of insulin in a cultured kidney epithelial cell line, *Am. J. Physiol.* **257:**C190–C196.

Dahl, D. C., Tsao, T., Duckworth, W. C., Frank, B. H., and Rabkin, R., 1990, Effect of bacitracin on retroendocytosis and degradation of insulin in cultured kidney epithelial cell line, *Diabetes* **39:**1339–1346.

Daugaard, H., Egfjord, M., and Olgaard, K., 1988, Metabolism of intact parathyroid hormone in isolated perfused rat liver and kidney, *Am. J. Physiol.* **254:**E740–E748.

Deen, W. M., Ueki, I. F., and Brenner, B. M., 1976, Permeability of renal peritubular capillaries to neutral dextrans and endogenous albumin, *Am. J. Physiol.* **231:**283–291.

Deen, W. M., Bridges, C. R., Brenner, B. M., and Myers, B. D., 1985, Heteroporous model of glomerular size selectivity: Application to normal and nephrotic humans, *Am. J. Physiol.* **249:**F374–F389.

Douglas, J. G., 1987, Angiotensin receptor subtypes of the kidney cortex, *Am. J. Physiol.* **253:** F1–F7.

Drakenberg, K., Östenson, C.-G., and Sara, V., 1990, Circulating forms and biological activity of intact and truncated insulin-like growth factor I in adult and neonatal rats, *Acta Endocrinol. (Copenhagen)* **123:**43–50.

Duckworth, W. C., Hamel, F. G., Liepnieks, J., Frank, B. H., Yagil, C., and Rabkin, R., 1988, High performance liquid chromatographic analysis of insulin degradation products from a cultured kidney cell line, *Endocrinology* **123:**2701–2708.

Duckworth, W. C., Hamel, F. G., Liepnieks, J., Peavy, D., Frank, B., and Rabkin, R., 1989, Insulin degradation products from perfused rat kidney, *Am. J. Physiol.* **256:**E208–E214.

Dzau, V. J., 1987, Implications of local angiotensin production in cardiovascular physiology and pharmacology, *Am. J. Cardiol.* **59:**59A–65A.

Emmanouel, D. S., Jaspan, J. B., Rubenstein, A. H., Huen, A. H.-J., Fink, E., and Katz, A. I.,

1978, Glucagon metabolism in the rat: Contribution of the kidney to the metabolic clearance rate of the hormone, *J. Clin. Invest.* **62:**6–13.

Emmanouel, D. S., Lindheimer, M. D., and Katz, A. I., 1980, Pathogenesis of endocrine abnormalities in uremia, *Endocr. Rev.* **1:**28–44.

Emmanouel, D. S., Goldwasser, E., and Katz, A. I., 1984a, Metabolism of pure human erythropoietin in the rat, *Am. J. Physiol.* **247:**F168–F176.

Emmanouel, D. S., Stavropoulos, T., and Katz, A. I., 1984b, Role of the kidney in metabolism of gonadotropins in rats, *Am. J. Physiol.* **247:**E786–E792.

Fawcett, J., and Rabkin, R., 1990, Isolated renal cortical endosomes degrade insulin, *J. Am. Soc. Nephrol.* **1:**698 (abstract).

Feria-Velasco, A., 1974, The ultrastructural bases of the initial stages of renal tubular excretion: A cytochemical study using horseradish peroxidase as a tracer, *Lab. Invest.* **30:** 190–200.

Foulkes, E. C., 1982, Tubular reabsorption of low molecular weight proteins, *Physiologist* **25:** 56–59.

Fuller, F., Porter, J. G., Arfsten, A. E., Miller, J., Schilling, J. W., Scarborough, R. M., Lewicki, J. A., and Schenk, D. B., 1988, Atrial natriuretic peptide clearance receptor, *J. Biol. Chem.* **263:**9395–9401.

Galaske, R. G., Van Liew, J. B., and Feld, L. G., 1979, Filtration and reabsorption of endogenous low-molecular-weight protein in the rat kidney, *Kidney Int.* **16:**394–403.

Ganapathy, V., and Leibach, F. H., 1986, Carrier-mediated reabsorption of small peptides in renal proximal tubule, *Am. J. Physiol.* **251:**F945–F953.

Gerbes, A. L., and Vollmar, A. M., 1990, Degradation and clearance of atrial natriuretic factors (ANF), *Life Sci.* **47:**1173–1180.

Goligorsky, M. S., and Hruska, K. A., 1986, Transcytosis in cultured proximal tubular cells, *J. Membr. Biol.* **93:**237–247.

Gorevic, P. D., Casey, T. T., Stone, W. J., DiRaimondo, C. R., Prelli, F. C., and Frangione, B., 1985, Beta-2 microglobulin is an amyloidogenic protein in man, *J. Clin. Invest.* **76:**2425–2429.

Hammerman, M. R., 1985, Interaction of insulin with the renal proximal tubular cell, *Am. J. Physiol.* **249:**F1–F11.

Harrison, J. F., and Barnes, A. D., 1970, The urinary excretion of lysozyme in dogs, *Clin. Sci.* **38:**533–547.

Hasegawa, K., Matsushita, Y., Inoue, T., Morii, H., Ishibashi, M., and Yamaji, T., 1986, Plasma levels of atrial natriuretic peptide in patients with chronic renal failure, *J. Clin. Endocrinol. Metab.* **63:**819–822.

Herington, A. C., Ymer, S., and Stevenson, J., 1986, Identification and characterization of specific binding proteins for growth hormone in normal human sera, *J. Clin. Invest.* **77:** 1817–1823.

Herrman, J., Simmons, R. E., Frank, B. H., and Rabkin, R., 1988, Differences in renal metabolism of insulin and cytochrome c, *Am. J. Physiol.* **254:**E419–E428.

Hjelle, J. T., Oparil, S., and Peterson, D. R., 1984, Subcellular sites of insulin hydrolysis in renal proximal tubules, *Am. J. Physiol.* **246:**F409–F416.

Hollister, A. S., Rodeheffer, R. J., White, F. J., Potts, J. R., Imada, T., and Inagami, T., 1989, Clearance of atrial natriuretic factor by lung, liver, and kidney in human subjects and the dog, *J. Clin. Invest.* **83:**623–628.

Hruska, K. A., Kopelman, R., Rutherford, W. E., Klahr, S., and Slatopolsky, E., 1975, Metabolism of immunoreactive parathyroid hormone in the dog: The role of the kidney and the effects of chronic renal disease, *J. Clin. Invest.* **56**:39–48.

Hruska, K. A., Korkor, A., Martin, K., and Slatopolsky, E., 1981, Peripheral metabolism of intact parathyroid hormone: Role of liver and kidney and the effect of chronic renal failure, *J. Clin. Invest.* **67**:885–892.

Inagami, T., 1989, Atrial natriuretic factor, *J. Biol. Chem.* **264**:3043–3046.

Jaspan, J. B., Mako, M. E., Kuzuya, H., Blix, P. M., Horwitz, D. L., and Rubenstein, A. H., 1977, Abnormalities in circulating beta cell peptides in chronic renal failure: Comparison of C-peptide, proinsulin and insulin, *J. Clin. Endocrinol. Metab.* **45**:441–446.

Johnson, V., and Maack, T., 1977, Renal extraction, filtration, absorption, and catabolism of growth hormone, *Am. J. Physiol.* **233**:F185–F196.

Just, M., and Habermann, E., 1973, Interactions of a protease inhibitor and other peptides with isolated brush border membranes from rat renal cortex, *Naunyn Schmiedebergs Arch. Pharmacol.* **280**:161–176.

Just, M., and Habermann, E., 1977, The renal handling of polybasic drugs. 2. In vitro studies with brush border and lysosomal preparations, *Naunyn Schmiedebergs Arch. Pharmacol.* **300**:67–76.

Just, M., Erdmann, G., and Habermann, E., 1977, The renal handling of polybasic drugs. 1. Gentamicin and aprotinin in intact animals, *Naunyn Schmiedebergs Arch. Pharmacol.* **300**:57–66.

Kashtan, C. E., Butkowski, R. J., Kleppel, M. M., First, M. R., and Michael, A. F., 1990, Posttransplant anti-glomerular basement membrane nephritis in related males with Alport syndrome, *J. Lab. Clin. Med.* **116**:508–515.

Kato, Y., and Braunstein, G. D., 1988, β-core fragment is a major form of immunoreactive urinary chronic gonadotropin in human pregnancy, *J. Clin. Endocrinol. Metab.* **66**:1197–1201.

Katsube, N., Schwartz, D., and Needleman, P., 1986, Atriopeptin turnover: Quantitative relationship between in vivo changes in plasma levels and atrial content, *J. Pharmacol. Exp. Ther.* **239**:474–479.

Katz, A. I., and Rubenstein, A. H., 1973, Metabolism of proinsulin, insulin, and C-peptide in the rat, *J. Clin. Invest.* **52**:1113–1121.

Kaysen, G. A., Myers, B. D., Couser, W. G., Rabkin, R., and Felts, J. M., 1986, Biology of disease: Mechanisms and consequences of proteinuria, *Lab. Invest.* **54**:479–498.

Kenny, A. J., and Stephenson, S. L., 1988, Role of endopeptidase-24.11 in the inactivation of atrial natriuretic peptide, *FEBS Lett.* **232**:1–8.

Kerjaschki, D., Sharkey, D. J., and Farquhar, M. G., 1984, Identification and characterization of podocalyxin—the major sialoprotein of the renal glomerular epithelial cell, *J. Cell Biol.* **98**:1591–1596.

Kuku, S. F., Jaspan, J. B., Emmanouel, D. S., Zeidler, A., Katz, A. I., and Rubenstein, A. H., 1976, Heterogeneity of plasma glucagon. Circulating components in normal subjects and patients with chronic renal failure, *J. Clin. Invest.* **58**:742–750.

Lee, J. C., Parthemore, J. G., and Deftos, L. J., 1977, Immunochemical heterogeneity of calcitonin in renal failure, *J. Clin. Endocrinol. Metab.* **45**:528–533.

Lefort, G. P., Stolk, J. M., and Nisula, B. C., 1986, Renal metabolism of the β-subunit of human choriogonadotropin in the rat, *Endocrinology* **119**:924–931.

Lemley, K. V, and Kriz, W., 1991, Anatomy of the renal interstitium, *Kidney Int.* **39**:370–381.

Leung, D. W., Spencer, S. A., Cachianes, G., Hammonds, R. G., Collins, C., Henzel, W. J., Barnard, R., Waters, M. J., and Wood, W. I., 1987, Growth hormone receptor and serum binding protein: Purification, cloning and expression, *Nature* **330**:537–543.

Maack, T., 1986, Renal clearance and isolated kidney perfusion techniques, *Kidney Int.* **30**: 142–151.

Maack, T., Johnson, V., Kau, S. T., Figueiredo, J., and Sigulem, D., 1979, Renal filtration, transport, and metabolism of low-molecular-weight proteins: A review, *Kidney Int.* **16**: 251–270.

Maack, T., Park, C. H., and Camargo, M. J. F., 1985, Renal filtration, transport, and metabolism of proteins, in: *The Kidney: Physiology and Pathophysiology* (D. W. Seldin and G. Giebisch, eds.), Raven Press, New York, pp. 1773–1803.

Maack, T., Suzuki, M., Almeida, F. A., Nussenzveig, D., Scarborough, R. M., McEnroe, G. A., and Lewicki, J. A., 1987, Physiological role of silent receptors of atrial natriuretic factor, *Science* **238**:675–678.

Maack, T., Almeida, F. A., Suzuki, M., and Nussenzveig, D. R., 1988, Clearance receptors of atrial natriuretic factor, *Contrib. Nephrol.* **68**:58–65.

Maratos-Flier, E., Kao, C.-Y. Y., Verdin, E. M., and King, G. L., 1987, Receptor-mediated vectorial transcytosis of epidermal growth factor by Madin–Darby canine kidney cells, *J. Cell Biol.* **105**:1595–1601.

Margulies, K. B., Cavero, P. G., Seymour, A. A., Delaney, N. G., and Burnett, J. C., Jr., 1990, Neutral endopeptidase inhibition potentiates the renal actions of atrial natriuretic factor, *Kidney Int.* **38**:67–72.

Markkanen, S. O., and Rajaniemi, H. J., 1979, Uptake and subcellular catabolism of human choriogonadotropin in the proximal tubule cells of rat kidney, *Mol. Cell. Endocrinol.* **13**: 181–190.

Markkanen, S., Töllikkö, K., Vanha-Perttula, T., and Rajaniemi, H., 1979, Disappearance of human [^{125}I] iodochorionic gonadotropin from the circulation in the rat: Tissue uptake and degradation, *Endocrinology* **104**:1540–1547.

Martin, K. J., Finch, J. L., Hruska, K., and Slatopolsky, E., 1987, Effect of biological activity of PTH on its peripheral metabolism in the rat, *Kidney Int.* **31**:937–940.

Marx, S. J., Woodward, C., Aurbach, G. D., Glossmann, H., and Keutmann, H. T., 1973, Renal receptors for calcitonin: Binding and degradation of hormone, *J. Biol. Chem.* **248**: 4797–4802.

Mason, D. Y., Howes, D. T., Taylor, C. R., and Ross, B. D., 1975, Effect of human lysozyme (muramidase) on potassium handling by the perfused rat kidney: A mechanism for renal damage in human monocytic leukemia, *J. Clin. Pathol.* **28**:722–727.

Meezan, E., and Freychet, P., 1982, Binding and degradation of ^{125}I-insulin by renal glomeruli and tubules isolated from rats, *Diabetologia* **22**:276–284.

Meezan, E., Pillion D. J., and Elgavish, A., 1988, Binding and degradation of ^{125}I-insulin by rat renal brush border membranes: Evidence for low affinity, high capacity insulin recognition sites, *J. Membr. Biol.* **105**:113–129.

Mehls, O., Tönshoff, B., Blum, W. F., Heinrich, U., and Seidel, C., 1990, Growth hormone and insulin-like growth factor I in chronic renal failure—Pathophysiology and rationale for growth hormone treatment, *Acta Paediatr. Scand.* **370**(Suppl.):28–34.

Mogensen, C. E., and Sølling, K., 1977, Studies on renal tubular protein reabsorption: Partial and near complete inhibition by certain amino acids, *Scand. J. Clin. Lab. Invest.* **37:** 477–486.

Muggia, F. M., Heinemann, H. O., Farhangi, M., and Osserman, E. F., 1969, Lysozymuria and renal tubular dysfunction in monocytic and myelomonocytic leukemia, *Am. J. Med.* **47:** 351–366.

Murthy, K. K., Thibault, G., Garcia, R., Gutkowska, J., Genest, J., and Cantin, M., 1986, Degradation of atrial natriuretic factor in the rat, *Biochem. J.* **240:**461–469.

Nakamura, R., Hayashi, M., Emmanouel, D. S., and Katz, A. I., 1986, Sites of insulin and glucagon metabolism in the rabbit nephron, *Am. J. Physiol.* **250:**F144–F150.

Nielsen, J. T., and Christensen, E. I., 1985, Basolateral endocytosis of protein in isolated perfused proximal tubules, *Kidney Int.* **27:**39–45.

Nielsen, J. T., Nielsen, S., and Christensen, E. I., 1986, Handling of lysozyme in isolated perfused proximal tubules, *Am. J. Physiol.* **251:**F822–F830.

Nielsen, S., and Nielsen, J. T., 1988, Influence of flow rate and perfused load on insulin absorption in isolated proximal tubules, *Am. J. Physiol.* **254:**F802–F812.

Nielsen, S., Nielsen, J. T., and Christensen, E. I., 1987, Luminal and basolateral uptake of insulin in isolated, perfused, proximal tubules, *Am. J. Physiol.* **253:**F857–F867.

Oparil, S., and Bailie, M. D., 1973, Mechanism of renal handling of angiotensin II in the dog, *Circ. Res.* **33:**500–507.

Petersen, J., Kitaji, J., Duckworth, W. C., and Rabkin, R., 1982, Fate of [^{125}I]insulin removed from the peritubular circulation of isolated perfused rat kidney, *Am. J. Physiol.* **243:**F126–F132.

Peterson, D. R., Oparil, S., Flouret, G., and Carone, F. A., 1977, Handling of angiotensin II and oxytocin by renal tubular segments perfused in vitro, *Am. J. Physiol.* **232:**F319–F324.

Peterson, D. R., Chrabaszcz, G., Peterson, W. R., and Oparil, S., 1979, Mechanism for renal tubular handling of angiotensin, *Am. J. Physiol.* **236:**F365–F372.

Peterson, D. R., Carone, F. A., Oparil, S., and Christensen, E. I., 1982, Differences between renal tubular processing of glucagon and insulin, *Am. J. Physiol.* **242:**F112–F118.

Pinter, G. G., 1988, Regulation of the interstitial volume in the renal cortex, *Contrib. Nephrol.* **68:**245–249.

Pitts, R. F., 1974, *Physiology of the Kidney and Body Fluids*, 3rd ed., Year Book Medical, Chicago.

Powell, D. R., Rosenfeld, R. G., Baker, B. K., Liu, F., and Hintz, R. L., 1986, Serum somatomedin levels in adults with chronic renal failure: The importance of measuring insulin-like growth factor I (IGF-1) and IGF-II in acid-chromatographed uremic serum, *J. Clin. Endocrinol. Metab.* **63:**1186–1192.

Pullman, T. N., Oparil, S., and Carone, F. A., 1975, Fate of labeled angiotensin II microinfused into individual nephrons in the rat, *Am. J. Physiol.* **228:**747–751.

Pullman, T. N., Carone, F. A., Oparil, S., and Nakamura, S., 1978, Effects of constituent amino acids on tubular handling of microinfused angiotensin II, *Am. J. Physiol.* **234:** F325–F331.

Rabkin, R., and Colwell, J. A., 1969, The renal uptake and excretion of insulin in the dog, *J. Lab. Clin. Med.* **73:**893–900.

Rabkin, R., and Kitabchi, A. E., 1978, Factors influencing the handling of insulin by the isolated rat kidney, *J. Clin. Invest.* **62:**169–175.

Rabkin, R., and Kitaji, J., 1983, Renal metabolism of peptide hormones, *Mineral Electrolyte Metab.* **9**:212–226.

Rabkin, R,. and Mahoney, C. A., 1987, Hormones and the kidney, in: *Diseases of the Kidney* (R. W. Schrier and C. W. Gottschalk, eds.), 4th ed., Little, Brown, Boston, pp. 309–355.

Rabkin, R., and Petersen, J., 1984, Peritubular uptake and processing of insulin, *Contrib. Nephrol.* **42**:38–48.

Rabkin, R., Simon, N. M., Steiner, S., and Colwell, J. A., 1970, Effect of renal disease on renal uptake and excretion of insulin in man, *N. Engl. J. Med.* **282**:182–187.

Rabkin, R., Pimstone, B. L., and Eales, L., 1973, Autoradiographic demonstration of glomerular filtration and proximal tubular absorption of growth hormone ^{125}I in the mouse, *Horm. Metab. Res.* **5**:172–175.

Rabkin, R., Unterhalter, S. A., and Duckworth, W. C., 1979a, Effect of prolonged uremia on insulin metabolism by isolated liver and muscle, *Kidney Int.* **16**:433–439.

Rabkin, R., Share, L., Payne, P. A., Young, J., and Crofton, J., 1979b, The handling of immunoreactive vasopressin by the isolated perfused rat kidney, *J. Clin. Invest.* **63**:6–13.

Rabkin, R., Gottheiner, T. I., and Fang, V. S., 1981, Removal and excretion of immunoreactive rat growth hormone by the isolated kidney, *Am. J. Physiol.* **240**:F282–F287.

Rabkin, R., Petersen, J., and Mamelok, R., 1982, Binding and degradation of insulin by isolated renal brush border membranes, *Diabetes* **31**:618–623.

Rabkin, R., Glaser, T., and Petersen, J., 1983, Renal peptide hormone metabolism, *The Kidney* **16**:25–29.

Rabkin, R., Ryan, M. P., and Duckworth, W. C., 1984, The renal metabolism of insulin, *Diabetologia* **27**:351–357.

Rabkin, R., Hirayama, P., Roth, R. A., and Frank, B. H., 1986, Effects of experimental diabetes on insulin binding by renal basolateral membranes, *Kidney Int.* **30**:348–354.

Rabkin, R., Yagil, C., and Frank, B., 1989, Basolateral and apical binding, internalization, and degradation of insulin by cultured kidney epithelial cells, *Am. J. Physiol.* **257**:E895–E902.

Reams, G., Villarreal, D., and Bauer, J. H., 1990, Intrarenal metabolism of angiotensin II, *Am. J. Physiol.* **258**:F1510–F1515.

Richards, M., Espiner, E., Frampton, C., Ikram, H., Yandle, T., Sopwith, M., and Cussans, N., 1990, Inhibition of endopeptidase EC 24.11 in humans: Renal and endocrine effects, *Hypertension* **16**:269–276.

Rippe, B., and Haraldsson, B., 1987, How are macromolecules transported across the capillary wall? *News in Physiol. Sci.* **2**:135–138.

Rosivall, L., Narkates, A. J., Oparil, S., and Navar, L. G., 1987, De novo intrarenal formation of angiotensin II during control and enhanced renin secretion, *Am. J. Physiol.* **252**:F1118–F1123.

Rutanen, E.-M., and Pekonen, F., 1990, Insulin-like growth factors and their binding proteins, *Acta Endocrinol. (Copenhagen)* **123**:7–13.

Sanders, P. W., Herrera, G. A., Chen, A., Booker, B. B., and Galla, J. H., 1988, Differential nephrotoxicity of low molecular weight proteins including Bence–Jones proteins in the perfused rat nephron in vivo, *J. Clin. Invest.* **82**:2086–2096.

Sara, V. R., and Hall, K., 1990, Insulin-line growth factors and their binding proteins, *Physiol. Rev.* **70**:591–614.

Saxena, R., Bygren, P., Butkowski, R., and Wieslander, J., 1989, Specificity of kidney-bound antibodies in Goodpasture's syndrome, *Clin. Exp. Immunol.* **78**:31–36.

Schardijn, G. H. C., and van Eps, L. W. S., 1987, β_2-microglobulin: Its significance in the evaluation of renal function, *Kidney Int.* **32**:635–641.

Seikaly, M. G., Arant, B. S., Jr., and Seney, F. D., Jr., 1990, Endogenous angiotensin concentrations in specific intrarenal fluid compartments of the rat, *J. Clin. Invest.* **86**: 1352–1357.

Shade, R. E., and Share, L., 1976, Metabolic clearance of immunoreactive vasopressin and immunoreactive [^{131}I]iodo vasopressin in the hypophysectomized dog, *Endocrinology* **99**: 1199–1206.

Shade, R. E., and Share, L., 1977, Renal vasopressin clearance with reductions in renal blood flow in the dog, *Am. J. Physiol.* **232**:F341–F347.

Sievertsen, G. D., Lim, V. S., Nakawatase, C., and Frohman, L. A., 1980, Metabolic clearance and secretion rates of human prolactin in normal subjects and in patients with chronic renal failure, *J. Clin. Endocrinol. Metab.* **50**:846–852.

Silbernagl, S., 1988, The renal handling of amino acids and oligopeptides, *Physiol. Rev.* **68**: 911–1007.

Silverman, M., Whiteside, C., and Trainor, C., 1984, Glomerular and postglomerular transcapillary exchange in dog kidney, *Fed. Proc.* **43**:171–179.

Simionescu, M., Ghitescu, L., Fixman, A., and Simionescu, N., 1987, How plasma macromolecules cross the endothelium, *News in Physiol. Sci.* **2**:97–100.

Simmons, R. E., Hjelle, J. T., Mahoney, C., Deftos, L. J., Lisker, W., Kato, P., and Rabkin, R., 1988, Renal metabolism of calcitonin, *Am. J. Physiol.* **254**:F593–F600.

Slatopolsky, E., Lopez-Hilker, S., Delmez, J., Dusso, A., Brown, A., and Martin, K. J., 1990, The parathyroid–calcitriol axis in health and chronic renal failure, *Kidney Int.* **38**(Suppl. 29):S41–S47.

Stephenson, S. L., and Kenny, A. J., 1987, Metabolism of neuropeptides: Hydrolysis of the angiotensins, bradykinin, substance P and oxytocin by pig kidney microvillar membranes, *Biochem. J.* **241**:237–247.

Stetler-Stevenson, M. A., Flouret, G., and Peterson, D. R., 1981, Handling of luteinizing hormone-releasing hormone by renal proximal tubular segments in vitro, *Am. J. Physiol.* **241**:F117–F122.

Sumpio, B. E., and Hayslett, J. P., 1985, Renal handling of proteins in normal and disease states, *Q. J. Med.* **57**:611–635.

Sumpio, B. E., and Maack, T., 1982, Kinetics, competition, and selectivity of tubular absorption of proteins, *Am. J. Physiol.* **243**:F379–F392.

Talor, Z., Emmanouel, D. S., and Katz, A. I., 1982, Insulin binding and degradation by luminal and basolateral tubular membranes from rabbit kidney, *J. Clin. Invest.* **69**:1136–1146.

Talor, Z., Emmanouel, D. S., and Katz, A. I., 1983, Glucagon degradation by luminal and basolateral rabbit tubular membranes, *Am. J. Physiol.* **244**:F297–F303.

Tisher, C. C., and Brenner, B. M., 1989, Structure and function of the glomerulus, in: *Renal Pathology with Clinical and Functional Correlations* (C. C. Tisher and B. M. Brenner, eds.), Lippincott, Philadelphia, pp. 92–110.

Tonolo, G., McMillan, M., Polonia, J., Pazzola, A., Montorsi, P., Soro, A., Glorioso, N., and Richards, M. A., 1988, Plasma clearance and effects of alpha-hANP infused in patients with end-stage renal failure, *Am. J. Physiol.* **254**:F895–F899.

Ugolev, A. M., and De Laey, P., 1973, Membrane digestion: A concept of enzymic hydrolysis on cell membranes, *Biochim. Biophys. Acta* **300**:105–128.

van Deurs, B., and Christensen, E. I., 1984, Endocytosis in kidney proximal tubule cells and cultured fibroblasts: A review of the structural aspects of membrane recycling between the plasma membrane and endocytic vacuoles, *Eur. J. Cell Biol.* **33**:163–173.

Venkatachalam, M. A., and Karnovsky, M. J., 1972, Extravascular protein in the kidney: An ultrastructural study of its relation to renal peritubular capillary permeability using protein tracers, *Lab. Invest.* **27**:435–444.

Vierhapper, H., Gasic, S., Nowotny, P., and Waldhausl, W., 1990, Renal disposal of human atrial natriuretic peptide in man, *Metabolism* **39**:341–342.

Vincent, C., Revillard, J. P., Galland, M., and Traeger, J., 1978, Serum β2-microglobulin in hemodialyzed patients, *Nephron* **21**:260–268.

Waldmann, T. A., Strober, W., and Mogielnicki, R. P., 1972, Renal handling of low molecular weight proteins. II. Disorders of serum protein catabolism in patients with tubular proteinuria, the nephrotic syndrome, or uremia, *J. Clin. Invest.* **51**:2162–2174.

Wall, D. A., and Maack, T., 1985, Endocytic uptake, transport, and catabolism of proteins by epithelial cells, *Am. J. Physiol.* **248**:C12–C20.

Ward, P. E., Erdos, E. G., Gedney, C. D., Dowben, R. M., and Reynolds, R. C., 1976, Isolation of membrane-bound renal enzymes that metabolize kinins and angiotensins, *Biochem. J.* **157**:643–650.

Ward, P. E., Schultz, W., Reynolds, R. C., and Erdos, E. G., 1977, Metabolism of kinins and angiotensins in the isolated glomerulus and brush border of rat kidney, *Lab. Invest.* **36**: 599–606.

Wehmann, R. E., and Nisula, B. C., 1981, Metabolic and renal clearance rates of purified human chronic gonadotropin, *J. Clin. Invest.* **68**:184–194.

Wehmann, R. E., Blithe, D. L., Flack, M. R., and Nisula, B. C., 1989, Metabolic clearance rate and urinary clearance of purified β-core, *J. Clin. Endocrinol. Metab.* **69**:510–517.

Weiss, J. H., Williams, R. H., Galla, J. H., Gottschall, J. L., Rees, E. D., Bhathena, D., and Luke, R. G., 1981, Pathophysiology of acute Bence–Jones protein nephrotoxicity in the rat, *Kidney Int.* **20**:198–210.

Whiteside, C. I., Lumsden, C. J., and Silverman, M., 1988, In vivo characterization of insulin uptake by dog renal cortical epithelium, *Am. J. Physiol.* **255**:E357–E365.

Wochner, R. D., Strober, W., and Waldmann, T. A., 1967, The role of the kidney in the catabolism of Bence–Jones proteins and immunoglobulin fragments, *J. Exp. Med.* **126**: 207–220.

Wolgast, M., 1985, Renal interstitium and lymphatacis, in: *The Kidney: Physiology and Pathophysiology* (D. W. Seldin and G. Giebisch, eds.), Raven Press, New York, pp. 497–517.

Woods, R. L., 1988, Contribution of the kidney to metabolic clearance of atrial natriuretic peptide, *Am. J. Physiol.* **255**:E934–E941.

Yagil, C., Frank, B. H., and Rabkin, R., 1988, Internalization and catabolism of insulin by an established renal cell line, *Am. J. Physiol.* **254**:C822–C828.

Yoshioka, T., Mitarai, T., Kon, V., Deen, W. M., Rennke, H. G., and Ichikawa, I., 1986, Role for angiotensin II in an overt functional proteinuria, *Kidney Int.* **30**:538–545.

Zapf, J., Hauri, C., Waldvogel, M., and Froesch, E. R., 1986, Acute metabolic effects and half-lives of intravenously administered insulinlike growth factors I and II in normal and hypophysectomized rats, *J. Clin. Invest.* **77**:1768–1775.

Chapter 13

Mechanisms for the Hepatic Clearance of Oligopeptides and Proteins

Implications for Rate of Elimination, Bioavailability, and Cell-Specific Drug Delivery to the Liver

Dirk K. F. Meijer and Kornelia Ziegler

1. SCOPE OF THE REVIEW

The role of the liver in the disposition of oligopeptides and proteins of various size, charge, and hydrophobicity is of paramount importance in relation to the following aspects:

- The organ can efficiently remove certain oligopeptides that entered the portal system after gastrointestinal absorption and consequently can be a major factor determining systemic bioavailability.
- If oligopeptides enter the general circulation or are parenterally administered, the clearance by various cell types in the liver can largely determine the residence time in the body of such agents.
- Oligopeptides can have major pharmacological effects on the liver circulation

Dirk K. F. Meijer • Department of Pharmacology and Therapeutics, University Centre for Pharmacy, Groningen, The Netherlands. *Kornelia Ziegler* • Department of Pharmacology and Toxicology, Justus-Liebig University, Giessen, Germany.

Biological Barriers to Protein Delivery, edited by Kenneth L. Audus and Thomas J. Raub. Plenum Press, New York, 1993.

and function of the cells. Some representatives of these peptides are involved in local communication between the cell types.

- Proteins can be used for cell-specific delivery of drugs to the various cell types in the liver with the aims of increasing the local therapeutic effects and of reducing the general side effects.

The fact that numerous biologically active peptides and monoclonals are produced and optimal formulations are sought for their desired therapeutic effects, has accelerated research aimed at a better understanding of the fundamental processes that define their kinetic profiles in the body.

Questions that arise are: What are the mechanisms of hepatic uptake? How can major first-pass elimination by the liver be prevented? And/or how can the residence time of peptides in the body be prolonged by manipulating hepatic clearance function? Such practical questions can only be answered if the fundamental processes that are responsible for hepatic uptake and degradation of proteins and oligopeptides are unraveled. In particular, the mechanisms of these processes in relation to the chemical structure of the ligands are relevant in this respect. Studies on structure-kinetic relationship in the hepatic disposition of oligo- and polypeptides are scarce. Such studies should enable proper predictions of the kinetic profiles of proteins and peptides *in vivo* on the basis of their physicochemical features. The purpose of this chapter is to review the nature of the membrane transport processes and metabolic pathways for oligopeptides and proteins in the various cell types in the liver in relation to their chemical properties. On the other hand, the basic knowledge of these processes can be favorably employed for site-specific drug delivery purposes and the achievements in this rapidly growing field will be discussed.

2. THE LIVER

2.1. Organizational Aspects

The liver plays a major role in the clearance of endogenous and exogenous (macro)molecules from the circulation. Many compounds from the intestinal tract, including various peptides and nutrients, reach the liver via the portal vein but normally fail to enter the systemic circulation to a considerable extent because they are efficiently removed by this organ (De Boer *et al.*, 1982; Pond and Tozer, 1984).

The liver not only synthesizes proteins but also has a function in their removal from the blood. In particular, structurally modified endogenous proteins, in addition to exogenous peptides, can be easily captured. Endogenous small peptides that are released from the antral and duodenal mucosa in response to various physiologic stimuli enter the portal circulation and traverse the liver before arriving in the systemic circulation en route to potential target tissues. A potential target that is primarily reached consequently is the liver itself.

   The liver is well equipped for this clearance function because of the abundant presence of transporting and endocytotic cells. The organ contains the major part of the total number of cells belonging to the reticuloendothelial system (RES) and is therefore equipped to phagocytose particulate and macromolecular material such as effete red blood cells, circulating tumor cells, microorganisms, colloidal material, immune complexes, aggregated proteins as well as monomolecular glycoproteins. The liver architecture guarantees an optimal exchange between blood and the sinusoidal liver cells and also with the parenchymal cells or hepatocytes. These cell populations are equipped with numerous mechanisms to endocytose soluble material including peptides. The blood, including the blood cells and other large particles, flows through the sinusoids in direct contact with the sinusoidal lining cells, i.e., Kupffer cells and endothelial cells (Jones and Summerfield, 1988; Brouwer *et al.*, 1988), while the hepatocytes are exposed to the blood plasma via the fenestrated endothelial lining and the space of Disse (see Fig. 1). The rate by which the liver removes a given substance from the blood depends on the nature of the substance, the rate of delivery to the liver as determined by hepatic blood flow (especially for highly extracted compounds), the transport capacity of the liver cell type involved in the clearance as well as the relative affinity of the substrate for the carrier- or receptor-proteins in the plasma membrane of the particular cell type.

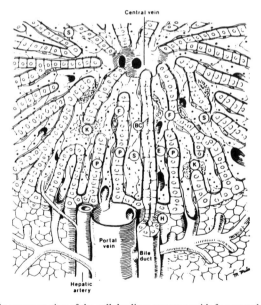

Figure 1.   Schematic representation of the cellular liver structure with fenestrated endothelial lining of the sinusoids (S) that receive blood from the hepatic artery and terminal portal venules. Bile is primarily excreted in the bile canaliculi (BC) draining into the bile ductules. Four cell types are indicated: parenchymal cells (hepatocytes) (P), endothelial cells (E), Kupffer cells (K), and fat-storing cells or lipocytes (F). Sinusoids finally drain their contents in the central veins.

Lipophilic molecules including very hydrophobic peptides [log P(octanol/H_2O) ≥ 2] can enter liver cells by passive diffusion through the plasma membrane (Ziegler *et al.*, 1988b). More polar substances like glucose, amino acids, bilirubin, bile salts, hydrophilic peptides, and charged drugs can only be taken up by carrier-mediated transport mechanisms (Meijer, 1989; Meijer *et al.*, 1990b; Klaassen and Watkins, 1984; Berk *et al.*, 1987). Relatively large peptides (> 10 kDa) most commonly undergo receptor-mediated endocytosis whereas smaller peptides can be taken up in the liver via various carrier-mediated membrane transport processes.

2.2. Functional Liver Structure

The hepatic artery delivers up to 25% of the total hepatic blood flow in the form of oxygen-rich blood, while the portal vein yields about 75% of the total hepatic blood flow and contains blood enriched in endogenous and exogenous components absorbed from the gastrointestinal tract (Fig. 1). Sinusoidal blood flows from the terminal portal venule (Campra and Reynolds, 1988) toward the terminal hepatic venule, creating different microenvironments around periportal (zone 1) and perivenous (zone 3) hepatocytes (Fig. 2). A zonal heterogeneity in hepatocyte functions is

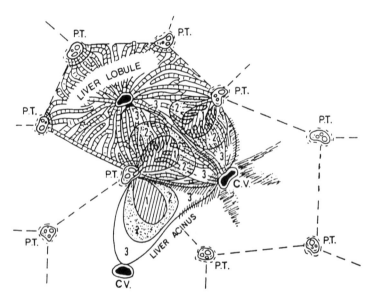

Figure 2. Schematic representation of the microcirculatory units of the liver: the acini. The liver acinus can be divided into three zones: zone 1 (with periportal cells arranged around the sinusoids), zone 2 (intermediate), and zone 3 (mostly the pericentral cells). The classical hexagonal liver lobule arranged around the central vein (CV) does not adequately represent a functional unit. PT, portal triad.

observed not only in ultrastructure, but also in metabolic and transport functions (reviewed in Gumucio and Chianale, 1988; Groothuis et al., 1985; Gumucio, 1989; Thurman and Kauffman, 1985). In sinusoidal blood, there is a zone 1-to-zone 3 gradient of substances efficiently taken up by hepatocytes (Groothuis et al., 1985; Gumucio, 1989). Assuming that hepatocytes along the acinus have identical transport characteristics, a similar gradient would be expected inside the cells. Tissue gradients of endogenous and exogenous compounds observed in various reports support this hypothesis (Jones et al., 1980; Groothuis et al., 1982; Braakman et al., 1987). Such gradients can be due to the different location of the cells regarding the incoming blood, as mentioned above, as well as due to intrinsic differences in uptake rates between zone 1 and zone 3 cells. The latter differences may be the indirect consequence of acinar gradients of oxygen, hormones, and other endogenous compounds (i.e., amino acids and bile acids) (Jungermann, 1986).

2.3. Cell Types of the Liver

The liver contains several cell types of which the arrangement in the hepatic acinus is depicted in Fig. 1, showing hepatocytes, endothelial cells, Kupffer cells, and fat-storing cells. Their contribution to the total liver volume is 78, 2.8, 2.1, and 1.4% (Ballet, 1990; Hendriks et al., 1987; Brouwer et al., 1988) and to the total liver cell volume 92.5, 3.3, 2.5, and 1.7%, respectively. About 15% of the total liver volume consists of intravascular space, the space of Disse (between endothelial lining and the hepatocytes), bile canaliculi and ductuli as well as lymphatic vessels. The space of Disse, although optically empty in the standard EM pictures, contains the components of the extracellular matrix such as collagens and glycosaminoglycans (predominantly heparan sulfates and fibronectin). These components may influence the elasticity of the fenestrated blood vessel wall and probably also the rate of exchange of substrates. Development of perisinusoidal fibrosis may create an extra diffusion barrier for compounds to be taken up in hepatocytes (Ballet, 1990).

The *endothelial cells* form a sieved wall within the sinusoids with fenestrations having a diameter of 50–150 nm (Brouwer et al., 1988). This arrangement of cells enables hepatocytes to exchange substances with plasma via the space of Disse. However, due to the limited size of the pores, there is a distinct selection of substances to be allowed through. Molecules exceeding 250 kDa cannot pass through the pores (Ballet, 1990; Brouwer et al., 1988) and, therefore, do not significantly interact with hepatocytes. This may partly protect the hepatocytes against direct injuries caused by bacteria and viruses but also limits the opportunities to deliver drugs in fair amounts to hepatocytes through inclusion in normal liposomes (Poznansky and Juliano, 1984).

Kupffer cells are located in the sinusoidal lumen at intersections of sinusoids and anchored to the endothelial cells by long processes. The preferential location of

Kupffer cells in zone 1 of the acinus suggests that blood entering the sinusoid is monitored for material that can be endocytosed. It appears that in analogy to hepatocytes, there is also a functional heterogeneity in the Kupffer cell population, for instance with regard to endocytic and tumoricidal capacities (Brouwer et al., 1988; Smit et al., 1991). Kupffer cells are part of the RES, and represent 80–90% of all resident macrophages in the body. They are involved in antigen presentation and processing, modulation of the immune response, as well as the uptake and metabolism of lipids and glycoproteins (Brouwer et al., 1988; Jones and Summerfield, 1988). Kupffer cells also modulate certain hepatocyte functions through the release of various mediators demonstrating a local and chemical communication between these cell types (Brouwer et al., 1988; Jones and Summerfield, 1988; Ballet, 1990). They are activated by agents such as α-interferon, endotoxins, platelet activating factor (PAF), phorbol esters, zymosan, and Ca^{2+} ionophores. After activation, they produce cytokines such as interleukin 1 and 6, tumor necrosis factor, α- and β-interferon as well as lipid mediators (PAF, leukotrienes, and prostaglandins) (Ballet, 1990).

2.4. Protein Synthesis in the Liver

The liver is the principal source of most of the plasma proteins such as albumin and acute-phase proteins. The synthetic activity is mainly situated in the parenchymal cells that exhibit a highly developed vectorial mechanism for sorting and secretion of the manufactured proteins. Certain amino acid sequences as sorting domains in the particular proteins play a role in intracellular routing in addition to glycosylation of the proteins in the endoplasmic reticulum and Golgi apparatus (McCullough and Tavil, 1987). Biosynthetic controls include the overall composition, the nature of the terminal sugar, as well as branching and microheterogeneity of the oligosaccharide chains of the particular glycoproteins. Transcriptional control of glycosyl transferases and glycosidases in the assembling process may, therefore, in principle provide an additional controlling system in intracellular routing and export.

It follows that synthesis, intracellular routing, and secretion of proteins is a complex phenomenon regulated at various levels which can be potentially influenced by multiple physiological and nutritional stimuli as well as by the presence of drugs. Secretion, endocytosis, and degradation of polypeptides in the liver is a dynamic equilibrium, the status of which may also influence clearance of exogenous peptides since competition may in principle occur at the levels of receptor recognition and degradative enzymes.

2.5. Degradation of Proteins and Peptides in the Liver

Catabolism of proteins in the liver involves multiple mechanisms in various cell types. In fact, synthesis and degradation may be closely linked phenomena in secretion-coupled degradation en route from the Golgi system to the cell surface. Such a coupling provides a major posttranslational route in controlling the *net* protein secretion as well as protein-precursor processing. Muscle and liver are the major sites for proteolysis in mammals and half-lives of hepatic proteins are the shortest, reflecting their high turnover rate and proteolytic capacity of the particular cells (Mayer and Doherty, 1985; Yamazaki and Larusso, 1989). Intracellular protein catabolism is composed of an extensive set of energy (ATP)-requiring processes that include lysosomal and nonlysosomal systems acting on "short"- and "long"-lived proteins, respectively. The particular digestive pathways may be classified into four categories (Yamazaki and Larusso, 1989): on the basis of the origin of the protein to be degraded (i.e., heterophagy or autophagy), on the basis of the site of digestion (lysosomal and nonlysosomal), on the basis of the regulatory status of the proteolytic process (basal proteolysis and stimulated proteolysis), as well as on the basis of the properties of the proteins to be degraded (i.e., short-lived and long-lived proteins).

Degradation of intracellular resident proteins is called autophagy and occurs predominantly by lysosomal digestion of cytoplasmic organelles and proteins. Large bodies are initially sequestered in autophagosomes without lysosomal enzymes and subsequently fuse with lysosomes (Mayer and Doherty, 1985; Yamazaki and Larusso, 1989). Soluble proteins are directly sequestered by the lysosomal membranes. Total cellular proteolysis cannot be completely blocked by inhibitors of lysosomal function. Other organelles such as nuclei, mitochondria, and ER as well as the cytoplasmic fraction display abundant proteolytic activities. The latter include Ca^{2+}-activated cysteine endopeptidases such as the calpains that specifically cleave substrates that give rise to physiological responses or regulate intracellular cell movements. In the so-called ubiquitin pathway, target proteins become covalently coupled to the 8.5-kDa protein ubiquitin in an ATP-requiring process. Lysine amino groups and/or N-terminal amino groups of the proteins to be degraded react with the carboxyl glycine of this small peptide and the particular poly-ubiquitin conjugates are proper substrates for cytosolic proteases. Proteolysis of these protein conjugates also requires ATP and regenerates ubiquitin with progressive enzymatic cleavage of the protein ligand (Mayer and Doherty, 1985; Yamazaki and Larusso, 1989). In liver, ubiquitin itself also undergoes proteolytic degradation via a cathepsin-like thiol protease. Small peptides that enter the cytosol by carrier-mediated processes or passive diffusion may, therefore, enter multiple degradative routes. Short-lived endogenous proteins including rate-limiting enzymes such as ornithine decarboxylase are degraded by the nonlysosomal systems. Long-lived proteins are mainly degraded by the lysosomal apparatus. The rate of lysosomal proteolysis may depend on physicochemical features like molecular weight, subunit size, charge, hydrophobicity-hydrophilicity balance, thermolability and conformation of the particular molecules among others. Proteins were previously regarded to be degraded faster when they were larger, more acidic, more hydrophobic, and had little oligosaccharide chains. It is more likely, however, that the half-life of the proteins is largely determined by the presence of specific amino acid sequences in the protein.

The responsible proteases can be inhibited by certain peptides such as leupeptin.

This tripeptide inhibits the cathepsins B, H, and L and causes a marked change in lysosomal density presumably by lysosomal constipation. Low lysosomal pH that is maintained by an ATP-driven electrogenic proton pump is essential for the function of such enzymes and therefore tertiary amines (chloroquine), inhibitors of the H^+ pump (quercetin, vanadate), and proton exchangers such as monensin will largely affect proteolytic activity of these organelles (Mayer and Doherty, 1985; Yamazaki and Larusso, 1989; see Chapter 3).

Proteolysis is started by endopeptidases that act on the middle part of the proteins. The major proteinase is cathepsin D with an intralysosomal concentration of 1 mM. Oligopeptides released by such enzymes are further degraded by exopepti- dases but the endopeptidase step is rate-limiting (McCullough and Tavil, 1987). The resulting amino acids and dipeptides reenter the metabolic pool of the cell.

It should be emphasized that the rate of glycoprotein catabolism in the liver can also be largely influenced by the rate of glycanolysis, i.e., the removal of the oligosaccharide moieties. The presence of the branched antennary sugar chains may "protect" the protein core from rapid breakdown since the prosthetic groups have to be degraded in a stepwise process first. The particular glucosidases and man- nosidases can be blocked by N-containing sugar derivatives, i.e., the glucosidase and mannosidase inhibitors such as castanospermine, and deoxynojirimicin, among others (Mayer and Doherty, 1985; McCullough and Tavil, 1987).

Glycoproteins that are taken up by receptor-mediated endocytosis generally enter a lysosomotropic route that will finally lead to proteolytic degradation in a hetero- phagic process. Some proteins can bypass this abortive route and may undergo trans- cytosis to the biliary pole of the cell (McCullough and Tavil, 1987) to be secreted or will be recirculated to the sinusoidal domain of the plasma membrane to reenter the bloodstream. It is important to realize that the relative contribution of those different intracellular routes in the total disposition can be largely dose-dependent and influenced by pathological conditions and presence of pharmacologically active com- pounds. In the case of insulin, apart from chloroquine-sensitive lysosomal degrada- tion, major endosomal proteolysis by an insulin-glucagon protease has also been described. The latter process can be inhibited by the antibiotic bacitracin (Burwen and Jones, 1990). Finally, internalization of peptides is not always a prerequisite for proteolysis: peptic hydrolases can be present at the cell surface as has been shown for degradation of certain dipeptides in the liver (Lochs et al., 1986; see Chapter 2).

3. MECHANISMS FOR HEPATIC UPTAKE OF PROTEINS AND OLIGOPEPTIDES

3.1. Carrier-Mediated Membrane Transport

For relatively small, polar or charged substrates, carrier-mediated membrane transport is necessary for substantial penetration into the cell (Berk et al., 1987;

Meijer, 1989; Smith, 1971; Meijer *et al.*, 1990b; Meier, 1988; Boyer, 1986) (see Fig. 3).

Carriers probably are single or grouped structural membrane proteins that form some sort of pores or undergo conformational changes that result in reallocation in the lipid bilayer by which a bound substrate is shuttled through this cellular barrier. It is not known whether this shuttling process involves passage through gated channels, flip-flop movements of single proteins, or positional changes of closely associated pairs of proteins (Sorrentino and Berk, 1988; Meier, 1988). Nevertheless, much progress was made in identification of potential carrier proteins by specific photo-affinity labeling (Kurz *et al.*, 1989), protein isolation, and determination of their amino acid and sugar composition (Berk *et al.*, 1987; Sorrentino and Berk, 1988; Meijer *et al.*, 1990b). Antibodies raised against such potential carrier proteins potentially can block transport and are employed to study cellular and tissue localization (Sorrentino and Berk, 1988; Levy, 1989). Some of the isolated carrier proteins have been reconstituted in proteoliposomes (Sottocasa *et al.*, 1982; Meier, 1988; Von Dippe and Levy, 1990; Ruetz *et al.*, 1988) and attempts are under way to sequence the proteins and identify the genetic material involved in their synthesis (Meier, 1988).

Passive, carrier-mediated transport involves recognition by a specific carrier

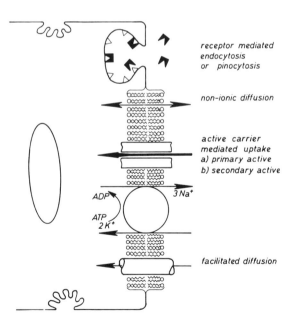

Figure 3. Mechanisms for membrane transport in the liver repsonsible for uptake of peptides, proteins, and other endogenous and exogenous compounds. Carrier-mediated transport can be facilitated, secondary active, and primary active, depending on type of energization. Endocytosis occurs with entirely different mechanisms.

on the basis of the physicochemical features of the substrate (Stein, 1986; Meijer, 1989) (see Fig. 3). Transport occurs along the (electro-) chemical concentration gradient. This process is saturable, displays competition between structurally related ligands, and transstimulations or accelerated exchange diffusion (Stein, 1986; Meier, 1988). The latter phenomena can be demonstrated in isolated cells or membrane vesicles by preloading them with a transportable agent and observing an increased *initial* rate of uptake of a chemically related compound under these conditions. This preloading effect is supposed to be due to reorientation of the carrier binding site to the outer site of the membrane, a process that is supposed to occur only in the drug-bound form (Stein, 1986).

Vectorial transport of charged compounds can lead to unequal concentration at each side of the membrane simply due to equilibration according to the membrane potential (Meijer, 1989). For instance, compounds with a net cationic charge can reach a fourfold higher intracellular concentration (Meijer, 1989; Meijer *et al.*, 1989) only due to the inside negative membrane potential of the hepatocyte (-30 to 40 mV) while anionic compounds can be driven out of the cell for instance at the canalicular level for the same reason (Weinman *et al.*, 1989; Meier, 1988). Real "uphill transport" against an electrochemical gradient requires energy-rich cosubstrates such as ATP (*primary active transport*) or may operate via existing gradients of organic ions (H^+, OH^-, HCO_3^-, Na^+, or K^+) driving the translocation process (Meier, 1988). Since such ion gradients can only be built up through the use of energy-rich compounds, such transfer of the organic solute via ion gradients is called *secondary active transport*. In the latter category, one finds clear examples of Na^+-linked transport of bile acids (Ruifrok and Meijer, 1982; Meier, 1988; Blitzer *et al.*, 1982; Zimmerli *et al.*, 1989), fatty acids (Stremmel, 1989; Potter *et al.*, 1989), and certain amino acids (Christensen, 1979). One or more Na^+ ions are supposed to be bound to the carrier together with the substrate. After translocation to the intracellular site where the Na^+ concentration is much lower, Na^+ dissociates from the carrier and the drug is released because affinity of the bile acid for the Na^+-free carrier is supposed to be reduced (Stein, 1986; Zimmerli *et al.*, 1989).

Primary active transport using ATP is definitely established for Na^+,K^+ transport (the electrogenic Na^+,K^+-ATPase pump system) and recently also for the removal of hydrophobic and cationic antineoplastic drugs from resistant tumor cells. This system operates via the so-called gp-170 multidrug resistance protein that is an ATPase too (Beck, 1987; Zamora *et al.*, 1988). This protein is also present in the canalicular pole of normal hepatocytes (Kamimoto *et al.*, 1989) and is a candidate for extrusion of lipophilic, mostly cationic drugs (Beck, 1987; Zamora *et al.*, 1988). Uptake of certain organic cations into the liver has been demonstrated to be very sensitive to depletion of ATP (Steen *et al.*, 1991). Also for the excretion of non-bile-salt anionic compounds such as glutathione conjugates, sulfates, glucuronides, and, in general, organic anions with two or more negative charges, an ATP-dependent multispecific transporter has recently been identified by several groups in the canalicular pole of the hepatocyte (Oude Elferink *et al.*, 1990; Ishikawa *et al.*, 1990;

Kobayashi *et al.*, 1990; Kamimoto *et al.*, 1989). There is also increasing evidence for ATP-dependent uptake of organic anions at the sinusoidal pole of the cell that may operate in the case of hydrophilic organic anions such as dibromosulfophalein (DBSP) and aniline naphthaline sulfonate (ANS), but interestingly not for more hydrophobic compounds such as BSP and Rose Bengal (Yamazaki *et al.*, 1990).

3.2. Multiplicity in Hepatic Transport Mechanisms

It is clear that all of the mentioned hepatic uptake processes in principle can contribute to some extent to the clearance of a ligand. However, the physicochemical features of the molecule such as charge, charge density, and distribution of charge, as well as the size and the hydrophilicity-lipophilicity balance of the drug molecule determine which process will be favored (Meijer, 1989; Smith, 1966). It is important to note here that for endogenous and exogenous organic compounds including peptides, multiple carrier-mediated mechanisms in the sinusoidal domain of the plasma membrane of liver parenchymal cells are operating in concert (see Fig. 4). At least three different processes are present for anionic compounds (Berk *et al.*, 1987; Meijer, 1989; Tiribelli *et al.*, 1990; Meier, 1988), four separate mechanisms have been described for endogenous and exogenous cationic compounds (Meijer, 1989; Meijer *et al.*, 1990b), and at least two have been described for uncharged compounds like the cardiac glycosides and steroidal hormones (Meijer, 1989; Smith, 1966; Klaassen and Watkins, 1984). Some of these carrier systems also accommodate small peptides (Frimmer and Ziegler, 1988). Especially at relatively low concentrations these processes may be quite specific for a certain molecule, but at higher concentrations there is overlapping substrate specificity: more than one of the above-mentioned carrier systems contribute to the net hepatic uptake of the drug. It is not excluded that, in addition, an aspecific (multiplicit) transport process is involved in the translocation of amphipathic molecules including small peptides (Kurz *et al.*, 1989; Frimmer and Ziegler, 1988). Such a system is hypothesized to accommodate all kinds of "bulky," hydrophobic drug molecules irrespective of charge. This idea is based on photoaffinity labeling of structural membrane polypeptides in the range of 50 kDa for a wide variety of drugs, showing transport interactions and mutual inhibition of photolabeling (Kurz *et al.*, 1989; Frimmer, 1982). Alternatively, these observations could be compatible with overlapping substrate specificity of a family of closely related carrier proteins that have a similar molecular mass but slightly different composition and net charge (Berk *et al.*, 1987; Boyer, 1986; Sorrentino and Berk, 1988; Meijer *et al.*, 1990b). It is clear, however, that independent processes must exist for bile acids, anionic dyes, and for uncharged steroids since they develop differently from birth (Klaassen and Watkins, 1984; Suchy *et al.*, 1989).

Some of the unexpected interactions observed during hepatic uptake between very different categories of organic compounds can be explained by the fact that drug

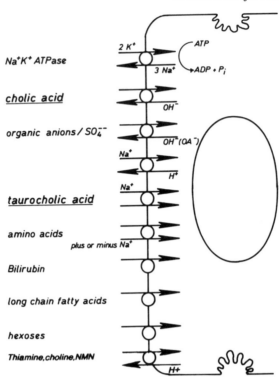

Figure 4. Transport systems for various types of endogenous organic compounds that have been identified in the sinusoidal domain of the hepatocyte plasma membrane. Small peptides can be accommodated by the bile acid carriers and possibly also by the cation carriers. Separate processes are indicated; however, the organic anion transporter may be identical to the bilirubin carrier. There is recent evidence that the organic cations thiamine, choline, and N-methyl noctinamide (NMN) as well as cationic drugs are taken up by dissimilar mechanisms.

molecules can be present in the body in multiple forms or, like certain peptides, expose both anionic and cationic groups. For instance, bile acids, at physiological pH, can be present as uncharged steroidal structures, but also in the anionic (dissociated) form, and can even be combined with Mg^{2+} or Ca^{2+} in a net positively charged complex (Meijer, 1989). Assuming that multiple mechanisms for anionic, cationic, and uncharged compounds exist, bile acids in these various forms could interfere with all of these processes and also be transported via such mechanisms, especially at relatively high concentrations (Vonk *et al.*, 1978). Consequently, mutual transport interactions between structurally unrelated agents should be anticipated even without existence of the hypothesized nonspecific transport system. On the other hand, a single channel-like structure for all of the above-mentioned categories of drugs could be envisioned, realizing that any organic solute could be accompanied by a different and unequal number of inorganic ions while passing the channel.

Consequently, transport would still show a certain substrate specificity as the consequence of interactions with the charged lining of the "pore" (Frimmer, 1982). There is evidence now that one taurocholate molecule is transported combined with more than one Na^+. Such a symport system is electrogenic (carrying net positive charge) and could therefore be helped by the negative membrane potential (Meijer et al., 1990b). Various types of oligopeptides display high affinity for these carriers due to the presence of charged groups and hydrophobic regions (see Sections 4.1– 4.9). Indirect evidence suggests that anionic compounds could be accompanied by Na^+ as well as Cl^- ions in a symport system (Frimmer, 1982) or exchange for OH^- or HCO_3^- ions (anion antiport) (Meier, 1988). Interactions of the complexes with the charged inner sites of the carrier pores during transport or the ability to pass charged barriers within the pore could explain such ion requirements (Berk et al., 1987).

Apart from interactions on the basis of charged groups, hydrophobic interactions may play an important role in the affinity of organic solutes for carrier systems in the sinusoidal membranes. Structure-kinetic studies with anionic and cationic compounds clearly indicate that, in contrast to the renal tubular secretion system, the liver parenchymal transport system favors transport of relatively hydrophobic (lipophilic) drugs. These are often compounds with a relatively high molecular mass and their bulky ring structures render them hydrophobic. Since the various carrier proteins mentioned above very likely possess hydrophobic binding sites, that largely determine the affinity for their substrates, it is not surprising that many cholephilic ligands and also certain peptides noncompetitively interact during hepatic uptake in spite of their different overall charge. Part of the observed interactions may alternatively be explained by a direct influence of the lipid bilayer that may indirectly affect carrier mobility and efficiency of carrier-mediated transport. Treatment of rats with ethinyl estradiol leads to a cholestatic condition and impaired bile acid transport that may be due to an increased rigidity of the sinusoidal plasma membrane. It has been shown that membrane fluidity and transport of bile acids and BSP can be restored by administration of certain detergents (Simon et al., 1980; Miccio et al., 1989).

4. CARRIER- AND NON-CARRIER-MEDIATED UPTAKE OF OLIGOPEPTIDES BY THE LIVER

4.1. Introduction

Peptides, in particular hormone modifications, superagonists and antagonists, and some peptide receptor blockers, are gaining importance in drug development (Hruby, 1985; Bertrams and Ziegler, 1991a; Gloff and Benet, 1990; Bocci, 1990; Wünsch, 1983) (see Fig. 5). Drug design based upon peptide structure, however, must take into consideration two problems: one is the prevention of proteolytic degradation. Efforts to circumvent the tendency of rapid degradation have succeeded in the development of enzymatically stable peptides. These compounds are confor-

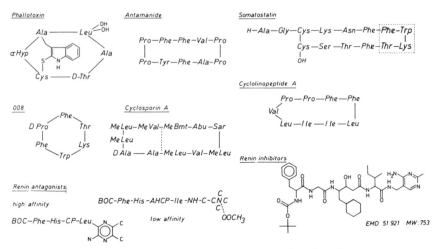

Figure 5. Chemical composition of some oligopeptides with cyclic or linear structures.

mationally restricted and often have only a partial peptide structure, especially in the case of linear renin-inhibiting peptides (Hui *et al.*, 1988). The other problem is that some of these proteolytically stable peptides also have a short half-life *in vivo* as a result of rapid hepatocellular elimination (Caldwell *et al.*, 1985).

The first step in hepatic elimination is the permeation of these substrates through the sinusoidal membrane of liver cells. Data are available for the molecular mechanism of hepatocellular uptake of certain small cyclic and linear peptides of pharmacological value or of toxicological importance, such as somatostatins (Wolfe *et al.*, 1987; Caldwell *et al.*, 1985), cyclosomatostatins (Kessler *et al.*, 1986), cyclolinopeptides (Tancredi *et al.*, 1989), cyclosporins (Borel *et al.*, 1976), antamanide (Wieland *et al.*, 1969), phallotoxins (Frimmer, 1975), linear renin-inhibiting peptides (Haber *et al.*, 1987) (Fig. 5), cholecystokinins (Gores *et al.*, 1986a,b, 1989; Hunter *et al.*, 1990; Doyle *et al.*, 1984), dipeptides (Gly-Ala; Gly-Leu; Gly-Pro; Gly-Sar) (Lochs *et al.*, 1986), cyclo(His-Pro) (Koch *et al.*, 1982) oligopeptides (Lombardo *et al.*, 1988), leupeptin (Leu-Leu-Arg) (Dennis and Aronson, 1985), α-amanitin (Kröncke *et al.*, 1986), gastrin-releasing peptide and gastrin (Doyle *et al.*, 1984; Wolfe *et al.*, 1987; Seno *et al.*, 1985) (Table I).

In comparative kinetic studies on isolated hepatocytes, some of the above small cyclic and linear peptides are taken up by energy-dependent carrier-mediated transport as well as by nonionic diffusion. The physicochemical properties of the individual compounds determine the mode of transport. For dipeptides and some oligopeptides (Lochs *et al.*, 1986; Lombardo *et al.*, 1988), clearance seems to be due to extracellular hydrolysis by enzymes located on the liver plasma membrane or released from the cytosol. From these studies, it was concluded that the liver does not possess a transport system for dipeptides. On the other hand, Mori and colleagues

Table I
Physiochemical Properties of Peptides

	Size (Da)	No. of amino acids	Lipophilicity[a]
Dipeptides	~220	2	0.000000014
Glutathione	307	3	—
Leupeptin	476	3	—
Leu-enkephalin	555	5	1.12
Met-enkephalin	573	5	0.031
Phalloidin	789	7	—
Bradykinin	1,060	9	—
Antidiuretic hormone	1,084	9	—
Cyclolinopeptide	1,048	9	—
Vasopressin	1,083	9	—
Oxytocin	1,007	9	—
Antamanide	1,147	10	—
Cyclosporin A	1,200	11	991
Substance P	1,347	11	0.275
Somatostatin (N-Tyr)	1,637	14	0.081
Endothelin	2,500	21	—
Secretin	3,052	27	—
VIP	3,327	28	—
Glucagon	3,485	29	0.063
Calcitonin	3,604	32	—
Cholecystokinin	3,716	33	—
β-Endorphin (N-Tyr)	3,540	31	0.014
Insulin	6,000	51	0.021
Lysozyme	14,307	129	—
γ-Interferon	25,000 (+G)[b]	166	—
α_1-Acid glycoprotein	41,000 (+G)	181	—
Ovalbumin	45,000 (+G)	400	—
Transferrin	79,000 (+G)	679	—
Albumin	65,000	575	—
IgA	160,000 (+G)	—	—
Fibronectin	250,000 (+G)	—	—
α_2-Macroglobulin	820,000 (+G)	1450	—

[a]Lipophilicity as determined by partition between n-octanol and aqueous buffers.
[b]+G = including oligosaccharide chains.

(Mori *et al.*, 1986, 1989) detected a highly specific receptorlike uptake mechanism for cyclo(His-Pro), a metabolite of thyrotropin-releasing hormone (TRH).

4.2. Peptide Transport by Bile Acid Carriers

Cyclosomatostatins (Ziegler *et al.*, 1990, 1991; Ziegler and Frimmer, 1986a), cyclolinopeptides (Ziegler *et al.*, 1990), phallotoxins (Petzinger and Frimmer, 1980), antamanide (Petzinger *et al.*, 1983), α-amanitin (Kröncke *et al.*, 1986), hydrophilic and hydrophobic linear renin-inhibiting peptides (Bertrams and Ziegler, 1989a,b; Seeberger and Ziegler, 1990a,b), and cholecystokinin octapeptide (Gores et al., 1986a,b, 1989; Hunter et al., 1990) enter liver cells by an energy-dependent carrier-mediated process. Since it is improbable that the liver possesses a specific transport system for various foreign peptides, the possible physiological substrates of these carrier systems were sought.

There exist several transport systems for endogenous substrates (Fig. 4) at the basolateral membrane of the hepatocytes (see Section 3.2). The transport systems for bile acids turned out to be important for the elimination of peptides. Sodium-dependent and independent transport systems were described for the uptake of conjugated and unconjugated bile acids, respectively. Eighty percent of the uptake of taurocholate is sodium-dependent, whereas only 50% of cholate uptake is mediated by an inwardly directed sodium gradient (Frimmer and Ziegler, 1988). These relative contributions, however, are concentration-dependent.

In addition, the transport of several of the peptides studied is inhibited by bromosulfophthalein, a putative foreign substrate of the bilirubin carrier (for review see Frimmer and Ziegler, 1988). This interference of uptake is due to a noncompetitive type of transport inhibition, suggesting that transport sites for bilirubin/BSP and peptides are not identical.

In earlier studies with phalloidin it could be shown by Frimmer and his group (Frimmer et al., 1980; Petzinger, 1981) that a transport system physiologically handling cholate is also responsible for the hepatospecific elimination of the toxin phalloidin. Several chemically different compounds such as cholecystographic agents, cardiac glycosides, and loop diuretics have been identified as further substrates of this carrier system (for review see Frimmer and Ziegler, 1988). It seems that this system is unable to discriminate between steroids and cyclopeptides. In 1985, the above transport system was termed multispecific transporter (Ziegler et al., 1985) (see Fig. 6).

On the basis of the presently collected data, it emerges that the transport system for cyclosomatostatins (Ziegler et al., 1991), cyclolinopeptides (Ziegler et al., 1990), antamanide (Petzinger et al., 1983), phalloidin (Frimmer and Ziegler, 1988), and cholate is identical. This interpretation is supported by mutual transport inhibition in isolated hepatocytes as well as isolated rat liver plasma membrane vesicles (see Table II). The compounds tested turned out to be mutually competitive transport inhibitors. In addition, these peptides also competitively inhibit sodium taurocholate cotransport and vice versa. The affinity of the above cyclopeptides to the multispecific cholate transporter, however, is tenfold higher than to the sodium taurocholate cotransporter. Furthermore, these cyclopeptides inhibit sodium-dependent as well as sodium-independent uptake of cholate and taurocholate. This is in contrast to the hydrophobic linear renin inhibitors (Bertrams and Ziegler, 1989a,b). The latter peptides have a higher affinity to the sodium taurocholate cotransporter than to the multispecific cholate transporter. They noncompetitively inhibit the sodium-independent part of taurocholate uptake, but competitively block the sodium-dependent part.

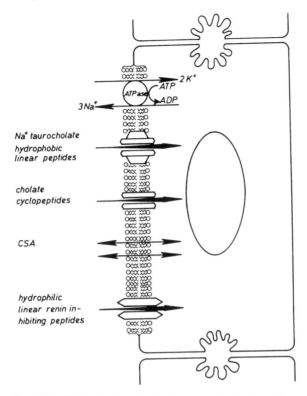

Figure 6. Overview of identified mechanisms for hepatocyte uptake of various classes of oligopeptides. Relatively hydrophobic peptides prefer the Na⁺-dependent bile acid carrier, cyclopeptides the multi-specific one, and hydrophilic linear peptides a separate route. Very lipophilic peptides like cyclosporin (CSA) enter by passive lipoid permeation.

Table II
Kinetic Properties of Hepatocellular Uptake

	Taurocholate	Cholate	Cyclosomatostatin	Linear peptides
K_m (μM)	19–57	58–74	5, 8	2
V_{max} (pmole/mg per min)	750–1300	570–1170	165	160
Energy dependent	+	+	+	+
Activation energy (kJ/mole)	121	108	37	60
Sodium dependent	+	+/−	−	−
Membrane potential dependent	+	+	+	+
Mutual competitive inhibition	+	+	+	+
Uptake into AS 30 D ascites cells	−	−	−	−
Identified binding proteins (kDa)	48–50, 54	48–50, 54	50, 54	50, 54
Functional molecular mass (kDa)	170	109	100	n.d.

At present, the identity of protein components of the multispecific cholate transporter with those of the sodium-dependent taurocholate cotransporter is not proven. But as discussed later, evidence exists that two different bile acid transport systems are located in the sinusoidal membrane of liver cells. One of them, responsible for the sodium-dependent uptake of taurocholate, also transports linear hydrophobic renin inhibitors and the other mediates the uptake of cholate and certain cyclopeptides. The two transport systems, however, show overlapping substrate specificities.

LaRusso and colleagues (Gores et al., 1986b, 1989) studied the processing of cholecystokinin by isolated liver cells. The authors described an active carrier-mediated transport system for cholecystokinin octapeptide, which could be inhibited by taurocholate. It was suggested that these peptides are taken up by a similar mechanism as discussed above.

4.3. Mechanisms and Energization of the Peptide Transporters

All cyclic as well as linear peptides tested in our studies (Frimmer, 1989; Bertrams and Ziegler, 1991a) noncompetitively inhibit the uptake of BSP and rifampicin, probably substrates of the bilirubin carrier in the liver. Other carrier systems existing in the liver cell membrane (Frimmer and Ziegler, 1988; Meier, 1988), for example those for amino acids, hexoses, cationic compounds, and long-chain fatty acids, were not influenced by the peptides and vice versa (Fig. 4).

Cyclosomatostatins (Bertrams and Ziegler, 1991a; Ziegler et al., 1990; Ziegler and Frimmer, 1986a) and some linear hydrophobic renin-inhibiting peptides (Bertrams and Ziegler, 1989a,b, 1991b), are taken up into liver cells in an energy-, temperature-, and concentration-dependent manner. Comparing the kinetics of the uptake of cholate, taurocholate, cyclosomatostatins, and hydrophobic linear renin inhibitors (Table II), it became evident that the affinity of the peptides for the transport system is higher than that of bile acids. In contrast, the V_{max} is higher for bile acids. At high substrate concentrations, a minor part of the uptake is mediated by diffusion. Permeability coefficients in the range of 8.1×10^{-6} cm/sec were calculated. The carrier-mediated part of the uptake is saturable as well as temperature- and energy-dependent. Activation energies in the range of 37–60 kJ/mole also indicate a carrier-mediated process. In the absence of oxygen or in the presence of metabolic inhibitors, uptake of cyclic and linear peptides is inhibited. These transport characteristics are also shared by bile acids.

4.4. Driving Forces for Hepatic Uptake of Peptides

The main difference in the transport characteristics of peptides and bile acids is the driving force needed for the uptake. Whereas 80% of the uptake of taurocholate and 40% of the uptake of cholate are driven by an inwardly directed sodium gradient, the uptake of the tested cyclopeptides as well as of the linear peptides is not sodium-dependent (Table II). The same phenomenon was described for the uptake of cholecystokinin octapeptide (Gores *et al.*, 1989; Hunter *et al.*, 1990). Sodium can be replaced by choline or lithium without any effect on uptake. When sodium is replaced by potassium, the uptake is reduced significantly. Since potassium also alters the membrane potential, it was inferred that the membrane potential could be the dominant driving force for the uptake of cyclopeptides and linear hydrophobic renin inhibitors. This suggestion is supported by anion replacement studies: the uptake of peptides is stimulated by making the inner side of the membrane more negative, whereas a more positive membrane potential inhibits the uptake of the above substrates. One must, however, realize that the cyclosomatostatins are neutral at physiological pH whereas the hydrophobic renin-inhibiting peptides are positively charged.

4.5. Structure-Kinetic Relationships of Peptides

At present, the structural requirements determining the affinity of the peptides to the bile acid transporters are not unequivocally known. Studies with various derivatives of cyclosomatostatins, cyclolinopeptides, and linear renin-inhibiting peptides were performed. The structure-activity relationship of peptide analogues on hepatocellular transport inhibition was investigated (Ziegler, 1989; Kessler *et al.*, 1987, 1988; Zanotti *et al.*, 1990).

Apart from the number of amino acids, the hydrophobicity and certain conformational characteristics are prerequisites for binding to bile acid carrier systems. The oligopeptide charge *per se* seems not to be a crucial factor determining extent of hepatic clearance. Cyclopeptides that are neutral at physiological pH, linear hydrophobic renin-inhibiting peptides that are positively charged, and the cholecystokinins that exist principally in the anionic form all are rapidly taken up in the liver especially if hydrophobicity reaches a certain threshold (Gores *et al.*, 1989; Hunter *et al.*, 1990). Amino acids and dipeptides have no affinity to the transporter, while some tripeptides have a low affinity (Bertrams and Ziegler, 1991b). Larger oligopeptides of up to 33 amino acids, as shown in the case of cholecystokinins, have a sevenfold lower affinity for these carrier systems compared with the endogenous bile acid substrates (Gores *et al.*, 1989; Hunter *et al.*, 1990). For the linear renin-inhibiting peptides tested, there is a nonlinear correlation between their hydrophobicity and their affinity for the transporter (Bertrams and Ziegler, 1991a). Hydrophilic linear renin-inhibiting peptides have a 200-fold lower affinity for the bile acid transporters compared with bile acids themselves. In addition, the type of transport inhibition turned out to be noncompetitive (Seeberger and Ziegler, 1990, 1993). Hydrophilic linear renin-inhibiting peptides are not substrates of bile acid carriers; nevertheless, they are probably taken up by some other kind of organic cation carrier system (Seeberger and Ziegler, 1990, 1993).

Since most linear peptides, up to a size of approximately 30 amino acids, are mixtures of many conformations, structurally rigid cyclic peptides, such as cyclosomatostatins, cyclolinopeptides, and antamanide, were used to elaborate the conformational prerequisites for affinity to bile acid transporters. Natural somatostatin has a low affinity for the bile acid transport system. A concentration of 200 μM is needed to inhibit bile acid transport by 50% (Ziegler *et al.*, 1985). Cyclic peptides derived from the active sequence of somatostatin are about 100-fold more active than natural somatostatin (Table III). The analogues possess a reduced conformational freedom owing to cyclization and exhibit increased enzymatic stability.

Interestingly, alteration of the direction of the peptide bonds, generating retro peptides, led to the development of analogues with higher activity (Ziegler *et al.*, 1985). Since both the peptides and their retro analogues have affinity for bile acid carriers, it is suggested that a small recognition area is responsible for binding to the bile acid carrier. The peptide with the highest affinity is a retro peptide with the sequence C(D-Pro-Phe-Thr-Lys-Trp-Phe), termed 008 (Fig. 5). The affinity is further increased when the Lys side chain is protected by a Z group (benzyloxycarbonyl). This is true for all cyclohexa- and pentapeptides synthesized. It points to the fact that a higher hydrophobicity of the analogues, due to the Z group, has effects on the affinity (Kessler *et al.*, 1987, 1988).

By replacing each amino acid in 008 by an alanine residue, the importance of the side chains of each amino acid for affinity to the bile acid carrier was defined. The conformation of the peptides was determined by NMR spectroscopy and shown to be very similar. All alanine-containing peptides inhibited bile acid uptake. All amino acids in the peptides turned out to be important for binding, since substitution of each of the aromatic amino acids led to a decreased affinity.

Table III
Chemical Structure and Inhibition of Uptake of Cholate and Taurocholate
of Cyclolinopeptide A (CLA) and Its Analogues[a]

Structure									Size (Da)	IC$_{50}$ Cholate	IC$_{50}$ Taurocholate
Pro	Pro	Phe	Phe	Leu	Ile	Ile	Leu	Val	1048	0.83 ± 0.19	69.75 ± 23.47
Ala	—	—	—	—	—	—	—	—	1022	1.22 ± 0.15	24.88 ± 5.87
—	Ala	—	—	—	—	—	—	—	1022	1.30 ± 0.80	11.36 ± 1.63
—	—	Ala	—	—	—	—	—	—	972	5.65 ± 0.94	25.21 ± 5.42
—	—	—	Ala	—	—	—	—	—	972	3.19 ± 0.81	40.84 ± 19.19
—	—	—	—	Aib	Aib	—	D-Ala	—	941	20.92 ± 1.81	58.46 ± 9.07
—	—	—	—	Gly					546	—	—
—	—	—	β-Ala	β-Ala					610	—	—
—	—	—	β-Ala	β-Ala					386	—	—

[a]Amino acid substitutions and/or chain shortening leads to dissimilar changes in affinity for the multispecific cholate carrier and the Na$^+$ dependent taurocholate carrier system.

The important structural elements common to all cyclopeptide analogues are two adjacent aromatic amino acids such as Phe-Phe or Phe-Trp and a neighboring Pro or D-Pro residue. The Phe residue in position 3 of cyclolinopeptides is the most important in this series. This sequence, however, is not alone sufficient for affinity to the transporter. In addition to sequential requirements, the spatial arrangement is important. The synthetic cyclic pentapeptides cyclo-Pro-Pro-Phe-Phe-Gly or -Ala do not inhibit bile acid uptake. The conformation of these analogues differ from that of bile acid transport-inhibiting cyclopeptides, indicating that a certain conformation is required for recognition by bile acid transporters. The amino acids surrounding the aromatic fragment have the task of fixing a certain conformation for the aromatic side chain.

Vasoactive intestinal peptide has an efficient first-pass clearance in the isolated perfused rat liver, while gastrin-releasing peptide (GRP-14 and GRP-27) and somatostatin (S-14 and S-28) are poorly cleared (Doyle *et al.*, 1984; Wolfe *et al.*, 1987; Seno *et al.*, 1985). Hepatic processing of cholecystokinin was studied in more detail (Gores *et al.*, 1986a,b, 1989). Small forms of cholecystokinin (CCK-4 and CCK-8) are extracted more efficiently than are large forms (e.g., CCK-33). Furthermore, hepatic recognition and uptake seem to be dependent upon the carboxyl-terminal tetrapeptide sequence (Trp-Met-Asp-Phe-NH$_2$).

The physicochemical determinants for hepatic clearance of CCK and related compounds were recently studied (Hunter *et al.*, 1990). A series of 13 tetrapeptides, including 8 analogues of the carboxyl-terminal tetrapeptide of CCK-8 but with different charge, hydrophobicity, and amino acid sequence, were investigated. Hydrophobicity of individual radiolabeled peptides was calculated or measured directly by determining their partitioning between octanol and an aqueous phase. First-pass hepatic extraction of the peptides in the isolated perfused rat liver varied from 4% to 86% and correlated significantly with hydrophobicity. Hydrophobic peptides with either positive, neutral, or negative charges were avidly extracted (30% to 86%) by the liver. Interestingly, first-pass clearance of hydrophobic peptides with similar charge varied with amino acid sequence. In contrast, the first-pass hepatic extraction of positively or negatively charged hydrophilic tetrapeptides was negligible ($<$ 10%). These results suggest that hydrophobicity and amino acid sequence, but not anionic or cationic nature, are the major determinants of hepatic extraction of CCK, and perhaps other small circulating peptides.

4.6. Carrier Proteins of the Peptide Uptake Systems

Protein components of these carrier systems have been identified by affinity- and photoaffinity-labeling with chemically reactive or photoreactive bile acid or cyclosomatostatin analogues (Kramer *et al.*, 1982; Wieland *et al.*, 1984; Ziegler, 1989; Ziegler and Frimmer, 1986b; Ziegler *et al.*, 1990, 1991). Proteins of 48–50, 52–54,

67, and 37 kDa were labeled with conjugated and unconjugated bile acid analogues in isolated rat liver cell plasma membranes (Table II) (Ziegler, 1989). Photoreactive cyclosomatostatin analogues bind to proteins of identical molecular masses (Ziegler, 1989). Binding of the photoreactive analogues can be prevented by preincubation with an excess of bile acids or with cyclosomatostatins and vice versa. Studies in other laboratories suggest that the protein subunit of 48–50 kDa is related to the sodium-dependent taurocholate cotransporter (Ananthanarayanan *et al.*, 1988), whereas the protein subunit of 52–54 kDa is involved in the binding of cholate (Kurz *et al.*, 1989). It was shown that cyclosomatostatins have affinity for both systems.

Binding proteins for linear hydrophobic peptides were identified by affinity chromatography on Affi-Gel-15 columns (Bio-Rad Laboratories, Richmond, Calif.) with a hydrophobic linear renin inhibitor as ligand (Sänger, 1989). Under these conditions, proteins of 48, 50, 54, and 58 kDa were also partially isolated (Table II). In contrast, no proteins could be eluted after chromatography of solubilized plasma membrane proteins from AS 30D ascites hepatoma cells (Sänger and Ziegler, 1990). This points again to the specificity of these proteins as components of the transport system.

4.7. Energy-Dependent Carrier-Mediated Peptide Transport Not Related to Bile Acid Carriers

In contrast to the hydrophobic linear renin inhibitors, hydrophilic modifications do not yield substrates for the bile acid carriers (Seeberger and Ziegler, 1990, 1993). These hydrophilic peptides noncompetitively inhibit the uptake of cholate and taurocholate and vice versa. They are transported into hepatocytes by an energy-dependent carrier- mediated mechanism, which is not related to bile acid, amino acid, bilirubin, long-chain fatty acid carriers but rather to cation transporters (see Meijer, 1989; Meijer *et al.*, 1990b). The physiological substrate of this transporter is not known at present; as competitive inhibitors, *d*-tubocurarine and ouabain have been found (Ziegler *et al.*, unpublished observation).

4.8. Peptide Uptake by Passive Lipoid Permeation

Cyclosporins (CSA) (Borel *et al.*, 1976), very hydrophobic cyclic peptides, permeate the membrane by nonionic passive diffusion (Ziegler *et al.*, 1988b). They noncompetitively inhibit the uptake of cholate (Ziegler and Frimmer, 1986c), whereas the uptake of taurocholate is blocked competitively (Kukongviriyapan and Stacey, 1988). In photoaffinity labeling studies, a photoreactive CSA analogue binds to proteins of 85, 54, 50, and 37 kDa (Ziegler and Frimmer, 1986c). This probably means that CSA interacts with bile acid transport systems, but that CSA is not transported by these carrier systems (Fig. 6).

4.9. Current Knowledge and Physiological Meaning of Peptide Clearance by the Liver

As described above, many pharmacologically interesting peptides can be rapidly taken up by the liver without earlier degradation. Depending on the physicochemical properties of the peptides (Table I), energy-dependent carrier-mediated transport processes, nonionic diffusion, or receptor-mediated endocytosis are responsible for their first-pass elimination (see Fig. 6).

Cyclosomatostatins, cholecystokinins, cyclolinopeptides, antamanide, phallotoxins, and hydrophobic linear renin inhibitors are taken up by liver cells using carrier-mediated mechanisms, which are related to the transport of bile acids. There is evidence for the conclusion that various cyclopeptides are substrates of the so-called multispecific cholate transporter, whereas hydrophobic linear renin inhibitors have higher affinity for the sodium taurocholate cotransporter (Table II, Fig. 6). In contrast, the very hydrophobic cyclosporins enter liver cells by nonionic diffusion, whereas the hydrophilic linear renin inhibitors are taken up an energy-dependent carrier-mediated mechanism, which is not related to bile acid carriers.

The overall physiological significance of peptide clearance by the liver is still unclear and remains somewhat speculative. Some investigators have proposed that peptide clearance by the liver may be important for nutrition by processing dipeptides from the portal circulation or from parenteral formulations into assimilable free amino acids (Lochs et al., 1986). Membrane hydrolases are the predominant factors in disposition of such dipeptides (see Chapter 2). Such very small peptides may not be good substrates for the "multispecific transporter," since the majority of possible dipeptide combinations are expected to be too hydrophilic to be effectively cleared by this mechanism. A second physiologically significant role might be in the clearance of circulating peptide hormones. Although a significant first-pass hepatic clearance of CCK-8 was found, CCK-33 was only minimally cleared (Hunter et al., 1990; Doyle et al., 1984). Other investigators found minimal first-pass clearance of S-14 and S-28, GRP-14 and GRP-27, and gastrin-17 (Wolfe et al., 1987; Seno et al., 1985). Thus, for many peptide hormones, particularly those whose chain length exceeds 12–14 amino acids, the liver does not appear to play a significant role in clearance from the plasma (Table I). Although the earlier-mentioned "multispecific transporter" consequently may lack physiological significance in the clearance of long peptide hormones, a number of biologically active, short peptide hormones that exist in the circulation are better substrates (e.g., vasopressin, oxytocin, bradykinin, CCK-8). The major importance of peptide processing by the hepatic bile acid type of transporters therefore may be pharmacologic in nature. In the case of CCK-8, the efficient clearance by the liver prevents spillover to the systemic circulation, also implying that the liver itself may represent a target organ for this peptide.

The transport systems for the uptake of phallotoxins (Petzinger and Frimmer, 1984; Frimmer and Ziegler, 1988), antamanide (Petzinger and Frimmer, 1984; Frimmer and Ziegler, 1988), cyclosomatostatins (Ziegler et al., 1990, 1991), cyclolinopeptides, and linear renin-inhibiting peptides (Bertrams and Ziegler, 1991a,b) are liver- and probably hepatocyte-specific. All substrates of these systems cannot be taken up by other cell types such as AS 30D ascites hepatoma cells or by fibroblasts (Table II).

5. MECHANISMS FOR HEPATIC CLEARANCE OF PROTEINS

5.1. Endocytosis and Phagocytosis

The liver is involved in the disposition of a variety of endogenous and exogenous macromolecules in the form of both particles and soluble species. Solid, often opsonized, particles are removed from the bloodstream by a process called *phagocytosis*. Soluble material, such as lipid particles, immune complexes, bacterial toxins, certain lysosomal enzymes as well as denatured or otherwise abnormal glycoproteins, can be handled by the liver by a versatile apparatus for receptor-mediated uptake and degradation (Forgac, 1988; Shepherd, 1989; Jones and Summerfield, 1988; Burwen and Jones, 1990; Sugiyama and Hanano, 1989).

Among the macromolecules that undergo receptor-mediated endocytosis are growth factors, hormones, transport proteins, proteins destined for degradation (α_2-macroglobulin), proteins to be transported across polarized cells (pIgA) as well as toxins and viruses. Lateral movement leads to clustering of receptor-substrate complex in a specialized coated part of the membrane (coated pits) inducing formation of coated vesicles within the cell (see Fig. 7 and Chapter 4). This process can be partially inhibited by hypotonic media and acidification of the cytosol (Burwen and Jones, 1990; Sugiyama and Hanano, 1989; see Chapter 4). A fundamental difference with carrier-mediated translocation of small organic and inorganic substrates is that following internalization the transported ligand remains segregated from the cytoplasm within vesicular structures. Receptor molecules possess a hydrophilic extracellular (ligand-binding) domain with various glycosylation sites, a hydrophobic membrane-spanning region, and a cytoplasmic tail. The intracellular part may exhibit protein kinase activity (e.g., in the case of insulin and epidermal growth factor) (Burwen and Jones, 1990; Sugiyama and Hanano, 1989). For some receptors (LDL, transferrin, asialoglycoprotein), internalization occurs even without ligand binding (constitutive endocytosis); for others, such as EGF and insulin,

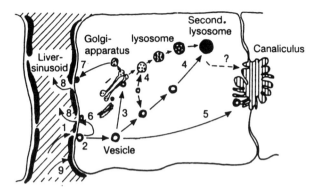

1. Binding to plasma membrane receptors
2. Endocytosis in hepatocytes
3. Vesicle transport to Golgi-system
4. Vesicle transport first and second lysosomes
5. Vesicle transport and diacytosis into bile
6. Recycling and diacytosis at plasma membrane
7. Diacytosis in Golgivesicle
8. Hepatocyte reflux of glycoproteins
9. Endocytosis in sinusoidal cells

Figure 7. Endocytosis of proteins in the hepatocyte and in sinusoidal cell types (black). Receptor-mediated endocytosis (1, 2) direct diacytosis (6), and indirect diacytosis via the Golgi system (3, 7) at the sinusoidal domain of the plasma membrane as well as transcytosis at the canalicular membrane (5) are indicated. Intracellular routing to Golgi system and lysosomes is depicted.

ligand-receptor formation is necessary (triggered endocytosis) (Burwen and Jones, 1990; Shepherd, 1989).

Endocytosis is an energy-requiring process in which the microfilament system is involved in the invagination process (Forgac, 1988; Kirshan, 1975; Jones and Summerfield, 1988; Burwen and Jones, 1990; Sugiyama and Hanano, 1989). The intracellular vesicles formed are uncoated through an uncoating ATPase and form endosomes or phagosomes that are acidified through an ATP-dependent proton pump (Yamashiro and Maxfield, 1984; Yamashiro *et al.*, 1983; see Chapter 3). This process is supported to be essential in the dissociation of the substrate and the receptor (Mellman *et al.*, 1986; Schwartz and Al-Awqati, 1986). From the endosomal compartment, vesicles containing the receptor molecules often recycle to the plasma membrane whereas the endocytosed ligand in the smooth endosomes is trafficked to the lysosomes (Geuze *et al.*, 1983). This acidic organelle contains a variety of aggressive enzymes to degrade proteins, lipids, and oligosaccharides, among others (De Duve and Wattiaux, 1966; see Chapter 3). Some vesicles escape this pathway and return to the plasma membrane. They exocytose their contents back to the plasma in a short-circuit process (Regoeczi *et al.*, 1985; Russel *et al.*, 1983) or diacytose the material at the other pole of the cell (Brown and Kloppel, 1989; Ahnen *et al.*, 1985) to the bile canaliculi (see Fig. 7). The signals for the intracellular sorting and routing remain to be defined (Geuze *et al.*, 1983; Stoorvogel *et al.*, 1989; Shepherd, 1989). Asialoglycoproteins (ASGP), transferrin, and epidermal growth factor (EGF) may be present in the same primary endosomes after internalization, yet ASGP is trafficked to lysosomes and after uncoupling of the ligand its receptor recycles to the plasma membrane. Transferrin recycles completely, still bound to its receptor, whereas much of the EGF receptor ends up in lysosomes associated with its ligand (Marti *et al.*, 1989). These different patterns of uncoupling and sorting may be related to maturation of the early endosomes (Stoorvogel *et al.*, 1989; Shepherd, 1989). Since the average pH of the vesicles initially drops rapidly from 7.4 to 6.3, followed by a further gradual acidification to 5.0, differences in the pH dependence of receptor-ligand dissociation may induce unequal sites of escape from the endosomes and thus unequal probability of recycling and/or modification of the proteins in the trans-Golgi reticulum (Stoorvogel *et al.*, 1989; Shepherd, 1989). Also, the sequence of changes in Ca^{2+} content, membrane potential, and membrane fluidity of the endosomes may underlie such specific receptor- and ligand-specific intracellular routing.

Ligand-receptor dissociation, endosome sorting, and transcellular movements can be largely affected by temperature, metabolic inhibitors, and weakly basic compounds that interfere with acidification of the endocytotic vesicles (NH_4Cl and tertiary amines such as chloroquine). Agents that dissipate H^+ gradients (nigericin and monensin), inhibitors of microfilament function (cytochalasin B) as well as disassemblers of the microtubular tubulin polymers (colchicine and vinblastine) consequently can also largely influence these processes (Forgac, 1988; van der Sluijs *et al.*, 1985; Ohare *et al.*, 1989; Burwen and Jones, 1990; Sugiyama and Hanano, 1989).

5.2. Receptor-Mediated Endocytosis in Various Cell Types of the Liver

A large number of receptor-mediated processes have been identified on hepato-cytes, endothelial cells, and Kupffer cells (Table IV). On hepatocytes one finds the ASGP receptor, recognizing galactose- and *N*-acetylgalactosamine-terminated gly-coproteins (Ashwell and Harford, 1982; Steer and Ashwell, 1986; Schwartz, 1984), a receptor for IgA (called secretory component) (Brown and Kloppel, 1989; Ahnen *et al.*, 1985), in addition to receptors for HDL particles (Havel and Hamilton, 1988; Goldstein and Brown, 1977), EGF (Marti *et al.*, 1989; Burwen *et al.*, 1984), transferrin (Forgac, 1988; Steer and Ashwell, 1986; Brown and Kloppel, 1989), hemopexin, chylomicron remnants, hemoglobin, insulin and peptide hormones (Burwen and Jones, 1990; Sugiyama and Hanano, 1989), as well as for lysosomal proteases (after complex formation with α_2-macroglobulins) (Sottrup-Jensen, 1987; Barret, 1982).

Receptors for immune and complement complexes (Fc and C3 receptors) and for fibronectin were detected on Kupffer cells; these also mediate endocytosis of

Table IV
Receptors for Endocytosis of Proteins in Cell Types of the Liver

Hepatocytes (Parenchymal cells) 78% of liver volume	Kupffer cells (Macrophages) 2.1% of liver volume	Endothelial cells 2.8% of liver volume
Asialoglycoprotein receptor (< 10 nm, galactose recognition)	Agalactoglycoprotein receptor (mannose recognition)	Agalactoglycoprotein receptor (mannose recognition)
Insulin receptor	Galactose-particle receptor (> 10 nm, galactose recognition)	Scavenger receptor (negatively charged proteins)
Epidermal growth factor receptor	LDL receptor	Sulfated polysaccharide receptor (chondroitin sulfate, heparin)
IgA receptor (polymeric IgA)	Macrophage enzyme receptor (tissue-derived enzymes)	Fc recptor (immune complexes)
Transferrin receptor (transferrin/Fe complexes)	α_2-Macroglobulin receptor (α_2-macroglobulin/protease complex)	
LDL receptor	Fibronectin receptor (opsonized material)	
HDL receptor	Fc receptor (immune complexes, opsonized particles) C3b receptor (complement factors)	

opsonized particulate material (Jones and Summerfield, 1988). In addition, a galactose-particle receptor is described, recognizing cells and other particles with a diameter exceeding 10 nm (Kolb-Bachofen et al., 1984) that expose galactose groups. As in other macrophages, a mannose/N-acetylglucosamine/fucose receptor has been identified (Stahl and Schlesinger, 1980) that not only recognizes terminal mannose and N-acetylglucosamine groups on glycoproteins but may also accommodate (neo-)glycoproteins with terminal glucose and fructose groups (Sano et al., 1990; Stahl, 1990; Smedsrød et al., 1990). Apart from this system, specific receptors for fucose-terminated glycoproteins (α_{1-3} fucose residues) were identified on parenchymal cells while on Kupffer cells a receptor binding Fuc-β-albumin was found distinct from the mannose/N-acetylglucosamine receptor (Jones and Summerfield, 1988). Kupffer cells, along with hepatocytes, can also endocytose lysosomal proteases by complex formation with circulating α_2-macroglobulin (Sottrup-Jensen, 1987; Barret, 1982). Finally, Kupffer cells exhibit receptors for positively charged (Bergmann et al., 1984) as well as negatively charged proteins (scavenger receptors) (Smedsrød et al., 1990). Although the above-mentioned receptors on Kupffer cells were once thought to be specific for this cell type, it is now agreed that many of them are also present on endothelial cells (Praaning Van Dalen et al., 1981; Smedsrød et al., 1990; van der Laan-Klamer et al., 1986). The endothelial cells also have receptors for LDL particles (Goldstein and Brown, 1977). Scavenger receptors may even be predominantly localized in the endothelial cells (Praaning Van Dalen et al., 1981; Smedsrød et al., 1990; van Berkel et al., 1986).

In the intact organism, the relative contribution of the various cell types to the clearance of a certain protein will be determined by the receptor density and the exposed surface of the cell types involved, by the acinar localization of the cells, by the presence of competing endogenous substrates and by the dose of the substrate administered. The latter factor should be seen in relation to receptor capacity and affinity (V_{max}/K_m ratio) of the endocytotic processes (van der Sluijs et al., 1985). These parameters are not necessarily constant since downmodulation of the receptors can occur with chronic administration of substrates (Sugiyama and Hanano, 1989; Weiss and Ashwell, 1989). Since various cell types may "compete" for the same substrate, dose dependence should be taken into account in measuring the relative clearance of macromolecules by the various cell types.

Receptor-mediated uptake of macromolecules requires a preceding selective binding of these substances to the plasma membrane of either the hepatocytes or the sinusoidal cell types. The binding is determined by a number of structural properties of the molecule such as size, charge, terminal sugar, and the capability of the molecule to form complexes with other circulating proteins such as α_2-macroglobulin, fibronectin, and immunoglobulins. Progress has been made in characterization of various receptor systems for endocytosis of glycoproteins at the functional and molecular level (see Table IV) (Forgac, 1988). Most of the receptor proteins have been isolated and purified to homogeneity using affinity chromatography and other protein separation techniques. Molecular mass, degree of glycosylation, amino acid composition as well as functional clustering of the receptor complexes determined by radiation inactivation have been reported (Forgac, 1988; Smedsrød et al., 1990; Stahl, 1990). The number of receptor complexes per cell and the maximal internalization rate of the particulate ligands were estimated from kinetic studies. For instance, each rat hepatocyte contains 250,000 dimeric ASGP surface receptors (subunits size ~ 48 kDa) with a K_a for asialoorosomucoid of 10^{-7} M. Each cell can maximally internalize 5×10^6 molecules/hr (Forgac, 1988; Ashwell and Harford, 1982). A portion of the receptor molecules may also be detected in intracellular pools as related to the processes of synthesis, repair, recycling, and degree of downmodulation (Weiss and Ashwell, 1989). It should be realized, however, that remarkable differences in receptor characteristics can occur between in vivo and in vitro (cell culture) conditions (Dunn et al., 1983).

5.3. Particle Endocytosis by Opsonin- and Receptor-Mediated Mechanisms

Experiments with monomers, dimers, and polymers of the enzymes ribonuclease A and lysozyme demonstrated that the rate of uptake of these enzymes by Kupffer cells is positively correlated with the size of these molecules (Jones and Summerfield, 1988; Bouma and Smit, 1988). If solid or particulate material is involved, such processes are called phagocytosis. The recognition of phagocytosable material occurs either by circulating opsonic factors like fibronectin or by the cell membrane receptors directly via carbohydrate interactions. An example of the latter is the galactose-particle receptor (Kolb-Bachofen *et al.*, 1984). Particles must have a minimum size in order to be phagocytosed. For colloidal particles < 10 nm, aggregation by agglutination is necessary. The process can be saturated and after maximal phagocytosis the capacity of the cells for further removal of particles can be blocked for more than 72 hr (Jones and Summerfield, 1988). This transient depression of phagocytic function is due to some kind of defect in membrane attachment and should be distinguished from a decrease in phagocytosis due to low fibronectin levels. The number of functional Kupffer cells rather than the receptor density on the cells seems to determine the total removal capacity (Jones and Summerfield, 1988). The rapid clearance of aggregated molecules has been shown, for instance, in the case of heat-aggregated serum albumin (Jones and Summerfield, 1988).

Endocytosis of aggregates and particles by the Kupffer cells, which is often related to thrombotic processes, can be mediated by a nonspecific opsonic serum protein (fibronectin). Fibronectin is distinct from other opsonic factors such as immunoglobulins and C3b, which exert opsonic activity in the process of bacterial phagocytosis. It is assumed that fibronectin binds to circulating aggregates and particles, and that this bound opsonic serum protein acts as a recognition site for a receptor on the Kupffer cells (Smedsrød *et al.*, 1990).

Kupffer cell function *in vivo* is usually tested with probes such as radioiodinated microaggregated human serum albumin. At a dose of 50 mg/kg body weight, clearance is so efficient that it reflects hepatic blood flow. At a 100-fold higher dose, the removal process is saturated and under such conditions is a measure of total Kupffer cell phagocytic function (Jones and Summerfield, 1988). IgG- or IgM-coated and ^{51}Cr-labeled erythrocytes can be used to test Fc receptor-mediated and C3-mediated clearance, respectively. Abnormalities in such functional tests with various liver diseases may, apart from portosystemic shunting, be due to elevated bile acid levels or circulation of abnormally immunoreactive immunoglobulins (Jones and Summerfield, 1988).

Complex formation of a circulating macromolecule with the plasma protein α_2-macroglobulin is a mechanism by which a group of enzymes (mainly proteases) are cleared from the circulation and taken up by hepatocytes and by Kupffer cells. The binding is in principle competitive and sterically hinders the enzymatic activity of the enclosed proteinase. Human α_2-macroglobulin is a protein of about 725 kDa and consists of four identical subunits. One molecule is able to bind two enzyme molecules (Sottrup-Jensen, 1987; Barret, 1982). Each subunit of the α_2-macroglobulin molecule possesses a "bait region" that is susceptible to limited proteolysis. When cleaved by a proteinase, the α_2-macroglobulin subunit changes shape in such a way that the proteinase molecule itself is irreversibly trapped. Proteinase larger than 90 kDa does not fit in, some smaller ones such as urokinase, kallikrein, and renin are excluded due to their specificity, while proteinases smaller than 10 kDa can more or less freely diffuse out of the closed trap. Complex formation decreases the half-life of α_2-macroglobulin drastically due to conformational change in the α_2-macroglobulin (Sottrup-Jensen, 1987; Barret, 1982). After trapping a proteinase, the molecule becomes more compact as demonstrated by pore-limited gel electrophoresis and dichroic spectra (Sottrup-Jensen, 1987; Barret, 1982). After internalization, the macroglobulin is proteolysed with the accompanying enzymes; however, the receptor recycles to the membrane and can be reutilized during various cycles (Kaplan and Keogh, 1983). Clearance of the particular complexes is not influenced by the presence of mannose-terminated glycosidases and it is very likely that independent processes are involved in the removal of these types of lysosomal material (Kaplan and Keogh, 1983). The uptake of enzyme-α_2-macroglobulin complexes by the Kupffer cells shows some similarity with the uptake of antigen-antibody complexes. In the latter case, the binding of antigen induces a conformational change in the Fc part of the antibody and this modified Fc part can now be recognized by receptors on the Kupffer cells (Givol et al., 1974; Smedsrød et al., 1990). Tumor-specific antibodies can be rapidly removed especially in tumor-bearing individuals through complex formation with circulating tumor antigens via Fc receptor systems.

5.4. Endocytosis of Polymeric IgA

In rat liver, IgA is efficiently transported from plasma into bile (Burwen and Jones, 1990; Brown and Kloppel, 1989; Ahnen et al., 1985). Especially polymeric IgA is taken up into hepatocytes after specific receptor recognition on the sinusoidal plasma membranes. In this case, the endocytosed material completely escapes the lysosomal route and reaches the opposite pole of the cell (transcytosis). This transport system is mediated by a plasma membrane glycoprotein called secretory component which is synthesized in the hepatocytes. After association of this secretory component with polymeric IgA, it is transported in a vesicular complex to the biliary pole of the hepatocyte followed by a discharge of the vesicles into the bile canicular lumen. Because of a protective coat, the vesicle contents escape the lysosomal system, and unchanged IgA, still complexed to secretory component, is excreted in bile. However, during passage through the cell, the IgA receptor itself is proteolytically cleaved by some extralysosomal degradation pathway. The ligand-binding domain, however, is spared and secreted together with IgA (Brown and Kloppel, 1989; Ahnen et al., 1985). IgA-antigen complexes could also be removed from the circulation in this manner and this may have an important role in limiting the systemic immune responses to antigens absorbed from the gut (Brown and Kloppel, 1989). In the human, IgA transport to bile may partly occur through transcytosis across the biliary epithelial cells in which secretory component is expressed as well (Brown and Kloppel, 1989). Interestingly, hepatic endocytosis and exocytosis of IgA in bile in humans may partly occur via interaction with the ASGP receptor system in hepatocytes rather than through secretory component in this cell type (Brown and Kloppel, 1989; Burwen and Jones, 1990).

5.5. Endocytosis of Proteins with a Net Positive Charge

Positively charged enzyme molecules are also recognized by the Kupffer cells of the liver. Experiments with the isozymes of lactate dehydrogenase and other tissue-derived enzymes demonstrated that the positive charge of the enzymes is an important feature in the rapid uptake by the Kupffer cells. Acetylation of amino groups of such enzymes was shown to result in loss of positive charge and decrease of plasma clearance rate (Bouma and Smit, 1988). Endothelial cells of several tissues bear anionic sites on their plasma membrane (Ghitescu and Fixman, 1984). Proteins with increased isoelectric points (due to modifications with positively charged moieties) may bind to these sites, and could subsequently be internalized (Bergmann et al., 1984). If human serum albumin is cationized with hexanediamine labeled with ^{125}I and subsequently injected into rats, the modified albumins are rapidly cleared from the plasma at a rate that increases with their isoelectric points (Bergmann et al., 1984). Protein-associated ^{125}I concentration is greatest in the liver, presumably in the sinusoidal cells. At the same time, positively charged proteins may also escape the general circulation by glomerular filtration since the negatively charged glomerulus favors passage of such proteins as has been demonstrated for albumin derivatives (Bergmann et al., 1984).

5.6. Receptors for Proteins with a Net Negative Charge

Various scavenger receptors for negatively charged proteins have been found on both macrophages and endothelial cells in the liver (Smedsrød et al., 1990; van Berkel et al., 1986; Jones and Summerfield, 1988; Goldstein and Brown, 1974; Praaning van Dalen et al., 1981). Such receptors are sensitive to polyanionic compounds such as dextran sulfate, fucoidin, polyinosinic acid, and poly-L-glutamic acid but not to the same extent (Horiuchi et al., 1987). LDL and albumin modified with anionic reagents were also shown to bind avidly to scavenger receptors followed by receptor-mediated endocytosis of the polyanionic molecules. The presence of numerous coated regions on the strategically oriented endothelial cells, that expose cationic residues, favors endocytosis of various connective tissue macromolecules (i.e., hyaluronan and chondroitin sulfate) as well as procollagen peptide (Smedsrød et al., 1990). Elevated levels of such macromolecules during liver disease may, apart from increased synthesis, result from a dysfunction of endothelial cells. A similar process may be responsible for the sinusoidal cell uptake of fluoresceinated proteins (van der Sluijs *et al.*, 1986) (see Fig. 9). Reaction of FITC with peptidyl lysine, for instance, in albumin, neutralizes the positive charge of the ε-amino group of this amino acid. Extra negative charges are provided by the carboxyl group of fluorescein and dissociation of an aromatic hydroxyl group in the dye. Maximally, a charge alteration of up to 3 results after reaction of 1 mole of lysine with 1 mole of FITC. The polyanion dextran-500-sulfate almost completely inhibited the internalization (van der Sluijs *et al.*, 1986; Meijer and van der Sluijs, 1989) and degradation of ^{125}I-labeled fluoresceinated albumin, indicating that a scavenger system is involved in the clearance process. If drugs with negatively charged groups are coupled to proteins, they may consequently alter the cell specificity of the particular carrier (van der Sluijs *et al.*, 1986; Meijer and van der Sluijs, 1989; Meijer *et al.*, 1990a). The mechanism by which such chemical derivatizations of proteins induce a scavenger type of clearance remains to be established (Haberland and Fogelman, 1985). Modification of BSA with malondialdehyde or maleic anhydride produces molecules recognized by scavenger receptors, a phenomenon that, apart from modification of the overall charge of the protein, is due to an altered conformation of the protein. This modification even persists after removal of the maleyl groups from the protein and in itself improves the interaction of the reacted albumin with the scavenger receptor (Haberland and Fogelman, 1985). Derivatization of functional groups in proteins with organic compounds generally increases hydrophobicity of the protein, a modification that may lead to an intensified interaction with hydrophobic regions on endothelial cells (Wright *et al.*, 1988).

The removal of nonenzymatically glycosylated proteins that are normally present in the plasma and especially during diabetic conditions can be blocked by aldehyde-modified proteins (Takata *et al.*, 1988). That such negatively charged proteins inhibit the clearance of glycosylated proteins may indicate the involvement of some sort of scavenger system in the removal of the latter products (Monnier, 1990; Vlassara *et al.*, 1989). Advanced glycosylation end products (AGE proteins) as well as particles with such proteins at their surface are scavenged by a high-affinity 90-kDa receptor identified on several types of macrophages which is instrumental in the removal of senescent macromolecules (Monnier, 1990). Internalization via the AGE protein receptor is accompanied by release of tumor necrosis factor and interleukin-1, substances that also upregulate this receptor in contrast to insulin, which induces a downregulation (Vlassara *et al.*, 1989).

Recent studies indicate that at least two different (more general) scavenger receptors are present in the liver: one on endothelial cells and one on Kupffer cells. Formaldehyde-treated albumin contains a monomeric form that is predominantly cleared by endothelial cells and a polymeric (aggregated) form that is mainly endocytosed by Kupffer cells (Jansen *et al.*, 1991b). A similar conclusion was reached on the basis of clearance of various forms of modified lipid particles (Bijsterbosch and van Berkel, 1990) and studies demonstrating that acetylated LDL and formaldehyde-treated albumin do not mutually interact (Horiuchi *et al.*, 1987).

Glutaraldehyde-treated albumin (Michalak and Bolger, 1989) also has a high affinity for hepatocytes and hepatitis B virus particles (surface antigen). Cross-linking of amino groups of lysine and carboxamide groups of glutamine in the albumin molecule may play a role (Thung *et al.*, 1989). This modification can also be produced by transglutaminase treatment (Thung *et al.*, 1989) and the products may provide an attachment point for viral entry into hepatocytes (Michalak and Bolger, 1989). Yet most of the injected glutaraldehyde-treated albumin is endocytosed by endothelial and Kupffer cells by their scavenger receptors (Wright *et al.*, 1988).

5.7. Galactosyl (Asialoglycoprotein) Receptors for Endocytosis in Hepatocytes

Many enzymes, acute-phase proteins, and most plasma proteins of the immune system are glycoproteins. The principal sugars forming the oligosaccharide chains of the glycoproteins are mannose, *N*- acetylglucosamine, galactose, and sialic acid (*N*-acetylneuraminic acid) (Forgac, 1988; Steer and Ashwell, 1986; Ashwell and Harford, 1982). The basic structure of the oligosaccharide of asialoorosomucoid, a glycoprotein used in many experimental studies, is depicted in Fig. 8. The basic concept is that the terminal sialic acid residues on the carbohydrate moieties of glycoproteins are essential for the normal survival of these agents in the circulation. A glycoprotein that is desialylated is cleared much more rapidly from the circulation

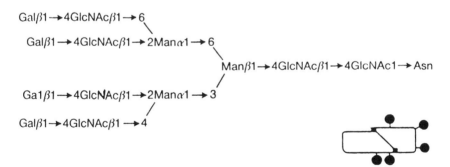

Figure 8. Characteristic structure of an oligosaccharide side chain of naturally occurring glycoproteins, such as α_1-acid glycoprotein (orosomucoid) and fetuin. Normally the major part of the galactose groups is connected to *N*-acetylneuraminic acid (sialic acid). However, microheterogeneity in the antennary structure and in sialic acid composition exists under pathological conditions and acute-phase reactions. The inset shows localization of five of these chains at the orosomucoid polypeptide moiety, having two domains connected with disulfide bridges.

than is the corresponding native glycoprotein (Steer and Ashwell, 1986; Schwartz, 1984; Ashwell and Harford, 1982). The galactosyl residues of the glycoproteins, exposed by desialylation, are the functional groups that mediate this specific and rapid uptake. These sugar groups are recognized by a specific receptor on the plasma membrane of the hepatocyte, designated as hepatic binding protein. The crucial role of the terminal galactosyl residues was demonstrated by showing a markedly decreased clearance after the galactose moiety is modified by treatment with galactose oxidase. Similar results have been obtained with a large number of plasma glycoproteins like orosomucoid, fetuin, lactoferrin, α_2-macroglobulin, and haptoglobulin (Steer and Ashwell, 1986; Schwartz, 1984; Ashwell and Harford, 1982).

The physiological role of the hepatic binding protein of the hepatocytes in the catabolism of circulating plasma glycoproteins remains a matter of speculation. Neuraminidase is widely distributed in the intact animal and it is certain that desialylation of glycoproteins occurs continuously *in vivo* (Irie and Tavassoli, 1986; Schwartz, 1984). Variable quantities of ASGPs have been detected in the plasma of normal subjects. It has been found that in patients with liver disorders like cirrhosis, the plasma concentration of ASGPs are appreciably higher than in normal subjects (Schwartz, 1984; Burwen and Jones, 1990; Steer and Ashwell, 1986; Ashwell and Harford, 1982).

The dimeric ASGP receptor or hepatic binding protein was isolated from rat liver, rabbit liver, and human liver. It has been identified as a water-soluble glycoprotein, in which at least 25% of the dry weight is composed of sialic acid, galactose, mannose, and glucosamine (Forgac, 1988; Stockert and Morell, 1983). This purified hepatic lectin is able to agglutinate erythrocytes, and induce mitogenesis and cytotoxicity of lymphocytes against hepatocytes (Stockert and Morell, 1983). The presence of Ca^{2+} is essential for the binding properties. The binding capacity of hepatic lectin depends upon the presence of two sialic acid-terminated oligosaccharide chains in each of the receptor protein moieties. The binding activity of the receptor is lost after neuraminidase treatment and is restored by resialylation (Forgac, 1988; Steer and Ashwell, 1986). It is assumed that the inactivation of the hepatic lectin after desialylation is caused by its binding to its own exposed galactosyl residues (Weiss and Ashwell, 1989; Steer and Ashwell, 1986). Downmodulation of the receptor can therefore occur through partial desialylation (Weiss and Ashwell, 1989). Recycling of the receptor following internalization takes about 5 min for each roundtrip. Since the half-life of the receptor is 20 hr, several hundred rounds of endocytosis are possible. The total process from binding of a ligand to the cells to proteolytic digestion may take up to 20 min. There is recent evidence, however, that at least two different intracellular pathways exist for internalized ASGP, of which one is particularly sensitive to colchicine. This indicates that in this pathway, microtubules are involved in the delivery to degradative compartments and therefore exhibit a degradation lag time of 15–20 min in contrast to the other pathway, which has a much shorter lag time (Clarke and Weigel, 1989). Degradation of ASGPs in lysosomes can be effectively and dose-dependently blocked by the protease inhibitor leupeptin (a tripeptide). Lysosomal degradation of fluoresceinated asialoorosomucoid can be followed by the biliary excretion of the fluorescein-lysine degradation product (see Fig. 9) since protein degradation is the rate-limiting step in the biliary output of the fluorescent material. ASGP receptor activity and expression are modulated by various physiological and pathological factors as well as by treatment with certain hormones and drugs. Binding affinity for the receptor is reduced in fetal and neonatal liver, in regenerating liver, in chemically induced hepatocarcinoma, and in senescence (Devirgiliis *et al.*, 1989). Treatment with estrogens changes intracellular routing. Instead of 3–5%, up to 25% of the ASGP dose is excreted in the bile whereas biliary output of IgA is reduced (Goldsmith *et al.*, 1987).

The distribution of the receptor system on the cell surfaces along the sinusoids is not well known. Hepatocytes are heterogeneous in their transport (Groothuis *et al.*, 1985) and metabolic (Gumucio and Chianale, 1988; Gumucio, 1989; Jungermann, 1986; Thurman and Kauffman, 1985) functions. Some earlier reports provided evidence for a heterogeneous acinar binding of asialoalkaline phosphatase (Hardonk and Scholtens, 1980) and EGF (see Burwen and Jones, 1990) in rat liver. The acinar distribution of the ASGP processing system was investigated more recently using

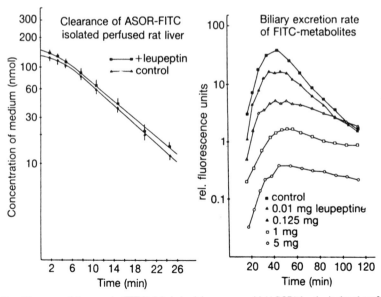

Figure 9. Clearance of fluorescein (FITC)-labeled asialoorosomucoid (ASOR) by the isolated perfused rat liver. The covalently bound fluorescein can only be released through proteolysis of the protein following uptake. The tripeptide leupeptin does not affect uptake into the liver but dose-dependently inhibits biliary excretion of fluorescein metabolites through inhibition of the rate-limiting proteolytic process in lysosomes (van der Sluys *et al.*, 1905).

[^{125}I]asialoorosomucoid in anterogradely and retrogradely perfused rat livers, by performing quantitative autoradiography (van der Sluijs *et al.*, 1988; Burwen and Jones, 1990). A gradient was observed descending from zone 1 to zone 3 cells upon perfusion in the normal direction. The zone 3-to-zone 1 gradient observed in the retrograde perfusions was significantly steeper than the 1-to-zone 3 gradient in anterograde perfusions, indicating a higher intrinsic activity of endocytosis in zone 3 (van der Sluijs *et al.*, 1988). Thus the "unfavorable" position of zone 3 cells in the acinus seems to be partly compensated for by a higher endocytotic activity. By virtue of this high intrinsic activity, perivenous cells may also have the opportunity to minimize spillover to the general circulation of desialylated glycoproteins produced by endothelial cells (van der Sluijs *et al.*, 1988; Irie and Tavassoli, 1986). Both differences in sinusoidal anatomy and intrinsic cellular function may underlie the normal zone 1-to-zone 3 gradient. It is not likely that zonal differences in fenestral diameter and the 30% higher porosity of the sinusoidal lining in zone 3 (Wisse *et al.*, 1985) have an influence on the uptake rates of the relatively small glycoprotein molecules (Meijer and van der Sluijs, 1989). Rather, the greater tortuosity and higher surface-to-volume ratio of the sinusoids in the periportal area (Gumucio and Chianale, 1988; Gumucio, 1983) tend to favor the uptake rate in zone 1, and the intrinsic endocytotic activity of zone 3 cells mentioned above may in fact have been somewhat underestimated (van der Sluijs *et al.*, 1988).

5.8. Galactosyl Receptors in Other Cell Types

Hepatic macrophages and endothelial cells also have been shown to contain an N-acetyl-D-galactosamine/D-galactose-recognizing particle receptor (Kolb-Bachofen *et al.*, 1984; van Berkel *et al.*, 1986). Early *in vitro* studies with isolated Kupffer cells revealed that this receptor can mediate adhesion of neuraminidase-treated erythrocytes and lymphocytes (Kolb-Bachofen *et al.*, 1984; Kolb-Bachofen, 1986). Although the sugar specificity of the Kupffer cell ASGP receptor appeared similar to the one on hepatocytes, these lectins do not seem to be immunologically related (Kolb-Bachofen *et al.*, 1984; Kolb-Bachofen, 1986). The preclustered arrangement of the Kupffer cell ASGP receptor in coated pits enables it to bind and endocytose ASGP-coated particulate material (with a size > 10 nm), which is refractory to internalization via the hepatocyte receptor. It is now clear that the galactose particle receptor can also mediate internalization of soluble ASGPs, although not as efficiently as the hepatocyte ASGP receptor (Praaning Van Dalen *et al.*, 1981; van Berkel *et al.*, 1986). Apart from the Kupffer cell ASGP receptor, other galactose-recognizing receptors have been identified in bone marrow and macrophages (Regoeczi *et al.*, 1980; Samlowski *et al.*, 1985). Regoeczi and colleagues (Regoeczi *et al.*, 1980) suggested that rabbit bone marrow tissue contains a receptor recognizing ASGPs with at least two biantennary asialoglycans, such as occur in human asialotransferrin type 1. Galactosylated bovine serum albumin containing more than 38 galactose molecules per molecule of protein, was bound to and internalized by endothelium of bone marrow (Kataoka and Tavassoli, 1985). The various galactosyl receptors in liver, macrophages, and extrahepatic endothelia should be more precisely characterized in terms of their binding requirements and sugar specificity.

5.9. Mannosyl Receptors for Endocytosis in Sinusoidal Cell Types

The ASGP and galactose particle recognition systems are not the sole mechanism by which glycoproteins are cleared from the circulation by the liver. A hepatic mannose/N-acetylglucosamine/fucose recognition system has been defined in various species (Jones and Summerfield, 1988; Burwen and Jones, 1990; Stahl, 1990). After intravenous infusion or local release, many lysosomal hydrolases are rapidly cleared from the circulation by the liver. Uptake of these enzymes can be abolished by pretreating them with sodium periodate, which oxidizes sugar residues, but not by intravenous administration of an excess of ASGP. In contrast, the administration of agalactoglycoproteins abolishes the uptake of these enzymes by the liver. This "selective" uptake is based on recognition of terminal N-acetylglucosamine, fucose, glucose, or mannose groups by the Kupffer cells and endothelial cells. This particular receptor has been identified by electron microscopic and autoradiographic studies (Hubbard *et al.*, 1979) and the protein has been isolated, purified, and chemically characterized (Stahl and Schwartz, 1986; Stahl and Schlesinger, 1980; Stahl, 1990). Binding and internalization studies indicate that the mannosyl receptor is both pinocytotic and phagocytotic, but it is not clear whether this involves the same population of receptors. There is recent evidence that the mannosyl receptor on endothelial cells may differ from that on Kupffer cells in that internalization rate in these cell types is different (Sano *et al.*, 1990). The mannosyl receptor is a 162-kDa glycoprotein with high-mannose chains that binds mannosylated BSA with a K_d of 5 nM. The mannosyl receptor recycles every 6–8 min so that, taking into account its half-life of several days, it recycles hundreds of times (Stahl, 1990). Receptor binding is Ca^{2+} dependent and optimal at pH 7.0, and decreases strongly as the pH is lowered to 5. The mannosyl receptor is not found on circulating monocytes but only on differentiated macrophages. Interferon-τ downregulates the receptors while glucocorticosteroids and vitamin D accelerate its synthesis (Stahl, 1990). The mannosyl receptor is, apart from the clearance of soluble proteins, also involved in the scavenging of microorganisms coated with mannose, glucose, etc. This cellular defense system operates in concert with the humoral system of circulating mannose-binding protein, which is considered to be an acute-phase protein (Stahl, 1990; Gordon and Rabinowitz, 1989). This protein is structurally related to the mannosyl receptor and very likely opsonizes organisms with exposed mannose or N-acetylglucosamine such as the human immunodeficiency virus (Stahl, 1990).

5.10. Influence of Sugar Density and Clustering on Endocytosis

Efficiency of endocytosis and further degradation of glycoproteins can be largely influenced by the number, density, and thus by the clustering of the sugar groups on the protein molecule. That sugar clustering is important can be based on earlier reports on naturally occurring glycoproteins containing the typical tetra-, tri-, bi-, or monoantennary oligosaccharide structure. In hepatocytes, triantennary ASGPs have a much higher affinity for the hepatic lectin than the bi- and monoantennary derivatives (Baenziger and Fiete, 1980; Conolly et al., 1982). The latter may partly undergo "short-circuit" exocytosis and after primary endocytosis are again released from the cells in intact form (Townsend et al., 1984; Evans, 1981; Regoeczi et al., 1985). This may lead to repeated cycles of endocytosis and exocytosis and an overall slow clearance of the macromolecule. Efflux of glycoproteins from the cells is also stimulated by high concentrations of bile salts (Russel et al., 1983), N-acetyl-galactose, and EGTA (Townsend et al., 1984) without destruction of the cells. Recycling of (asialo-) glycoproteins was suggested to be related to some kind of repair through resialylation in the Golgi system, a process that may occur only at a low concentration range of such glycoproteins (Evans, 1981; Regoeczi et al., 1985).

Of the natural mannose-terminated glycoproteins, only the tetraantennary structures containing proteins are endocytosed (Jones and Summerfield, 1988; Taylor et al., 1987; Maynard and Baenziger, 1981). Receptor affinity of complex oligosaccharides and efficiency of receptor-mediated endocytosis is much higher than for neoglycoproteins with a similar number of monosaccharides. A high density of the sugar molecules and thereby the possibility of binding to several receptor sites clustered in coated pits may explain the more efficient triggering of endocytosis. The presence of specific sugar groups on a protein molecule can also inhibit lysosomal proteolysis (Jones and Summerfield, 1988; Sano et al., 1990; Taylor et al., 1987). Density and clustering of the sugar groups and microheterogeneity of the oligosaccharide chains will therefore largely affect cellular degradation rate. Consequently, the composition as well as the geometry of the glycoprotein structure will influence the overall involvement of certain cells in the disposition of a particular glycoprotein.

In principle, neoglycoproteins prepared by reductive amination yield at random glycosylations in the protein molecule (Schwartz and Gray, 1977). When a critical galactosyl density is exceeded, such glycoproteins become substrates for the ASGP receptor in hepatocytes. At least 13 moles of galactose groups per mole of albumin is necessary for binding to hepatocyte membranes (Vera et al., 1984). In general, the equilibrium binding constant increases with sugar density. Neoglycoproteins prepared by reductive amination, however, clearly lack the clustered galactose groups present at the penultimate part of the oligosaccharide structures of naturally occurring glycoproteins. This may largely influence internalization rate as observed for tracer doses of ^{125}I-labeled asialoorosomucoid and ^{125}I-labeled lactosaminated albumin with 25 sugars (L_{25}HSA) (van der Sluijs et al., 1985, 1986, 1987, 1988). Half-time for internalization in perfused rat liver was 3.4 min for asialoorosomucoid and 34.9 min for the lactosaminated serum albumin. Conceivably, neoglycoproteins which are prepared by reductive amination (Schwartz and Gray, 1977) or amidination (Stowell and Lee, 1980) with simple sugars have a lower affinity for the ASGP receptor. On the other hand, neoglycoproteins prepared through thiophosgene activation of p-aminophenyl sugars (Kataoka and Tavassoli, 1985) will contain extra negative charge and this may lead to partial capturing by scavenger-like receptors on nonparenchymal cell types (Jansen et al., 1991a). The extent of endocytosis of such glycoproteins by nonparenchymal cells is probably inversely related to the relative affinity of the particular neoglycoprotein for the ASGP receptor in the hepatocytes (Meijer and van der Sluijs, 1989). Fiume et al. (1987b) prepared Ara-AMP conjugates of [^{3}H]-L_{30}HSA and demonstrated that, upon injection in rats, the concentration in hepatocytes and that in sinusoidal cells were about equal. The authors compared the distribution of the conjugates with that of the nonderivatized carriers and showed that conjugation of L_{30}HSA with 12 molecules of Ara-AMP resulted in a 7-fold increase in sinusoidal cell uptake, whereas this was only 1.5-fold for the high-affinity hepatocyte-specific asialoorosomucoid carrier.

In studies on the disposition of lipoproteins (van Berkel et al., 1986), attempts were made to decrease plasma low-density lipoprotein (LDL) concentrations by modifying the particles with tris-galactosyl-cholesterol (TGC), in order to reshape them to suitable ligands for the hepatocyte ASGP receptor. The design of this molecule was based on the consideration that the hepatocyte ASGP receptor binds and internalizes triantennary structures with a significantly higher efficiency than bi- or monoantennary glycoproteins (Ashwell and Harford, 1982; Baenziger and Fiete, 1980; Conolly et al., 1982). After injection of these modified LDL particles, a higher clearance was found. The Kupffer cell population was shown to be predominantly responsible for this increase in uptake and subsequent degradation. Similar experiments were done with TGC-modified high-density lipoprotein (HDL), but in this case the ligand was almost exclusively taken up in hepatocytes (van Berkel et al., 1986). This differential intrahepatic targeting was achieved by the selective interaction of the large (23 nm) LDL particles with the Kupffer cells. Liposomes in which lactosylceramide (Roerdink et al., 1988; Soriano et al., 1983) or other galactolipids were incorporated revealed a strikingly stimulated plasma clearance and hepatic uptake in the rat in vivo. Isolation of various cell types showed that this increase largely occurred in Kupffer cells through recognition by the N-acetyl-D-galactosamine/D-galactose particle receptor (Bijsterbosch and van Berkel, 1989). The smaller (10 nm) galactose-derivatized HDL particles preferentially bind to the hepatocyte ASGP receptor. Another possibility of changing the hepatic fate of LDL is by modifying the apoprotein β recognition part of LDL with sugars to gain extra cell specificity (Attie et al., 1980; Bijsterbosch and van Berkel, 1990). A relatively low degree of lactosylation (around 60 lactose/LDL) directs most of the particles to parenchymal cells. At a lactosylation degree of 300, the main site of uptake is in the Kupffer cells despite their much smaller total cellular volume (Bijsterbosch and van Berkel, 1990).

The disadvantage of this LDL preparation similar to liposomes is that the endothelial lining of blood vessels cannot be passed and application in drug delivery is restricted to blood cells and cell types in direct contact with blood.

6. HEPATIC CLEARANCE OF THERAPEUTIC PROTEINS AND OLIGOPEPTIDES: AN INTEGRATION OF CONCEPTS

6.1. Chemical and Physiological Factors

Many potential therapeutic proteins are either under investigation or have been introduced in the clinic. Modification of the structure of peptidyl lead compounds with the aim of changing the pharmacological activity may at the same time largely affect the pharmacokinetic behavior of the particular molecules. Such modifications include amino acid substitution, deletions, and additions, as well as cyclization, drug conjugation, and (de)glycosylation. In this manner, size, overall charge, charge distribution, spatial conformation as well as hydrophilicity-hydrophobicity balance can be altered with major consequences for the distribution and elimination processes that determine their fate in the body. For instance, even natural and recombinant proteins can largely differ in their biological disposition. Among the class of therapeutic proteins are enzymes, hormones, hormone releasing factors, interferons, hybrid proteins, monoclonal antibodies, muramyl peptides, and immunosuppressive peptides such as cyclosporin. The selected peptides given in Table I demonstrate that major differences in amino acid number, overall conformation, hydrophobicity, and molecular mass occur in this category of agents. It remains a challenge to predict the pharmacokinetic profiles of the largely varying peptides on the basis of such structural and physicochemical properties (see Section 4.5 and Fig. 10). Questions to be answered are: What is their distribution volume *in vivo*? What are the physiological barriers for penetration into organs and tissues? And what is the relative contribution of systemic degradation, hepatic and renal clearance to their removal from the body? As dealt with in the preceding sections, a number of physicochemical and physiological factors may determine the pharmacokinetics of oligopeptides and proteins *in vivo* (see Table V):

a. Are the available compounds to be injected in the monomeric, polymeric, or aggregated form or are they mixtures of these forms?

b. What is the source of the particular peptide (species, natural or recombinant, presence of chemical coating, isotopes, or fluorescent labels)?

c. If introduced into the bloodstream, does the injected peptide remain in the same state or is it effectively opsonized (antigens, monoclonal antibodies, proteases)?

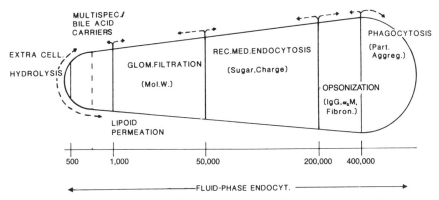

Figure 10. Clearance mechanisms for oligopeptides and proteins in the intact organism. Dependent on size, charge, sugar recognition, aggregation to particles, and formation of complexes with opsonication factors, peptides can be filtered in the kidneys, or hydrolyzed in the blood or at cell surfaces, or recognized by carrier- or receptor-mediated transport systems. Molecular weight as indicated on the x axis roughly determines the clearance mechanisms. However, the indicated mechanisms largely overlap and are dependent not only on size but also on functional groups (sugars), charge, lipophilicity, and vulnerability for circulating or fixed proteases. Fluid-phase endocytosis in principle occurs over the entire molecular weight range.

d. If such compounds are taken up by endocytotic or carrier-mediated mechanisms, are endogenous inhibitors present under physiological and pathological conditions?

e. Related to c and d: is the pharmacokinetic profile time- dependent (presence in the plasma of bile acids, acute-phase proteins, drugs)?

f. Is disposition of the type of peptide known to be largely species-dependent?

g. Is the compound stable in the bloodstream and does the bioassay used specifically determine functionally active peptide?

h. Does the molecular size and lipophilicity of the peptide predict major affinity for the carrier-mediated uptake mechanisms in the liver?

i. Does the peptide contain oligosaccharide chains important for endocytosis and, if so, what is (are) the terminal sugar(s)?

j. Is the peptide small enough to undergo glomerular filtration and is urinary excretion masked by extensive tubular reabsorption?

k. Are polyanionic or polycationic features present that invite adsorptive endocytosis?

Some small physiological oligopeptides (see Table I) that are often locally released may undergo extremely rapid extracellular hydrolysis in the bloodstream (Fig. 10). Gastroenteric as well as many artificial oligopeptides with 3–12 amino acids may exhibit high affinity for various carrier-mediated uptake systems in the hepatocytes,

Table V

Factors in Clearance of Proteins and Peptides

Physicochemical factors
  Molecular size (glomerular filtration in kidney, carrier/receptor-mediated clearance in liver)
  Net charge and charge distribution (glomerular filtration, reabsorption in kidney, endocytocsis in liver)
  Hydrophilicity/hydrophobicity balance (affinity for carriers in liver)
  Glycosylation (terminal sugar, number/density sugars, endocytosis)
  Amino acid composition (substitution, deletions, enantiomeric form)
Biopharmaceutical factors
  Monomeric/polymeric, aggregated form?
  Conformation of the protein (cyclization, branching)
  Source of the protein, preparation method (species, cell type, recombinant/natural)
  Re-(de-) glycosylation
  Chemical protectants (hybrid proteins, coatings, protein conjugates)
Physiological factors
  Opsonization (IgG, fibronection, α_2-macroglobulin)
  Species differences in kinetics
  Extracellular degradation (bloodstream, cell sufaces)
  Endogenous competing substrates (receptor/carrier level, proteolytic enzymes)

especially when they display sufficient hydrophobicity (Hunter *et al.*, 1990) (see Section 4.2). Very lipophilic proteins may even be rapidly removed by passive lipoid diffusion into the cells of the eliminating organs followed by proteolytic degradations (see Section 4.8). It should be emphasized, however, that no simple rules exist here with regard to relation between parameters such as partition coefficients and rate of membrane permeation of peptides. *n*-Octanol/water partition, for instance, seems to be a poor predictor in the case of peptides because of the influence of solvated amide bonds and the presence of terminal charges on zwitterionic peptides (Ho *et al.*, 1990). Interestingly, some hydrophilic peptides also have been shown to be avidly taken up in hepatocytes via a saturable cation transport mechanism (Seeberger and Ziegler, 1990, 1993). Oligopeptides with 10 to 30 amino acids may possess a relatively low affinity for the above-mentioned carrier systems in the liver and, consequently, may be cleared predominantly by the kidneys (see Section 4.5) (Takakura *et al.*, 1990). If (roughly speaking) the molecular size of proteins exceeds 50 kDa, glomerular filtration is no longer possible (Takakura *et al.*, 1990) and such substrates can, in principle, only be removed by endocytotic mechanisms either by substrate-specific receptors (insulin) or by more general receptors (sugar recognizing) (see Section 5.2). If a critical number of exposed sugar groups (mannose, galactose, fucose, *N*-acetylglucosamine, *N*-acetylgalactosamine, or glucose) is exceeded, receptor-mediated endocytotic processes in the liver provide an efficient removal system (Fig. 10). In addition, complexation with α_2-macroglobulin, or with immunoglobulins may also lead to rapid removal of soluble proteins (see Sections 5.3 and 5.4). Proteins modified with simple aldehydes or bifunctional reactants may be phagocytosed either after opsonization, through absorptive endocytosis, or by scavenger systems depending on their overall charge, size, and aggregation state (see Sections 5.5 and 5.6). Proteins that are stable in the bloodstream often have a homogeneous charge distribution and/or protecting oligosaccharide side chains (sialic acid). Such proteins can have a relatively long half-life in the circulation since only fluid-phase endocytosis may determine their turnover rate (Fig. 10). Finally, protein aggregates or protein-derived nanoparticles may be phagocytosed by opsonization factors (fibronectin, immunoglobulins) or endocytosed by galactose- or mannose-recognizing particle receptors (Kolb-Bachofen et al., 1984; Kolb-Bachofen, 1986; Stahl, 1990).

The interaction of these complex peptide structures with the wealth of circulating factors and membrane-bound recognition sites makes it a difficult task to predict pharmacokinetic behavior on the basis of chemical structure and physicochemical features. Nevertheless, considerable progress was made in structure-kinetic studies (see Section 4.5) while at the same time cell biological knowledge increased spectacularly. As mentioned in the preceding sections, a few categories of peptides seem to enjoy special interest with regard to the liver: proteins abnormal in charge and glycosylation patterns, hydrophobic small peptides, certain immune proteins and immune complexes as well as hormones and cytokines that encounter specific and functional receptors on the surface of the liver cells. Receptor-mediated endocytosis of insulin and EGF are well-known examples of the latter (Sugiyama and Hanano, 1989; Burwen and Jones, 1990). The authors do not pretend to exhaustively deal with the hepatic disposition of the many individual therapeutic oligopeptides investigated, yet a number of interesting examples in hepatic disposition are worthwhile mentioning.

1. The group of interferons (17–23 kDa) was intensively studied by Bocci (1987, 1990), who found that the uptake rate of the natural IFN-α (unglycosylated) was negligible. Clearance was much larger for the IFN-β (glycosylated) especially if desialylated, suggesting an important role for sugar signals. Interestingly, they showed that unglycosylated recombinant IFN-β, IFN-α, and also interleukins were very rapidly removed by the liver via unknown mechanisms. Possibly a scavenger system for cationic proteins may play a role. The cationic character of such products may also help renal filtration although the presence of dimeric and trimeric forms may negatively influence the renal contribution. Taken together, plasma decay of these recombinant modalities in patients may be surprisingly rapid. It should be emphasized here that both elimination and distribution of interferons can be largely species dependent (Gloff and Benet, 1990).

2. Another interesting aspect is that clearance of therapeutic proteins may be largely influenced by endogenous factors either in the normal or in the pathological state. The rapid clearance of various forms of the antineoplastic enzyme asparaginase (130–180 kDa) very likely occurs through scavenging by the macrophages of the RES including the Kupffer cells in the liver (Bocci, 1987). The half-life of such proteins is

greatly prolonged when the RES is blocked by overloading with phagocytosable materials (Jones and Summerfield, 1988) or for instance through prior infection with a virus that destroys specific populations of these hepatic cells (Smit *et al.*, 1991). Another example of kinetic perturbations in pathological conditions is the clearance of monoclonal antibodies and immunotoxins. The half-life of most normal immunoglobulins ranges from 5 to 25 days (Gloff and Benet, 1990; Bocci, 1990). That of F(ab) fragments is up to 35 times shorter mainly through renal filtration and catabolism (Covell *et al.*, 1986). The uptake of labeled monoclonal antibodies in the liver is receptor-mediated and saturable. Consequently, it can be decreased by previous injection of an excess of unlabeled antibody. Monoclonal antibody clearance by the liver is markedly different in normal and tumor-bearing hosts if the antibody is tumor-specific. Important is the observation that clearance of tumor-related antibodies is positively correlated with the tumor size and antigen content of the tumor (Shea *et al.*, 1989). The reason seems to be that much of the liver uptake is due to formation of antigen-antibody complexes that are rapidly removed by the Fc receptor on the various cell types (see Section 5.3). It is important to note that the kinetics of such specific proteins under pathological conditions may be largely different from those found in normal individuals.

3. In view of the importance of bile acid carrier systems in the removal of small peptides (see Section 4.2) as well as the influence of high bile salt concentration on receptor-mediated endocytosis (Russel *et al.*, 1983), it can be anticipated that hepatic clearance of certain oligopeptides and proteins may be altered during postprandial bile acid absorption and during cholestatic diseases. Virtually no data are available at present, but the possible competitive interactions during carrier-mediated uptake could lead to time-dependent changes in the bioavailability of highly cleared small peptides (see Section 4.5) and also in largely decreased clearance of glycoproteins during cholestatic conditions (Russel *et al.*, 1983). Indeed, marked elevation of diagnostic enzymes and acute-phase proteins is seen in such patients (Burwen and Jones, 1990). Finally, clearance of mannose-terminated glycoproteins can be largely decreased in diabetic patients as a consequence of the elevated glucose levels (Jones and Summerfield, 1988).

6.2. Manipulations of the Hepatic Clearance of Peptides

In the case that liver first-pass of oligopeptides is impractically high or if efficient hepatic clearance largely limits the residence time of peptides in the body, one could consider procedures to counteract these processes. Very different approaches have been taken:

- Carbohydrate-mediated clearance by the liver of certain immunotoxins can be clearly decreased by deglycosylation (Blakey and Thorpe, 1988). Also, enzymatic reglycosylation and addition of sialic acid can be used.

- Clearance of interferons can be reduced by binding to albumin and monoclonal antibodies or by enclosing the drug in liposomes or erodable polymers (Tomlinson, 1990).
- Conjugation with hydrophilic polymers such as polyethylene glycols and dextrans may, apart from increasing the half-life, also reduce immunogenicity (see Chapter 4). Examples are interleukins and asparaginase (Tomlinson, 1990).
- Proteins can be derivatized to so-called hybrid proteins to alter their residence time in the body. Immunotoxins (Poznansky and Juliano, 1984) are well-known examples but also toxins covalently bound to growth factors and hormones have been widely employed in experimental work (Blakey and Thorpe, 1988; Tomlinson, 1990).
- Receptor- and carrier-mediated removal processes could be temporarily blocked by nonactive and nontoxic competing substrates. Biodegradable polycarbohydrates (Smedsrød et al., 1990) and inactive isomers could, in principle, be considered for such purposes.
- Decrease in hydrophobicity and changes in essential amino acid sequences in oligopeptides could reduce affinity for hepatocyte carriers (if compatible with effectivity). This will not always lead to major successes since multiple alternative carrier systems are present in these cells (see Section 4.5).

7. PROTEIN TRANSPORT IN RELATION TO DRUG TARGETING IN THE LIVER

7.1. Introduction

Since glycoproteins represent a wide variety of macromolecules which, depending on their charge and carbohydrate structure, can be more or less specifically recognized by certain tissues and cell groups within these tissues, they potentially can be used as drug carriers targeting drugs to their targets in the body (Poznansky and Juliano, 1984; Monsigny et al., 1988; Meijer and van der Sluijs, 1989; Shen, 1987; Fallon and Schwartz, 1985). A number of important advantages are provided: drug concentrations at crucial sites could be attained which would never have been produced by normal administration. Furthermore, unwanted effects at other sites could be reduced and elimination from the body considerably retarded. A stable linkage of a drug (or an enzyme) to glycoproteins may completely change the normal distribution pattern of the parent compound which is now mainly dictated by the properties of the carrier molecules. Especially covalent binding should be used, provided that the drug retains its activity or is released after degradation of the carrier in lysosomes.

Drug targeting to the liver is performed for diagnostic and therapeutic purposes. Liver disorders or liver metabolism can be influenced by various agents: agents for antiviral, antiparasitic, antineoplastic, or antilipidemic effects, drugs counteracting liver fibrosis, or hepatotoxic reactions, and finally modified enzymes to correct genetic enzyme deficiencies (Table VI).

7.2. Potential Carrier Systems for Liver Targeting

Both soluble and particle carrier materials have been employed for the purpose of drug targeting to the various cell types in the liver. Relevant in this respect is the endothelial barrier: particles > 100–150 nm cannot pass the fenestrae (pores) in the endothelial lining (Brouwer et al., 1988). This is true for most of the normal sized liposomes, and such carriers can only reach Kupffer cells, pit cells, and endothelial cells. Small liposomes and HDL particles can, in principle, be used for delivery of lipophilic (pro-)drugs to hepatocytes and lipocytes (van Berkel et al., 1986, 1987; Roerdink et al., 1988). Such agents can also be enclosed in the lipid core of LDL particles or in liposomes and be delivered to the sinusoidal cell types (van Berkel et al., 1986, 1987). After degradation of the carrier, such drugs could locally act in these cells or be slowly released from the cells to reach the other cell types in the liver. Reuptake of the parent drugs in these cells, however, may not be efficient enough to reach therapeutic concentrations, and at the same time prevent toxic levels in extrahepatic tissues.

The choice of such carriers therefore is determined by a number of considerations (Poznansky and Juliano, 1984; Meijer and van der Sluijs, 1989; Wu, 1988; Shen, 1987):

- What is the cost of the chosen carrier, taking into account that clinical application is the final goal?
- Can the targeted cell type actually be reached (physiological barriers, rapid removal by RES)?
- Does the carrier allow sufficient drug loading (taking into account drug potency and the capacity for carrier-mediated endocytosis)?
- Is the carrier to a large extent cell-specific in the whole organism (reaction with opsonic factors, or removal by competing cell types) and is specificity unaltered after linkage of the drug?
- Is the carrier biodegradable (proper release rate of the drug and RES toxicity of the carrier itself)?
- Are degradation products of the carrier material nontoxic?
- Can major removal by the RES system be prevented (proper residence time for therapeutic effects)?
- Do immunogenic reactions occur after chronic administration?
- Is the drug released in its active form and at an adequate rate (therapeutic effect, leakage from the tissue)?
- Is the targeted receptor pathway also expressed in pathological conditions and is it influenced by other drugs?

Table VI
Targeting to the Liver with Polypeptides

Drugs/agents	Carrier	Experimental method	Reference
Antiviral agents			
Ara-AMP	Lactosaminated HSA	Mouse *in vivo*	Fiume *et al.* (1986b)
	Galactosyl-poly-L-lysine	Mouse *in vivo*	Fiume *et al.* (1986a)
	Lactosaminated HSA	Patients (hepatitis B)	Fiume *et al.* (1988b)
Trifluorothymidine	Asialofetuin	Mouse *in vivo*	Fiume *et al.* (1982)
Antiparasitic agents			
Primaquine	Asialofetuin	Mouse *in vivo*	Trouet *et al.* (1982a)
Pepstatin	Asialofetuin	Rat *in vivo*	Furuno *et al.* (1983)
Antineoplastic agents			
Abrin A chain	IgG (tumor-specific antibody)	Guinea pig *in vivo*	Hwang *et al.* (1984)
Daunomycin	IgG (anti-alphafeto-protein AB)	Mouse *in vitro*, patients	Hirai *et al.* (1983), Belles-Isles and Page (1988)
	Lactosaminated HSA	Mouse *in vivo*	Schneider *et al.* (1984)
Diphtheria toxin A	Asialofetuin	Cultured rat hepatocytes	Cawley *et al.* (1981)
	Asialoorosomucoid	Cultured rat hepatocytes	Chang and Kulberg (1982)
Methotrexate	Poly-L-lysine	Mouse *in vivo*	Arnold (1985)
Mitomycin C	IgG (anti-alphafeto-protein AB)	Rat *in vivo*	Kato *et al.* (1983)
Ricin	Asialofetuin	Cultured rat hepatocytes	Herschmann *et al.* (1984)
	Epidermal growth factor	Mouse *in vivo*	Herschmann *et al.* (1984)
Trenimon	IgG	Mouse *in vivo*	Ghose *et al.* (1982)
Diagnostic agents			
Fluorescein	Asialoorosomucoid	Isolated rat liver	van der Sluijs *et al.* (1985)
Technetium-99m	Galactosyl-HSA	Rat/rabbit	Vera *et al.* (1985)
	Galactosyl-HSA	Human	Stadalnik *et al.* (1985)
Radioscanning agents	IgG (hepatoma AB)	Rat *in vivo*, patients	Wu (1988)
Agents affecting lipid metabolism			
LDL/HDL	Tris-galactosyl-cholesterol		van Berkel *et al.* (1986)
LDL	Lactosaminated form	Isolated hepatocytes	Attie *et al.* (1980)
Antitoxicants			
N-Acetylcysteine	Asialofetuin	Hep G_2 cells	Wu *et al.* (1985)
Folinic acid	Asialofetuin	Hep G_2 cells	Wu (1988)
Uridine monophosphate	Asialofetuin/poly-L-lysine	Isolated hepatocytes	Wu *et al.* (1988)
Genes			
CAT	Asialoorosomucoid-poly-L-lysine	Hep G_2 cells, rat *in vivo*	Wu and Wu (1988), Wu *et al.* (1989), (1990)
HSA	Asialoorosomucoid-poly-L-lysine	Rat *in vivo* (Nagase)	Wu *et al.* (1991)
LDL receptor	Asialoorosomucoid-poly-L-lysine	Rabbit (Watanabe)	Wilson *et al.* (1989)

7.3. Lysosomal Release of Active Drug from the Protein Carrier

The final goal of any drug targeting concept is to make the active drug available in the cell type of interest. Drugs such as doxorubicin that are covalently linked via their NH_2 groups to the COOH groups of the carrier protein can only be slowly released in their active form. Introduction of amino acid spacers between the drug and the protein can improve release rate and such spacers can be used to manipulate the cellular concentrations (Trouet et al., 1982a,b; Shen, 1987; Schneider et al., 1984). Drugs with functional COOH groups or isothiocyanate moieties can be linked to lysine-NH_2 groups of proteins but the final degradation product will contain at least one amino acid (lysine) since this linkage cannot be split from the drug by the lysosomal enzymes (van der Sluijs et al., 1985). FITC covalently coupled to asialoorosomucoid is quantitatively taken up by the parenchymal cells of the liver. After degradation, the dye is released but still bound to lysine and after conjugation with glucuronic acid it is excreted into bile (van der Sluijs et al., 1985). The excretion rate of the fluorescent material in bile reflects the rate-limiting lysosomal degradation of the glycoprotein carrier (van der Sluijs et al., 1985) (Fig. 9). Uncoupling of the drug can also occur too fast: trifluorothymidine, covalently linked to the hepatotropic carrier asialofetuin, is so rapidly released from lysosomes in the hepatocytes that substantial amounts escape to the general circulation and reach nontarget tissues such as bone marrow (Fiume et al., 1982). Depression of bone marrow by the redistributed drug was exhibited in otherwise healthy animals but not in animals with experimental hepatic necrosis. This phenomenon was explained by a reduced lysosomal release rate of the drug in the liver of the latter animals (Fiume et al., 1982).

An important factor determining the rate of release from lysosomes is probably the extent of protonation of the delivered drug. Basic drugs will be largely protonated at the relatively low internal pH of this organelle (pH 5–6) and, consequently, will undergo persistent storage and relatively slow release. In contrast, acidic drugs will be present mainly in the undissociated form and therefore will easily leave the lysosomal compartment. This was shown for the protease inhibitor pepstatin which was administered covalently coupled to asialofetuin (Furuno et al., 1983). The carboxylic tripeptide was properly delivered to hepatocyte lysosomes but after carrier degradation it was rapidly excreted into bile. However, in spite of this rapid release, intralysosomal levels remained high enough for inhibition of cathepsin during 6 hr after injection, demonstrating the extreme potency of the pepstatin molecule.

7.4. Preparation of (Glyco)protein–Drug Conjugates

Glycoproteins represent excellent objects for the targeting of drugs: the structure can be modified with regard to the protein backbone as well as the functional sugar groups. The relatively small molecular mass does not exclude passage through the vascular endothelium (Poznansky and Juliano, 1984; Bodmer and Dean, 1988;

Schneider *et al.*, 1984) and immunogenicity after chronic administration is limited (Fiume *et al.*, 1987a). Naturally occurring plasma proteins such as orosomucoid (α_1-acid glycoprotein) and fetuin were often used. They exhibit a relatively high affinity for the ASGP receptor probably due to the clustered arrangement of the antennary oligosaccharide side chains (see Section 5.10) but after drug conjugation may display immunogenic properties. Some artificial carrier systems mimic this geometric organization of sugar groups. Plasma proteins such as albumin and apoprotein B as well as certain enzymes can be randomly derivatized with sugar groups. Well-known methods are the reductive amination using boronhydride and lactose. In this reaction the aldehyde moiety of glucose is reacted with ϵ-NH_2 groups of lysine. Up to 60 lysine molecules can be linked to lactose in this manner. It should be realized that the nitrogen atom in the protein-sugar linkage can still become protonated and that consequently no positive charge is lost. Such glycosylated albumin molecules may lose their normal neutral-base transition properties and loss of flexibility in the protein molecule due to sugar derivatization can therefore occur. Other methods to connect sugars include the thioglycoside method (Stowell and Lee, 1980) and coupling via thiophosgene activation of amino sugars (Monsigny *et al.*, 1988; Kataoka and Tavassoli, 1985). Especially the last method may lead to negatively charged proteins since the nitrogen atom in the linking moiety cannot be protonized in contrast to the lysine-NH_2 groups. This charge modification is further increased by subsequent covalent coupling of drugs to the remaining NH_2 groups. Drugs with acidic functional groups will amplify this problem (see Section 5.5).

Alternatively, drug peptide carriers can be designed with a simpler structure than the naturally occurring plasma proteins. Various types of polylysines with different charge can be tailor-made and even provided with clustered sugar groups (Pompipom *et al.*, 1984; Arnold, 1985; Shen, 1987; Monsigny *et al.*, 1988; Fiume *et al.*, 1986a).

From all of these considerations, it follows that the design of proper drug targeting preparations requires an integrated scientific approach on the cell biological level as well as on the level of whole animal testing. The step from experimental work in animals to that in man and finally to the clinic is still large and implies special problems with regard to chronic cellular toxicity and immunogenicity (Poznansky and Juliano, 1984; Poste, 1984). Application in man also implies large-scale preparations of chemically well-defined drug-carrier modalities and the design of biopharmaceutically standardized dosage forms.

7.5. Current Achievements in Drug Targeting to the Liver with Protein Carriers

7.5.1. INTRODUCTION

The liver probably is the most systematically studied organ in the field of drug targeting (Table VI). Quite a few examples can be found in which successful and selective delivery of drug to this organ was achieved *in vivo* (Wu, 1988; Fallon and Schwartz, 1989; Stahl, 1990; Meijer *et al.*, 1990a) (see Table VI). It is of importance to note that the idea of drug targeting in fact turned around the usual rules in drug design and development: very toxic agents that seemed to be shelved by pharmacologists became serious candidates again for therapeutic application. A vast number of chemical methods were developed for a proper preparation of drug-macromolecular conjugates.

7.5.2. TARGETING OF ANTIVIRAL AGENTS

Fiume and co-workers coupled 5–15 molecules of adenine arabinoside monophosphate (Ara-AMP) to lactosaminated albumin containing 20–50 terminal galactose groups per albumin molecule (Fig. 11). The phosphoamide bond, linking the phosphate group about equally to ε-NH_2 groups of lysine and imidazole nitrogens of histidine (Fiume *et al.*, 1988a), is acid-labile and apart from hydrolysis in acidic endosomal and lysosomal compartments, it can be proteolytically cleaved to the monophosphates. The carbodiimide coupling reaction should be performed at pH 7.5 to prevent production of polymeric and poorly soluble material (Fiume *et al.*, 1986b, 1988a). The drug-protein conjugate is delivered not only to hepatocytes where hepatitis B virus mainly replicates, but also to sinusoidal cell types (Fiume *et al.*,

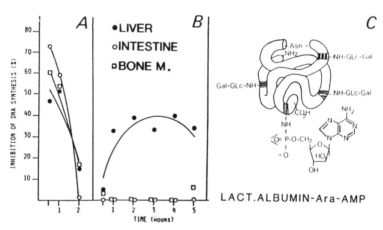

Figure 11.   Increased efficiency of 9-β-D-arabino-furanosyl adenine monophosphate (Ara-AMP) on the hepatic level and decreased side effects on intestinal mucosa and bone marrow through coupling to human serum albumin containing 20–50 lactose molecules per mole protein (C). Ectromelia viral hepatitis causes increased (mainly viral) DNA synthesis and is quantified by [³H]thymidine incorporation 24 hr after infection. The lactosylated albumin-Ara-AMP conjugate in a threefold-lower dose shows a prolonged inhibitory pattern (B) compared with Ara-AMP itself (A), but does not suppress DNA synthesis in the other tissues as is the case for uncoupled Ara-AMP. (According to Fiume *et al.*; see References list.)

1987b). This kinetic behavior is probably due to electronegativity of the albumin carrier in which a large part of the positively charged groups have been used for sugar and drug derivatization. Since in the initial stages of viral infection sinusoidal cell types are attacked (a process that destroys the endothelial barrier) (Fiume *et al.*, 1982), this broad hepatic distribution is not necessarily inadequate. This particular drug formulation reached the clinical stage and a preliminary study showed promising results in HBV-infected patients (Fiume *et al.*, 1988b). More definite studies are awaited. It is of importance to note here that using such carriers, phosphorylated drugs that normally cannot penetrate the cell membrane, can be delivered intracellularly. Anyway, the liberated Ara-AMP is retained within the cells long enough to permit effective phosphorylation to the di- and triphosphate forms and consequently systemic toxicity is reduced.

In contrast to the parent drug, the Ara-AMP conjugate with lactosaminated albumin is almost devoid of side effects in bone marrow and intestinal mucosa (Fiume *et al.*, 1982, 1986b) (see Fig. 11). Another positive aspect is that the conjugate does not seem to evoke major immunogenic reactions after chronic administration, if homologous albumin is used (Fiume *et al.*, 1987a). Interesting alternatives to this carrier might be antibodies directed to HBV surface antigens (Shouval *et al.*, 1982) or albumin treated with glutaminase or malondialdehyde. The latter proteins display a high affinity for receptors on the virus particle as well as the hepatocytes (Michalak and Bolger, 1989; Thung *et al.*, 1989).

7.5.3. TARGETING OF ANTINEOPLASTIC AGENTS AND ANTITOXICANTS

These include cytostatic agents such as daunorubicin, vindesine, methotrexate, cyclophosphamide, trenimon, and toxins such as ricin, abrin, gelonin as well as diphtheria toxins (Blakey and Thorpe, 1988; Wu, 1988; Meijer and van der Sluijs, 1989; Fallon and Schwartz, 1989). As carriers, (neo-)glycoproteins (Trouet *et al.*, 1982a; Monsigny *et al.*, 1988; Bodmer and Dean, 1988), tumor-specific antibodies (Ghose *et al.*, 1982), poly-L-lysine (Arnold, 1985; Ryser and Shen, 1986; Shen, 1987; Pompipom *et al.*, 1984), and various other synthetic polymers were used. It should be realized here that antibodies *per se* [e.g., against HBVsAg (Shouval *et al.*, 1982) and alpha-fetoprotein (Hirai *et al.*, 1983)] can be effective in the *in vitro* and *in vivo* lysis of hepatoma cells. Conjugates of antibodies and lactosaminated albumin with daunorubicin increased survival of animals with implanted tumors (Schneider *et al.*, 1984; Trouet *et al.*, 1982a) and in some cases of cancer patients (Schneider *et al.*, 1984; Wu, 1988). ASGPs were linked to bacterial toxins yielding extremely potent drugs that inhibit protein synthesis in the hepatoma cells at a concentration of 10^{-11} M (Cawley *et al.*, 1981; Chang and Kullberg, 1982; Simpson *et al.*, 1982). Since in some but not all hepatomas the density of the particular receptor is considerably decreased, this approach is apt to fail in such cases (Wright, 1989).

The fact that in some hepatoma cells there is a relative lack of ASGP receptors can be used to rescue noncancer cells from the intoxication of high doses of methotrexate (Wu, 1988). This was accomplished by coupling folinic acid to ASGPs and combining this with the antineoplastic drug. Only the receptor-negative cells are killed by the combination. It can be concluded from these data that the rescue factor, after release from the carrier, does not redistribute rapidly to the cancer cells. This targeting concept was later expanded to prevent hepatoxicity of acetaminophen (Wu et al., 1985) and galactosamine (Wu et al., 1988a) through covalent coupling of N-acetylcysteine and uridine monophosphate, respectively, to ASGPs. It is assumed that in vivo, the particular antitoxicant preparations are equally delivered to all the hepatocytes. However, studies on the acinar distribution of injected ASGPs revealed zonal heterogeneity (van der Sluijs et al., 1988; Burwen and Jones, 1990) with relatively low concentrations in the pericentral cells. This predicts that the drug may only be adequately presented to the pericentral cells if the portal concentration of the carrier exceeds the K_m for receptor-mediated uptake. Since the affinity of most galactose-terminated neoglycoproteins for the receptor as well as the hepatic extraction is considerably lower than for the naturally occurring ASGPs, a major heterogeneity in distribution of the neoglycoproteins is not anticipated (van der Sluijs and Meijer, 1991).

7.5.4. TARGETING OF GENES

The process of receptor-mediated endocytosis can in principle also be employed for delivery of small fragments of DNA to certain cell types. The option here is to correct genetic deficiencies in the production of essential proteins such as peptide hormones, plasma proteins, and membrane receptors. The delivered plasmid should be taken up in the cell and integrated in the cellular genome. Transcriptions to mRNA and translation into polypeptides can be monitored to detect the cellular expression of the targeted gene. Gene targeting has been attempted for insulin, using a proinsulin gene included in small liposomes (Nicolau and Cudd, 1989). In order to test the suitability of the targeted gene product, foreign "reporter" genes such as the bacterial gene coding for chloramphenicol acetyltransferase (CAT) are often used. In vitro, genes can be quite easily introduced in cultured cells by microinjection or Ca^{2+} phosphate-mediated permeation of the plasma membrane. Fusion with plasmid-containing liposomes (Nicolau and Cudd, 1989) or cells and cellular delivery by viruses (Wilson, 1986) are also employed. Evidently, it is much more difficult to perform this in the intact organism. Recent developments in gene targeting to the liver in vivo indicate that persistent expression of genes might be achieved (Wilson et al., 1989; Wu and Wu, 1988; Wu et al., 1988b, 1989). Receptor-mediated endocytosis is used to introduce the gene, in spite of the fact that usually most of the material is trafficked to the lysosomes. The crucial question here is how part of the endocytosed plasmid escapes this degradation route and reaches the cell nucleus instead. There is evidence for some interiorized ligands that a small part of the endocytotic vesicles may become associated with the Golgi system, may recycle to the plasma membrane, or can undergo transcytosis (Russel *et al.*, 1983; Townsend *et al.*, 1984; see Chapters 4 and 11). For a ligand such as EGF, vectorial transport to the cell nucleus normally is a minor route but during regeneration (following hepatectomy) complete perturbation of cellular routing is observed in the sense that much less of the ligand is trafficked to lysosomes and much more to the cell nucleus (Marti *et al.*, 1989). Whether this is true for other glycoproteins remains to be studied. Nevertheless, integration of targeted DNA and also its expression can be greatly enhanced by prior partial hepatectomy (Wu *et al.*, 1988b, 1989) probably as a consequence of rapid cellular replication. Also agents that induce hyperplasia of the liver can be used to improve integration of the foreign DNA. Pretreatment of rats with the hypolipidemic agent nafenopin, which induces a marked liver growth (Meijer *et al.*, 1977), leads to persistent expression of the CAT gene injected *in vivo* (Wu *et al.*, 1990).

The prime technical item is how to design a suitable carrier for targeting DNA to organs *in vivo*; the complex should be sufficiently stable in the bloodstream, noncovalent binding of DNA should be preferred, and the DNA-carrier complex should be water-soluble. Wu *et al.* (1990) used the basic polypeptide polylysine, which strongly binds DNA (see Fig. 12). The polylysine matrix was then covalently coupled to lactosaminated albumin to obtain hepatocyte specificity (Wu, 1988). Direct galactosylation of polylysine negatively influences its binding affinity for DNA while the use of cluster glycosides with spacer arms is in principle possible though expensive (Wu, personal communication). Depending on the size of the plasmid, polylysine polymers of different length can be employed. Careful titration is necessary to keep the complex in solution. This concept was applied to target a foreign gene (CAT) combined with various promoters to the liver and clear expression of the gene was demonstrated (Wu and Wu, 1988) especially after partial hepatectomy (Wu *et al.*, 1988b). Excess asialoorosomucoid prevents DNA targeting to the liver probably through competition with the ASGP receptors. Both viral promoters and human promoter genes can be used. Recently, gene targeting was used to correct genetic analbuminemia in the so-called Nagase rat model (Wu *et al.*, 1989) and cholesterolemia due to an LDL receptor deficiency in the rabbit (Wilson *et al.*, 1989). A gene coding for human serum albumin came to expression in rats within 48 hr after injection, raising the albumin concentration in the plasma from zero to 18 mg/100 ml. Persistent expression was seen until 4 weeks postinjection (Fig. 12).

A gene coding for the LDL receptor protein was targeted to the rabbit liver *in vivo*, resulting in a 25–50% decrease in cholesterol plasma concentration lasting for 2 days after injection of the plasmid polylysine-lactosaminated albumin preparation. In both cases, expression was controlled by detecting the particular mRNAs as well as the proteins themselves. These preliminary observations may mark the beginning of a new era in the manipulation of cell biology *in vivo* (Monsigny *et al.*, 1988; Fallon and Schwartz, 1985; Meijer and van der Sluijs, 1989). Applications in corrections of genetic disorders such as enzyme deficiencies, receptor deficiencies as well as *in vivo*

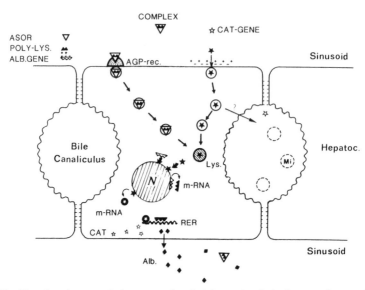

Figure 12. Targeting of genes to the hepatocyte. In principle, two interiorization procedures can be used: cationization of the plasma membrane leading to adsorptive endocytosis of the DNA fragments [indicated for the chloramphenicol acetyl transferase (CAT) gene] or noncovalent binding of the plasmid to poly-L-lysine, covalently coupled to the asialoglycoprotein asialoorosomucoid (ASOR) (indicated for the albumin gene). After receptor-mediated or adsorptive endocytosis, an unknown part of the endocytosed material escapes the abortive lysosomal route and the DNA becomes integrated in the cellular genome. Persistent expression was detected by mRNAs and synthesis of the particular proteins. (See Wu *et al.* in References list.)

production of pharmacologically active polypeptides represent novel options in the field of cell-specific drug delivery.

REFERENCES

Ahnen, D. J., Brown, W. R., and Kloppel, T. M., 1985, Secretory component: The polymeric immunoglobulin receptor. What's in it for the gastroenterologist and hepatologist? *Gastroenterology* **89:**667–682.

Ananthanarayanan, M., Von Dippe, P., and Levy, D., 1988, Identification of the hepatocyte Na$^+$-dependent bile acid transport protein using monoclonal antibodies, *J. Biol. Chem.* **263:**8338–8343.

Arnold, L. J., 1985, Polylysine-drug conjugates, *Methods Enzymol.* **112:**270–285.

Ashwell, G., and Harford, J., 1982, Carbohydrate specific receptors in the liver, *Annu. Rev. Biochem.* **51:**531–544.

Attie, A. D., Pittman, R. C., and Steinberg, D., 1980, Metabolism of native and of lactosylated human low density lipoprotein, evidence for two pathways for catabolism of exogenous proteins in rat hepatocytes, *Proc. Natl. Acad. Sci. USA* **77:**5923–5927.

Baenziger, J. U., and Fiete, D., 1980, Galactose and N-acetylgalactosamine specific endocytosis of glycopeptides by isolated rat hepatocytes, *Cell* **22**:611–620.

Ballet, F., 1990, Hepatic circulation—Potential for therapeutic intervention, *Pharmacol. Ther.* **47**:281–328.

Barret, A. J., 1982, α2-Macroglobulin, in: *Methods in Enzymology* (L. Lorand, ed.), Academic Press, New York, pp. 442–455.

Beck, W. T., 1987, The cell biology of multiple drug resistance, *Biochem. Pharmacol.* **36**: 2879–2884.

Begg, E. J., Atkinson, H. C., Jeffery, G. M., and Taylor, N. W., 1989, Individualised aminoglycoside dosage based on pharmacokinetic analysis is superior to dosage based on physician intuition at achieving target plasma drug concentrations, *Br. J. Clin. Pharmacol.* **28**:137–141.

Belles-Isles, M., and Page, M., 1988, In vitro activity of daunomycin–anti-alphafoetoprotein conjugate on mouse hepatoma cells, *Br. J. Cancer* **41**:841–842.

Bergmann, P., Kacenelenbogen, R., and Vizaet, A., 1984, Plasma clearance, tissue distribution and catabolism of cationized albumins with increasing isoelectric points in the rat, *Clin. Sci.* **67**:35–43.

Berk, P. D., Potter, B. J., and Stremmel, W., 1987, Role of plasma membrane ligand binding proteins in the hepatocellular uptake of albumin-bound organic anions, *Hepatology* **7**: 165–176.

Bertrams, A., and Ziegler, K., 1989a, Uptake of renin inhibitors into liver cells, *Biol. Chem. Hoppe Seyler* **370**:875.

Bertrams, A., and Ziegler, K., 1989b, Uptake of renin inhibitors into liver cells: Evidence for an active transport related to the sinusoidal bile acid transport, *Naunyn Schmiedebergs Arch. Pharmacol.* **340**:R65.

Bertrams, A., and Ziegler, K., 1991a, Hepatocellular uptake of peptides by bile acid transporters, *Biochim. Biophys. Acta* **1091**:337–348.

Bertrams, A., and Ziegler, K., 1991b, New substrates of the multispecific bile acid transporter in liver cells, *Biochim. Biophys. Acta* **1073**:213–220.

Bijsterbosch, M. K., and Van Berkel, T. J. C., 1989, The galactose-specific receptor on K-cells; the effect of the extent of lactosylation on uptake, in: *Cells of the Hepatic Sinusoid*, Volume 2 (E. Wisse, D. L. Knook, and K. Decker, eds.), Kupffer Cell Foundation, Rijswijk, pp. 130–131.

Bijsterbosch, M. K., and Van Berkel, T. J. C., 1990, Uptake of lactosylated low-density lipoprotein by galactose-specific receptors in rat liver, *Biochem. J.* **270**:233–239.

Blakey, D. C., and Thorpe, P. E., 1988, Prevention of carbohydrate-mediated clearance of ricin-containing immunotoxins by the liver, in: *Immunotoxins* (A. E. Frankel, ed.), Kluwer Academic Publishers, Boston, pp. 457–473.

Blitzer, B. L., Ratoosh, S. L., Donovan, C. B., and Boyer, J. L., 1982, Effects of inhibitors of Na^+-coupled ion transport on bile acid uptake by isolated rat hepatocytes, *Am. J. Physiol.* **243**:48–53.

Bocci, V., 1987, Metabolism of protein anticancer agents, *Pharmacol. Ther.* **34**:1–49.

Bocci, V., 1990, Catabolism of therapeutic proteins and peptides with implications for drug delivery, *Adv. Drug Deliv. Rev.* **4**:149–169.

Bodmer, J. L., and Dean, R. T., 1988, Drug and enzyme targeting, *Methods Enzymol.* **112**: 298–306.

Borel, J. F., Feurer, C., Gubler, H. U., and Stahelin, H., 1976, Biological effects of cyclosporin A, *Agents Actions* :468–475.

Bouma, I. M. W., and Smit, M. J., 1988, Elimination of enzymes from plasma in the rat, *Clin. Enzymol.* **6:**111–119.

Boyer, J. L., 1986, Mechanisms of bile secretion and hepatic transport, in: *Physiology of Membrane Disorders*, 2nd ed. (T. E. Andreoli, J. F. Hoffman, D. D. Fanestil, and S. G. Schultz, eds.), Plenum Medical, New York, pp. 609–636.

Braakman, I., Groothuis, G. M. M., and Meijer, D. K. F., 1987, Acinar redistribution and heterogeneity in transport of the organic cation rhodamine B in rat liver, *Hepatology* **7:** 849–855.

Brouwer, A., Wisse, E., and Knook, D. L., 1988, Sinusoidal endothelial cells and perisinusoidal fat-storing cells, in: *The Liver: Biology and Pathobiology*, 2nd ed. (I. M. Arias, W. B. Jakoby, H. Popper, D. Schachter, and D. A. Shafritz, eds.), Raven Press, New York, pp. 665–682.

Brown, W. R., and Kloppel, T. M., 1989, The liver and IgA: Immunological, cell biological and clinical implications, *Hepatology* **9:**763–784.

Burwen, S. J., and Jones, A. L., 1990, Hepatocellular processing of endocytosed proteins, *J. Electron Microsc. Tech.* **14:**140–151.

Burwen, S. J., Barker, M. E., Goldman, I. S., Hradek, G. T., Raper, S. E., and Jones, A. L., 1984, Transport of epidermal growth factor by rat liver, *J. Cell Biol.* **99:**1259–1265.

Caldwell, L. J., Parr, A., Beihn, R. M., Agha, B. J., Mlodozeniec, A. R., Jay, M., and Digenis, G. A., 1985, Drug distribution and biliary excretion pattern of cyclic somatostatin analogs: A comparison of ^{14}C labeled drug and a ^{131}I iodinated drug analog, *Pharm. Res.* **2:**80–83.

Campra, J. L., and Reynolds, T. B., 1988, The hepatic circulation, in: *The Liver: Biology and Pathobiology*, 2nd ed. (I. M. Arias, W. B. Jakoby, H. Popper, D. Schachter, and D. A. Shafritz, eds.), Raven Press, New York, pp. 911–930.

Cawley, D. B., Simpson, D. L., and Herschman, H. R. 1981, Asialoglycoprotein receptor mediates the toxic effects of an asialofetuin diphtheria toxin fragment A conjugate on cultured rat hepatocytes, *Proc. Natl. Acad. Sci. USA* **78:**3383–3387.

Chang, T. M., and Kullberg, D. W., 1982, Studies on the mechanism of cell intoxication by diphtheria toxin fragment A-asialoorosomucoid hybrid toxins, *J. Biol. Chem.* **257:**12563–12572.

Christensen, H. N., 1979. Exploiting amino acid structure to learn about membrane transport, *Adv. Enzymol.* **49:**41–101.

Clarke, B. L., and Weigel, P. H., 1989, Differential effects of leupeptin, monensin and colchicine on ligand degradation mediated by the two asialoglycoprotein receptor pathways in isolated rat hepatocytes, *Biochem. J.* **262:**277–284.

Conolly, D. T., Townsend, R. R., Kawaguchi, R., Bell, K., and Lee, Y. C., 1982, Binding and endocytosis of cluster glycosides by rabbit hepatocytes: Evidence for a short circuit pathway that does not lead to degradation, *J. Biol. Chem.* **257:**939–945.

Covell, D. G., Barbet, J., Holton, O. D., Black, C. D. V., Parker, R. J., and Weinstein, J. N., 1986, Pharmacokinetics of monoclonal immunoglobulin G_1, F(ab')$_2$, and Fab' in mice, *Cancer Res.* **46:**3969–3978.

Creighton, T. E., 1990, Protein folding, *Biochem. J.* **270:**1–16.

De Boer, A. G., Moolenaar, F., De Leede, L. G. J., and Breimer, D. D., 1982, Rectal drug administration: Clinical pharmacokinetic considerations, *Clin. Pharmacokinet.* **7:** 285–311.

De Duve, C., and Wattiaux, R., 1966, Functions of lysosomes, *Annu. Rev. Physiol.* **28:** 435–492.

Dennis, P. A., and Aronson, N. N., 1985, Metabolism of leupeptin by rat liver, *Arch. Biochem. Biophys.* **240:**768–776.

Devirgiliis, L. C. Bruscalupi, G., and Dini, L., 1989, Modulation of asialoglycoprotein binding activity in livers of pregnant or estrogen-treated rats, *Biosci. Rep.* **9:**701–707.

Doyle, J. W., Wolfe, M. M., and Mcguigan, J. E., 1984, Hepatic clearance of gastrin and cholecystokinin peptides, *Gastroenterology* **87:**60–68.

Dunn, W. A., Wall, D. A., and Hubbard, A. L., 1983, Use of isolated, perfused liver in studies of receptor-mediated endocytosis, *Methods Enzymol.* **98:**225–241.

Evans, W. H., 1981, Membrane traffic at the hepatocytes sinusoidal and canalicular surface domains, *Hepatology* **5:**452–458.

Fallon, R. J., and Schwartz, A. L., 1985, Receptor mediated endocytosis and targeted drug delivery, *Hepatology* **5:**899–901.

Fallon, R. J., and Schwartz, A. L., 1989, Receptor-mediated delivery of drugs to hepatocytes, *Adv. Drug Deliv. Rev.* **4:**49–65.

Fiume, L., Bussi, C., Mattioli, A., Balboni, P. G. Barbanti-Brodano, G., and Wieland, T., 1982, Hepatocyte targeting of antiviral drugs coupled to galactosyl-terminating glycoproteins, in: *Targeting of Drugs* (G. Gregoriades, J. Senior, and A. Trouet, eds.), Plenum Press, New York, pp. 1–17.

Fiume, L., Bassi, B., Busi, C., Mattioli, A., Spinosa, G., and Faulstich, H., 1986a, Galactosylated poly (L-lysine) as a hepatotropic carrier of ara-AMP, *FEBS Lett.* **203:**203–206.

Fiume, L., Bassi, B., Busi, C., Mattioli, A., and Spinosa, G., 1986b, Drug targeting in antiviral chemotherapy. A chemically stable conjugate of 9-beta-D-arabinofuranosyladenine 5′ monophosphate with lactosaminated albumin accomplishes a selective delivery of the drug to liver cells, *Biochem. Pharmacol.* **35:**967–972.

Fiume, L., Busi, C., Preti, P., and Spinosa, G., 1987a, Conjugates of ara-AMP with lactosaminated albumin: A study of their immunogenicity in mouse and rat, *Cancer Drug Deliv.* **4:**145–150.

Fiume, L., Mattioli, A., and Spinosa, G., 1987b, Distribution of a conjugate of 9-beta-D-arabinofuranosyladenine monophosphate [Ara-AMP] with lactosaminated albumin in parenchymal and sinusoidal cells of rat liver, *Cancer Drug Deliv.* **4:**11–16.

Fiume, L., Bassi, B., and Bongini, A., 1988a, Conjugates of 9-beta-D-arabinofuranosyladenine 5′-monophosphate (Ara-AMP) with lactosaminated albumin. Characterization of the drug-carrier bonds, *Pharm. Acta Helv.* **63:**137–139.

Fiume, L., Bonino, F., Mattioli, A., Chiaberge, E., Cerenzia, M. R. T., Busi, C., Brunetto, M. R., and Verme, G., 1988b, Inhibition of hepatitis B virus replication by ARA-AmP conjugated with lactosaminated serum albumin, *Lancet* **2:**13–15.

Forgac, M., 1988, Receptor-mediated endocytosis, in: *The Liver: Biology and Pathobiology* (I. M. Arias, W. B. Jakoby, H. Popper, D. Schachter, and D. A. Shafritz, eds.), Raven Press, New York, pp. 207–225.

Frimmer, M., 1975, Phalloidin, a membrane specific toxin, in: *Pathogenesis and Mechanism of Liver Cell Necrosis* (D. Keppler, ed.), MTP Press, Lancaster, U.K., pp. 163–173.

Frimmer, M., 1982, Organotropism by carrier-mediated transport, *Trends Pharmacol. Sci.* **3:**395–397.

Frimmer, M., 1989, Uptake of foreign cyclopeptides by liver cells, in: *Hepatic Transport of*

Organic Substances (E. Petzinger, R. H. K. Kinne, and H. Sies, eds.), Springer-Verlag, Berlin, pp. 309–316.

Frimmer, M., and Ziegler, K., 1988, The transport of bile acids in liver cells, *Biochim. Biophys. Acta* **947**:75–99.

Frimmer, M., Petzinger, E., and Ziegler, K., 1980, Protective effect of anionic cholecystographic agents against phalloidin on isolated hepatocytes by competitive inhibition of the phallotoxin uptake, *Naunyn Schmiedebergs Arch. Pharmacol.* **13**:85–89,

Furuno, K., Miwa, N., and Kato, K., 1983, Receptor mediated introduction of pepstatin-asialofetuin conjugates into lysosomes of rat hepatocytes, *J. Biochem.* **93**:249–256.

Geuze, H. J., Slot, J. W., Strous, G. J. A. M., Lodish, H. F., and Schwartz, A. L., 1983, Intracellular site of asialoglycoprotein receptor ligand uncoupling: Double label immuno electron microscopy during receptor mediated endocytosis, *Cell* **32**:1277–1287.

Ghitescu, J., and Fixman, A., 1984, Surface charge distribution on the endothelial cell of liver sinusoids, *J. Cell Biol.* **99**:639–647.

Ghose, T., Blair, A. H., Vaugham, K., and Kulkarni, P., 1982, Antibody-directed drug targeting in cancer therapy, in: *Targeted Drugs* (E. P. Goldberg and J. Wiley, eds.), Plenum Press, New York, pp. 1–22.

Givol, D., Pecht, I., Hochman, J., Schlessinger, J., and Steinberg, I. Z., 1974, Conformation changes in the Fab and Fc of the antibody as a consequence of antigen binding, in: *Progress in Immunology II*, Volume 1, North-Holland, Amsterdam, pp. 39–48.

Gloff, C. A., and Benet, L. Z., 1990, Pharmacokinetics and protein therapeutics, *Adv. Drug Deliv. Rev.* **4**:359–386.

Goldsmith, M. A., Jones, A. L., Underdown, B. J., and Schiff, J. M., 1987, Alterations in protein transport events in rat liver after estrogen treatment, *Am. J. Physiol.* **253**:G195–G200.

Goldstein, J. L., and Brown, M. S., 1974, Binding and degradation of low density lipoproteins by cultured human fibroblasts: Comparison of cells from normal subject and from a patient with homozygous familial hypercholesterolemia, *J. Biol. Chem.* **249**:5153–5162.

Goldstein, J. L., and Brown, M. S., 1977, The low density lipoprotein pathway and its relation to atherosclerosis, *Annu. Rev. Biochem.* **46**:879–930.

Gordon, S., and Rabinowitz, S., 1989, Macrophages as targets for drug delivery, *Adv. Drug. Deliv. Rev.* **4**:27–47.

Gores, J., LaRusso, N. F., and Miller, L. J., 1986a, Hepatic processing of cholecystokinin peptides I, *Am. J. Physiol.* **250**:344–349.

Gores, J., Miller, L., and LaRusso, N. F., 1986b, Hepatic processing of cholecystokinin peptides II, *Am. J. Physiol.* **250**:350–356.

Gores, J., Kost, L. J., Miller, L. J., and La Russo, N. J., 1989, Processing of cholecystokinin by isolated liver cells, *Am. J. Physiol.* **257**:G242–G248.

Groothuis, G. M. M., Hardonk, M. J., Keulemans, K. T. P., Nieuwenhuis, P., and Meijer, D. K. F., 1982, Autoradiographic and kinetic demonstration of acinar heterogeneity of taurocholate transport, *Am. J. Physiol.* **243**:455–462.

Groothuis, G. M. M., Hardonk, M. J., and Meijer, D. K. F., 1985, Hepatobiliary transport of drugs: Do periportal and perivenous hepatocytes perform the same job? *Trends Pharmacol. Sci.* **6**:322–327.

Gumucio, J. J., 1983, Functional and anatomical heterogeneity in the liver acinus: Impact on transport, *Am. J. Physiol.* **244**:578–582.

Gumucio, J. J., 1989, The coming of age from the description of a biological curiosity to a partial understanding of its physiological meaning and regulation, *Hepatology* **9:**154–160.

Gumucio, J. J., and Chianale, J., 1988, Liver cell heterogeneity and liver function, in: *The Liver: Biology and Pathobiology*, (I. M. Arias, W. B. Jakoby, H. Popper, D. Schachter and D. A. Shafritz, eds.), Raven Press, New York, pp. 932–947.

Haber, E., Hui, K. Y., Carlson, W. D., and Bernatowicz, M. S., 1987, Renin inhibitors: A search for principles of design, *J. Cardiovasc. Pharm.* **7:**54–58.

Haberland, M. E., and Fogelman, A. M., 1985, Scavenger receptor mediated recognition of maleyl bovine plasma albumin and the demaleylated protein in human monocyte macrophages, *Proc. Natl. Acad. Sci. USA* **82:**2693–2697.

Hardonk, M. J., and Scholtens, H. B., 1980, A histochemical study about the zonal distribution of the galactose-binding protein in rat liver, *Histochemistry* **69:**289–297.

Havel, R. J., and Hamilton, R. C., 1988, Hepatic lipoprotein receptors and intracellular lipoprotein catabolism, *Hepatology* **8:**1689–1704.

Hendriks, H. F. J., Brouwer, A., and Knook, D. L., 1987, The role of hepatic fat storing (stellate) cells in retinoid metabolism, *Hepatology* **7:**1368–1371.

Herschmann, H. R., Cawley, D., and Simpson, D. L., 1984, Toxic conjugates of epidermal growth factor and asialofetuin, in: *Receptor-Mediated Targeting of Drugs* (G. Gregoriades, G. Poste, J. Senior, and A. Trouet, eds.), Plenum Press, New York, pp. 27–51.

Hirai, H., Tsakada, Y., Koji, T., Ishii, N., Kaneda, H., and Kasai, Y., 1983, Attempts of treatment of hepatoma with antibody to alpha-fetoprotein, *Protides Biol. Fluids Proc. Colloq.* **31:**357–365.

Ho, N. F. H., Day, J. S. Barsuhn, C. L., Burton, P. S. and Raub, T. J., 1990, Biophysical model approaches to mechanistic transepithelial studies of peptides, *J. Controlled Release* **11:** 3–24.

Horiuchi, S., Takata, K., Murakami, M., and Morino, Y., 1987, Receptor-mediated endocytosis of aldehyde-modified proteins by sinusoidal liver cells, *J. Protein Chem.* **6:**191–205.

Hruby, V. J., 1985, Design of peptide hormone and neurotransmitter analogues, *Trends Pharmacol. Sci.* **6:**259–262.

Hubbard, A. L., Wilson, G., Ashwell, G., and Stukenbrok, H., 1979, An electron microscopic autoradiographic study of carbohydrate recognition systems in the rat liver, *J. Cell Biol.* **83:**47–64.

Hui, K. Y., Holtzman, E. J., Quinones, M. A., Hollenberg, N. K., and Haber, E., 1988, Design of renin inhibitory peptides, *J. Med. Chem.* **31:**1679–1686.

Hunter, E. B., Powers, S. P., Kost, L. J., Pinon, D. I., Miller, L. J., and LaRusso, N. F., 1990, Physicochemical determinants in hepatic extraction of small peptides, *Hepatology* **12:** 76–82.

Hwang, K. M., Foon, K. A., Cheung, P. H., Pearson, J. W., and Oldham, R. K., 1984, Selective antitumor effect on L10 hepatocarcinoma cells of a potent immunoconjugate composed of the A chain of abrin and a monoclonal antibody to a hepatoma-associated antigen, *Cancer Res.* **44:**4578–4586.

Inoue, M., Hirata, E., Morino, Y., Nagase, S., Roy Chowdhury, J., Roy Chowdhury, N., and Arias, I. M., 1985, The role of albumin in the hepatic transport of bilirubin: Studies in mutant analbuminemic rats, *J. Biochem.* **97:**737–743.

Irie, S., and Tavassoli, M., 1986, Liver endothelium desialylates ceruloplasmin, *Biochem. Biophys. Res. Commun.* **140:**94–100.

Ishikawa, T., Müller, M., Klünemann, C., Schaub, T., and Keppler, D., 1990, ATP-dependent primary active transport of cysteinyl leukotrienes across liver canalicular membrane. Role of the ATP-dependent transport system for glutathione S-conjugates, *J. Biol. Chem.* **265:** 19279–19286.

Jansen, R. W., Molema, G., Ching, T. L., Oosting, R., Harms, G., Moolenaar, F., Hardonk, M. J., and Meijer, D. K. F., 1991a, Hepatic endocytosis of various types of mannose-terminated albumins; what is important, sugar recognition, net charge or the combination of these features. *J. Biol. Chem.* **266:**3343–3348.

Jansen, R. W., Molema, G., Harms, G., Kruijt, J. K., Van Berkel, T. J. C., Hardonk, M. J., and Meijer, D. K. F., 1991b, Formaldehyde treated albumin contains monomeric and poly-meric forms that are differently cleared by endothelial and Kupffer cells of the liver: Evidence for scavenger receptor heterogeneity, *Biochem. Biophys. Res. Commun.* **180:** 23–32.

Jones, A. L., Hradek, G. T., Renston, R. H., Wong, K. Y., Karlaganis, G., and Paumgartner, G. 1980, Autoradiographic evidence for hepatic lobular concentration gradient of bile acid derivative, *Am. J. Physiol.* **238:**233–237.

Jones, E. A., and Summerfield, J. A., 1988, Kupffer cells, in: *The Liver: Biology and Pathobiology*, 2nd ed. (I. M. Arias, W. B. Jakoby, H. Popper, D. Schachter and D. A. Shafritz, eds.), Raven Press, New York, pp. 683–703.

Jungermann, K., 1986, Functional heterogeneity of periportal and perivenous hepatocytes, *Enzyme* **35:**161–180.

Kahan, B. D., 1989, Pharmacokinetics and pharmacodynamics of cyclosporine, *Transplant. Proc.* **21:**9–15.

Kamimoto, Y., Gatmaitan, Z., Hsu, J., and Arias, I. M., 1989, The function of Gp 170, the multidrug resistance gene product in rat liver canalicular membrane vesicles, *J. Biol. Chem.* **264:**11693–11698.

Kaplan, J., and Keogh, E. A., 1983, Studies in the physiology of macrophage receptors for α-macroglobulin protease complexes, *Ann. N.Y. Acad. Sci.* **420:**442–455.

Kataoka, H., and Tavassoli, M., 1985, Identification of lectin-like substances recognizing galactosyl residues of glycoconjugates in the plasma membrane of marrow sinus endo-thelium, *Blood* **65:**1165–1171.

Kato, Y., Tsukada, Y., Hara, T., and Hirai, H., 1983, Enhanced antitumor activity of mitomycin C conjugated with anti-alphafetoprotein antibody by a novel method of conjugation, *J. Appl. Biochem.* **5:**313–319.

Kessler, H., Gehrke, M., Haupt, A., Klein, M., Müller, A., and Wagner, K., 1986, Common structural features for cytoprotection activities of somatostatin, antamanide and related peptides, *Klin. Wochenschr.* **64:**74–78.

Kessler, H., Haupt, A., Frimmer, M., and Ziegler, K., 1987, Synthesis of a cyclic retro analog of somatostatin suitable for photoaffinity labeling, *Int. J. Peptide Protein Res.* **29:** 621–628.

Kessler, H., Schudock, M., Ziegler, K., and Frimmer, M., 1988, Peptide conformations (1): Synthesis and structure activity relationships of side chain modified peptides of cyclo(D-Pro-Phe-Thr-Lys-Trp-Phe), *Int. J. Peptide Protein Res.* **32:**183–193.

Kirshan, A., 1975, Rapid flow cytofluorometric analysis of mammalian cell cycle by pro-pidium iodide staining, *J. Cell Biol.* **66:**188–193.

Kitamura, T., Jansen, P., Hardenbrook, C., Kamimoto, Y., Gatmaitan, Z., and Arias, I. M.,

1990, Defective ATP-dependent bile canalicular transport of organic anions in mutant (TR⁻) rats with conjugated hyperbilirubinemia, *Proc. Natl. Acad. Sci. USA* **87**:3557–3561.

Klaassen, C. D., and Watkins, J. B., 1984, Mechanisms of bile formation, hepatic uptake and biliary excretion, *Pharmacol. Rev.* **36**:1–67.

Kobayashi, K., Sogame, Y., Hara, H., and Hayashi, K., 1990, Mechanisms of glutathione S-conjugate transport in canalicular and basolateral rat liver plasma membranes, *J. Biol. Chem.* **265**:7737–7741.

Koch, Y., Battini, F., and Peterkofsky, A., 1982, [³H]cyclo-(histidyl-proline) in rat tissues: Distribution, clearance, and binding, *Biochem. Biophys. Res. Commun.* **104**:823–829.

Kolb-Bachofen, V., 1986, Mammalian lectins and their function—A review, *Lectins* **V**:197–206.

Kolb-Bachofen, V., Schlepper-Schafer, J., Roos, P., Hulsmann, D., and Kolb, H., 1984, GalNac/Gal specific rat liver lectins: Their role in cellular recognition, *Biol. Cell* **51**:219–226.

Kramer, W., Bickel, Y., Bascher, H. P., Gerok, W., and Kurz, G., 1982, Bile-salt binding polypeptides in plasma membranes of hepatocytes revealed by photoaffinity labeling, *Eur. J. Biochem.* **129**:13–24.

Kröncke, K. D., Fricker, G., Meier, P. J., Gerok, W., and Kurz, G., 1986, Alpha-amanitin uptake into hepatocytes, *J. Biol. Chem.* **261**:12562–12567.

Kukongviriyapan, V., and Stacey, N. H., 1988, Inhibition of taurocholate transport by cyclosporin A in cultured rat hepatocytes, *J. Pharmacol. Exp. Ther.* **247**:685–689.

Kurz, G., Muller, M., Schramm, U., and Gerok, W., 1989, Identification and function of bile salt binding polypeptides of hepatocyte membrane, in: *Hepatic Transport of Organic Substances* (R. K. H. Kinne, E. Petzinger, and H. Sies, eds.), Springer-Verlag, Berlin, pp. 207–278.

Levy, D., 1989, Characterization of the bile acid transport system in normal and transformed hepatocytes using monoclonal antibodies, in: *Hepatic Transport of Organic Substances* (R. K. H. Kinne, E. Petzinger, and H. Sies, eds.), Springer-Verlag, Berlin, pp. 279–285.

Lochs, H., Morse, E. L., and Adibi, S. A., 1986, Mechanism of hepatic assimilation of dipeptides, *J. Biol. Chem.* **261**:14976–14981.

Lombardo, Y. B., Morse, E. L., and Adibi, S. A., 1988, Specificity and mechanism of influence of amino acid residues on hepatic clearance of oligopeptides, *J. Biol. Chem.* **263**:12920–12926.

McCullough, A. J., and Tavil, A. S., 1987, Hepatic protein metabolism: Basic and applied biochemical clinical aspects, in: *The Liver Annual* (I. M. Arias, M. Frenkel, and J. H. P. Wilson, eds.), Elsevier, Amsterdam, pp. 49–90.

Marti, U., Burwen, S. J., and Jones, D. L., 1989, Biological effects of epidermal growth factor, with emphasis on the gastro-intestinal tract and liver: An update, *Hepatology* **9**:126–138.

Mayer, R. J., and Doherty, F., 1985, Intracellular protein catabolism: State of the art, *FEBS Lett.* **198**:181–193.

Maynard, Y., and Baenziger, J. U., 1981, Oligosaccharide specific endocytosis by isolated rat hepatic reticuloendothelial cells, *J. Biol. Chem.* **256**:8063–8068.

Meier, P. J., 1988, Transport polarity of hepatocytes, *Semin. Liver Dis.* **8**:293–307.

Meijer, D. K. F., 1989, Transport and metabolism in the hepatobiliary system, in: *Handbook of Physiology: The Gastrointestinal System*, Volume III (J. Shulz, G. Forte, and B. B. Rauner, eds.), Oxford University Press, London, pp. 717–758.

Meijer, D. K. F., and van der Sluijs, P., 1989, Covalent and noncovalent protein binding of drugs: Implications for hepatic clearance, storage, and cell-specific drug delivery, *Pharm. Res.* **6:**105–118.

Meijer, D. K. F., Vonk R. J., Keulemans, K., and Weitering, J. G., 1977, Hepatic uptake and biliary excretion of dibromosulphthalein, albumin dependence, influence of phenobarbital and nafenopin pretreatment and the role of Y- and Z-protein, *J. Pharmacol. Exp. Ther.* **202:**8–21.

Meijer, D. K. F., Mol. W., Müller, M., and Kurz, G., 1989, Mechanisms for hepatobiliary transport of cationic drugs studied with the intact organ and on the membrane level, in: *Proceedings of Life Sciences: Hepatic Transport of Organic Substances* (E. Petzinger, R. K. H., Kinne, and H. Sies, eds.), Springer-Verlag, Berlin, pp. 344–367.

Meijer, D. K. F., Jansen, R. W., and Molema, G., 1990a, Design of cell-specific drug targeting preparations for the liver: Where cell biology and medicinal chemistry meet, in: *Trends in Drug Research* (V. Claassen, ed.), Elsevier, Amsterdam, pp. 303–332.

Meijer, D. K. F., Mol, W. E. M., Muller, M., and Kurz, G., 1990b, Carrier-mediated transport of drugs in the hepatic distribution and elimination of drugs, with special reference to the category of organic cations, *J. Pharmacokinet. Biopharm.* **18:**35–70.

Mellman, I., Fuchs, R., and Helenius, A., 1986, Acidification of the endocytic and exocytic pathways, *Annu. Rev. Biochem.* **55:**663–700.

Miccio, M., Orzes, N., Lunazzi, G. C., Gazzin, B., Corsi, R., and Tiribelli, C., 1989, Reversal of ethinylestradiol-induced cholestasis by epomediol in rat. The role of liver plasma-membrane fluidity, *Biochem. Pharmacol.* **38:**3559–3563.

Michalak, T. I., and Bolger, G. T., 1989, Characterization of the binding sites for glutaraldehyde-polymerized albumin on purified woodchuck hepatocyte plasma membranes, *Gastroenterology* **96:**153–166.

Mohri, T., Uesugi, K., and Kamisaka, A., 1985, Buculome N-glucuronide: Purification and identification of a major metabolite of buculome in rat bile, *Xenobiotica* **15:**615–621.

Mol, W. E. M., Muller, M., Kurz, G., and Meijer, D. K. F., 1992, Characterization of a hepatic uptake system for organic cations with a photoaffinity label of procainamidethobromide, *Biochem Pharmacol.* **43:**2217–2226.

Molteni, L., 1979, Dextrans as drug carriers, in: *Drug Carriers in Biology and Medicine* (G. Gregoriades, ed.), Academic Press, New York, pp. 107–125.

Monnier, V. M., 1990, Nonenzymatic glycosylation, the Maillard reaction and the aging process, *J. Gerontol.* **45:**B105–B111.

Monsigny, M., Roche, A. C., and Midoux, P., 1988, Endogenous lectins and drug targeting, *Ann. N.Y. Acad. Sci.* **551:**399–414.

Mori, M., Yamada, M., Yamagushi, M., Suzuki, M., Oshima, K., Kobayashi, I., and Kobayashi, S., 1986, Cyclo(His-Pro), a metabolite of thyrotropin-releasing hormone, *Biochem. Biophys. Res. Commun.* **134:**443–451.

Mori, M., Yamada, M., Iriuchijima, T., Murakami, M., and Kobayashi, S., 1989, Characterization and solubilization of cyclo(His-Pro) binding from rat liver membranes, *Proc. Soc. Exp. Biol. Med.* **191:**108–112.

Nicolau, C., and Budd, A., 1989, Liposomes as vesicles, in: *Critical Reviews in Therapeutic Drug Carrier Systems*, Vol. 6, CRC Press, Chicago, pp. 239–271.

Ohare, K. B., Hume, I. C., Scarlett, L., Chytry, V., Kopeckova, P., Kopecek, J., and Duncan, R., 1989, Effect of galactose on interaction of N(2-hydroxypropyl)methacrylamide copolymers with hepatoma cells in culture—Preliminary application to an anticancer agent, daunomycin, *Hepatology* **10**:207–214.

Oude Elferink, R. P. J., Ottenhoff, R., Liefting, W. G. M., Schoenmaker, B., Groen, A. K., and Jansen, P. L. M., 1990, ATP-dependent efflux of GSSG and GS-conjugate from isolated rat hepatocytes, *Am. J. Physiol.* **258**:G699–G706.

Petzinger, E., 1981, Competitive inhibition of the uptake of demethylphalloin by cholic acid in isolated hepatocytes. Evidence for a transport competition rather than a binding competition, *Naunyn Schmiedebergs Arch. Pharmacol.* **316**:345–349.

Petzinger, E., and Frimmer, M., 1980, Comparative studies on the uptake of ^{14}C-bile acids and ^{3}H-demethylphalloin in isolated rat liver cells, *Arch Toxicol.* **44**:127–135.

Petzinger, E., and Frimmer, M., 1984, Driving forces in hepatocellular uptake of phalloidin and cholate, *Biochim. Biophys. Acta* **778**:539–548.

Petzinger, E., Joppen, C., and Frimmer, M., 1983, Common properties of hepatocellular uptake of cholate, iodipamide and antamanide, as distinct from the uptake of bromosulfophthalein, *Naunyn Schmiedebergs Arch. Pharmacol.* **322**:174–179.

Pompipom, M. M., Bugianesi, R. L., Robbins, J. C., Doebben, T. W., and Shen, T. Y., 1984, Receptor mediated targeting of drugs, in: *Receptor-Mediated Targeting of Drugs* (G. Gregoriades, J. Poste, J. Senior, and A. Trouet, eds.), Plenum Press, New York, pp. 53–71.

Pond, S. M., and Tozer, T. N., 1984, First pass elimination: Basic concepts and clinical consequences, *Clin. Pharmacokinet.* **9**:1–25.

Poste, G., 1984, Drug targeting in human cancer chemotherapy, in: *Receptor-Mediated Targeting of Drugs* (G. Gregoriades, S. Poste, J. Senior, and A. Trouet, eds.), Plenum Press, New York, pp. 1–25.

Potter, B. J., Sorrentino, D., and Berk, P. D., 1989, Mechanisms of cellular uptake of fatty acids, *Annu. Rev. Nutr.* **9**:253–270.

Poznansky, M. J., and Juliano, R. L., 1984, Biological approaches to the controlled delivery of drugs: A critical review, *Pharmacol. Rev.* **36**:277–336.

Praaning Van Dalen, D. P., Brouwer, A., and Knook, K. L., 1981, Clearance capacity of rat liver Kupffer, endothelial, and parenchymal cells, *Gastroenterology* **81**:1036–1044.

Regoeczi, E., Chindemi, P. A., Hatton, M. W. C., and Berg, L. R., 1980, Galactose-specific elimination of human asialotransferrin by bone marrow in the rabbit, *Arch. Biochem. Biophys.* **205**:76–84.

Regoeczi, E., Charlwood, P. A., and Chindemi, P., 1985, The effects of cytotropic compounds on the resialylation of human asialotransferrin type 3 in the rat, *Ex. Cell Res.* **157**:495–503.

Roerdink, F. H., Daemen, T., Bakker-Woudenberg, I. A. J. M., Horm, G., Crommelin, D. J. A., and Scherphof, G. L., 1988, Therapeutic utility of liposomes, in: *Drug Delivery Systems, Fundamentals and Techniques* (P. Johnson and J. G. Lloyd-Jones, eds.), VCH, New York, pp. 66–80.

Ruetz, S., Hugentobler, G. and Meier, P. J., 1988, Functional reconstitution of the canalicular bile salt transport system of rat liver, *Proc. Natl. Acad. Sci. USA* **85**:6147–6151.

Ruifrok, P. G., and Meijer, D. K. F., 1982, Sodium coupled uptake of taurocholate by rat liver plasma membrane vesicles, *Liver* **2**:28–34.

Russel, F. G. M., Weitering, J. G., Oosting, R., Groothuis, G. M. M., Hardonk, M. J., and Meijer, D. K. F., 1983, Influence of taurocholate on hepatic clearance and biliary excretion of asialo intestinal alkaline phosphatase in the rat in vivo and in isolated perfused rat liver, *Gastroenterology* **85**:225–234.

Ryser, H. J. P., and Shen, W. C., 1986, Drug-poly(lysine) conjugates: Their potential for chemotherapy and for the study of endocytosis, in: *Targeting of Drugs with Synthetic Systems* (G. Gregoriades, J. Senior, and G. Poste, eds.), Plenum Press, New York, pp. 103–121.

Samlowski, W. E., Braaten, B. A., and Daynes, R. A., 1985, Characterization of the in vitro interaction of PNA[hi] lymphocytes with the bone marrow and hepatic asialoglycoprotein receptors, *Cell. Immunol.* **95**:1–14.

Sänger, U., 1989, Isolation of protein components of the hepatocellular multispecific bile acid transport system by affinity chromatography using a linear peptide with renin inhibitory activity as ligand, *Naunyn Schmiedebergs Arch. Pharmacol.* **340**:86.

Sänger, U., and Ziegler, K., 1990, Partial identification and isolation of binding proteins for linear renin inhibiting peptides in isolated rat liver plasma membranes by affinity chromatography, *Biol. Chem. Hoppe Seyler* **371**:813.

Sano, A., Taylor, M. E., Leaning, M. S., and Summerfield, J. A., 1990, Uptake and processing of glycoproteins by isolated rat hepatic endothelial and Kupffer cells, *J. Hepatol.* **10**:211–216.

Schneider, Y. J., Abarca, J., Aboud-Pirak, E., Baurain, E., Ceulemans, F., Deprez-De Campeneere, D., Lesur, B., Masquelier, C., Otte-Schlachmuylder, C., Rolin van Swieten, D., and Trouet, A., 1984, Drug targeting in human cancer chemotherapy, in: *Receptor-Mediated Targeting of Drugs* (G. Gregoriades, S. Poste, J. Senior, and A. Trouet, eds.), Plenum Press, New York, pp. 1–25.

Schwartz, A. L., 1984, The hepatic asialoglycoprotein receptor, *CRC Crit. Rev. Biochem.* **16**:207–223.

Schwartz, B. A., and Gray, G. R., 1977, Proteins containing reductively aminated disaccharides: Synthesis and characterization, *Arch. Biochem. Biophys.* **181**:542–549.

Schwartz, G. J., and Al-Awqati, Q., 1986, Regulation of transepithelial proton transport by exocytosis and endocytosis, *Annu. Rev. Physiol.* **48**:153–161.

Seeberger, A. and Ziegler, K., 1990, Characterization of the hepatocellular uptake of the hydrophilic linear peptide EMD 56133, *Biol. Chem. Hoppe Seyler* **371**:749.

Seeberger, A., and Ziegler, K., 1993, Characterization of the hepatocellular transport of hydrophilic linear peptides with renin-inhibiting activity, *Naunyn Schmiedebergs Arch. Pharmacol,* in press.

Seno, M., Seino, Y., Takemura, Y., Nishi, S., Ishida, H., Kitano, N., and Imura, H., 1985, Comparison of somatostatin-28 and somatostatin-14 clearance by the perfused rat liver, *Can. J. Physiol. Pharmacol.* **63**:62–67.

Shea, C. R., Chen, N., Wimberly, J., and Hasan, T., 1989, Rhodamine dyes as potential agents for photochemotherapy of cancer in human bladder carcinoma cells, *Cancer Res.* **49**:3961–3965.

Shen, T. Y., 1987, Preferential membrane permeation and receptor recognition in drug targeting, in: *Bioreversible Carriers in Drug Design, Theory and Application* (B. Roche, ed.), Pergamon Press, Elmsford, N.Y., pp. 214–225.

Shen, W. C., and Ryser, H. J. P., 1981, cis-Aconityl spacer between daunomycin and macromolecular carriers: A model of pH-sensitive linkage by releasing drug from a lysosomotropic conjugate, *Biochem. Biophys. Res. Commun.* **102**:1048–1054.

Shepherd, V. L., 1989, Intracellular pathways and mechanisms of sorting in receptor-mediated endocytosis, *Trends Pharmacol. Sci.* **10**:458–462.

Shouval, D., Wands, J., Zurawski, V. R., Isselbacker, K. J., and Shafrik, D. A., 1982, Selective binding and complement mediated lysis of human hepatoma cells (PLC/PRF15) in culture by monoclonal antibodies to hepatitis B surface antigen, *Proc. Natl. Acad. Sci. USA* **77**: 650–654.

Silverstein, S. C., Steinman, R. M., and Cohn, Z. A., 1977, Endocytosis, *Annu. Rev. Biochem.* **46**:469–722.

Simon, F. R., Gonzales, M., Sutherland, E., Accatino, L., and Davis, R. A., 1980, Reversal of estradiol-induced bile secretory failure with Triton WR 1339, *J. Clin. Invest.* **65**:851–860.

Simpson, D. L., Cawley, D. B., and Herschmann, H., 1982, Killing of cultured hepatocytes by conjugates of asialofetuin and epidermal growth factor linked to the A chain of ricin or diphtheria toxin, *Cell* **29**:469–473.

Smedsrød, B., Pertoft, H., Gustafson, S., and Laurent, T. C., 1990, Scavenger function of the liver endothelial cells, *Biochem. J.* **266**:313–327.

Smit, M. J., Duursma, A. M., Koudstaal, J., Hardonk, M. J., and Bouma, J. M. W., 1991, Infection of mice with lactate dehydrogenase-elevating virus destroys the subpopulation of Kupffer cells involved in receptor-mediated endocytosis of lactate dehydrogenase and other enzymes, *Hepatology* **12**:1192–1200.

Smith, R. L., 1966, The biliary excretion and enterohepatic circulation of drugs and other organic compounds, *Prog. Drug Res.* **9**:299–360.

Smith, R. L., 1971, Excretion of drugs in bile, in: *Handbook of Experimental Pharmacology* (B. Brodie and J. G. Gilette, eds.), Springer-Verlag, Berlin, pp. 354–389.

Soriano, P., Dijkstra, J., Legrand, A., Spanjer, H., Londos, D., Roerdink, F., Scherphof, G., and Nicolau, C., 1983, Targeted and non-targeted liposomes for in vivo transfer to rat liver cells of a plasmid containing the preproinsulin I gene, *Proc. Natl. Acad. Sci. USA* **80**: 7128–7131.

Sorrentino, D., and Berk, P., 1988, Mechanistic aspects of hepatic bilirubin uptake, *Semin. Liver Dis.* **8**:119–136.

Sottocasa, G. L., Baldini, G., Sandri, G., Lunazzi, G., and Tiribelli, C., 1982, Reconstitutions in vitro of sulfobromophthalein transport by bilitranslocase, *Biochim. Biophys. Acta* **685**: 123–128.

Sottrup-Jensen, L., 1987, α_2-Macroglobulin and related thiol ester plasma proteins, in: *The Plasma Proteins*, Volume 5, 2nd ed. (F. W. Putman, ed.), Academic Press, New York, pp. 192–291.

Stadalnik, R. C., Vera, D. R., Woodle, E. S., Turdeau, W. L., Porter, D. A., Ward, R. E., Krohn, K. A., and O'Grady, L. F., 1985, Technetium-99m NGA functional hepatic imaging: Preliminary clinical experience, *J. Nucl. Med.* **26**:1233–1242.

Stahl, P. D., 1990, The macrophage mannose receptor—Current status, *Am. J. Respir. Cell Mol. Bio.* **2**:317–318.

Stahl, P. D., and Schlesinger, P. H., 1980, Receptor-mediated pinocytosis of mannose/N-acetylglucosamine-terminated glycoproteins and lysosomal enzymes by macrophages, *Trends Biochem. Sci.* **5**:194–196.

Stahl, P., and Schwartz, A. L., 1986, Receptor mediated endocytosis, *J. Clin. Invest.* **77**: 657–662.

Steen, H., Oosting, R., and Meijer, D. K. F., 1991, Mechanisms for uptake of cationic drugs in the liver. A study with tributylmethylammonium (TBuMA) concerning substrate specificity and potential driving forces, *J. Pharmacol.* **258**:537–543.

Steer, C. J., and Ashwell, G., 1986, Hepatic membrane receptors for glycoproteins, in: *Progress in Liver Diseases*, Volume VIII (H. Popper and F. Schafner, eds.), Grune & Stratton, New York, pp. 99–123.

Stein, W. D., 1986, *Transport and Diffusion across Cell Membranes*, Academic Press, New York.

Stockert, R. J., and Morell, A. G., 1983, Hepatic binding protein: The galactose specific receptor of mammalian hepatocytes, *Hepatology* **3**:750–757.

Stoorvogel, W., Geuze, H. J., Griffith, J. M., Schwartz, A. L., and Strous, G. J., 1989, Relation between intracellular pathways of the receptors for transferrin, asialoglycoprotein and mannose 6-phosphate in human hepatoma cells, *J. Cell Biol.* **108**:2137–2148.

Stowell, C. P., and Lee, Y. C., 1980, Neoglycoproteins: The preparation and application of synthetic glycoproteins, *Adv. Carbohydr. Chem. Biochem.* **37**:225–281.

Stremmel, W., 1989, Mechanisms of hepatic fatty acid uptake, *J. Hepatol.* **9**:374–382.

Strous, G. J., 1989, Relation between intracellular pathways of the receptors for transferrin, asialoglycoprotein and mannose 6-phosphate in human hepatoma cells, *J. Cell Biol.* **108**:2137–2148.

Suchy, F. J., Ananthanarayanan, M., and Bucuvalas, J. C., 1989, The ontogeny of hepatic bile acid transport, in: *Hepatic Transport of Organic Substances* (R. K. H. Kinne, E. Petzinger, and H. Sies, eds.), Springer-Verlag, Berlin, pp. 257–265.

Sugiyama, Y., and Hanano, M., 1989, Receptor-mediated transport of peptide hormones and its importance in the overall hormone disposition in the body, *Pharm. Res.* **6**:192–202.

Takakura, Y., Fujita, T., Hashida, M., and Sezaki, H., 1990, Disposition characteristics of macromolecules in tumor-bearing mice, *Pharm. Res.* **7**:339–346.

Takata, K., Horiuchi, S., Araki, N., Shiga, M., Saitoh, M., and Morino, Y., 1988, Endocytic uptake of nonenzymatically glycosylated proteins is mediated by a scavenger receptor for aldehyde-modified proteins, *J. Biol. Chem.* **263**:14819–14825.

Tancredi, T., Zanotti, G., Rossi, F., Benedetti, E., Pedone, C., and Temussi, P. A., 1989, Comparison of the conformations of cyclolinopeptide A in the solid state and in solution, *Biopolymers* **28**:513–523.

Taylor, M. E., Leaning, M. S., and Summerfield, J. A., 1987, Uptake and processing of glycoproteins by rat hepatic mannose receptor, *Am. J. Physiol.* **252**:E690–E698.

Thung, S. N., Wang, D., Fasy, T. M., Hood, A., and Gerber, M. A., 1989, Hepatitis B surface antigen binds to human serum albumin cross-linked by transglutaminase, *Hepatology* **9**:726–730.

Thurman, R. G., and Kauffman, F. C., 1985, Sublobular compartmentation of pharmacological events (SCOPE): Metabolic fluxes in periportal and pericentral regions of the liver lobule, *Hepatology* **5**:144–151.

Tiribelli, C., Lunazzi, G. C., and Sottocasa, G. L., 1990, Biochemical and molecular aspects of the hepatic uptake of organic anions, *Biochim. Biophys. Acta* **1031**:261–275.

Tomlinson, E., 1990, Control of the biological dispersion of therapeutic proteins, in: *Protein Design and the Development of New Therapeutics and Vaccines* (J. B. Hook and G. Poste, eds.), Plenum Press, New York, pp. 331–357.

Townsend, R. R., Wall, D. A., Hubbard, A. L., and Lee, A. C., 1984, Rapid release of galactose-terminated ligands after endocytosis by hepatic parenchymal cells. Evidence for a role of carbohydrate structure in the release of internalized ligand from receptor, *Proc. Natl. Acad. Sci. USA* **81:**466–470.

Trouet, A., Baurain, R., Deprez-De Campeneere, D., Masquelier, M., and Pirson, P., 1982a, Targeting of antitumor and antiprotozoal drugs by covalent linkage to protein carriers, in: *Targeting of Drugs* (G. Gregoriades, J. Senior, and A. Trouet, eds.), Plenum Press, New York, pp. 19–35.

Trouet, A., Masquelier, M., Baurain, R., and Deprez-De Campeneere, D., 1982b, A covalent linkage between daunorubicin and proteins that is stable in serum and reversible by lysosomal hydrolysis as required for a lysosomotropic drug-carrier conjugate: In vitro and in vivo studies, *Proc. Natl. Acad. Sci. USA* **79:**626–629.

Tsuji, A., Terasaki, T., Tamai, I., Nakashima, E., and Takanosu, K., 1985, A carrier-mediated transport system for benzylpenicillin in isolated rat hepatocytes, *J. Pharm. Pharmacol.* **37:**55–57.

Tsuji, A., Terasaki, T., Takanosu, K., Tamai, I., and Nakashima, E., 1986, Uptake of benzylpenicillin, cefpiramide and cefazolin by freshly prepared rat hepatocytes. Evidence for a carrier-mediated transport system, *Biochem. Pharmacol.* **35:**151–158.

Unadkat, J. D., Bartha, F., and Sheiner, L. B., 1986, Simultaneous modeling of pharmaco-kinetics and pharmacodynamics with nonparametric kinetic and dynamic models, *Clin. Pharmacol. Ther.* **40:**86–93.

van Berkel, T. J. C., Kruyt, J. K., Harkes, L., Nagelkerke, J. F., Spanjer, H., and Kempen, H. J. M., 1986, Receptor-dependent targeting of native and modified lipoproteins to liver cells, in: *Site-Specific Drug Delivery* (E. Tomlinson and S. S. Davis, eds.), Wiley, New York, pp. 49–68.

van Berkel, T. J. C., Dekker, D. J., Kruyt, J. K., and Van Eyk, H. G., 1987, Internalization in vivo of transferrin and asialotransferrin with liver cells, *Biochem. J.* **243:**715–722.

van der Laan-Klamer, S. M., Harms, G., Atmosoerodjo, J. E., Meijer, D. K. F., Hardonk, M. J., and Hoedemaker, P. J., 1986, Studies on the mechanism of binding and uptake of immune complexes by various cell types of rat liver in vivo, *Scand J. Immunol.* **23:** 127–133.

van der Sluijs, P., and Meijer, D. K. F., 1991, Limitations on the specificity of targeting asialoglycoprotein-drug conjugates to hepatocytes, in: *Targeted Diagnosis and Therapy of Liver Diseases: Cell-Surface Receptors and Liver-Directed Agents* (G. Y. Wu and C. H. Wu, eds.), Dekker, New York, pp. 235–264.

van der Sluijs, P., Oosting, R., Weitering, J. G., Hardonk, M. J., and Meijer, D. K. F., 1985, Biliary excretion of FITC metabolites after administration of FITC-labeled asialo-orosomucoid as a measure of lysosomal proteolysis, *Biochem. Pharmacol.* **34:**1399–1405.

van der Sluijs, P., Bootsma, H. P., Postema, B., Moolenaar, F., and Meijer, D. K. F., 1986, Drug targeting to the liver with lactosylated albumins. Does the glycoprotein target the drug, or is the drug targeting the glycoprotein? *Hepatology* **6:**723–728.

van der Sluijs, P., Postema, B., and Meijer, D. K. F., 1987, Lactosylation of albumin reduces uptake rate of dibromosulfophthalein in perfused rat liver and dissociation rate from albumin in vitro, *Hepatology* **4:**688–695.

van der Sluijs, P., Braakman, I., Meijer, D. K. F., and Groothuis, G. M. M., 1988, Heterogeneous acinar localization of the asialoglycoprotein internalization system in rat hepatocytes, *Hepatology* **8:**1521–1529.

Vera, D. R., Krohn, K. A., Stadalnik, R. C., and Schneider, P. O., 1984, Tc-99 galactosyl-neoglycoalbumin: In vitro characterization of receptor mediated binding, *J. Nucl. Med.* **25**:779–787.

Vera, D. R., Stadalnik, R. C., and Krohn, K. A., 1985, Technetium 99m galactosylneoglyco-albumin: Preparation and preclinical studies, *J. Nucl. Med.* **26**:1157–1167.

Vlassara, H., Moldawar, L., and Chan, B., 1989, Macrophage monocyte receptor for non-enzymatically glycosylated proteins is upregulated by cachectin tumor necrosis factor, *J. Clin. Invest.* **84**:1813–1820.

Von Dippe, P., and Levy, D., 1990, Reconstitution of the immunopurified 49-kDa sodium-dependent bile acid transport protein derived from hepatocyte sinusoidal plasma membranes, *J. Biol. Chem.* **265**:14812–14816.

Vonk, R. J., Van Doorn, A. B. D., Mulder, G. J., and Meijer, D. K. F., 1978, The influence of bile salts on hepatocellular transport, *Falk Symp.* **26**:121–126.

Vonk, R. J., Danhof, M., Coenraads, T., Van Doorn, A. B. D., Keulemans, K., Scaf, A. H. J., and Meijer, D. K. F., 1979, Influence of bile salts on hepatic transport of dibromo-sulphthalein, *Am. J. Physiol.* **237**:524–534.

Weinman, S. A., Graf, J., and Boyer, J. L., 1989, Voltage driven taurocholate dependent secretion in isolated hepatocytes couplets, *Am. J. Physiol.* **256**:826–852.

Weiss, P., and Ashwell, G., 1989, The asialogylcoprotein receptor, properties and modulation by ligand, in: *Alpha₁-Acid Glycoprotein: Genetics, Biochemistry, Physiological Functions and Pharmacology* (P. Bauman, C. B. Eap, W. E. Muller, and J. P. Tillement, eds.), Liss, New York, pp. 169–184.

Wieland, T., Lüben, G., Fösel, J., Ottenheym, H., De Vries, S. X., Konz, W., Prox, A., and Schmid, J., 1969, Antamanide: Discovery, isolation, structure elucidation and synthesis, *Angew. Chem. Int. Ed. Engl.* **7**:204–208.

Wieland, T., Nassal, M., Kramer, W., Fricker, G., Bickel, U., and Kurz, G., 1984, Identity of hepatic membrane transport systems for bile salts, phalloidin and antamanide by photo-affinity labeling, *Proc. Natl. Acad. Sci. USA* **81**:5232–5236.

Wilson, G., 1986, Gene therapy: Rationale and realisation, in: *Site-Specific Drug Delivery* (E. Tomlinson and S. S. Davis, eds.), Wiley, New York, pp. 149–164.

Wilson, J. M., Wu, C. H., Wu, G. Y., Roy Chowdhury, N., Epstein, A., Waltman, R., and Roy Chowdhury, J., 1989, Temporary amelioration of hyperlipidemia in LDL receptor deficient rabbits by targeted gene delivery, *Hepatology* **10**:631.

Wisse, E., De Zanger, R. B., Charels, K., Van der Smissen, P., and McCuskey, R. S., 1985, The liver sieve: Considerations concerning the structure and function of endothelial fenestrae, the sinusoidal wall and the space of Disse, *Hepatology* **5**:683–692.

Wolfe, M. M., Doyle, J. W., and Mcguigan, J. E., 1987, Hepatic clearance of somatostatin and gastrin-releasing peptide, *Life Sci.* **40**:335–342.

Wright, T. L., 1989, Targeted therapy for hepatic cancer: Good in theory, problematic in practice, *Hepatology* **9**:657–658.

Wright, T. L., Roll, F. J., Jones, A. L., and Weisiger, R. A., 1988, Uptake and metabolism of polymerized albumin by rat liver, *Gastroenterology* **94**:443–452.

Wu, G. Y., 1988, Targeting in diagnosis and therapy, in: *The Liver: Biology and Pathobiology* (I. M. Arias, W. B. Jakoby, H. Popper, D. Schachter, and D. A. Shafritz, eds.), Raven Press, New York, pp. 1303–1313.

Wu, G. Y., and Wu, C. H., 1988, Evidence for targeted gene delivery to Hep G2 hepatoma cells in vitro, *Biochemistry* **27**:887–892.

Wu, G. Y., Wu, C. H., and Rubin, M. I., 1985, Acetaminophen hepatotoxicity and targeted rescue: A model for specific chemotherapy of hepatocellular carcinoma, *Hepatology* **5**: 709–713.

Wu, G. Y., Keegan-Rogers, V., Franklin, S., Midford, S., and Wu, C. H., 1988, Targeted antagonism of galactosamine toxicity in normal hepatocytes *in vitro*, *J. Biol. Chem.* **263**: 4719–4723.

Wu, G. Y., Wilson, J. M., and Wu, C. H., 1989, Targeting genes: Delivery and persistent expression of a foreign gene driven by mamalian regulatory elements *in vivo*, *J. Biol. Chem.* **264**:16985–16987.

Wu, G. Y., Tangco, M. V., and Wu, C. H., 1990, Targeted gene delivery: Persistence of foreign gene expression achieved by pharmacological means, *Hepatology* **12**:871 (Abstract).

Wu, G. Y., Wilson, J. M., Shalaby, F., Grossman, M., Shafritz, D. A., and Wu, C. H. 1991. Receptor-mediated gene delivery *in vivo*. Partial correction of genetic analbuminemia in nagase rats. *J. Biol. Chem.* **266**:14338–14362.

Wünsch, E., 1983, Peptide factors as pharmaceuticals: Criteria for application, *Biopolymers* **22**:493–505.

Yamashiro, D. J., and Maxfield, F. R., 1984, Acidification of endocytic compartments and the intracellular pathways of ligand and receptors, *J. Cell. Biochem.* **26**:231–246.

Yamashiro, D. J., Fluss, S. R., and Maxfield, F. R., 1983, Acidification of endocytic vesicles by an ATP dependent proton pump, *J. Cell Biol.* **97**(3):929–934.

Yamazaki, K., and Larusso, N. F., 1989, The liver and intracellular digestion: How liver cells eat, *Hepatology* **10**:877–886.

Yamazaki, M., Suzuki, H., Sugiyama, Y., Iga, T., and Hanano, M., 1990, Uptake of organic anions by isolated rat hepatocytes. A classification in terms of ATP dependency, *Proc. Third Int. Congr. Mathematical Modelling of Liver Excretory Processes*, Tokyo, pp. 322–328.

Zamora, J. M., Pearce, H. L., and Bock, W. T., 1988, Physical chemical properties shared by compounds that modulate multidrug resistance in human leukemic cells, *Mol. Pharmacol.* **33**:454–462.

Zanotti, G., Rossi, F., Di Blasio, B., Pedone, C., Benedetti, E., Ziegler, K., and Tancredi, T., 1990, Structure activity relationship in cytoprotective peptides, in: *Peptides* (J. E. Rivier and G. K. Marshall, eds.), Pierce Chemical Co., Rockford, Ill, pp. 118–119.

Ziegler, K., 1989, Identification of carrier proteins in hepatocytes by (photo)affinity labels derived from foreign cyclopeptides, in: *Hepatic Transport of Organic Substances* (E. Petzinger, R. K. H. Kinne, and H. Sies, eds.), Springer-Verlag, Berlin, pp. 317–326.

Ziegler, K., and Frimmer, M., 1986a, Molecular aspects of cytoprotection by modified somatostatins, *Klin. Wochenschr.* **64**:87–89.

Ziegler, K., and Frimmer, M., 1986b, Identification of cyclosporin binding sites in rat liver plasma membranes, isolated hepatocytes, and hepatoma cells by photoaffinity labeling using [^{3}H]cyclosporin-diaziridine, *Biochim. Biophys. Acta* **855**:147–156.

Ziegler, K., and Frimmer, M., 1986c, Cyclosporin A and a diaziridine derivative inhibit the hepatocellular uptake of cholate, phalloidin and rifampicin, *Biochim. Biophys. Acta* **855**: 136–142.

Ziegler, K., Frimmer, M., Kessler, H., Damm, I., Eiermann, V., Koll, S., and Zarbock, J., 1985, Modified somatostatins as inhibitors of a multispecific transport system for bile acids and phallotoxins in isolated hepatocytes, *Biochim. Biophys. Acta* **845:**86–93.

Ziegler, K., Frimmer, M., Kessler, H., and Haupt, A., 1988a, Azidobenzamido-008, a new photosensitive substrate for the multispecific bile acid transporter of hepatocytes, *Biochim. Biophys. Acta* **945:**263–272.

Ziegler, K., Polzin, G., and Frimmer, M., 1988b, Hepatocellular uptake of cyclosporin A by simple diffusion, *Biochim. Biophys. Acta* **938:**44–50.

Ziegler, K., Frimmer, M., Müllner, S., and Fasold, H., 1989, Bile acid binding proteins in hepatocellular membranes of newborn and adult rats. Identification of transport proteins with azidobenzamidotaurocholate, *Biochim. Biophys. Acta* **980:**161–168.

Ziegler, K., Pedone, C., and Kessler, H., 1990, Common structural features of cyclopeptides that inhibit bile acid uptake, *Biol. Chem. Hoppe Seyler* **371:**754–755.

Ziegler, K., Lins, W., and Frimmer, M., 1991, Hepatocellular transport of cyclosomatostatins: Evidence for a carrier system related to the multispecific bile acid transporter, *Biochim. Biophys. Acta* **1061:**287–296.

Zimmerli, B., Valantinas, J., and Meier, P. J., 1989, Multispecificity of Na$^+$-dependent taurocholate uptake in basolateral (sinusoidal) rat liver plasma membrane vesicles, *J. Pharmacol. Exp. Ther.* **250:**301–308.

Chapter 14

The Immune System as a Barrier to Delivery of Protein Therapeutics

Michael W. Konrad

1. INTRODUCTION

Our immune system recognizes infecting organisms and protein toxins as foreign, and by inactivating them often protects us from harmful effects. However, a protein therapeutic may also be seen as foreign. The result of being perceived as foreign can be benign, neutralize the activity of the therapeutic, or generate an adverse physiological response that may even be lethal. The protein may be intrinsically foreign because it has been obtained from another species and has a different amino acid sequence, or it may be perceived as foreign due to a more subtle difference, such as being denatured or aggregated. Low-molecular-weight drugs generally do not initiate an immune reaction unless they become conjugated to protein, as occasionally occurs with penicillin.

The purpose of this chapter is to summarize the structure of the immune system, methods used to study it, past experiences with protein therapeutics, and suggest ways to predict and minimize immune reactions. However, the specificity and intensity of an immune reaction is defined by the individual genetic makeup of the host as well as the chemical and physical structure of the potential immunogen. Thus, there must always be a fundamental uncertainty in the prediction of an immune response among the genetically heterogeneous human population. In addition, we will see that many factors other than the actual structure of the protein therapeutic influence immunogenicity.

Michael W. Konrad • The deVlaminck Institute, Lafayette, California 94549.

Biological Barriers to Protein Delivery, edited by Kenneth L. Audus and Thomas J. Raub. Plenum Press, New York, 1993.

2. STRUCTURE OF THE IMMUNE SYSTEM

A foreign substance (the immunogen) can interact with the immune system in different ways and produce both quantitatively and qualitatively different responses. In order to have a common vocabulary for later discussions, it will be useful to briefly outline some parts of what is now known about the immune system. Readers familiar with immunology may want to skip to Section 5, while those wanting to learn more about the subject will want to consult an authoritative text. An up-to-date encyclopedic treatment is *Fundamental Immunology* (Paul, 1989), while perhaps a more philosophical and entertaining account is presented in *Immunology: The Science of Self–Nonself Discrimination* (Klein, 1982).

2.1. Antibodies and B Cells

There are two main branches of the immune system which react in distinct ways with the immunogen, but normally also interact with each other to produce a complete response. The humoral branch is responsible for the production of antibodies, proteins with molecular masses from 150 to 750 kDa that bind specifically to the immunogen or antigen. The structural unit of antibodies, or immunoglobulins, is a dimer, each containing a light (L) chain of 22 kDa and a heavy (H) chain of 53 kDa. The four polypeptide chains are held together by disulfide bridges to form a Y-shaped structure, with the two arms of the Y containing both an L and an H chain, while the stem of the Y is the extension of the two longer H chains. The end sections of the arms form two identical binding (the variable or V) regions which are responsible for the specific interaction with the immunogen or antigen. The stem of the antibody (the constant or C region) is characteristic of the immunoglobulin class, and mediates interactions with other components of the immune system. The immunogen is defined as the substance that has initiated the immune response, while the molecule that binds to antibody is an antigen. The distinction between immunogen and antigen may in some cases be difficult or meaningless.

There are five major classes of antibodies. IgM is normally the first to appear in an immune response, and consists of five basic units plus a shorter chain for a combined size of about 750 kDa. Typically, the same cells which produce IgM switch to produce IgG later in the immune response. IgG, with a molecular mass of 150 kDa, has the basic unit structure. IgE antibodies are present in serum at the lowest concentration of all classes, but are very important since they generate allergic reactions through interaction between their characteristic constant region and mast cells, which then release histamine. IgA antibodies are produced in the intestine and other mucosal tissue in reaction to ingested and inhaled substances (see Chapter 5). IgD facilitates B cell activation. IgG, IgM, and IgA have each been divided into subclasses.

Antibodies are produced by those B cells (bone marrow derived) that present membrane-bound immunoglobulins on their exterior surface which can bind to the immunogen. It is this binding that provides one of the main signals driving the cells to proliferate, differentiate into plasma cells, and produce and secrete large amounts of immunoglobulin. At birth there exists a great diversity of B cells in every individual, each producing antibodies with different V regions, so that there usually are many that bind at least weakly to the immunogen. During the growth and differentiation of B cells that bind immunogen, somatic rearrangement of antibody genes normally results in a switch to production of antibodies of the IgG class. If there is repeated exposure to immunogen, but it is not present in excess amount, those multiplying B cell clones which produce antibodies with high affinity can compete with lower-affinity-antibody-producing clones. This selective process gradually produces antibody of progressively higher affinity but typically lower diversity.

The half-life of the typical IgG antibody is about 20 days. B cells live only a short time if not stimulated by antigen, but if they have been stimulated to differentiate into plasma cells, they secrete antibody for a few weeks. Once stimulated, they also may live for a longer time as B memory cells.

2.2. T Cells

The second major arm of the immune system is represented by T cells (thymus-processed). These cells also recognize the immunogen, but by a completely different mechanism than used by B cells. The immunogen must be first processed, typically by macrophage cells, into fragments which are then exposed on the cell surface while bound to a receptor specified by the major histocompatibility gene complex (MHC). This histocompatibility receptor complex can then participate in the formation of a larger complex with a second, different receptor (the T cell receptor) on the surface of specific T cells. T cell receptors are homologous in structure to the immunoglobulin receptor on B cells, and like them are present in great diversity so that there is likely to be some that bind to at least some fragment of the immunogen. The T cell receptor is analogous in function since it plays a major role in the activation of the specific T cell. However, the interaction with and activation of T cells by the immunogen differs from B cell activation in three fundamental ways. The immunogen must be processed to produce a fragment which is then recognized as foreign; the fragment must interact in a generic way to form a complex with the MHC protein, and this complex must be recognized by a specific T cell receptor. This fundamental difference in recognition pathway can mean, for example, that a foreign substance may be a good B cell immunogen but a poor T cell antigen, and different parts of the immunogen may be recognized in the B and T cell response.

The full recognition of an immunogen is thus not an automatic or default reaction. It requires the presence of B cell receptors that bind immunogen, an MHC receptor that can bind fragments of processed immunogen, and T cell receptors that can recognize these bound fragments. It should not be surprising then that only certain parts of a foreign substance may be recognized as nonself, or that some foreign proteins are not detected at all. In particular, since the MHC complex is very heterogeneous (genetically polymorphic) in the human population, the extent of immune response to a specific immunogen varies greatly among the population.

T cells fall into three basic functional classes. The helper T cells participate in activation of specific B cells by the local secretion of growth and differentiation factors. The switch by B cells from secretion of IgM to IgG generally requires T cell help. T cells can live for years, so T memory cells are often the main participant in the secondary or amnestic response to immunogen. A second class of T cells kill target cells that are recognized by exposed antigen on their surface. A third class of T cells specifically suppress the activity of other T cells.

However, in order for the immune system to be beneficial, it must also distinguish self from nonself. The recognition of self, e.g., tolerance, is complex and often not complete or absolute. Earlier models of tolerance proposed that all immune cells that reacted with self molecules were clonally eliminated during embryonic development. While such negative selection is certainly involved in the production of tolerance, it is now known that the suppression of autoimmune responses in normal adults is also due to active mechanisms with T suppressor cells playing an important role. Thus, tolerance is not irreversible.

3. ASSAYS

3.1. Antigen-Specific Antibody

In development, production, and administration of a protein therapeutic, the goal will usually be to minimize immunogenicity. However, some immune reaction will almost certainly occur. The magnitude of the response in humans can generally not be predicted using animal models and, in fact, immune reaction to a human protein will usually make chronic toxicological studies impossible or meaningless (Teelmann et al., 1986). Thus, immunological monitoring during clinical trials is highly desirable. In the United States, monitoring immunogenicity of a protein therapeutic will almost certainly be required by the Food and Drug Administration, although the details of the program and evaluation of results will depend on the therapeutic, disease, and patient population.

Total levels of antigen-specific antibodies are easily followed with an enzyme-linked immunosorbent assay (ELISA). In one variant of this assay, the antigen, in this case the protein therapeutic, is used to coat a plastic surface, dilutions of patient serum are incubated on the plastic, unbound antibody is removed by washing, and any bound, antigen-specific antibody is detected by incubation with enzyme-labeled anti-human antibody. The ELISA, first used to follow antibodies during viral infection (Voller *et al.*, 1976), is now a standard method for this purpose. A variety of polyclonal antibodies produced in animals such as the rabbit and goat, and both polyclonal and monoclonal murine antibodies to specific classes and even subclasses of human immunoglobulins are commercially available. Using this method, one can follow the production of specific IgM and IgG and determine if a complete immune response has occurred.

Most of the modern work on serum IgE levels in patients sensitive to allergens has used the radioallergosorbent technique (RAST) (Wide *et al.*, 1967). The antigen (allergen) is covalently attached to a bead or disk and incubation with patient serum allows specific IgE to bind which is then detected with radioactively labeled anti-human IgE. The ELISA can also be used for the same purpose (Sepulveda *et al.*, 1979). However, the utility of measuring and using the serum IgE levels as an index of the severity of an allergic response is questionable. A functional skin test is perhaps more clinically relevant (Platts-Mills *et al.*, 1981).

3.2. Neutralizing Antibody

Most protein therapeutics will have biological activity that can be measured in an *in vitro* or *in vivo* assay. Thus, it is possible to test for the presence of antibodies in patient serum that can neutralize this activity.

There are three main characteristics of the serum that will have a major role in determining the magnitude of neutralization: concentration of specific antibody, affinity of this antibody to antigen, and the site on the antigen at which the antibody binds (the epitope). If neutralization is detected, it indicates that at least some fraction of the antibody population can bind at or near the active site and block activity. Thus, a high antibody concentration alone does not necessarily mean that there will be any ability of the serum to neutralize activity. The presence of neutralizing activity in such an assay does not necessarily mean that all therapeutic activity will be abrogated in the patient, although a correlation is usually seen. In some studies, patient serum has been tested for the ability to neutralize or block the signal in an immunoassay for the therapeutic. However, the presence or absence of this kind of blocking will in general have no relationship to the ability of the antibody to neutralize biological activity.

There is a general problem that the immunochemist will often have in interpreting assay results of serum from patients receiving a protein therapeutic. If the therapeutic is only slightly immunogenic, the antibody levels will be low, but more importantly, the antibodies will likely be very heterogenous and of low affinity. This is in contrast to typical "hyperimmune," high-affinity antibody preparations that are used in the laboratory for analytical purposes, which have been obtained from animals immunized repeatedly over a long period of time with preparations that are highly immunogenic by design. The difference in the shape of dilution curves when low- and high-affinity antisera to diphtheria toxin is used to neutralize toxin activity has been extensively documented and interpreted (Chase, 1977, particularly p. 305). A fundamental point in understanding these curves is that even a homogeneous antibody preparation must be described by two parameters, the concentration and the affinity. In some assays and conditions, both concentration and affinity affect the dilution curve shape and, thus, there is no single number or parameter that completely describes the antibody preparation and can unambiguously be defined as the antibody titer. A high-affinity antibody may effectively neutralize activity in a stoichiometric manner, so that if a fixed concentration of serum neutralizes 50% of toxin activity, the same serum dilution would eliminate essentially all activity if half as much toxin were used in the assay. In such a case, it might seem appropriate to describe the titer of the serum in units of toxin that 1 ml of serum could inactivate. Conversely, a low-affinity antibody might need to be present in great molar excess to neutralize 50% of activity if the antibody–toxin equilibrium dissociation constant is greater than the toxin concentration in the assay. In such a case, adding half the toxin to a second assay of that serum dilution would again result in half the activity being inactivated. In this case, it would seem appropriate to describe the neutralization titer of the serum as the serum dilution which neutralizes half the toxin activity, since this now would be the parameter characteristic of the serum and independent of the amount of toxin used. This could also be described as a case where dilution curves of a standard, high-affinity antibody and the unknown, low-affinity serum do not have the same shape and thus cannot be superimposed by a simple translation along the dilution axis in order to define an unambiguous titer.

This same problem was encountered in another context when an attempt was made to quantitate insulin antibodies from patients with different histories (Koch *et al.*, 1989). High-avidity (apparent association constant) antibodies were found in a patient with antibody-mediated insulin resistance induced by administration of animal insulin, while low-avidity antibodies were seen in an asymptomatic patient with autoantibodies to endogenous insulin. A liquid-phase assay measuring the amount of radiolabeled insulin that could be captured by antibodies selectively detected high-avidity antibodies and the dilution curves for high- and low-avidity antibodies had very different shapes. However, the solid-phase ELISA was far better and more equally efficient in detecting both high- and low-avidity antibodies, and thus, shape of the dilution curves were similar. Since biological neutralization assays must often be done in a liquid format, it may be difficult to directly compare antibodies of greatly different affinity.

3.3. T Cell Activity

T cell response to immunogen is important, and indeed necessary for a complete immunological response. The *in vitro* stimulation by immunogen of proliferation in lymphocyte cultures from patients can be used to measure T cell involvement in an immune response. In addition, the ability to isolate T cell clones after growth of patient lymphocytes in media containing IL-2 and immunogen has enabled identification of the specific epitopes recognized by this arm of the immune system (Nell *et al.*, 1985; Miller *et al.*, 1987; Naquet *et al.*, 1988). While such studies are playing an increasing role in adding to our knowledge of the immune response to protein therapeutics, the methodology is at present more complicated than simple assays for antibodies and is not typically a routine part of clinical monitoring.

4. ANTITOXIN SERUM: THE FIRST PROTEIN THERAPEUTIC

Antibodies specific for bacterial toxins, and able to neutralize their clinical effects, can be produced in a variety of nonhuman species (Ratner, 1943). Horse antidiphtheria serum was used successfully before the turn of the century. As would be expected, repeated use of such preparations produces an intense immune reaction. Unfractionated horse serum contains many proteins that are seen as foreign by the human immune system, the major component being equine serum albumin. However, even purified IgG antitoxin is seen as foreign since the amino acid sequence of the constant region of the horse immunoglobulin molecule is different from human immunoglobulin.

The clinical profile of toxicity seen after administration of antitoxin serum was often called serum sickness (Ratner, 1943; von Pirquet and Schick, 1951). This syndrome can include fever, skin lesions, arthralgia, lymphadenopathy, gastrointestinal distress, and proteinuria, and is most likely to occur when large amounts (hundreds of milligrams) of foreign protein are administered. This cannot only generate an intense reaction, but the large amount of antigen and antibody that are present together in the serum have the potential to form large amounts of antibody–antigen complexes (immune complexes) which mediate much of the toxicity. The size and characteristics of immune complexes depend on many factors, including the valence of the antigen, the antibody-to-antigen ratio, antibody affinity, and the absolute concentrations of antibody and antigen (Mannik, 1980). The size of the complex, which when large is known as an immune lattice, determines how rapidly the complex is removed from the circulation and the tissue and organ where the immune complex is likely to be deposited. Large complexes are removed by the Kupffer cells of the liver, but this system has a finite capacity and can thus be saturated. Large complexes also are deposited in the kidney, specifically in the glomerular capillary wall (see Chapter 12). The immune complexes in the kidney are phagocytosed by bone marrow-derived monocytes, and the lysosomal enzymes (see Chapter 13) released in this process may be part of the cause of glomerular damage that is characteristic of this event (Cochrane and Koffler, 1973; Mannik, 1980). Such immune complexes can also activate the complement system which then results in tissue damage. Several quantitative studies of serum sickness have been carried out in animal models (e.g., Wilson and Dixon, 1971). It was recently observed after administration of horse serum containing antithymocyte antibodies to patients with bone marrow failure (Gilliland, 1984; Lawley *et al.*, 1984).

5. INSULIN: A PARADIGM

5.1. Introduction and History

Much of our knowledge of the response of the immune system to a pure protein has derived from the use of insulin. This is partially due to the extent of clinical experience, extending over almost 70 years. The first diabetic human received pancreatic extracts containing insulin early in 1922, about one year after Banting and Best started their work and less than two weeks after the first public report of the hypoglycemic effects of these extracts in dogs. In June of 1922, the Eli Lilly Company had completed the first large-scale purification of insulin and by 1923 had produced an amount sufficient to treat 10,000 patients (Forsham, 1982). Immune reaction to insulin at the site of injection was first described in 1925 (Lawrence, 1925) and the induction of insulin neutralizing activity in the serum of a patient who had received repeated injections was reported in 1938 (Banting *et al.*, 1938).

Insulin is a relatively small protein of about 6 kDa, with two polypeptides of 21 and 30 amino acids linked by disulfide bonds. This simplicity, commercial availability in large quantities, and medical importance often caused it to be the object of study by early protein chemists. Sanger received the Nobel Prize for determination of the amino acid sequence of insulin which was completed in 1953.

5.2. Local Immune-Mediated Toxicities

Local cutaneous reactions may consist of a burning or itching sensation, swelling, erythema, induration, or wheal formation. The reaction may start a few minutes after the injection and rapidly decrease, or be delayed for 8 hr and continue for several days (Kahn and Rosenthal, 1979). The immediate type of reaction is associated with and mediated by IgE while the more delayed reactions are mediated by IgG (Galloway and Bressler, 1978). While quite rare with current purified insulin preparations, the destruction of subcutaneous fat (lipoatrophy) at the site of injection was not uncommonly seen 3–6 months after the initiation of therapy (Van Haeften, 1989). Biopsy of such lesions reveals deposits of IgM, IgG, and complement, suggesting that the tissue destruction is an example of the Arthus reaction (Reeves *et al.*, 1980). In this reaction, a variety of activated cells such as granulocytes and neutrophils are attracted by immune complexes and cause tissue necrosis through release of proteases and other mediators of inflammation.

5.3. Systemic Immune-Mediated Toxicities

5.3.1. INSULIN METABOLISM AND ELIMINATION

Insulin is synthesized and secreted into the blood by the beta cells of the pancreas. The blood then must flow through the liver before it joins the main peripheral circulation. During this single pass, about 40 to 50% of the insulin is removed by insulin receptor-mediated endocytosis (Field, 1972). Once past the liver, insulin is similarly removed by binding to insulin receptors in remaining tissue. Since insulin has a molecular mass of only 6 kDa, free insulin is removed from the blood entering the kidney by glomerular filtration (Brenner *et al.*, 1981). However, less than 2% of this filtered insulin appears in the urine since it is reabsorbed and degraded by cells lining the proximal convoluted tubules of the kidneys. An approximately equal amount of insulin is removed by the postglomerular peritubular capillaries in the kidney and some insulin reacts with and is removed by insulin receptors. Thus, though glomerular filtration is 130 ml/min in man (Renkin and Gilmore, 1973), the total renal clearance of insulin is about 200 ml/min (Rubenstein and Spitz, 1968). Typically, 40 U of insulin is produced in 24 hr by the pancreas, the kidney removes 6–8 U, and the serum insulin concentration is 20 mU/ml (1 U $\simeq$ 36 μg or 6 nmole).

The relative importance of insulin receptor-mediated clearance and nonspecific (mostly renal) clearance is suggested by the amounts of normal and mutant insulin in the serum of an individual heterozygous in the insulin gene. The mutant insulin contained a serine-for-phenylalanine substitution at position 24 in the B chain, had one-sixth the activity of normal insulin in stimulating glucose oxidation in an *in vitro* assay, but was present in 20-fold excess of the normal insulin in patient serum (Shoelson *et al.*, 1983). This is consistent with specific receptor-mediated metabolism being the major route of clearance. However, quantitative aspects of insulin distribution and metabolism can change when the concentration of blood insulin changes. As an example, the arterial–venous concentration across the kidney increases as the concentration of free insulin increases. Thus, the fraction of insulin removed by receptor-specific versus molecular size-dependent renal mechanisms will be dependent on insulin concentration (Rubenstein and Spitz, 1968).

5.3.2. INSULIN–ANTIBODY COMPLEXES

The properties of the antibodies to insulin will vary depending on patient genetic background, onset of disease, species of insulin administered, and many other factors. However, there are some common characteristics that are useful in under-standing the effects of antibodies. Typically, human antibody–insulin complexes are not precipitable, are predominantly univalent, and form and dissociate in a reversible manner. Using serum from patients with high antibody titers, plots of bound/free versus bound insulin at equilibrium are strongly biphasic with dissociation constants typically 10^{-7} and 10^{-9} M for the two antibody species. Rapid dilution of complexes demonstrates half-lives of 4 and 140 min (Berson and Yalow, 1959).

5.3.3. INSULIN RESISTANCE

Insulin resistance is a functional term used to describe the failure of a patient to respond to a normal dose of insulin. Any patient requiring more than 60 U/day of insulin is resistant, although it may not be considered significant until more than 200 U/day is required, and half of resistant patients require more than 1000 U/day (Kahn and Rosenthal, 1979). Chronic insulin resistance can be due to the failure of tissue to respond to a normal concentration of free insulin, caused for example by a decrease in the effective number or affinity of insulin receptors, or can be the result of binding and inactivation of a fraction of administered insulin by neutralizing antibodies. Resistance is at least logically distinct from increased insulin consumption due to the increased metabolic requirements of obese patients (Federlin et al., 1980). The frequency of appearance of resistance caused by antibodies is not high, with one group reporting that between 1940 and 1960 only 0.01% of all insulin-requiring diabetics were resistant (Kahn and Rosenthal, 1979), while another estimate was 0.1% (Federlin et al., 1980).

A requirement for an increase in daily insulin dose might not by itself seem of great consequence, since the cost of insulin is not high. However, variation of insulin concentration in the blood is normally the mechanism by which the blood glucose level is kept constant. Thus, in a nondiabetic, insulin concentrations rapidly increase after a large meal when glucose is entering the blood and must be metabolized. Insulin levels are low early in the morning before breakfast when glucose must be conserved in order to maintain a constant blood glucose level. Since a significant fraction of insulin is removed in each pass of the blood through the body, insulin levels respond to changes in the rate of insulin release by the pancreas in minutes. Abnormally low glucose levels cause loss of consciousness and shock, and chronically high glucose levels eventually result in a characteristic spectrum of tissue and organ damage. Blood glucose levels in a diabetic are controlled by a combination of planned meals to produce a constant pattern of caloric intake, and scheduled insulin injections of specific formulations which release insulin into the blood to precisely meet metabolic need (Galloway and Bressler, 1978). However, insulin bound by antibody has a molecular size of at least 160 kDa, and will thus not be removed by the kidney. Insulin bound to neutralizing antibody will not be removed by receptor-mediated endocytosis. Since insulin–antibody binding is reversible, these antibody complexes represent a reservoir which buffers the insulin concentration and thus complicates, or in extreme cases prevents, establishment of a satisfactory concentration–time profile for insulin (Van Haeften, 1989).

5.3.4. DEPOSITION OF IMMUNE COMPLEXES

The average size of antibody–insulin complexes in the peripheral circulation is typically small. However, medium-sized complexes can become more common in late-stage diabetes, at least partially because of impaired phagocytic function which normally removes them, and the deposition of these complexes is the likely cause of severe microangiopathy seen in late-stage diabetes (Iavicoli *et al.*, 1982).

5.4. Causes of Insulin Immunogenicity

When insulin is obtained from bovine pancreas, there are obvious reasons for it to be immunogenic, since the amino acid sequence of bovine insulin differs from that of human at positions 8 and 10 in the A chain, and at position 30 in the B chain in the C terminus. Indeed, procine insulin, which differs in sequence from human only at B30, is generally less immunogenic. However, a major factor in the immunogenicity of early preparations of insulin, purified by acid extraction and crystallization, was the presence of protein impurities. These included glucagon, pancreatic polypeptide, vasoactive intestinal peptide, and somatostatin (Reeves, 1981). Proinsulin, which contains the C peptide in which about half of the amino acids are different from human C peptide, was also a contaminant. However, when crystallized insulin is purified by size exclusion chromatography, on Sephadex G-50 for example, a single peak component is obtained which is 99% insulin, containing 5–10% desamido insulin. A second, ion-exchange chromatography yields "single component" or "monocomponent" insulin which has greatly reduced immunogenicity (Hansend *et al.*, 1981).

There are many other factors which affect the degree of immunogenicity of insulin preparations. These include the solubility, pH, and presence of agents such as zinc and protamine, intended to retard the dissolution of insulin into the circulation and produce an extended insulin–time profile (Reeves, 1981). Storage of insulin preparations at 5°C can result in deamidation and polymerization (Fisher and Porter, 1981) and both changes would be expected to increase immunogenicity. The delivery of insulin, particularly in infusion devices, results in shear rates proportional to the rate of delivery. High shear produced in this way accelerates self-association into oligomers and macromolecular aggregates (Sato *et al.*, 1983), which would increase immunogenicity. High titers of insulin antibodies are often induced when insulin administration has been reinitiated after a period without treatment (Van Haeften, 1989).

5.5. Use of Human Insulin

It is of course now possible to produce human proteins using genetic engineering techniques and human insulin was one of the first pharmaceutical targets of this technology (Johnson, 1983). As described in the previous section, there are many factors other than the amino acid sequence that are very important in determining the immunogenicity of a specific therapeutic protocol using insulin. However, it is generally observed that immunological reactions are less frequent and of less severity when human insulin is used (Fireman *et al.*, 1982; Fineberg *et al.*, 1983; Schernthaner *et al.*, 1983; Wilson *et al.*, 1985; Di Mario *et al.*, 1986). This does not mean that commercial preparations of insulin with the human amino acid sequence are not immunogenic, or even that there are always detectable differences in immunogenicity between human and porcine insulin. In one study in which three types of insulin preparations were administered to diabetics who had received no other insulin previously, insulin antibodies developed in 11/31 or 35% of patients given human insulin, in 25/41 or 61% of patients receiving monocomponent porcine insulin, and 16/21 or 76% of patients receiving a mixture of purified porcine and bovine insulins (Fireman *et al.*, 1982). While the number of patients is admittedly small in this one trial, the same order in the degree of immunogenicity has been seen in other studies. Even if human insulin were not less immunogenic than porcine, it may be reassuring to have an unlimited supply that is not linked to the vagaries of the use of pork as food.

The benefit of a change from animal insulin to human is not necessarily the same as starting with the human type. In some studies a decrease in antibody levels was seen after a switch to human insulin (Fineberg *et al.*, 1982; Maneschi *et al.*, 1982; Spijker *et al.*, 1982; Grammer and Roberts, 1989), but in other cases no improvement was seen (Walhausl *et al.*, 1981; Clark *et al.*, 1982; Small and Lerman, 1984; Di Mario *et al.*, 1986), and in one case an allergic reaction was induced after administration of one human insulin formulation (Monotard) but not by another (Actrapid) (Silverstone, 1986). It is clear that there is often extensive cross-reactivity in the ability of antibodies from diabetic patients induced by one species of insulin to bind to another (Galloway and Bressler, 1978; Grammer *et al.*, 1984, 1985, 1987), that there is considerable variation among patients, and that decreases in antibody levels after a change from animal to human insulin, in those patients in which it occurs, typically require several months.

5.6. Etiology of Diabetes: An Autoimmune Disease?

There is a broad spectrum of clinical conditions that involve an inability to adequately control serum glucose levels, but only a fraction of patients are unable to produce any insulin and require chronic insulin administration. The sudden onset of inability to produce insulin [type I diabetes mellitus or insulin-dependent diabetes mellitus (IDDM)], which has a frequency of about 1 out of 300 under the age of 20 in North America, has a strong genetic component (Nerup and Lernmark, 1981; Stiller *et al.*, 1984). It is associated with the HLA-B8 and HLA-B15 markers and may possibly be a sequela of viral infection (Onodera *et al.*, 1981). Of importance for the present discussion is the fact that autoantibodies specific to beta islet cell membranes, insulin receptors, and insulin can often be found in the serum even before the onset of diabetic symptoms, and certainly before administration of insulin (Nerup and Lernmark, 1981). Clones from individual B cells can also be derived from IDDM patients before insulin treatment that produce high-affinity antibody (K_d 10^{-7} M) against insulin (Casali *et al.*, 1990). In another study (Keller, 1990), antibodies to human insulin were found in 6/9 prediabetic patients but in 0/12 normal subjects. However, T cell cultures were stimulated to proliferate by human insulin in 8/9 of the prediabetic patients, but were also been in 4/12 of the normal subjects. This is an example of the fact that immunological tolerance is neither absolute nor irreversible. While it could be argued that diabetes is a special case, individuals have been found to produce autoantibodies to a number of autologous proteins.

6. FACTOR VIII

Hemophilia A is a hereditary, X chromosome-linked disease in which a deficiency of the blood coagulation pathway is due to either the lack of factor VIII or the presence of an inactive protein. In about 5 to 20% of patients receiving human factor VIII as replacement therapy, serum inhibitors to coagulation activity develop, with the frequency being highest in the group with the most severe deficiency (Shapiro, 1979; Roberts and Cromartie, 1984). These inhibitors have been shown to be neutralizing IgG antibodies, usually IgG4, the rarest of the four IgG subclasses, and typically containing only kappa light chains (Hultin *et al.*, 1977; Shapiro, 1979; Hoyner *et al.*, 1984).

Physiologically, factor VIII deficiency is very different from diabetes, since even a complete lack of activity is not immediately lethal, and an excess has no adverse effect. Factor VIII is also very different structurally from insulin, since it is a protein of 2332 amino acids, which circulates in the blood in noncovalent association with von Willebrand factor, an even larger multimeric glycoprotein. Even with its large size, the neutralizing antibodies found in hemophiliacs appear to bind at two rather restricted segments of the factor VIII chain, one at the amino-terminal and the other near the C-terminal end (Fulcher *et al.*, 1985). The immune complexes present after administration of factor VIII contain less than three and possibly one antibody molecule, perhaps because there is only one epitope per antigen molecule (Shapiro, 1979). While some hemophiliacs produce an inactive protein which has immunological cross-reactivity with native factor VIII, such material is absent in most severe deficiencies (Shapiro, 1979; Roberts and Cromartie, 1984). DNA sequence analysis of the factor VIII gene has revealed that mutations occur within introns to produce sequences that may cause incorrect RNA processing, while others generate premature stop codons that should result in the production of factor VIII chains without C-terminal segments (Gitschier *et al.*, 1985). In these hemophiliacs it is thus possible that immunological tolerance is absent for the epitopes on native factor VIII molecules. There does seem to be a nonrandom association between HLA genotype and factor VIII deficiency, but the pattern is complex (Shapiro, 1984). An intriguing insight into one mechanism for tolerance is obtained by the observation of a spontaneous recovery from an autoimmune inhibition of factor VIII. The recovery was found to be the result of the induction in the patient of anti-idiotype antibodies which neutralized the inhibitor activity of the factor VIII antibodies (Sultan *et al.*, 1987).

The neutralizing activity in serum can be measured and, if low, can be overcome by merely increasing the amount of factor VIII administered (Strauss, 1969), although the additional expense may not be negligible. In about half of patients treated in this way, increases in inhibitor titers are induced, while in others titers may remain fairly constant (Allain and Frommel, 1976; Rizza and Matthews, 1982). In one case, the administration of massive doses of factor VIII to a highly resistant patient resulted in a dramatic and progressive decrease in neutralizing titer which appeared to have induced tolerance (Brackmann and Gormsen, 1977). The human antibodies often cross-react only weakly with porcine factor VIII and, thus, in an inversion of the strategy used with insulin therapy, the substitution of porcine for human factor VIII can be an effective treatment (Kernoff *et al.*, 1984). The use of an activated prothrombin-complex concentrate (FEIBA) to bypass the need for factor VIII also shows promise (Roberts, 1981; Sjamsoedin *et al.*, 1981).

7. INTERFERONS AND LYMPHOKINES

7.1. Introduction

Interferons are a class of proteins which have the ability to protect cells from being killed by viral infection. The α and β interferons have the greatest antiviral activity, while γ interferon appears to function mainly as a lymphokine in mediating aspects of the immune response. Alpha interferons, IFNα, consist of several closely related species with molecular masses in the range of 15 to 20 kDa. Beta interferon, IFNβ, has a similar amino acid sequence, but is heavily glycosylated. Native and recombinant human interferons have been used mainly as a cancer therapeutic, although they have also shown efficacy against hepatitis and opportunistic infections associated with AIDS.

Lymphokines are proteins which are produced by lymphocytes and modulate the activity of other lymphocytes. While over ten lymphokines have been identified and cloned, we will only discuss interleukin-2 (IL-2). IL-2 has approximately the same molecular mass as IFNα or IFNβ, 13 kDa (Taniguchi *et al.*, 1983). It was originally described as a T cell growth factor, but has subsequently been found to act on several cell types and to alter cellular function as well as stimulate growth. IL-2 has been used mainly as a cancer therapeutic, and has induced partial and complete tumor regression, particularly in renal cell carcinoma and melanoma (Rosenberg *et al.*, 1984; Rosenberg, 1988, 1990).

7.2. Interferon Alpha

The first observation of antibodies to interferon was made in a patient with nasopharyngeal carcinoma successfully treated with native IFNβ (Vallbracht *et al.*, 1981). Neutralizing activity appeared only after administration of the IFNβ preparation. Cells cultured from the patient were able to produce IFNβ which was neutralized by his serum to the same extent as the exogenous IFNβ (Vallbracht *et al.*, 1982). Thus, it did not appear that a lack of tolerance to IFNβ was a result of a genetic deficiency in ability to produce IFNβ. Antibody to IFNα was found to be present in another patient before administration of native IFNα (Mogensen *et al.*, 1981) and it was speculated that this neutralizing activity could have been involved in the etiology of varicella zoster for which the IFNα had been administered.

The first trial of recombinant IFNα was described in 1982 and employed 16 patients (Gutterman *et al.*, 1982). Neutralizing activity was not seen in any of the patients before therapy, but appeared in 3 as a result of IFNα administration. Seven of the sixteen showed partial tumor regression, and none of the antibody-positive patients were in this group. In a group of 653 evaluable cancer patients receiving recombinant IFN-α2A (Roferon-A, produced by Hoffman–La Roche, Inc.), 160 (26%) were found to have developed neutralizing antibodies to the recombinant interferon, with less than 1% having neutralizing activity before treatment (Itri *et al.*, 1987). The incidence of antibody induction ranged from a high of 44% in 88 patients with renal cell carcinoma to a low of 4% in 82 patients with leukemia. The duration of treatment was only slightly longer for antibody-positive patients. There was no significant difference in the frequency of partial or complete remission between the antibody-negative group (112 of 465 or 24%), and the antibody positive group (43 of 152 or 28%). The median duration of response was the same for the two groups. A later report provided an update on a slightly expanded number of patients, including this group, and provided data on the time from the start of treatment to the detection of antibodies (Itri *et al.*, 1989). The median was 2.1 months, range 0.9 to 15.2, for 96 renal cell carcinoma patients, and was 11.7, range 3.8 to 29.7, for 18 leukemia patients. No deleterious clinical or laboratory effects had been found to be associated with the development of neutralizing activity to this interferon (Jones and Itri, 1986).

It may seem paradoxical that in these trials there is not a strong negative correlation between clinical response and the development of neutralizing activity, especially since the appearance of neutralizing activity in the sera of patients typically does coincide with a decrease in IFN-induced fatigue and anorexia, a return to normal WBC counts and liver function tests (Quesada *et al.*, 1986). However, the use of IFN in treating cancer is different in several respects to the administration of insulin to diabetics or factor VIII to hemophiliacs. IFN is not administered to correct an obvious deficiency; for most cancer types it induces regression in a small fraction of patients treated; the mechanism of the anticancer action is either unknown or at least controversial; the kinetics of anticancer activity related to IFN administration are usually not well known since tumor regression is typically monitored at infrequent intervals; and the dose, schedule, and length of treatment vary greatly.

It is possible that antibodies that are able to neutralize IFN activity *in vitro* are not effective *in vivo*, perhaps because an important target cell population is not accessible to antibodies. It is possible that *in vivo* toxicities are neutralized but *in vivo* anticancer activity is not. However, there are also other explanations for the apparent lack of correlation between neutralizing activity and response. The necessary anticancer activity of IFN may often occur before neutralizing antibodies appear. In fact, relapse of tumor regression sometimes occurs coincidently with the development of neutralizing antibodies (Quesada *et al.*, 1985). In addition, some cancer types and treatments that have been administered before IFN administration, directly or indirectly inhibit activity of the immune system, and thus diminish the ability of the patient to make antibody to IFN. As described previously, the frequency of appearance of neutralizing activity and the frequency of tumor regression differ by at least a factor of ten among patient populations with different malignancies. In fact, neutralizing activity is most common in cancer types with high response rates and this positive correlation could be confounding a negative effect of neutralizing activity on response when large patient groups composed of many cancer types are considered as a whole. Thus, the most useful information can only be obtained from experience with large numbers of patients and one has to be careful not to force comparison of disparate clinical groups. Thus, the cautious conclusion would be that neutralizing activity defined by an *in vitro* assay may well inhibit the anticancer activity of IFN, but that in patients receiving an initial IFN treatment, this inhibition is not obvious in the overall frequency of response.

The frequency of induction of neutralizing antibodies has been reported to be only 13 of 537 patients, 2.4% after administration of IFN-α2B (Intron A, produced by Schering Corp.), much lower than the 25% seen after administration of recombinant IFN-α2A. However, the assay for neutralizing activity against IFN-α2b was not based on a bioassay for IFN activity, but rather was defined as the ability of patient sera to block an immunoassay for IFNα (Protzman *et al.*, 1984). Mechanistically, the inhibition of antibody binding in an immunoassay has no necessary relationship to inhibition of binding to a cell receptor in a bioassay. It was subsequently demonstrated that the immunoassay seriously underestimates the frequency of sera with biologically neutralizing activity and that there is no reason to believe that there is a significant difference is frequency of neutralizing activity after administration of the two IFNα preparations (Itri *et al.*, 1987).

7.3. Interferon Beta

In one clinical trial in which each patient received recombinant IFNβ (Betaseron, produced by Cetus Corp.), by the i.v. or i.m. route, IFN-specific antibodies were induced in 24 of 30 patients (Konrad *et al.*, 1987). However, although the antibody concentrations were quite high in some patients, no neutralizing activity could be detected. Even in patients with high antibody levels, the pharmacokinetics of IFNβ did not seem to be altered, although immune complexes containing IFNβ might be expected to be cleared more slowly than an IFN monomer. This suggests that this IFNβ may normally be present as a larger complex even in the absence of antibodies. In a subsequent trial in which the IFN was given only by the i.v. route, antibodies were seen in only 2 of 36 patients. While the numbers of patients is small, this is still a dramatic illustration of the influence of mode of administration on immunogenicity.

This IFNβ had been intentionally altered by genetic engineering techniques to contain a serine in place of cysteine at position 17 (Mark *et al.*, 1984). However a potentially greater difference between the recombinant and native protein was the result of expression of the IFNβ gene in bacteria, since the resulting protein was missing the glycosylation of the human material. When the native IFNβ gene was expressed in Chinese hamster ovary (CHO) cell culture, glycosylated protein was produced with a specific activity of 1.2×10^9 IU/mg protein (McCormick *et al.*, 1984). This might appear to be higher than the value of "greater than 10^8 IU/mg" given for the nonglycosylated material (Khosrovi, 1984), although a direct comparison of the two interferons was not described in either study. It should be noted that the spectrum of carbohydrate structures in IFNβ is a function of the animal cell type used for its expression, although the material produced by CHO cells is very similar to that produced by human fibroblasts (Kagawa *et al.*, 1988). In one report, removal of carbohydrate from glycosylated IFNβ using glycopeptidase F resulted in an insoluble protein (Conradt *et al.*, 1987), which would certainly be expected to increase immunogenicity. However, this does not prove that the nonglycosylated IFNβ produced in bacteria is insoluble.

7.4. Interleukin-2

Native IL-2 produced in animal cells was used in early clinical trials whereas recombinant IL-2 produced by bacterial fermentation has been used in later trials. A mutant IL-2 (Wang *et al.*, 1984) with a cysteine residue replaced by a serine (Proleukin, produced by Cetus Corp.) was found to induce nonneutralizing antibodies in approximately half of the patients receiving the protein by the i.v. route (Allegretta *et al.*, 1986; Kolitz *et al.*, 1987). However, when this IL-2 was administered subcutaneously, either alone (Whitehead *et al.*, 1990) or in combination with

IFNβ (Krigel *et al.*, 1988; Paolozzi *et al.*, 1989), neutralizing antibodies developed in approximately 30% of patients and this route of administration was discontinued. Early results indicated that these antibodies did not bind to native IL-2, but studies of additional patients revealed that patient serum could also neutralize native IL-2 (unpublished data). There is no evidence that the amino acid substitution is the cause of the immunogenicity of this protein; however, as has been demonstrated by the study of insulin antibodies, the cross-reactivity of patient serum also does not exclude the immunogenicity of the substitution.

8. MONOCLONAL ANTIBODIES AND IMMUNOTOXINS

Antibodies have a great apparent potential as protein therapeutics. They can be used to supplement the body's immune defenses against viral, bacterial, and parasitic disease, to neutralize protein mediators of the immunoinflammatory cascade, and to image and deliver toxins to malignant cells (Larrick, 1989; Ahmad and Law, 1990). However, one of the major problems in the use of antibodies, particularly the repeated administration of a monoclonal antibody, is the rapid appearance in the patient of antibodies to the administered antibody. In one cancer patient receiving murine monoclonal antibody specific to Leu-1, an antigen present on the surface of lymphoma cells, free human IgG specific to the murine antibody appeared at day 18 (Levy and Miller, 1983). At this time, 210 mg of murine monoclonal antibody had been administered in six doses. However, it is likely that antimurine antibody was present by day 14, but merely bound to monoclonal antibody, since at that time there was no detectable free murine monoclonal in the blood. This dosing protocol and the pattern of appearance of antibody to the administered antibody is quite typical of the use of murine monoclonal antibodies in the treatment of cancer. However, it should be noted that even after appearance of neutralizing activity, the concentration of the target Leu-1-positive lymphoma cells started to fall at day 18 and continued to decrease until day 40, when it was about 5% of the pretreatment level.

Even though the total doses of monoclonal antibodies are often in the range of several hundred to a thousand milligrams, the toxicities are usually mild, and are generally not those associated with the serum sickness syndrome (Levy and Miller, 1983). Immunotoxins are antibodies to which cytotoxic toxins have been chemically linked. A favorite protein for this purpose has been the A chain of the plant toxin ricin, although, as one would expect, ricin is immunogenic in humans (Godal *et al.*, 1983). Toxicities seen after administration of the murine monoclonal ricin A chain immunotoxin Xomaxyme-mel, were a transient fall in serum albumin, weight gain associated with edema, malaise, fatigue, myalgia, and fever (Spitler *et al.*, 1987).

The development of human antibodies to the murine constant region of the antibody chain is to be expected, since the human constant regions have different amino acid sequences, and thus the murine antibody is seen as foreign. However, the variable region of the monoclonal, which binds to the target antigen and is thus essential for the therapeutic effect, is also not present in the human patient, and is thus seen as foreign. The murine monoclonal antibody OKT3 is specific to antigen associated with the human T cell receptor and has been used as a general immuno-suppressive agent in transplant patients. Antibodies to the variable region, anti-idiotype antibodies, are seen after administration of OKT3 (Colvin *et al.*, 1982) even during administration of other immunosuppressive drugs such as azathioprine and corticosteroids (Jaffers *et al.*, 1986), and can completely neutralize the therapeutic effectiveness of the antibody (Chatenoud *et al.*, 1986). Anti-idiotype responses have also been seen after administering murine monoclonal antibodies 791T/36, specific to colorectal antigen (Rowe *et al.*, 1985), 9.2.27 and T101, specific to leukemia antigen (Schroff *et al.*, 1985; Shawler *et al.*, 1985), and Xomaxyme-mel, specific to melanoma antigen (Mischak *et al.*, 1990). After administration of the monoclonal CO17-1A, specific to colorectal antigen, both anti-idiotype and anti-anti-idiotype antibodies are sequentially induced as a cascade response (Wettendorff *et al.*, 1989).

It is now possible to produce chimeric antibody molecules, with the variable region derived from a murine monoclonal antibody and the constant region from a human antibody (Morrison *et al.*, 1984; Brown *et al.*, 1987; Nishimura *et al.*, 1987; Riechmann *et al.*, 1988). While antibodies do not appear to be generated against the human constant region, antibodies specific to the variable region do appear (LoBuglio *et al.*, 1989). The fact that the immune response is restricted to the variable region after administration of human or human chimeric monoclonal antibodies suggests that one might be able to evade effects of this immune response by successively administering a series of antibodies with variable regions that differ sufficiently so as to not be neutralized by the anti-idiotype antibodies of the previous antibody, but still able to bind to a part of the target molecule or cell. The extent to which this is possible, both technically and economically, remains to be seen.

9. PREDICTING IMMUNOGENICITY

9.1. Covalent Structure: Amino Acid Sequence and Glycosylation

When considering the immunogenicity of a nonhuman protein therapeutic, it may be natural to first focus on the difference in the amino acid sequence between the native human protein and the therapeutic. The following groups have made important recent contributions to an understanding of the relationship between antigenicity and sequence, but there has been so much important work in this field that this list might be considered a stochastic sample: Niman *et al.* (1983), Benjamin *et al.* (1984), Westhof *et al.* (1984), Berzofsky (1985), Novontny *et al.* (1986), Margalit *et al.* (1987), Grey *et al.* (1989). The following differences between native and test protein are likely to result in increased immunogenicity, although these "rules" are merely generalizations and quantitative predictions are not possible:

- A difference in a large number of amino acids
- A replacement of one type of an amino acid for another, e.g., polar for nonpolar, acidic for basic, hydrophobic for hydrophilic
- A covalent change that is likely to alter secondary structure, e..g, introduction of a proline into an α-helical region
- Changes in parts of the protein that are on the surface of the molecule, accessible to large probes, and flexible (appear to have a high temperature in the crystal structure)

Most protein therapeutics are likely to be proteins that are normally present in serum and essentially all serum proteins except albumin are glycosylated. However, the structure of the carbohydrate in a glycoprotein may be difficult to determine. There often is not a single species but a spectrum of related structures, and different tissues (Rademacher *et al.*, 1988) or cell cultures (Conradt *et al.*, 1987) can produce a different spectrum of carbohydrate structures. The role and importance of the carbohydrate in defining biological activity and immunogenicity must certainly vary among serum proteins, and any general statement concerning its importance would be controversial at best. Glycosylation is known to modulate biological activity of erythropoietin (Goto *et al.*, 1988) and choriogonadotropin (Manjunath and Sairam, 1982). As mentioned in Section 7.3, glycosylation is likely to have a role in the solubility of IFNβ and thus perhaps in its immunogenicity.

9.2. Everything Else

Noncovalent alterations in structure, such as denaturation and aggregation, generally increase immunogenicity. However, characterization and even definition and description of noncovalent structure can be difficult. Even if such altered protein has reduced biological activity, the average specific activity of the entire protein preparation is a blunt tool in revealing a minor, inactive species, since many bioassays are not very accurate, and the specific activity of maximally active native material may be poorly determined. Gel exclusion chromatography of the final, formulated therapeutic, under conditions as close to physiological as possible, is one tool that can reveal aggregation. However, such characterization is not commonly published. Many of the details of purification and formulation, which might suggest deviations from native structure, are regarded as proprietary by pharmaceutical companies. The very fact that a therapeutic is generally prepared and stored in a very concentrated and nonnative form, e.g., lyophilized, and then administered over a short time at one site of the body may induce changes in the physical state of even completely native material that would never occur *in vivo*.

As we have seen from discussion of exemplary protein therapeutics, there are factors other than protein structure that determine appearance and intensity of an immune reaction. These include route and schedule of administration, immunosuppression due to disease or drug administration, and autoimmunity or lack of tolerance associated with etiology of the need for the therapeutic. In addition, there is usually going to be considerable variation in the immune response in a human population which is some cases can be shown to be linked to MHC genotype.

10. SUMMARY AND CONCLUSION

This review has used five groups of therapeutic proteins as examples: antitoxin animal serum, insulin, factor VIII, interferons and lymphokines, and monoclonal antibodies. The nature of immune reactions to them and the clinical consequences of these reactions differ greatly. The differences in clinical toxicities are not just the result of differences in the chemical and physical structure of the proteins, but are also due to the therapeutic role of the protein.

Insulin is perhaps the therapeutic most sensitive to immune reactions. It often needs to be administered daily over almost an entire lifetime and either an increase or a decrease of activity, or even an alteration of the time profile of activity can have adverse consequences. At the other extreme, a monoclonal antibody used for tumor imaging may have to be given once and thus an immune response can have no adverse effect.

There are not many general techniques to reduce the immune response to a foreign protein that are effective and clinically feasible. Chemically induced immunosuppression, which is usually required after an organ transplant, is generally too toxic to be used to control immune reaction to a protein therapeutic. One technique that has shown promise is the covalent attachment of polyethylene glycol to the protein (Abuchowski et al., 1977; Savoca et al., 1979; Wilkinson et al., 1987; Katre, 1990; Meyers et al., 1991).

The safest strategy in producing a protein therapeutic is to make it as similar as possible to the normal human counterpart, as is attempted with the manufacture of human insulin. However, there may be fundamental limitations on the extent this can be done, as in the case of a monoclonal antibody where the variable region must be foreign to the patient. It may be more expensive to produce native material, as in the case of IFNβ. All protein therapeutics are likely to produce an immune response if sensitive enough techniques are used to detect the response. The problem for the group producing the therapeutic is then to predict, i.e., guess, the degree of immunogenicity under a certain set of conditions and attempt to evaluate the clinical and commercial impact of this immunogenicity. If a mutant IFN induces tumor regression in a malignant cancer that is refractory to all other therapies and results in long-term survival and improvement in quality of life, it may be rather insignificant that it induces antibodies. At the other extreme, if one brand of human growth hormone has the human amino acid sequence, while a second brand has an extra methionine at the N-terminus, the first brand may well have at least a marketing advantage, even though no unequivocal proof exists that the nonhuman sequence produces a clinically significant immune reaction. As stated in a recent, more popularized review of this subject, "when proteins meet the immune system, the results can be complicated indeed" (Konrad, 1989). The decision of how much and what kind of immunogenicity is acceptable is also complex.

REFERENCES

Abuchowski, A., McCoy, J. R., Palczuk, N. C., van Es, T., and Davis, F. F., 1977, Effect of covalent attachment of polyethylene glycol on immunogenicity and circulating life of bovine liver catalase, *J. Biol. Chem.* **252**:3582–3586.

Ahmad, A., and Law, K., 1990, Recombinant targeted proteins for biotherapy, *Mol. Biother.* **2**: 67–73.

Allain, J.-P., and Frommel, D., 1976, Antibodies to factor VIII. V. Pattern of immune response to factor VIII in hemophilia A, *Blood* **47**:973–982.

Allegretta, M., Atkins, M. B., Dempsey, R. A., Bradley, E. C., Konrad, M. W., Childs, A., Wolfe, S. N., and Mier, J. W., 1986, The development of anti-interleukin-2 antibodies in patients treated with recombinant human interleukin-2 (IL-2), *J. Clin. Immunol.* **6**:481–490.

Banting, F. G., Franks, W. R., and Gairns, S., 1938, Anti-insulin activity of serum of insulin-treated patient, *Am. J. Psychiatry* **95**:562–566.

Benjamin, D. C., Berzofsky, J. A., East, I. J., Gurd, F. R. N., Hannum, C., Leach, S. J., Margoliash, E., Michael, J. G., Miller, A., Prager, E. M., Reichlin, M., Sercarz, E. E., Smith-Gill, S. J., Todd, P. E., and Wilson, A. C., 1984, The antigenic structure of proteins: A reappraisal, *Annu. Rev. Immunol.* **2**:67–101.

Berson, S. A., and Yalow, R. S., 1959, Quantitative aspects of the reaction between insulin and insulin-binding antibody, *J. Clin. Invest.* **38**:1996–2016.

Berzofsky, J. A., 1985, Intrinsic and extrinsic factors in protein antigenic structure, *Science* **229**:932–940.

Brackmann, H. H., and Gormsen, J., 1977, Massive factor-VIII infusion in haemophiliac with factor-VIII inhibitor, high responder, *Lancet* **2**:933.

Brenner, B. M., Ichikawa, I., and Dean, W. M., 1981, Glomerular filtration, in: *The Kidney* (B. M. Brenner and F. C. Rector, eds.), Saunders, Philadelphia. pp. 289–327.

Brown, B. A., Davis, G. L., Saltzgaber-Muller, J., Simon, P., Ho, M.-K., Shaw, P., Stone, B. A., Sands, H., and Moore, G. P., 1987, Tumor-specific genetically engineered murine/human chimeric monoclonal antibody, *Cancer Res.* **47**:3577–3583.

Casali, P., Nakamura, M., Ginsberg-Fellner, F., and Notkins, A. L., 1990, Frequency of B cells committed to the production of antibodies to insulin in newly diagnosed patients with insulin-dependent diabetes mellitus and generation of high affinity human monoclonal IgG to insulin, *J. Immunol.* **144**:3741–3747.

Chase, M. W., 1977, Neutralization reactions, in: *Methods in Immunology and Immunogenicity* (C. A. Williams and M. W. Chase, eds.), Academic Press, New York, pp. 275–398.

Chatenoud, L,. Baudrihaye, M. F, Chkoff, N., Kreis, H., Goldstein, G., and Bach, J.-F, 1986,

Restriction of the human in vivo immune response against the mouse monoclonal antibody OKT3, *J. Immunol.* **137**:830–838.

Clark, A. J., Wiles, P. G., Leifer, J. M., Knight, G., Adeniyi-Jones, R. O., Watkins, P. J., Ward, J. D., MacCuish, A. C., Keen, H., and Jones, R. H., 1982, A double-blind crossover trial comparing human insulin (recombinant DNA) with animal insulins in the treatment of previously insulin-treated diabetic patients, *Diabetes Care* **5**(Suppl. 2):127–134.

Cochrane, C. G., and Koffler, D., 1973, Immune complex disease in experimental animals and man, *Adv. Immunol.* **16**:185–264.

Colvin, R. B., Cosimi, A. B., Burton, R. C., Kurnick, J. T., Struzziero, C., Goldstein, G., and Russell, P. S., 1982, Anti-idiotype antibodies in patients treated with murine monoclonal antibody OKT3, *Fed. Proc.* **41**:363.

Conradt, H. S., Egge, H., Peter-Katalinic, J., Reiser, W., Siklosi, T., and Schaper, K., 1987, Structure of the carbohydrate moiety of human interferon-β secreted by a recombinant Chinese hamster ovary cell line, *J. Biol. Chem.* **262**:14600–14605.

Di Mario, U., Arduini, P., Tiberti, C., Lombardi, G., Pietravalle, P., and Andreani, D., 1986, Immunogenicity of biosynthetic human insulin: Humoral immune response in diabetic patients beginning in insulin treatment and in patients previously treated with other insulins, *Diabetics Res. Clin. Pract.* **2**:317–324.

Federlin, K., Velcovsky, H. G., and Maser, E., 1980, Clinical aspects of immunity to insulin, in: *Basic and Clinical Aspects of Immunity to Insulin* (K. Keck and P. Erb, eds.), de Gruyter, Berlin, pp. 203–218.

Field, J. B., 1972, Insulin extraction by the liver, in: *Handbook of Physiology*, American Physiological Society, Washington, D.C., pp. 505–513.

Fineberg, S. E., Galloway, J. A., Fineberg, N. S., and Rathbun, M. J., 1982, Immunological improvement resulting from the transfer of animal insulin-treated diabetic subjects to human insulin (recombinant DNA), *Diabetes Care* **5**(Suppl. 2):107–111.

Fineberg, S. E., Galloway, J. A., Fineberg, N. S., Rathbun, M. J., and Hufferd, S., 1983, Immunogenicity of recombinant DNA human insulin, *Diabetologia* **25**:465–469.

Fireman, P., Fineburg, S. E., and Galloway, J. A., 1982, Development of IgE antibodies to human (recombinant DNA), porcine, and bovine insulins in diabetic subjects, *Diabetes Care* **5**(Suppl. 2):119–125.

Fisher, B. V., and Porter, P. B., 1981, Stability of bovine insulin, *J. Pharm. Pharmacol.* **33**:203–206.

Forsham, P. H., 1982, Milestones in the 60-year history of insulin (1922–1982), *Diabetes Care* **5**(Suppl. 2):1–3.

Fulcher, C. A., Mahoney, S. G., Roberts, J. R., Kasper, C. K., and Zimmerman, T. S., 1985, Localization of human factor FVIII inhibitor epitopes to two polypeptide fragments, *Proc. Natl. Acad. Sci. USA* **82**:7728–7732.

Galloway, J. A., and Bressler, R., 1978, Insulin treatment in diabetes, *Med. Clin. North Am.* **62**:663–680.

Gilliland, B. C, 1984, Serum sickness and immune complexes, *N. Engl. J. Med.* **311**:1435–1436.

Gitschier, J., Wood, W. I., Tuddenham, E. G. D., Shuman, M. A., Goralka, T. M., Chen, E. Y., and Lawn, R. M., 1985, Detection and sequence of mutations in the factor VIII gene of hemophiliacs, *Nature* **315**:427–430.

Godal, A., Fodstad, O., and Pihl, A., 1983, Antibody formation against the cytotoxic proteins abrin and ricin in humans and mice, *Int. J. Cancer* **32**:515–521.

Goto, M., Akai, K., Murakami, A, Hashimoto, C., Tsuda, E., Ueda, M., Kawanishi, G., Takahashi, N., Ishimoto, A., Chiba, H., and Sasaki, R., 1988, Production of recombinant human erythropoietin in mammalian cells: Host-cell dependency of the biological activity of the cloned glycoprotein, *Biotechnology* **6**:67–72.

Grammer, L. C., and Roberts, M., 1989, Specificity of IgE antibody against various insulins in a patient with anaphylaxis to beef-pork insulin but not to human (rDNA) insulin, *Clin. Exp. Allergy* **19**:551–553.

Grammer, L. C., Metzger, B. E., and Patterson, R., 1984, Cutaneous allergy to human (recombinant DNA) insulin, *J. Am. Med. Assoc.* **251**:1459–1460.

Grammer, L. C., Roberts, M., and Patterson, R., 1985, IgE and IgG antibody against human (recombinant DNA) insulin in patients with systemic insulin allergy, *J. Lab. Clin. Med.* **105**:108–113.

Grammer, L. C., Roberts, M., Buchanan, T. A., Fitzsimons, R., Metzger, B. E., and Patterson, R., 1987, Specificity of immunoglobin E and immunoglobin G against human (recombinant DNA) insulin in human insulin allergy and resistance, *J. Lab. Clin. Med.* **109**: 141–146.

Grey, H. M., Sette, A., and Buus, S., 1989, How T cells see antigen, *Sc. Am.* Nov.:56–64.

Gutterman, J. U., Fine, S., Quesada, J., Horning, S. J., Levine, J. F., Alexanian, R., Bernhardt, L., Kramer, M., Spiegel, H., Colburn, W., Trown, P., Merigan, T., and Dziewanowski, Z., 1982, Recombinant leukocyte A interferon: Pharmacokinetics, single-dose tolerance, and biologic effects in cancer patients, *Ann. Intern. Med.* **96**:549–556.

Hansend, B., Nielsen, J. H., and Welinder, B., 1981, Immunogenicity of insulin in relation to its physio-chemical properties, in: *Basic and Clinical Aspects of Immunity to Insulin* (K. Keck and P. Erb, eds.), de Gruyter, Berlin, pp. 335–352.

Hoyner, L. W., Gawryl, M. S., and de la Fuente, B., 1984, Immunological characterization of factor VIII inhibitors, in: *Factor VIII Inhibitors* (L. W. Hoyner, ed.), Liss, New York, pp. 73–85.

Hultin, M. B., London, F. S., Shapiro, S. S., and Yount, W. J., 1977, Heterogeneity of factor VIII antibodies: Further immunochemical and biological studies, *Blood* **49**:807–817.

Iavicoli, M., di Mario, U., Pozzilli, P., Canalese, J., Ventriglia, L., Galfo, C., and Andreani, D., 1982, Impaired phagocytic function and increased immune complexes in diabetics with severe microangiopathy, *Diabetes* **31**:7–11.

Itri, L. M., Campion, M., Dennin, R. A., Palleroni, A. V., Gutterman, J. U., Groopman, J. E. and Trown, P. W., 1987, Incidence and clinical significance of neutralizing antibodies in patients receiving recombinant interferon alfa-2a by intramuscular injection, *Cancer* **59**: 668–674.

Itri, L. M., Sherman, M. I., Palleroni, A. V., Evans, L. M., Tran, L.-L., Campion, M., and Chizzonite, R., 1989, Incidence and clinical significance of neutralizing antibodies in patients receiving recombinant interferon-α_{2a}, *J. Interferon Res.* **9**(Suppl. 1):S9–S15.

Jaffers, G. J., Fuller, T. C., Cosimi, A. B., Russell, P. S., Winn, H. J., and Colvin, R. B, 1986, Monoclonal antibody therapy. Anti-idiotypic and non-anti-idiotypic antibodies to OKT3 arising despite intense immunosuppression, *Transplantation* **41**:572–578.

Johnson, I. S., 1983, Human insulin from recombinant DNA technology, *Science* **219**: 632–637.

Jones, G. J., and Itri, L. M., 1986, Safety and tolerance of recombinant interferon alfa-2a (Roferon-A) in cancer patients, *Cancer* **57**:1709–1715.

Kagawa, Y., Takasaki, S, Utsumi, J., Hosoi, K., Shimizu, H., Kochibe, N., and Kobata, A., 1988, Comparative study of the asparagine-linked sugar chains of natural human interferon-β1 and recombinant human interferon-β1 produced by three different mammalian cells, *J. Biol. Chem.* **263**:17508–17515.

Kahn, C. R., and Rosenthal, A. S., 1979, Immunological reactions to insulin: Insulin allergy, insulin resistance, and the autoimmune insulin syndrome, *Diabetes Care* **2**:283–295.

Katre, N. V., 1990, Immunogenicity of recombinant IL-2 modified by covalent attachment of polyethylene glycol, *J. Immunol.* **144**:209–213.

Keller, R. J., 1990, Cellular immunity to human insulin in individuals at high risk for the development of type I diabetes mellitus, *J. Autoimmun.* **3**:321–327.

Kernoff, P. B. A., Thomas, N. D., Lilley, P. A., Matthews, K. B., Goldman, E., and Tuddenham, E. G. D., 1984, Clinical experience with polyelectrolyte-fractionated porcine factor VIII concentrate in the treatment of hemophiliacs with antibodies to factor VIII, *Blood* **63**:31–41.

Khosrovi, B., 1984, The production, characterization, and testing of a modified recombinant human interferon beta, in: *Interferon: Research, Clinical Application, and Regulatory Consideration* (K. C. Zoon, P. D. Noguchi, and T.-Y. Liu, eds.), Elsevier, Amsterdam, pp. 89–100.

Klein, J., 1982, *Immunology: The Science of Self–Nonself Discrimination*, Wiley, New York.

Koch, M., Sodoyez, J. C. Sodoyez-Goffaux, F., Dozio, N., Di Silvio, L. S., and Kurtz, A. B., 1989, Is quantitative assessment of insulin-antibodies and autoantibodies feasible? *Diabetologia* **32**:774–778.

Kolitz, J. E., Welte, K., Wong, G. Y., Holloway, K., Merluzzi, V. J, Engert, A., Bradley, E. C. Konrad, M. W., Polivka, A., Gabrilove, J. L., Sykora, K. W., Miller, G. A., Fiedler, W., Krown, S., Oettgen, H. F., and Mertelsmann, R., 1987, Expansion of activated T-lymphocytes in patients treated with recombinant interleukin 2, *J. Biol. Response Mod.* **6**: 412–429.

Konrad, M., 1989, Immunogenicity of proteins administered to humans for therapeutic purposes, *Trends Biotechnol.* **7**:175–179.

Konrad, M. W., Childs, A. L., Merigan, T. C., and Borden, E. C., 1987, Assessment of the antigenic response in humans to a recombinant mutant interferon beta, *J. Clin. Immunol.* **7**:365–375.

Krigel, R. L., Padavic-Shaller, K. A., Rudolph, A. R., Litwin, S., Konrad, M., Bradley, E. C., and Comis, R. L., 1988, A phase I study of recombinant interleukin 2 plus recombinant β-interferon, *Cancer Res.* **48**:3875–3881.

Larrick, J. W., 1989, Potential of monoclonal antibodies as pharmacological agents, *Pharm. Rev.* **41**:539–557.

Lawley, T. J., Bielory, L., Gascon, P., Yancey, K. B., Young, N. S., and Frank, M. M., 1984, A prospective clinical and immunological analysis of patients with serum sickness, *N. Engl. J. Med.* **311**:1407–1413.

Lawrence, R. D., 1925, Local insulin reactions, *Lancet* **1**:1125–1126.

Levy, R., and Miller, R. A., 1983, Tumor therapy with monoclonal antibodies, *Fed. Proc.* **42**: 2650–2656.

LoBuglio, A. F., Wheeler, R. H., Trang, J., Haynes, A., Rogers, K., Harvey, E. B., Sun, L., Ghrayeb, J., and Khazaeli, M. B. 1989, Mouse/human chimeric monoclonal antibody in man: Kinetics and immune response, *Proc. Natl. Acad. Sci. USA* **86**:4220–4224.

McCormick, F., Trahey, M., Innis, M., Dieckmann, B., and Ringold, G., 1984, Inducible expression of amplified human beta interferon genes in CHO cells, *Mol. Cell. Biol.* **4:** 166–172.

Maneschi, F., Fineberg, S. E., and Kohner, E. M., 1982, Successful treatment of immune-mediated insulin resistance by human insulin (recombinant DNA), *Diabetes Care* **5** (Suppl. 2):175–179.

Manjunath, P., and Sairam, M. R., 1982, Biochemical, biological, and immunological properties of chemically deglycosylated human choriogonadotropin, *J. Biol. Chem.* **257:** 7109–7115.

Mannik, M., 1980, Physiochemical and functional relationships of immune complexes, *J. Invest. Dermatol.* **74:**333–338.

Margalit, H., Spouge, J. L., Cornette, J. L., Cease, K. B., Delisi, C., and Berzofsky, J. A., 1987, Prediction of immunodominant helper T cell antigenic sites from the primary sequence, *J. Clin. Immunol.* **138:**2213–2229.

Mark, D. F., Lu, S. D., Creasey, A. A., Yamamoto, R., and Lin, L. S., 1984, Site-specific mutation of the human fibroblast interferon gene, *Proc. Natl. Acad. Sci. USA* **81:**5662–5666.

Meyers, F. J., Goodman, G., Paradise, C., Konrad, M., and Scudder, S. A., 1991, A phase I study including pharmacokinetics of pegylated interleukin 2 (PEG IL-2), *Clin. Pharmacol. Ther.* **49:**307–313.

Miller, G. G., Pollack, M. S., Nell, L. J., and Thomas, J. W., 1987, Insulin-specific human T cells. Epitope specificity, major histocompatibility complex restriction, and alloreactivity to a diabetes-associated haplotype, *J. Immunol.* **139:**3622–3629.

Mischak, R. P., Foxall, C., Rosendorf, L. L., Knebel, K., Scannon, P. J., and Spitler, L. E., 1990, Human antibody responses to components of the monoclonal antimelanoma antibody ricin A chain immunotoxin Xomazyme-mel, *Mol. Biother.* **2:**104–109.

Mogensen, K. E., Daubas, P., Gresser, I., Sereni, D., and Varet, B., 1981, Patient with circulating antibodies to α-interferon, *Lancet* **2:**1227–1228.

Morrison, S. L., Johnson, M. J., Herzenberg, L. A., and Oi, V. T., 1984, Chimeric human antibody molecules: Mouse antigen-binding domains with human constant region domains, *Proc. Natl. Acad. Sci. USA* **81:**6851–6855.

Naquet, P., Ellis, J., Tibensky, D., Kenshole, A., Singh, B, Hodges, R., and Delovitch, T. L., 1988, T cell autoreactivity to insulin in diabetic and related non-diabetic individuals, *J. Immunol.* **140:**2569–2578.

Nell, L. J., Virta, V. J., and Thomas, J. W., 1985, Recognition of human insulin in vitro by T cells from subjects treated with animal insulins, *J. Clin. Invest.* **76:**2070–2077.

Nerup, J., and Lernmark, A., 1981, Autoimmunity in insulin-dependent diabetes mellitus, *Am. J. Med.* **70:**135–141.

Niman, H. L., Houghten, R. A., Walker, L. E., Reisfeld, R. A., Wilson, I. A., Hogle, J. M., and Lerner, R. A., 1983, Generation of protein-reactive antibodies by short peptides is an event of high frequency: Implications for the structural basis of immune recognition, *Proc. Natl. Acad. Sci. USA* **80:**4949–4953.

Nishimura, Y., Yokoyama, M., Araki, K., Ueda, R., Kudo, A., and Watanabe, T., 1987, Recombinant human–mouse chimeric monoclonal antibody specific for common acute lymphocytic leukemia antigen, *Cancer Res.* **47:**999–1005.

Novontny, J., Handschumacher, M., Haber, E., Bruccoleri, R. E., Carlson, W. B., Fanning, D. W., Smith, J. A., and Rose, G. D., 1986, Antigenic determinants in proteins coincide with surface regions accessible to large probes (antibody domains), *Proc. Natl. Acad. Sci. USA* **83:**226–230.

Onodera, T., Toniolo, A., Ray, U. R., Jenson, A. B., Knazek, R. A., and Notkins, A. L., 1981, Virus-induced diabetes mellitus, *J. Exp. Med.* **153:**1457–1473.

Paolozzi, F., Zamkoff, K., Doyle, M., Konrad, M., Bradley, E. C., Rudolph, A., Newman, N., Gullo, J., Scalzo, A., and Poiesz, B., 1989, Phase I trial of recombinant interleukin-2 and recombinant β-interferon in refractory neoplastic diseases, *J. Biol. Response Mod.* **8:** 122–139.

Paul, W. E., 1989, *Fundamental Immunology*, 2nd ed., Raven Press, New York.

Platts-Mills, T. A. E., Chapman, M. D., and Tovey, E., 1981, Radioimmunoassays in allergy, in: *Immunoassays for the 80s* (A. Voller, A. Bartlett, and D. Bidwell, eds.), University Park Press, Baltimore, pp. 289–311.

Protzman, W. P., Jacobs, S. L., Minnicozzi, M., Oden, E. M., and Kelsey, D. K., 1984, A radioimmunologic technique to screen for antibodies to α-2 interferon, *J. Immunol. Methods* **75:**317–323.

Quesada, J. R., Rios, A., Swanson, D., Trown, P., and Gutterman, J. U., 1985, Antitumor activity of recombinant-derived interferon alpha in metastatic renal cell carcinoma, *J. Clin. Oncol.* **3:**1522–1528.

Quesada, J. R., Moshe, T., Rios, A., Kurzrock, R., and Gutterman, J. U., 1986, Clinical toxicity of interferons in cancer patients: A review, *J. Clin. Oncol.* **4:**234–243.

Rademacher, T. W., Parekh, R. B., and Dwek, R. A., 1988, Glycobiology, *Annu. Rev. Biochem.* **57:**785–838.

Ratner, B., 1943, *Allergy, Anaphylaxis and Immunotherapy*, Williams & Wilkins, Baltimore.

Reeves, W. G., 1981, Antibody production during insulin therapy: Patterns of response and clinical sequels, in: *Basic and Clinical Aspects of Immunity to Insulin* (K. Peck and P. Erb, eds.), de Gruyter, Berlin, pp. 219–235.

Reeves, W. G., Allen, B., and Tattersall, R. B., 1980, Insulin-induced lipoatropy: Evidence for an immune pathogenesis, *Br. Med. J.* **280:**1500–1503.

Renkin, E. M., and Gilmore, J. P., 1973, Glomerular filtration, in: *Handbook of Physiology* (J. Field, ed.), Section 8, American Physiological Society, Washington, D.C.

Riechmann, L., Clark, M., Waldmann, H., and Winter, G., 1988, Reshaping human antibodies for therapy, *Nature* **332:**323–327.

Rizza, C. R., and Matthews, J. M., 1982, Effect of frequent factor VIII replacement on the level of factor VIII antibodies in haemophiliacs, *Br. J. Haematol.* **52:**13–24.

Roberts, H. R., 1981, Hemophiliacs with inhibitors: Therapeutic options, *N. Engl. J. Med.* **305:** 757–758.

Roberts, H. R., and Cromartie, R., 1984, Overview of inhibitors to factor VIII and IX, in: *Factor VIII Inhibitors* (L. W. Hoyer, ed.), Liss, New York, pp. 1–18.

Rosenberg, S. A., 1988, Immunotherapy of patients with advanced cancer using interleukin-2 alone or in combination with lymphokine activated killer cells, in: *Important Advances in Oncology* (V. T. DeVita, Jr., S. Hellman, and S. A. Rosenberg, eds.), Lippincott, Philadelphia, pp. 217–257.

Rosenberg, S. A., 1990, Adoptive immunotherapy for cancer, *Sci. Am.* **May:**62–69.

Rosenberg, S. A., Grimm, E. A., McGrogan, M., Doyle, M., Kawasaki, E., Koths, K., and Mark, D. F., 1984, Biological activity of recombinant human interleukin-2 produced in *E. coli*, *Science* **223:**1412–1415.

Rowe, R. E., Pimm, M. V., and Baldwin, R. W., 1985, Anti-idiotype antibody responses in cancer patients receiving a murine monoclonal antibody, *IRCS Med. Sci.* **13**:936–937.

Rubenstein, A. H., and Spitz, I., 1968, Role of the kidney in insulin metabolism and excretion, *Diabetes* **17**:161–169.

Sato, S., Ebert, C. D., and Kim, S. W., 1983, Prevention of insulin self-association and surface adsorption, *J. Pharm. Sci.* **72**:228–232.

Savoca, K. V., Abuchowski, A., van Es, T., Davis, F. F., and Palczuk, N. C., 1979, Preparation of a non-immunogenic arginase by the covalent attachment of polyethylene glycol, *Biochim. Biophys. Acta* **578**:47–53.

Schernthaner, G., Borkenstein, M., Fink, M., Mayr, W. R., Menzel, J., and Schober, E., 1983, Immunogenicity of human insulin (Novo) or pork monocomponent insulin in HLA-DR-typed insulin-dependent diabetic individuals, *Diabetes Care* **6**(Suppl. 1):43–48.

Schroff, R. W., Foon, K. A,. Beatty, S. M., Oldham, R. K., and Morgan, A. C., Jr., 1985, Human anti-murine immunoglobulin responses in patients receiving monoclonal antibody therapy, *Cancer Res.* **45**:879–885.

Sepulveda, R., Longbottom, J. L., and Pepys, J., 1979, Enzyme linked immunosorbent assay (ELISA) for IgG and IgE antibodies to protein and polysaccharide antigens of *Aspergillus fumigatus*, *Clin. Allergy* **9**:359–371.

Shapiro, S. S., 1979, Antibodies to blood coagulation factors, *Clin. Haematol.* **8**:207–214.

Shapiro, S. S., 1984, Genetic predisposition to inhibitor formation, in: *Factor VIII Inhibitors* (L. W. Hoyer, ed.), Liss, New York, pp. 45–55.

Shawler, D. L., Bartholomew, R. M., Smith, L. M., and Dillman, R. O., 1985, Human immune response to multiple injections of murine monoclonal IgG, *J. Immunol.* **135**:1530–1535.

Shoelson, S., Fickova, M., Haneda, M., Nahum, A., Musso, G., Kaiser, E. T., Rubenstein, A. H., and Tager, H., 1983, Identification of a mutant human insulin predicted to contain a serine-for-phenylalanine substitution, *Proc. Natl. Acad. Sci. USA* **80**:7390–7394.

Silverstone, P., 1986, Generalized allergic reaction to human insulin, *Br. Med. J.* **292**: 933–934.

Sjamsoedin, L. J. M., Heijnen, L., Mauser-Bunschoten, E. P., van Geijlswijk, J. L., van Houwelingen, H., van Asten, P., and Sixma, J. J., 1981, The effect of activated prothrombin-complex concentrate (FEIBA) on joint and muscle bleeding in patients with hemophilia A and antibodies to factor VIII, *N. Engl. J. Med.* **305**:717–721.

Small, P., and Lerman, S., 1984, Human insulin allergy, *Ann. Allergy,* **53**:39–41.

Spijker, A. J., Poortman, J., Thijssen, H. H., and Erkelens, D. W., 1982, Decrease of circulating insulin antibodies in two patients treated with continuous subcutaneous infusion of human insulin, *Diabetes Care* **5**(Suppl. 2):171–174.

Spitler, L. E., del Rio, M., Khentigan, A., Wedel, N. I., Brophy, N. A., Miller, L. L., Harkonen, W. S., Rosendorf, L .L., Lee, H. M., Mischak, R. P., Kawahata, R. T., Stoudemire, J. B., Fradkin, L. B., Bautista, E. E., and Scannon, P. J., 1987, Therapy of patients with malignant melanoma using a monoclonal antimelanoma antibody–ricin A chain immunotoxin, *Cancer Res.* **47**:1717–1723.

Stiller, C. R., Dupre, J., Gent, M., Jenner, M. R., Keown, P. A., Laupacis, A., Martell, R., Rodger, N. W., Graffenried, B. v., and Wolfe, B. M. J., 1984, Effects of cyclosporine immunosuppression in insulin-dependent diabetes mellitus of recent onset, *Science* **223**: 1362–1367.

Strauss, H. S., 1969, Acquired circulating anticoagulants in hemophilia A, *N. Engl. J. Med.* **281**:866–873.

Sultan, Y., Rossi, F., and Kazatchkine, M. D., 1987, Recovery from anti-VIII:C (anti-hemophilic factor) autoimmune disease is dependent on generation of antiidiotypes against anti-VIII:C autoantibodies, *Proc. Natl. Acad. Sci. USA* **84**:828–831.

Taniguchi, T., Matsui, H., Fugita, T., Takaoka, C., Kashima, N., Yoshimoto, R., and Hamuro, J., 1983, Structure and expression of a cloned cDNA for human interleukin-2, *Nature* **302**: 305–308.

Teelmann, K., Hohbach, C., Lehmann, H., and Group, T. I. W., 1986, Preclinical safety testing of species-specific proteins produced with recombinant DNA-techniques, *Arch. Toxicol.* **59**:195–200.

Vallbracht, A., Treuner, J., Flehmig, B., Joester, K. E., and Niethammer, D., 1981, Interferon-neutralizing antibodies in a patient treated with human fibroblast interferon, *Nature* **289**: 496–497.

Vallbracht, A., Treuner, J., Manncke, K. H., and Niethammer, D., 1982, Autoantibodies against human beta interferon following treatment with interferon, *J. Interferon Res.* **1**: 107–110.

Van Haeften, T. W., 1989, Clinical significance of insulin antibodies in insulin-treated diabetic patients, *Diabetes Care* **12**:641–648.

Voller, A., Bidwell, D., and Bartlett, A., 1976, Microplate enzyme immunoassays for the immunodiagnosis of virus infections, in: *Manual of Clinical Immunology* (N. Rose and H. Feldman, eds.), American Society for Microbiology, Washington, D. C., p. 506.

von Pirquet, C., and Schick, B., 1951, *Serum Sickness*, Williams & Wilkins, Baltimore.

Walhausl, W. K., Kastner, G., Komjati, M., and Bratusch-Marrain, P., 1981, Studies on the biologic actions of biosynthetic human insulin in vitro and in diabetic man, *Diabetes Care* **4**:205–208.

Wang, A., Lu, S., and Mark, D., 1984, Site-specific mutagenesis of the human interleukin-2 gene: Structure function analysis of the cysteine residues, *Science* **224**:1431–1434.

Westhof, E., Altschuh, D., Maras, D., Bloomer, A. C., Mondragon, A., Klug, A., and Van Regenmortel, M. H. V., 1984, Correlation between segmental mobility and the location of antigenic determinants in proteins, *Nature* **311**:123–126.

Wettendorff, M., Iliopoulos, D., Tempero, M., Kay, D., DeFreitas, E., Koprowski, H., and Herlyn, D., 1989, Idiotypic cascades in cancer patients treated with monoclonal antibody CO17-1A, *Proc. Natl. Acad. Sci. USA* **86**:3787–3791.

Whitehead, R. P., Ward, D., Hemingway, L., Hemstreet, G. P., III, Bradley, E., and Konrad, M., 1990, Subcutaneous recombinant interleukin 2 in a dose escalating regimen in patients with metastatic renal cell adenocarcinoma, *Cancer Res.* **50**:6708–6715.

Wide, L., Bennich, H., and Johansson, S. G. O., 1967, Diagnosis of allergy by an *in vitro* test for allergen antibodies, *Lancet* **2**:1105.

Wilkinson, I., Jackson, C.-J. C., Lang, G. M., Holford-Strevens, V., and Sehon, A. H., 1987, Tolerance induction in mice by conjugates of monoclonal immunoglobulins and mono-methoxypolyethylene glycol, *J. Immunol.* **139**:326–331.

Wilson, C. B., and Dixon, F. J., 1971, Quantitation of acute and chronic serum sickness in the rabbit, *J. Exp. Med.* **134**:7s–18s.

Wilson, R. M., Douglas, C. A., Tattersall, R. B., and Reeves, W. G., 1985, Immunogenicity of highly purified bovine insulin: A comparison with conventional bovine and highly purified human insulins, *Diabetologia* **28**:667–670.

V

Tissue Barriers

Chapter 15

Extravasation and Interstitial Transport in Tumors

Rakesh K. Jain and Laurence T. Baxter

1. INTRODUCTION

The use of high-molecular-weight agents such as proteins has been of increasing interest for cancer detection and treatment since the development of genetic engineering and hybridoma technology. These agents include monoclonal antibodies (conjugated with radionuclides, toxins, cytokines, or enzymes), growth factors, biological response modifiers, and enzymes. The use of cells, such as lymphokine-activated killer cells or tumor-infiltrating lymphocytes, is also being investigated. The potent toxicity of some of these agents toward cancer cells *in vitro* has ignited hopes for a "magic bullet." Although these agents show great potential, results in clinical studies have not been so positive.

The successful use of proteins and other macromolecules for cancer treatment has been hindered by inadequate and nonuniform distribution within tumors. In many cases this suboptimal response in tumors may be attributed to tumor physiology. In particular, there are physiological barriers to transport which may render the most toxic agent ineffective. There are several transport steps encountered by any molecule or cell injected into the circulation: (1) distribution within the vascular space, (2) transport across the vascular wall, (3) transport across the interstitium, and possibly (4) uptake by the cancer cell (Fig. 1). For subcutaneous injection or sustained release, transport across the interstitium is the major barrier, with removal of drug by

Rakesh K. Jain and Laurence T. Baxter • Department of Radiation Oncology, Harvard Medical School, Steele Laboratory, Massachusetts General Hospital, Boston, Massachusetts 02114.

Biological Barriers to Protein Delivery, edited by Kenneth L. Audus and Thomas J. Raub. Plenum Press, New York, 1993.

441

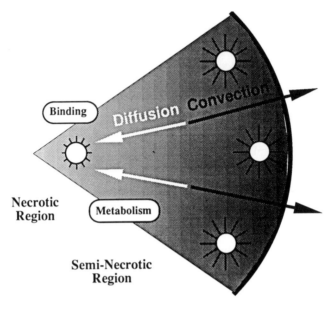

**Well Vascularized
Region**

Figure 1. Transport in tumors. This figure is a cross section of a spherical tumor. Filtration of fluid and macromolecules is lowest in the center where the interstitial pressure is highest. There is a convective flow of fluid and solute radially outward, resulting in a concentration gradient. This gradient leads to a diffusional flux inward. The macromolecule may also bind to the cell and/or be metabolized.

lymphatic and blood vessels. A molecule also may bind specifically or non-specifically to proteins or tissue components, or be metabolized or physically entrapped.

For the past ten years we have been studying the transport of proteins and macromolecules in tumors and the role of these physiological barriers. Through experimental techniques and mathematical models, we have studied extravasation and interstitial diffusion and convection. Through this we have determined the impact of various parameters on the uptake and distribution of drugs within tumors. In this chapter we review the role of extravasation and interstitial transport on drug delivery, describe ways to quantify these effects, and discuss the implications of physiological barriers, with an emphasis on cancer detection and treatment.

2. EXTRAVASATION

2.1. Starling's Law of Fluid Filtration

Tumor vasculature is comprised of both vessels from the preexisting normal tissue and new vessels from an angiogenic response initiated by the cancer cells. The distribution of blood vessels is nonuniform, and may vary greatly from tumor to tumor. Some tumors exhibit peripheral vascularization, in which a virtually avascular core is surrounded by a shell of well-vascularized tissue. Such tumors develop a necrotic core as they grow due to central hypoxia. Other tumors may be centrally vascularized, for which drug distributions are expected to be quite different. A given tumor may have several nodes, each showing different patterns and stages of vascularization (Jain and Ward-Hartley, 1984; Jain, 1988).

An important difference between normal and tumor vessels is that tumor vessels are dilated, tortuous, and lack a continuous basement membrane. Arteriovenous shunts and intervenular connections are more prevalent in tumor vessels. Blood vessels created in response to cancer cells also lack the smooth muscle cells surrounding normal arteries, and so many of the tumor vessels may be treated as giant capillaries. The perfusion rates may vary greatly in different regions of a tumor, especially when necrotic or semi-necrotic regions are present.

The filtration of fluid across a blood vessel wall is described mathematically by Starling's law, which states that the rate of filtration per unit area of blood vessel (J_v/S, ml/cm^2 per sec) is proportional to the difference in hydrostatic (p) and osmotic (π) pressures (mm Hg) inside and outside the blood vessel. The proportionality constant is the vascular hydraulic conductivity, L_p (cm/mm Hg per sec); the product of the hydraulic conductivity and the surface area is referred to as the capillary filtration coefficient (CFC). Since the vascular wall is semipermeable, only a fraction of the osmotic pressure results in fluid flow; this fraction is the osmotic reflection coefficient, σ. It is close to 0 for small molecules and nearly 1 for macromolecules. Albumin is the most abundant plasma protein, with σ approximately 0.8 in mesentery (Curry *et al.*, 1976) and 0.9 in subcutaneous tissue (Ballard and Perl, 1978). Starling's law may be written as:

$$J_v/V = L_p[(p_v - p_i) - \sigma(\pi_v - \pi_i)] \tag{1}$$

where the subscripts v and i are for the vascular and interstitial pressures, respectively.

2.2. Transcapillary Exchange

There are many proposed pathways for the extravasation of solute molecules, including: (1) transport through endothelial cells; (2) lateral membrane diffusion;

(3) convection and diffusion through interendothelial junctions; (4) endothelial fenestrae; and (5) vesicular transport (Renkin *et al.*, 1977; Jain, 1987a; see Chapter 10) (Fig. 2). The vesicular transport theory has been disputed by the work of Frokjaer-Jensen (1980) and Bundagaard *et al.* (1983). For macromolecules the most likely dominant mode of transport in nonfenestrated capillaries is the interendothelial junctions. Through these junctions there are two modes of transport. The convective transport, often dominant for macromolecules, is dependent on a pressure difference between the vascular and interstitial spaces. The diffusive transport occurs due to a concentration gradient.

Efforts have been made to quantify the transport of solutes across vessel walls, most often yielding an effective permeability coefficient, P_{eff}, defined as the solute flux (J_s/S) divided by the concentration difference between the plasma and interstitial space:

$$J_s/S = P_{eff}(C_p - C_i) \tag{2}$$

The effective vascular permeability has been determined by a variety of techniques. Curry (1984) measured P_{eff} in large capillaries of frogs using an optical method, by injecting a solute in a single perfused capillary and measuring the solute concentration outside the vessel. This method has not yet been used for tumor vessels.

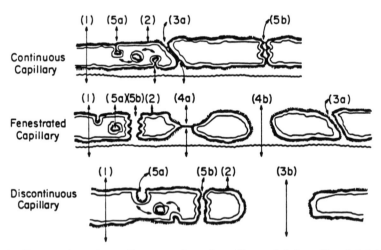

Figure 2. Transvascular exchange. Transport pathways in capillary endothelium. (1) endothelial cell; (2) lateral membrane diffusion; (3) interendothelial junctions—(a) narrow, (b) wide; (4) endothelial fenestrae—(a) closed, (b) open; (5) vesicular transport—(a) transcytosis, (b) transendothelial channels. Note that water and lipophilic solutes share pathways (1), (3), and (4). Lipophilic solutes may use pathway (2) as well. Hydrophilic solutes and macromolecules use pathways (3) and (4). Macromolecules may also follow pathway (5). (From Jain, 1987a, with permission.)

An alternate microscopic approach uses an exteriorized thin tissue or window preparation to monitor fluorescent molecules. Mesentery, the hamster cheek pouch, rabbit ear chambers, and dorsal skin windows have all been used effectively. Baxter *et al.* (1987) and Baxter and Jain (1988) have analyzed superfusate concentration data in the hamster cheek pouch to determine vascular permeability (P) and interstitial diffusion (D) coefficients. Gerlowski and Jain (1986) have developed a method to determine P and D by video image analysis of the fluorescent field around blood vessels in the rabbit ear chamber. No exteriorization or superfusion was required with this method. They found that tumor vessels were approximately seven times as permeable as normal vessels to dextran of 150 kDa. Other studies using macroscopic methods have shown a greater permeability for macromolecules in tumor vessels than in several normal vessels (e.g., Dewey, 1959; Song and Levitt, 1970; O'Conner and Bale, 1984).

Yuan *et al.* (1993) have measured the microvascular permeability to macromolecules in a human colon cancer xerograft in SCID mice. They have found values that are consistent with those of Gerlowski and Jain (1986): that the tumor vessels are an order of magnitude more permeable than the host vessels.

A potential problem with all of these methods is that they measure the *effective* vascular permeability, and do not distinguish between convection and diffusion. An analysis of solute flux which recognizes these two pathways yields (Patlak *et al.*, 1963):

$$J_s/S = (J_v/S)(1 - \sigma)C_p + P(C_p - C_i)[\text{Pe}/(e^{\text{Pe}} - 1)] \qquad (3)$$

where J_v/S is the fluid flux (from Starling's law), σ is the osmotic reflection coefficient, P is the true (diffusive) vascular permeability coefficient, and Pe is the Péclet number [$= J_v(1 - \sigma)/PS$], the ratio of convective to diffusive flux. From Eqs. (2) and (3), the effective permeability coefficient is given by:

$$P_{\text{eff}} = P\{[\text{Pe}/(e^{\text{Pe}} - 1)] + \text{Pe}/(1 - C_i/C_p)\} \qquad (4)$$

Thus, the effective permeability coefficient, calculated by the experiments described above, is a function of the permeability coefficient (P), concentration difference, and fluid filtration rate, which is itself a function of the hydraulic conductivity and the difference between plasma and interstitial pressures. In tumors the difficulty in comparing values of P_{eff} is complicated by the large differences in interstitial pressure from tumor to tumor and within the same tumor (Boucher *et al.*, 1990).

Rutili (1978) and Rippe and Haraldsson (1987) have shown for macromolecules larger than albumin that convection is the primary pathway for exchange through pores. When this is the case, the solute flux will be approximately proportional to the plasma concentration alone, and the effective permeability coefficient [$P_{\text{eff}} = J_s/S/(C_p - C_i)$] will be much greater than the diffusive permeability coefficient, P.

This makes the elevated interstitial pressure in tumors an important factor in drug delivery.

2.3. Interstitial Pressure

As discussed above, interstitial pressure plays an important role in determining uptake and the value of P_{eff} for macromolecules. The interstitial pressure in tumors has long been known to be greater than in normal tissues (Young *et al.*, 1950). Wiig *et al.* (1982) and Misiewicz (1986) showed that the pressure is greater in the center of the tumor than in the periphery. The high pressure serves to decrease the fluid filtration and convective solute flux from blood vessels in solid tumors, and so reduce the effective permeability coefficient.

Baxter and Jain (1989) developed a mathematical model to describe the interstitial pressure profiles in tumors. They found that the interstitial pressure could be uniform over large portions of the tumor, with a steep pressure gradient at the tumor periphery. This was due to the fact that for alymphatic tumors, fluid transport over macroscopic distances is slower than filtration or absorption through the blood vessel wall. The ratio of resistances to fluid flow was incorporated in a dimensionless parameter, $\alpha^2 = (R^2 L_p S)/(VK)$, where R (cm) is the tumor radius, S/V (cm^{-1}) is the exchange vessel surface area per unit volume, and K (cm^2/mm Hg sec^{-1}) is the interstitial hydraulic conductivity. The maximum interstitial pressure was equal to the effective vascular pressure, $p_e = p_v - \sigma(\pi_v - \pi_i)$. The model was extended to incorporate necrosis, heterogeneous blood perfusion, and lymphatics (Baxter and Jain, 1990, 1991a). The predicted pressure profiles were confirmed experimentally by Boucher *et al.* (1990) and shown to be driven by the microvascular pressure (Boucher *et al.*, 1992). Figure 3 shows some of their experimental pressure data with the model's best fit using two parameters, p_e and α^2. The estimated parameters agreed well with values calculated from the literature. The effect of interstitial pressure on macromolecular uptake will be discussed quantitatively in Section 4.1.1.

3. INTERSTITIAL TRANSPORT

3.1. Diffusion

After the molecule has extravasated, it must traverse the interstitium. The interstitial space fraction in tumors may be very large relative to normal tissues (Jain, 1987b). The interstitium is a gel-like region comprised mainly of a collagen and elastic fiber network. Within this network, polysaccharides (glycosaminoglycan and

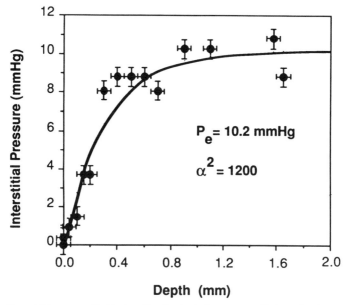

Figure 3. Interstitial pressure distribution. Here is a typical pressure profile in a subcutaneous mammary adenocarcinoma. The cirles (●) represent data points, while the solid line is the theoretical profile (Baxter and Jain, 1989) fitted for two parameters: the effective vascular pressure, p_e (10.2 mm Hg), and the hydraulic conductivity ratio, α^2. The error bars represent estimates of the precision of pressure and depth measurements. The error in depth measurements was obtained from the variation in the location of the surface between insertion and withdrawal of the micropipette.

proteoglycans) play a dominant role in creating resistance to fluid and molecular transport. In several tumor tissues, the collagen content is higher than for normal tissues, while the polysaccharide content is lower. This is thought to decrease the resistance to transport in tumor interstitium.

The two transport mechanisms within the interstitium are diffusion and convection. The diffusive flux is proportional to concentration gradients, while the convective flux is proportional to the interstitial fluid velocity. The diffusion coefficient, D (cm^2/sec), is the constant relating diffusive solute flux to the concentration gradient. The value of D will depend on both the structure of the interstitium and the size, shape, and charge of the solute.

Figure 4 shows the relationship between molecular weight (or Stokes–Einstein radius) and the effective diffusion coefficient for several macromolecules in water, tumor, and normal tissue. Diffusion coefficients are highest in water and lowest in normal tissues. The diffusivity decreases approximately as a power-law function of the molecular weight. The shape and charge of proteins also affect the value of D, and there are differences from tissue to tissue and between tumors.

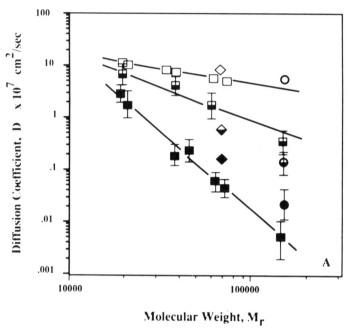

Molecular Weight, M_r

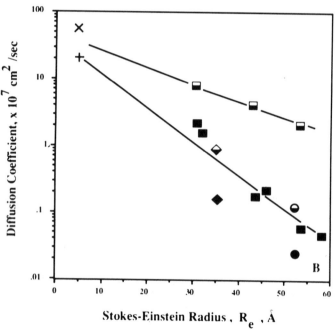

Stokes-Einstein Radius, R_e , Å

There is a problem with the diffusion of proteins and other macromolecules in large solid tumors. In a homogeneously perfused and uniformly extravasating tissue, material need only travel an intercapillary distance on the order of 100 μm to achieve a uniform distribution throughout the tissue. However, in tumors which have regions of low or no extravasation (e.g., necrotic or high-pressure areas), the solute must traverse a macroscopic distance on the order of the tumor radius in order to achieve a uniform distribution. For an IgG antibody the characteristic time scale for diffusion ($L^2/4D$) is 30 min over 100 μm but 7 months over a distance of 1 cm. Similarly for a Fab fragment (50 kDa), the time scales are 10 min and 2 months for 100 μm and 1 cm, respectively. The concentration profiles due to diffusion are shown in Fig. 5 for IgG and sodium fluorescein (Clauss and Jain, 1990).

3.2. Convection

The proportionality constant relating interstitial fluid velocity (u, cm/sec) to the interstitial pressure gradient is the interstitial hydraulic conductivity, K (cm^2/mm Hg per sec). As fluid moves through the interstitium, it will carry along solute molecules. The time required for convection over a length, L, is given by u/L. When there is a pressure gradient within a tumor, fluid will flow from areas of high pressure (center) to low pressure (periphery). Unless there is drainage within the tumor by lymphatics, this fluid will escape into the surrounding tissue or fluid. Butler *et al.* (1975) measured this fluid loss from isolated tumor preparations (rat mammary carcinomas), obtaining a radially outward velocity of 0.1 to 0.2 μm/sec at the periphery of a 1-cm tumor. This value is estimated to be an order of magnitude greater than the velocity in subcutaneous tumors (Jain and Baxter, 1988). At this maximum velocity, the time required for any solute to travel 100 μm and 1 cm are 8 min and 13 hr, respectively. However, the model of Jain and Baxter (1988) shows that the velocity away from the periphery is very much smaller than this, and is zero in central necrotic regions. It should also be noted that convection will always push molecules *away* from the center of a tumor, frequently opposing a slow diffusive flux toward the center.

One method of quantifying the interstitial transport parameters is by the use of transparent thin tissue preparations, such as the rabbit ear chamber. The spatial and

---

Figure 4. Molecular weight dependence of diffusivity. (A) The effective diffusion coefficient, D, has been plotted as a function of molecular weight for dextrans (Nugent and Jain, 1984; Gerlowski and Jain, 1986), albumin (Nugent and Jain, 1984) and IgG (Clauss and Jain, 1990) in water, normal and tumor tissue. Symbols: □, dextran, aqueous; ◇, bovine serum albumin, aqueous; ○, rabbit IgG, aqueous; □, dextran, tumor; ◇, bovine serum albumin, tumor; ○, rabbit IgG, tumor; ■, dextran, normal tissue; ◆, bovine serum albumin, normal tissue; ●, rabbit IgG, normal tissue. (B) The effective diffusion coefficient plotted versus the Stokes–Einstein radius. Symbols as in A plus: ×, sodium fluorescein, tumor; +, sodium fluorescein, normal tissue. (From Clauss and Jain, 1990, with permission.)

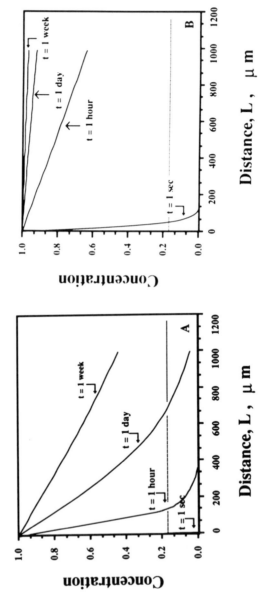

Figure 5. Interstitial diffusion. The concentration of IgG (A) and sodium fluorescein (B) in the tumor interstitium are shown as a function of position and time, using a simple one-dimensional model for diffusion from a constant source. Most of the sodium fluorescein has diffused 1 mm within 1 hr, while the fraction of IgG which as diffused 1 mm in 1 hr is negligible. (From Clauss and Jain, 1990, with permission.)

temporal variation in the fluorescence intensity due to labeled molecules is digitized and recorded and fit by a diffusion equation. Fox and Wayland (1979) have found diffusion coefficients in rat mesentery, and Nugent and Jain (1984), Gerlowski and Jain (1986), and Clauss and Jain (1990) have reported diffusion coefficients in granulation and tumor tissue for macromolecules such as albumin, IgG, and dextrans.

Recently, Chary and Jain (1987, 1989) have adapted a fluorescence recovery after photobleaching (FRAP) technique for use *in vivo* to measure the velocity and diffusion coefficients in the interstitium of tumor and granulation tissue. The recovery of fluorescence intensity following bleaching by a strong but brief laser pulse is monitored. Translation of the spot yields the velocity, while the spreading out of the intensity yields a diffusion coefficient. Typical velocity profiles in the interstitium are shown in Fig. 6 (Chary and Jain, 1989).

3.3. Binding

The binding of solutes to plasma proteins serves to increase the residence time in the bloodstream (unless the protein is being actively removed). However, binding in the extravascular space has a different influence. It is the specific binding of molecules such as monoclonal antibodies to tumor-associated antigens which has led to intense interest and study. There are at least three important aspects of extravascular binding. The first is similar to binding in the vascular space, i.e., an increase in the residence time in the extravascular space due to a decreased rate of reabsorption of solute into the bloodstream. The reduction in the reabsorption rate is proportional to the ratio of bound solute to free solute in the extravascular space. This ratio is determined by the binding affinity ($K_a = k_f/k_r$, the ratio of forward to reverse rate constants) and the number of sites available for binding. Presently, these binding parameters have only been measured *in vitro*, but Kaufman and Jain (1990) are developing the FRAP technique to estimate kinetic parameters *in vivo*.

The second aspect of binding is drug effectiveness. Many enzymes, proteins, or macromolecules must be internalized by a cell to cause the desired effect. In order to be internalized (and/or metabolized), the solute must first bind to the cell surface. The third role of binding is its effect on interstitial transport. A result of high-affinity binding which is often undesirable is a reduction in the effective diffusivity. When solute molecules bind tightly to the interstitium or to cells, they are not free to be transported by diffusion or convection. This effect may lead to a strongly heterogeneous perivascular distribution of the solute, whereby all extravasated material remains in a small region immediately outside the blood vessel. Cells far away from blood vessels are not affected. Modeling of the effect of binding of monoclonal antibodies and their fragments has been done by Weinstein *et al.* (1987), Fujimori *et al.* (1989), and Baxter and Jain (1991a,b).

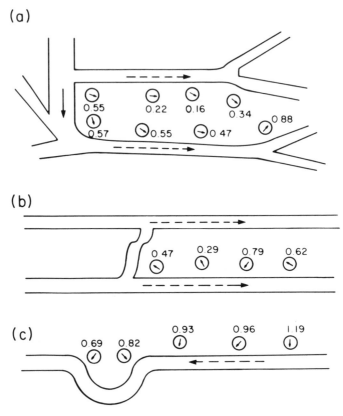

Figure 6. Interstitial velocity profiles. Representative regions in the microcirculation. Circles represent locations of fluorescence photobleaching experiments. The arrows inside the circles represent the direction of the interstitial fluid velocity at these locations. The nearby values show magnitudes of the velocity in μm/sec. (a) An area where interstitial flow parallels blood flow in the vessels. (b) Interstitial flow is opposite prevailing blood flow. (c) Fluid is absorbed from the interstitium into a postcapillary venule. (From Chary and Jain, 1989, with permission.)

4. TRANSPORT PROPERTIES OF TUMORS

4.1. Interstitial Properties

4.1.1. INTERSTITIAL PRESSURE

Figure 3 shows one set of experimental data for interstitial pressure in a tumor. The unusual finding was that the pressure was nearly uniform throughout the center of the tumor, with a very steep pressure gradient at the tumor periphery. Pressure drops of over 30 mm Hg were found in distances less than 1 mm from the surface of animal and human tumors (Boucher *et al.*, 1990, 1991; Roh *et al.*, 1991; Gutmann *et al.*, 1992; Less *et al.*, 1992). The elevated interstitial pressure in the center of the tumor has equilibrated with the effective vascular pressure ($p_v - \sigma\Delta\pi$), resulting in zero net filtration of fluid from the blood vessels. Our model shows that the convective solute flux, dominant for macromolecules in normal tissue, is severely reduced by the high interstitial pressure. Extravasation is then determined chiefly by the diffusive component of the vascular permeability coefficient (see Section 5.2). If interstitial pressure were low, as in most normal tissues, the uptake of proteins would be as uniform as the vascular density.

4.1.2. HYDRAULIC CONDUCTIVITY

The mathematical model of Baxter and Jain (1989) for interstitial fluid transport yields a characteristic distance at the tumor periphery over which the interstitial pressure gradient is significant. The distance (cm) over which the pressure rises from minimum to 63% of maximum is given by: $\sqrt{KV/L_pS}$. Using typical values for the hydraulic conductivities and surface area in a tumor, a length scale of 300 μm is obtained. The result is that throughout the majority of an alymphatic tumor mass larger than 1 mm in diameter, the interstitial pressure will be essentially in equilibrium with the effective vascular pressure, yielding no net filtration of fluid, and greatly reduced extravasation of solute.

4.2. Perfusion Characteristics

4.2.1. BLOOD SUPPLY

For macromolecules, the blood perfusion rate within nonnecrotic regions of a tumor does not play an important role in determining uptake. This happens because so little material extravasates in a single pass through the vasculature. The uptake is not limited by *blood flow rate*, but rather by the *surface area* available for exchange, by affecting the permeability–surface area (*PS*) product or the capillary filtration coefficient (CFC). This is in contrast to the flow rate-limited extravasation of small solutes. Increasing the blood vessel surface area by a factor of two while decreasing the blood flow rate would virtually double the macromolecular extravasation rate. A heterogeneous blood vessel distribution results in a correspondingly heterogeneous solute concentration, while the interstitial pressure would not be greatly affected.

4.2.2. LYMPHATICS

The lack of adequate lymphatic drainage in tumors is one important factor leading to elevated interstitial pressure. Any fluid which does extravasate is not able to be reabsorbed by nearby lymphatic vessels and must instead be transported a macroscopic distance through the interstitium to a region of lower pressure. If lymphatics were present in a tumor, whether residual vessels from the host tissue or formed by the growing tumor, they would serve to reduce the interstitial pressure and remove solute from the tissue. The effects of lymphatics in tumors are discussed further in Baxter and Jain (1990).

4.3. Role of Necrosis

Because there is virtually no filtration in the center of a uniformly perfused tumor, the presence of a necrotic core (or alternately, the lack of blood vessels) does not affect the magnitude or distribution of interstitial pressure. Only if the necrosis grows to the point of leaving a very thin viable tumor periphery will the interstitial pressure be reduced. This is borne out by the simulations of Baxter and Jain (1990).

On the other hand, necrosis does have a great effect on the distribution of macromolecular solutes. There is no source for extravasation in a necrotic core, and interstitial diffusion over macroscopic length scales is too slow to achieve a significant concentration in large necrotic regions. This may require a two-step approach for effective treatment of the whole tumor. There may be a possible benefit for necrosis in that there is likewise no *drainage* in a necrotic region, and so such a region may be used as a reservoir for material useful after the solute has been cleared from the plasma (Baxter and Jain, 1990).

4.4. Role of Tumor Size

For large solid tumors the tumor radius does not have a significant *direct* effect on the concentration or pressure profiles. The steep pressure gradients are confined to a very thin layer at the tumor periphery (see Section 4.1.2). Instead, the tumor radius has indirect effects by contributing to the blood perfusion rate, percent necrosis, and maximum interstitial pressure. Micrometastases (on the order of 100 μm), however, are small enough that large pressures do not build up in the interstitium. The time scale for interstitial diffusion in these tumors is minutes instead of months; thus, a more uniform uptake and distribution is expected.

5. TRANSPORT PROPERTIES OF PROTEINS

5.1. Diffusion Coefficients

The relationship between molecular weight and the diffusion coefficient was shown in Fig. 4. The time required for a solute to move a given distance is inversely proportional to the diffusion coefficient. The time required for most proteins to be distributed evenly between nearby capillaries is on the order of tens of minutes. However, for macromolecules the size of antibodies, the time scale for diffusion over a *macroscopic* distance (e.g., 1 cm) is on the order of months. For this reason the uptake and distribution of proteins is not very sensitive to the interstitial diffusion coefficient. Material extravasating at the periphery would take months to diffuse into the center of the tumor if there were no extravasation in the center.

5.2. Permeability Coefficients

   The effective vascular permeability coefficient plays an important role in determining the uptake of proteins. The characteristic time scale for extravasation is $(P_{eff}S/V)^{-1}$ sec. Doubling the effective permeability coefficient or surface area would double the rate of extravasation. In organs such as the kidney, liver, and lung, permeability is highest (see Chapter 10), while the permeability of blood vessels in the brain is low (see Chapter 11).

   . As discussed earlier, the effective vascular permeability coefficient in tumors may be greatly reduced by the elimination of the convective flux due to elevated interstitial pressure. In such cases the uptake is essentially determined by the diffusive component of vascular permeability. This is illustrated in Fig. 7, which shows simulations for the distribution of IgG in a tumor of 1 cm radius as a function of time (Jain and Baxter, 1988). In Fig. 7A the effective permeability is reduced to 10% of its expected value due to reduced transcapillary convection. Figure 7B shows the concentration profile resulting when P_{eff} is reduced to 1% of the value expected with low interstitial pressure, while in Fig. 7C it is assumed that the diffusive component is zero, allowing extravasation of solute only where there is fluid filtration. The only parameter changed in these simulations is the value of P, the diffusive permeability coefficient. Other parameter values used in these simulations may be found in Jain and Baxter (1988).

   The importance of the vascular permeability, P, and the distinction between P and P_{eff} is quite evident from Fig. 7. When P is zero (or extremely low), then there is virtually no material present in the center of the tumor, where interstitial pressure is high and fluid filtration negligible. For a value of $P = 5.7 \times 10^{-8}$ cm/sec (corresponding to 10% of the P_{eff} value measured by Gerlowski and Jain, 1986) the interstitial concentration approaches the plasma concentration after 72 hr. In addition, there is more material in the center of the tumor because solute is washed out of the periphery by a radial convective flow. For a value of P between 5.7×10^{-9} and 5.7×10^{-10} cm/sec, intermediate results are obtained. Reports of very low central concentrations of antibodies (e.g., Jones *et al.*, 1986) and experiments by Rippe and Haraldsson (1987) suggest that the diffusive permeability may be a small fraction of the effective permeability measured in tumor preparations with low interstitial pressures. The differences in the profiles in Fig. 7 demonstrate the

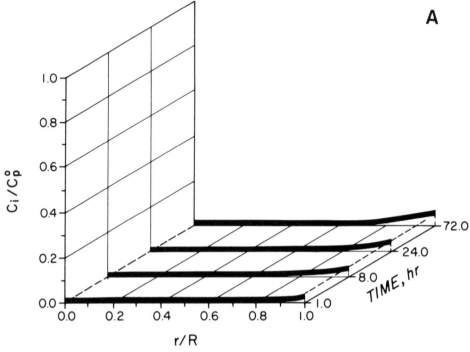

Figure 7. Interstitial concentration of IgG. The dimensionless interstitial concentration $[C_i/C_p\,(t=0)]$ is plotted as a function of radial position and time during 72 hr of continuous infusion in a tumor of 1 cm radius. The effect fo the diffusive permeability coefficient, P, is demonstrated. (A) $P = 5.7 \times 10^{-8}$; (B) $P = 5.7 \times 10^{-9}$; (C) $P = 0$ cm/sec. Center: $r/R = 0$; edge: $r/R = 1$. (From Jain and Baxter, 1988, with permission.)

need to determine the magnitude of P in the absence of convective transcapillary exchange.

The vascular hydraulic conductivity, L_p, has been measured in a variety of tissues. However, only recently has L_p been measured in tumors; Sevick and Jain (1991) obtained a value of 2.2 ml/min per mm Hg per 100 g, which is 10 to 1000 times higher than those found in several other normal tissues, and similar to the value in glomerular capillaries.

5.3. Binding and Metabolism Parameters

In Section 3.3 the effect of binding on delivery of proteins was discussed from a physical standpoint. The three factors of extravascular binding were to increase tissue

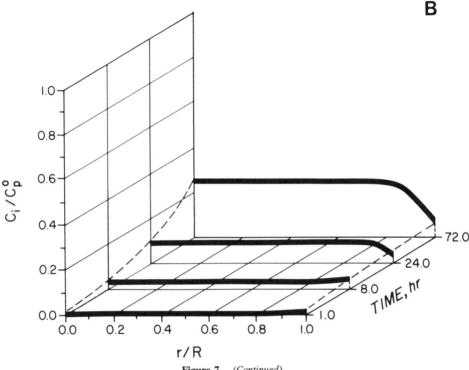

Figure 7. (*Continued*)

residence time, produce an effect on cells, and change the distribution of solute around blood vessels. The role of metabolism is cell and solute specific, besides a general effect of reducing the total amount of original material in the body. The effect on tissue concentration levels is greatest at late times, when there is little material left in the plasma to replenish extravascular regions.

One measure of the benefit of binding is the specificity ratio (SR), which is the ratio of the average concentration of a binding molecule to the average concentration of an equivalent nonbinding molecule. The effect of the binding affinity on the specificity ratio may be seen in Fig. 8, where the calculated values of the SR for an IgG molecule and its Fab fragment are shown as a function of the binding affinity (Baxter and Jain, 1991a). A large binding affinity leads to a large SR, which increases with time as the solute is cleared more rapidly from tissues to which it does not bind. Equivalent results were obtained by raising the association rate constant, decreasing the dissociation rate constant, or increasing the number of antibody binding sites.

If, however, the binding affinity is too high, then the distribution of proteins around individual blood vessels may become very nonuniform (Fujimori *et al.*, 1989; Baxter and Jain, 1991b). It may be desirable to use a molecule with a moderate affinity

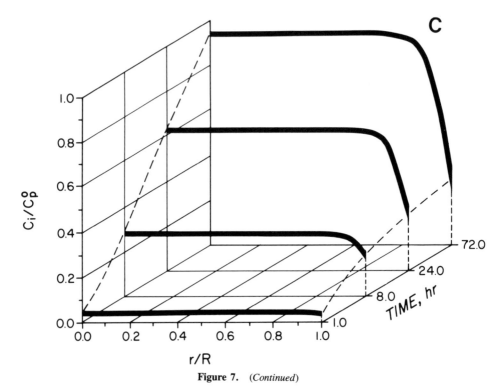

Figure 7. (*Continued*)

to achieve a high SR while minimizing the heterogeneity around a blood vessel. The dosage (or initial plasma concentration) is important in determining whether the interstitial antibody concentrations are above or below saturation levels. Dosages that achieve interstitial concentrations near saturation are most efficient.

5.4. Molecular Weight

The effect of molecular weight on the diffusion coefficient was seen in Fig. 4. However, the more important effect of molecular weight is on the permeability coefficient. Small molecules extravasate more readily than larger molecules. In addition, a greater fraction of the transcapillary solute flux is due to diffusion for smaller molecules. In trying to optimize the uptake in a tissue, the plasma clearance rate must also be considered. Small molecules are cleared from the body much more rapidly than large molecules (see Chapters 12 and 13). The molecular weight may also determine in which normal organs the drug will accumulate, i.e., the liver, kidney, or spleen.

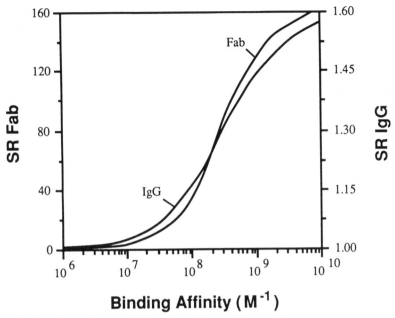

Figure 8. Specificity ratios versus binding affinity. The specificity ratio (average concentration throughout a tumor of a binding molecule divided by the average concentration of an equivalent nonbinding molecule) is plotted as a function of the binding affinity for IgG and Fab, 48 hr after injection. The maximum concentration of bound material is 1.18×10^{-8} M and 2.35×10^{-8} M, respectively.

6. SUMMARY AND IMPLICATIONS

6.1. Summary

There are many physiological and physicochemical factors which influence the extravasation and interstitial transport of proteins and macromolecules. In normal tissues the effective vascular permeability coefficient, blood vessel surface area, interstitial diffusion coefficient, and plasma clearance rate all play important roles. In tumors there is an additional factor or physiological barrier. The elevated interstitial pressure in tumors, due in part to a lack of lymphatic drainage, greatly reduces the driving force for filtration and extravasation and may lead to heterogeneities in macromolecular concentration levels (see Table I). The steep interstitial pressure gradient predicted by Jain and Baxter (1988) was seen experimentally in the periphery of large solid tumors (Boucher *et al.*, 1990).

Specific extravascular binding of solute molecules may also be important in increasing concentration levels of proteins. High binding affinities were seen to

Table I
Physiological Advantages and Problems in the Delivery of Macromolecules to Tumors[a]

Advantages
Relatively high degree of specificity of antibodies for tumor-associated antigens
Relatively large vascular permeability, interstitial diffusion coefficient, and hydraulic conductivity
Long plasma clearance times
Useful for targeting smaller drugs
Problems
Heterogeneous blood supply
Elevated interstitial pressure
Fluid loss from periphery
Large distances in the interstitium
Large affinity and heterogeneous binding
Possible immunological response
Metabolism

[a]Adapted from Jain (1990).

increase the total concentration levels as compared to nonbinding solutes. This makes a high affinity most effective unless uniform perivascular concentrations are required. Higher concentrations may be achieved (at the cost of lower SR) by using continuous infusion of the solute instead of a single bolus injection.

6.2. Implications for Drug Delivery

These results have significant implications for delivery of macromolecules to tumor tissue. In order to improve drug delivery in the presence of these physiological barriers, one must either overcome the barriers, exploit the transport barriers, or avoid them.

Attempts to overcome barriers due to extravasation and interstitial transport may include the following strategies. (1) Optimize kinetic parameters. The schedule and dosage of delivery should be chosen to operate near saturation level for as long as possible. The maximum affinity should be sought such that the heterogeneity around individual blood vessels is acceptable. When the solute is rapidly metabolized, the development of a system with slower metabolism may be more important than achieving the highest possible binding affinity. (2) Modulate pressures. Modifying the vascular hydrostatic or osmotic pressures may have a beneficial effect on delivery. The uptake of chemotherapeutic agents by carcinoma cells in the peritoneal cavity was increased significantly when the drugs were injected intraperitoneally in solutions of low osmolalities (Stephen et al., 1990). This strategy may be limited by side effects such as edema or a lack of selectivity in affecting tumor vasculature. The interstitial pressure in tumors might be reduced by radiation (Roh *et al.*, 1991), or mechanical or chemical means. This would in part reestablish transcapillary filtration. (3) Increase the vascular permeability. Low-molecular-weight agents have high permeability coefficients, but lack selectivity for cancer cells. Vasoactive agents may be used to increase the diffusive component of extravasation; however, the response may only be a transient one. Strategies based on modifying the diffusive permeability coefficient require more study on the permeability characteristics of tumor vasculature with and without fluid filtration. For subcutaneous injections, changes in the kinetic or interstitial parameters should be targeted rather than permeability or pressure.

Hyperthermia is a classic attempt to exploit transport limitations in tumors. The low perfusion rates and vascular densities found in some parts of tumors which reduce mass transfer also limit the ability of the tumor to remove harmful heat. As with other methods of cancer treatment, hyperthermia alone is not completely selective and may harm the surrounding normal tissue. A second method to exploit a transport barrier is to target the necrotic core itself. Epstein *et al.* (1988) developed antibodies which bind to intracellular components and are accessible only to cells with damaged cell membranes. For this method to be effective, the necrotic region must be small, or have some functional blood vessels present to deliver the antibodies. An alternate strategy for exploiting the high pressure would be to further *increase* the vascular and/or interstitial pressure with a view to collapsing the weakened tumor vessels. This strategy has not been tested.

Finally, one can seek to develop strategies which avoid these physiological barriers. Targeting the tumor vasculature itself for destruction completely avoids interstitial transport considerations. Routes of injection other than intravenous or intraarterial avoid the bloodstream and the corresponding permeability problem. For example, Weinstein *et al.* (1983) used a lymphatic pathway for detection of lymph node metastases. Schlom *et al.* (1990) cite the use of intracavitary administration of monoclonal antibodies, the use of interferons to upregulate antigen expression, and the possibility of using prior radiation to improve antibody uptake. The use of continuous infusion or controlled drug release avoids the decrease in plasma concentration, but possibly with adverse effect to normal tissues. For lower-molecular-weight agents, the vascular permeability and interstitial diffusion rates are high, but again there is frequently a lack of selectivity for cancer cells.

6.3. Two-Step Approaches

The limitations of high-molecular-weight agents suggest approaches to therapy which use a second low-molecular-weight solute in two stages. In the first phase, macromolecules such as monoclonal antibodies are administered to achieve selectivity, allowing long times or high concentrations of a non-toxic agent to overcome

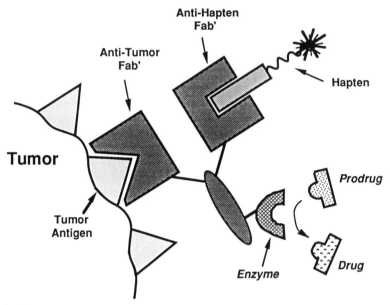

Figure 9. Bifunctional antibody approach. One example of adjuvant therapy with a lower-molecular-weight agent. One antibody binding site is tumor specific, and the other binds to the hapten, which is a toxin or radionuclide. Alternately, an enzyme may be conjugated to the antibody to convert a prodrug to a more toxic form locally. (Adapted from Stickney *et al.*, 1989.)

physiological barriers. A small solute may then be administered in the second phase to utilize the specificity gained by the macromolecules.

We have extended our mathematical analyses to study two novel therapies based on these two-phase approaches (Yuan *et al.*, 1991; Baxter *et al.*, 1992). Bifunctional antibodies (BFA) have two different binding sites: one for a tumor-associated antigen, and the second for the low-molecular-weight solute (hapten) (Fig. 9). Blood vessels are much more permeable to hapten, which carries a therapeutic agent. The local binding, ideally only at the site of BFA concentration, helps to retain the hapten at the desired location.

The second, related approach is antibody-derived enzyme–prodrug therapy (ADEPT). In this strategy, a prodrug which is itself relatively nontoxic, is injected some time after the antibody. An enzyme which is not found in normal tissues is conjugated to the antibody. The prodrug will be converted to its toxic form where the enzyme is located. Thus, a toxic species will be *produced* locally at the desired site. Controlled drug delivery devices for the prodrug may be useful to maintain high drug levels *in situ*.

ACKNOWLEDGMENTS. This work was supported by grants from the National Science Foundation, National Institutes of Health, American Cancer Society to R.K.J., and an NSF postdoctoral fellowship (1990–91) to L.T.B. and Humboldt U.S. Senior Scientist Award (1990–91) to R.K.J.

REFERENCES

Ballard, K., and Perl, W., 1978, Osmotic reflection coefficients of canine subcutaneous adipose tissue endothelium, *Microvasc. Res.* **16**:224–236.

Baxter, L. T., and Jain, R. K., 1988, Vascular permeability and interstitial diffusion in superfused tissues: A two-dimensional model, *Microvasc. Res.* **36**:108–115.

Baxter, L. T., and Jain, R. K., 1989, Transport of fluid and macromolecules in tumors. I. Role of interstitial pressure and convection, *Microvasc. Res.* **37**:77–104.

Baxter, L. T., and Jain, R. K., 1990, Transport of fluid and macromolecules in tumors. II. Role of heterogeneous perfusion and lymphatics, *Microvasc. Res.* **40**:246–263.

Baxter, L. T., and Jain, R. K., 1991a, Transport of fluid and macromolecules in tumors. III. Role of binding and metabolism, *Microvasc. Res.* **41**:5–23.

Baxter, L. T., and Jain, R. K., 1991b, Transport of fluid and macromolecules in tumors. IV. A microscopic model of the perivascular distribution, *Microvasc. Res.* **41**:252–272.

Baxter, L. T., Jain, R. K., and Svensjö, E., 1987, Vascular permeability and interstitial diffusion of macromolecules in the hamster cheek pouch: Effect of vasoactive drugs, *Microvasc. Res.* **34**:336–348.

Baxter, L. T., Yuan, F., and Jain, R. K., 1992, Pharmacokinetic analysis of the perivascular distribution of bifunctional antibodies and haptens: comparison with experimental data, *Cancer Res.* **52**:5838–5844.

Boucher, Y., and Jain, R. K., 1992, Microvascular pressure is the principal driving force for interstitial hypertension in solid tumors: Implications for vascular collapse, *Cancer Res.* **52**:5110–5114.

Boucher, Y., Baxter, L. T., and Jain, R. K., 1990, Interstitial pressure gradients in tissue-isolated and subcutaneous tumors: Implications for therapy, *Cancer Res.* **50**:4478–4484.

Boucher, Y., Kirkwood, J., Opacic, D., Desantis, M., and Jain, R. K., 1991, Interstitial hypertension in superficial metastatic melanomas in humans, *Cancer Res.* **51**:6691–6694.

Bundagaard, M., Hagman, P., and Crone, C., 1983, The three-dimensional organization of plasmalemmal vesicular profiles in the endothelium of rat heart capillaries, *Microvasc. Res.* **25**:358–368.

Butler, T. P., Grantham, F. H., and Gullino, P. M., 1975, Bulk transfer of fluid in the interstitial compartment of mammary tumors, *Cancer Res.* **35**:3084–3088.

Chary, S. R., and Jain, R. K., 1987, Analysis of diffusive and convective recovery of fluorescence after photobleaching–Effect of uniform flow field, *Chem. Eng. Commun.* **55**:235–249.

Chary, S. R., and Jain, R. K., 1989, Direct measurement of interstitial diffusion and convection of albumin in normal and neoplastic tissues using fluorescence photobleaching, *Proc. Natl. Acad. Sci. USA* **86**:5385–5389.

Clauss, M. A., and Jain, R. K., 1990, Interstitial transport of rabbit and sheep antibodies in normal and neoplastic tissues, *Cancer Res.* **50**:3487–3492.

Curry, F. E., 1984, Mechanics and thermodynamics of transcapillary exchange, in: *Handbook of Physiology*, Section 2 (E. M. Renkin and C. C. Michel, eds.), American Physiological Society, Bethesda, pp. 309–374.

Curry, F. E., Mason, J. C., and Michel, C. C., 1976, Osmotic reflection coefficients of capillary walls to low molecular weight hydrophilic solutes measured in single perfused capillaries of the frog mesentery, *J. Physiol. (London)* **261**:319–336.

Dewey, W. C., 1959, Vascular–extravascular exchange of I-131 plasma proteins in the rat, *Am. J. Physiol.* **197**:423–431.

Epstein, A. L., Chen, F. M., and Taylor, C., 1988, A novel method for the detection of necrotic lesions in human cancers, *Cancer Res.* **48**:5842–5848.

Fox, J. R., and Wayland, H., 1979, Interstitial diffusion of macromolecules in the rat mesentery, *Microvasc. Res.* **18**:255–276.

Frokjaer-Jensen, J., 1980, Three-dimensional organization of plasmalemmal vesicles in endothelial cells. An analysis by serial sectioning of frog mesenteric capillaries, *J. Ultrastruct. Res.* **73**:9–20.

Fujimori, K., Covell, D. G., Fletcher, J. E., and Weinstein, J. N., 1989, Modeling analysis of the global and microscopic distribution of immunoglobulin G, F(ab')$_2$, and Fab in tumors, *Cancer Res.* **49**:5656–5663.

Gerlowski, L. E., and Jain, R. K., 1986, Microvascular permeability of normal and neoplastic tissues, *Microvasc. Res.* **31**:288–305.

Gutmann, R., Leunig, M., Feyh, J., Goetz, A. E., Messmer, K., Kastenbauer, E., and Jain, R. K., 1992, Interstitial hypertension in head and neck tumors in patients: Correlation with tumor size, *Cancer Res.* **52**:1993–1995.

Jain, R. K., 1987a, Transport of molecules across tumor vasculature, *Cancer Metastasis Rev.* **6**:559–594.

Jain, R. K., 1987b, Transport of molecules in the tumor interstitium: A review, *Cancer Res.* **47**:3039–3051.

Jain, R. K., 1988, Determinants of tumor blood flow: A review, *Cancer Res.* **48**:2641–2658.

Jain, R. K., 1990, Physiological barriers to delivery of monoclonal antibodies and other macromolecules in tumors, *Cancer Res. (Suppl.)* **50**:814s–819s.

Jain, R. K., and Baxter, L. T., 1988, Mechanisms of heterogeneous distribution of monoclonal antibodies and other macromolecules in tumors: Significance of interstitial pressure, *Cancer Res.* **48**:7022–7032.

Jain, R. K., and Ward-Hartley, K. A., 1984, Tumor blood flow: Characterization, modifications and role in hyperthermia, *IEEE Trans. Sonics Ultrason.* **SU-31**:504–526.

Jones, P. L., Gallagher, B. M., and Sands, H., 1986, Autoradiographic analysis of monoclonal antibody distribution in human colon and breast tumor xenografts, *Cancer Immunol. Immunother.* **22**:139–143.

Kaufman, E. N., and Jain, R. K., 1990, Quantification of transport and binding parameters using FRAP: Potential for *in vivo* applications, *Biophys. J.* **58**:873–885.

Less, J. R., Posner, M. C., Boucher, Y., Borochovitz, D., Wolmark, N., and Jain, R. K., 1992, Interstitial hypertension in human breast and colorectal tumors, *Cancer Res.* **52**:6371–6374.

Misiewicz, M. A., 1986, Microvascular and interstitial pressure in normal and neoplastic tissues, M.S. thesis, Carnegie Mellon University, Pittsburgh.

Nugent, L. J., and Jain, R. K., 1984, Extravascular diffusion in normal and neoplastic tissues, *Cancer Res.* **44**:238–244.

O'Conner, S. W., and Bale, W. F., 1984, Accessibility of circulating immunoglobulin G to the extravascular compartment of solid rat tumors, *Cancer Res.* **44**:3719–3723.

Patlak, C. S., Goldstein, D. A., and Hoffman, J. F., 1963, The flow of solute and solvent across a two-membrane system, *J. Theor. Biol.* **5:**426–442.

Renkin, E. M., Watson, P. D., Sloop, C. H., Joyner, W. M., and Curry, F. E., 1977, Transport pathways for fluid and large molecules in microvascular endothelium of the dog's paw, *Microvasc. Res.* **14:**205–214.

Rippe, B., and Haraldsson, B., 1987, Fluid and protein fluxes across small and large pores in the microvasculature. Application of two-pore equations, *Acta. Physiol. Scand.* **131:**411–428.

Roh, H. D., Boucher, Y., Kaliniki, S., Buchsbaum, R., Bloomer, W. D., and Jain, R. K., 1991, Interstitial hypertension in cervical carcinomas in humans: Possible correlation with tumor oxygenation and radiation response, *Cancer Res.* **51:**6695–6698.

Rutili, G., 1978, Transport of macromolecules in subcutaneous tissue by FITC-dextrans, Ph.D. thesis, University of Upsaliensis, Uppsala, Sweden.

Schlom, J., Hand, P. H., Greiner, J. W., Colcher, D., Shrivastrav, S., Carrasquillo, J. A., Reynolds, J. C., Larson, S. M., and Raubitschek, A., 1990, Innovations that influence the pharmacology of monoclonal antibody guided tumor targeting, *Cancer Res. (Suppl.)* **50:** 820s–827s.

Sevick, E. M., and Jain, R. K., 1991, Measurement of capillary filtration coefficient in a solid tumor, *Cancer Res.* **51:**1352–1355.

Song, C. W., and Levitt, S. H., 1970, Effect of x-irradiation on vascularity of normal tissues and experimental tumor, *Radiology* **94:**445–447.

Stephen, R. L., Novak, J. M., Jensen, E. M., Kablitz, C., and Buys, S. S., 1990, Effect of osmotic pressure on uptake of chemotherapeutic agents by carcinoma cells, *Cancer Res.* **50:**4704–4708.

Stickney, D. R., Slater, J. B., Kirk, G. A., Ahlem, C., Chang, C., and Frincke, J. M., 1989, Bifunctional antibody: ZCE/CHA-indium-111-BLEDTA-IV clinical imaging in colorectal carcinoma, *Antibody Immunoconjugates Radiopharm.* **2:**1–13.

Weinstein, J. N., Steller, M. A., Keenan, A. M., Covell, D. G., Key, M. E., Sieber, S. M., Oldham, R. K., Hwang, K. M., and Parker, R. J., 1983, Monoclonal antibodies in the lymphatics: Selective delivery to lymph node metastases of a solid tumor, *Science* **222:** 423–426.

Weinstein, J. N., Eger, R. R., Covell, D. G., Black, C. D. V., Mulshine, J., Carrasquillo, J. A., Larson, S. M., and Keenan, A. M., 1987, The pharmacology of monoclonal antibodies, *Ann. N.Y. Acad. Sci.* **507:**199–210.

Wiig, H., Tveit, E., Hultborn, R., Reed, R. K., and Weiss, L., 1982, Interstitial fluid pressure in DMBA-induced rat mammary tumors, *Scand. J. Clin. Lab. Invest.* **42:**159–164.

Young, J. S., Lumsden, C. E., and Stalker, A. L., 1950, The significance of the "tissue pressure" of normal testicular and of neoplastic (Brown–Pearce carcinoma) tissue in the rabbit, *J. Pathol. Bacteriol.* **62:**313–333.

Yuan, F., Baxter, L. T., and Jain, R. K., 1991, Pharmacokinetic considerations in two-step approaches using bifunctional and enzyme-conjugated antibodies, *Cancer Res.* **51:**3119–3130.

Yuan, F., Leunig, M., Berk, D., and Jain, R. K., 1993, Microvascular permeability of albumin in human tumor xenograft LS174T in dorsal skin fold chamber of SCID mice, *Microvasc. Res.* (in press).

Chapter 16

Tissue Barriers

Diffusion, Bulk Flow, and Volume Transmission of Proteins and Peptides within the Brain

Conrad E. Johanson

1. OVERVIEW

Targeted cells in the brain, i.e., neurons and glia, can be selectively modulated by certain extracellular proteins and polypeptides. In some pathophysiological situations, specific modulation of particular groups of cells within the central nervous system (CNS) by proteins and hormones could be of considerable benefit to patients (Olson *et al.*, 1991). Exogenous proteins, when viewed as pharmacologic agents in the bloodstream, must circumvent either the blood–brain barrier (cerebral capillary wall) or the blood–cerebrospinal fluid (CSF) barrier (choroid plexus and arachnoid membrane) in order to gain access to the specialized extracellular fluid of the brain (Johanson, 1989b). Chapter 11 by Broadwell deals with strategies for delivering proteins across these impeding barrier systems. This chapter, however, deals not with circumvention of the CNS vascular barriers by proteins, but rather with the distributive movements within brain parenchyma of a macromolecule once it has gained access to the cerebral extracellular fluid. A knowledge of the structure and function of

Conrad E. Johanson • Cerebrospinal Fluid Research Laboratory, Department of Clinical Neurosciences, Program in Neurosurgery, Brown University/Rhode Island Hospital, Providence, Rhode Island 02902.

Biological Barriers to Protein Delivery, edited by Kenneth L. Audus and Thomas J. Raub. Plenum Press, New York, 1993.

extracellular pathways is fundamental in understanding how proteins move from one region to another in the CNS.

2. TWO MAJOR EXTRACELLULAR FLUIDS INSIDE THE CENTRAL NERVOUS SYSTEM

There are two main extracellular compartments in CNS: the large-cavity CSF and the cerebral interstitial fluid (ISF). The CSF and ISF have comparable chemical composition because they are formed by similar secretory mechanisms and because they are more or less in diffusion equilibrium with each other across the rather permeable ependymal and pia–glial membranes which physically separate these two extracellular fluids (Cserr and Patlak, 1991). The ependymal lining is interposed between the brain parenchyma and the interior ventricular CSF, whereas the pia–glial membrane lies between the brain substance and the exteriorly located CSF in the subarachnoid spaces.

2.1. Large-Cavity Cerebrospinal Fluid (CSF)

A significant portion of the CSF lies in the four ventricular cavities of the brain, i.e., the two lateral, the third and the fourth ventricles. The other sites that contain CSF are the central canal of the spinal cord, and the subarachnoid spaces that overlie both the cerebral hemispheres and the spinal cord. The total volume of CSF in an adult human is approximately 150 ml. Thus, peptides and proteins administered in relatively small injectate volumes intraventricularly would be substantially diluted in concentration before reaching target cells within the brain. Reabsorptive transport mechanisms in choroid plexus that clear peptides (Huang, 1982) and proteins (Van Deurs, 1976, 1978) from CSF would further diminish the concentration of these substances in CSF.

2.2. Cerebral Interstitial Fluid (ISF)

The other extracellular fluid, i.e., interstitial fluid, intimately surrounds neurons and glia as their microenvironment. The ISF exists as a thin film, about 200 Å in thickness, encompassing the cellular elements of nervous tissue. The ISF comprises about 15–20% of the wet weight of the brain (Levin *et al.*, 1970; Johanson, 1989a; Nicholson and Rice, 1991). Accordingly, a 1-kg brain, as in adult humans, contains at least 150 g (or ml) of ISF. Thus, mature human brain is composed of some 300 ml or more of total extracellular fluid, i.e., approximately 150 ml ISF plus 150 ml CSF. This

300 ml of fluid would seem to be close to the upper limit for the volume of distribution of an extracellularly distributed substance, such as a high-molecular-weight protein administered into brain ventricles. For smaller species, such as rats and other laboratory mammals, similar parameters are applicable for pharmacokinetic modeling. That is, the total extracellular fluid volume of CSF plus ISF is about 20–25% of the weight of brain and spinal cord.

3. DYNAMICS OF THE EXTRACELLULAR FLUID

Because proteins and hormones are carried by currents of extracellular fluid from one region of the CNS to many other sites, it is important to identify pathways for bulk fluid movements in brain and the forces that propel them. An overview of the extracellular conduits is schematized in Fig. 1. The extracellular fluid dynamics in the CNS is a complex interplay among: (1) CSF flow through the ventriculo-subarachnoid space–venous sinus pathway, and flow along sleeves of subarachnoid space (envelop-

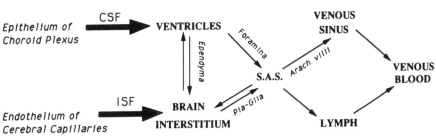

Figure 1. Pathways for the flow of extracellular fluid and proteins in the central nervous system (CNS). Cerebrospinal fluid (CSF) is derived from constituents of plasma ultrafiltrate in the plexuses, by an active secretory process occurring in the choroid epithelia. The nascent CSF containing proteins of plasma and choroidal origin percolates downward through the ventricular system and eventually passes out of the fourth ventricle, via foramina, into the subarachnoid space in the cisterna magna. From this large cistern, the CSF continues to flow in the subarachnoid space (S.A.S.) overlying the cerebral hemispheres and is finally reabsorbed across arachnoid villi in the dura mater. Simultaneous with CSF formation at the blood–CSF barrier is the production of interstitial fluid (ISF) by endothelial cells of the blood–brain barrier. ISF is thought to be secreted across the capillary wall into the brain interstitial space. Once formed, ISF undergoes bulk flow exteriorly across the pia–glial membrane into S.A.S. and interiorly into the ventricles across the ependymal lining. Net diffusion of ions and macromolecules can occur in either direction (bidirectional arrows), according to the prevailing concentration gradient, across the highly permeable brain–CSF interfaces. The fluid in the S.A.S. is a mixture of CSF and ISF. Subarachnoid fluid containing proteins and polypeptides drains from the brain by two major pathways, into venous sinuses by pinocytotic-like transport across villus-like structures and into lymphatic capillaries (nose and eye) that reach the deep cervical lymph nodes. The schema depicts transport interfaces by italic type, and the extracellular conduits by boldface type. See Spector and Johanson (1989) for illustration of sagittal view of the CNS showing spatial relationships among the various anatomical compartments.

ing certain cranial nerves) into lymphatic drainage systems of nose and eye, (2) ISF drainage along routes in brain parenchyma, and (3) CSF–ISF exchange across the ependymal wall, in either direction according to locally prevailing hydrostatic and osmotic pressure gradients (Cserr and Patlak, 1991).

3.1. CSF: Two Major Drainage Routes

3.1.1. VENTRICULO-SUBARACHNOID SPACE–VENOUS SINUS PATHWAY

The CSF proper is the conduction medium for the large-cavity pathway that originates in the lateral ventricles. Reflecting the integrity of the blood–CSF barrier, the adult CSF concentration of protein is normally two to three orders of magnitude less than in plasma. The main source of CSF, indeed of most extracellular fluid in CNS, is the choroid plexus tissue which is located in all four cerebroventricles (Spector and Johanson, 1989). As much as 80–90% of fluid generated in the brain originates from the choroid plexuses, whereas the balance of 10–20% is derived extrachoroidally (see Section 3.2). The nascent CSF moves from lateral to the third ventricle, then through the Sylvian aqueduct into the fourth ventricle. Foramina in the walls of the fourth ventricle allow fluid to escape into the cisterna magna and other parts of the subarachnoid system. Hydrostatic pressure gradients continue to drive the subarachnoid CSF, across valve-like structures in the arachnoid villi, out into the venous sinus. These "valves" are large enough to permit protein passage, but only in one direction, i.e., from CSF into blood (Butler, 1989). Thus, once a protein or peptide has flowed into the sinus, it has essentially been cleared from the brain.

3.1.2. CSF OUTFLOW INTO DEEP CERVICAL LYMPH

A considerable amount of CSF in mammals drains along the outside of cranial nerves down to submucosal tissue of the nose (Cserr and Patlak, 1991) and eye (Erlich *et al.*, 1986). In rats (Harling-Berg, 1989) and rabbits (Bradbury and Westrop, 1983), the flow of CSF along the olfactory nerve has been mapped. CSF flows via extensions of the subarachnoid space around the nerve, through the cribriform plate, and eventually reaches the nasal submucosa. Proteins and peptides can then readily drain from the submucous spaces of the nose (and eye) into afferent lymphatics.

Serum albumin has been used to trace CSF outflow to lymph. Radioiodinated serum albumin (RISA) drains slowly, but extensively, from the CNS to deep cervical lymph. Up to one half of RISA injected into CSF or brain can be recovered in lymph (Yamada *et al.*, 1991). Protein clearance from CSF is faster than from ISF. The half-time for turnover of CSF is 2 to 3 hr, while that for turnover of ISF (i.e., clearance of protein) is 6 to 18 hr.

3.2. ISF: Brain Parenchyma Distributive Pathways

The main source of cerebral ISF is ion secretion from the cerebral endothelium rather than metabolic water generation from brain cell metabolism (Cserr and Patlak, 1991). Rates of ISF production can be estimated from rates of albumin clearance from tissue (Szentistvanyi *et al.*, 1984; Yamada *et al.*, 1991). Calculated values for endothelial production of ISF at the blood–brain barrier are considerably less in magnitude than those directly measured for epithelial manufacturing of CSF by choroid plexuses. Notwithstanding the smaller capacity of the cerebral endothelium to secrete ISF, the continual albeit slow elaboration of ISF is sufficient to drive the bulk flow of ISF from brain into either ventricular or subarachnoid CSF. Such convective movements of fluid have implications for carriage of protein (native or exogenous) among regions of CNS.

3.2.1. BULK FLOW OF ISF

The narrow intercellular clefts between brain cells offer considerable resistance to bulk flow of ISF and therefore the substances dissolved in it; thus, the hydrodynamic conductance through these diminutive extracellular clefts in most regions is too low to accommodate appreciable flow of ISF via this interstitial pathway (Fenstermacher and Patlak, 1976). For bulk flow of proteins to occur throughout the cerebral interstitium, relatively large extracellular channels are required. Such a requirement is met by the extensive network of brain extracellular channels that exist: (1) between fiber tracts in white matter, (2) in the subependymal spaces close to the ventricles, and (3) as perivascular (Virchow–Robin) sleeves of CSF that follow large vessels penetrating from the subarachnoid space deep into brain substance. This anatomical triad of extracellular channel networking constitutes multiple pathways for convectively translocating proteins among various regions of brain and CSF (Cserr and Patlak, 1991). The brain ISF that drains into the CSF flows with the latter into blood or lymph (Section 3.1 and Fig. 1).

3.2.2. DIFFUSION THROUGH ISF

One experimental approach for evaluating diffusion versus bulk flow of substances has been to inject extracellular tracers of various sizes into brain, and then to quantitate the rate of efflux of the tracers from the site of administration. Cserr (1981) microinjected radiolabeled tracers directly into caudate nucleus, and found that the tracers (ranging from 69 to 0.9 kDa) were removed from tissue according to a single rate constant in spite of the five-fold difference in diffusion coefficients. Thus, the 69-kDa protein was removed from the site of injection at the same rate as the 0.9-kDa polyethylene glycol. Such a finding is consistent with removal of tracers by bulk flow. Obviously, some diffusion of proteins must occur in the cerebral interstitium; however, in regions where sizable extracellular pathways exist, the bulk flow or convective process must be the predominant distributive mechanism. The relatively permeable ependymal and pia–glial membranes that separate brain from CSF allow rapid diffusional exchange of ions and macromolecules between CSF and the ISF in superficial portions of the underlying nervous tissue.

3.3. CSF–ISF Exchange across the Ependymal Wall

It has been long appreciated that polysaccharides (e.g., inulin) and proteins (e.g., horseradish peroxidase) can readily permeate the ependymal lining (Brightman and Reese, 1969), i.e., sometimes referred to as the brain–CSF interface. The ependymal lining is one layer of cells connected to each other in most ventricular regions by gap junctions rather than zonulae occludentes. For nearly a century now, investigators have known that an intraventricularly administered substance has access to brain parenchyma. Thus, by bulk flow as well as diffusive fluxes, there is virtually free exchange of solutes and water across the "ependymal wall," the direction of movement being determined by local gradients for hydrostatic and osmotic pressures between CNS compartments and blood, and by chemical potential gradients for solutes between CSF and ISF.

3.3.1. BRAIN-TO-CSF DRAINAGE OF ISF

Normally, pressure relationships are such that brain ISF drains into large-cavity CSF either in the ventricles or in the subarachnoid compartment over the hemispheres (Cserr and Patlak, 1991). This conclusion was reached from dye distribution experiments. The patterns of staining suggest that the brain tissue hydrostatic pressure is usually greater than that in CSF. The CSF content of endogenous protein is normally low. Consequently, potentially deleterious proteinaceous materials leaking across the blood–brain barrier or resulting from cellular destruction, can move not only by bulk flow but also by diffusion, down respective concentration gradients from ISF to CSF, for subsequent elimination via fluid flow into venous sinuses.

3.3.2. CSF-TO-BRAIN FLUXES

Under pathologic conditions, ions, water, and proteins (e.g., horseradish peroxidase) can move from ventricular CSF into brain. A reversal of the hydrostatic pressure gradient between CSF and ISF results in retrograde flow of fluid from ventricles into brain substance. This can occur in hydrocephalus when CSF pressure is substantially elevated. Retrograde flow of CSF into brain also occurs when the osmotic pressure of blood increases (e.g., hypernatremia) or when there is severe hypotension. In response to hyperosmolar plasma, a reduction in ISF volume (Cserr et al., 1991; Nicholson and Rice, 1986) together with flow reversal have been documented by quantitative analysis of the kinetics of tracer penetration from CSF to brain (Pullen and Cserr, 1984; Pullen et al., 1987; Rosenberg et al., 1980).

Even when CSF pressure is normal, there is undoubtedly diffusive permeation of solutes from CSF to brain. For example, ^{36}Cl and ^{22}Na are actively transported rapidly from blood to CSF across the choroid plexuses, but more slowly across the blood–brain barrier (Smith and Rapoport, 1986; Murphy and Johanson, 1989). Therefore, shortly after intravenous or intraperitoneal administration of these tracers, there is a concentration (activity) gradient, directed from CSF to brain, which promotes net diffusion of ^{36}Cl and ^{22}Na from ventricular CSF into adjacent brain tissue. Similarly, earlier analyses of tissue concentration profiles of macromolecules, e.g., inulin (5.5 kDa) had shown that even large nonelectrolytes diffuse relatively unrestricted down concentration gradients from CSF into brain tissue (Rall et al., 1962).

Is protein diffusion from choroid plexus–CSF into brain rapid and extensive enough to be of functional significance for neurons and glia? While diffusional exchange between CSF and brain is rapid for areas immediately adjacent to CSF, it can be anticipated that diffusion of macromolecules is slow (many hours to days) for regions deep within tissue (Cserr, 1971). The relatively long diffusion times should not be a problem for proteins and other substances needed in very small amounts, e.g., transthyretin (Schreiber and Aldred, 1990) and micronutrients (Spector, 1989). If CSF concentrations of these important substances are set by the choroid plexus, then eventually they will diffuse out to the entire brain; furthermore, if proteinaceous materials are not significantly reabsorbed across the blood–brain barrier as they are diffusing in brain interstitium, their concentrations in ISF may become equal to those in CSF within a few days or less.

Overall, the findings in many types of experiments indicate that there is a dynamic exchange of proteins as well as water, ions, nonelectrolytes across the ependymal and pia–glial linings, respectively, of the interior and exterior surfaces of the brain. The exchange is potentially bidirectional, and can occur by bulk flow as well as diffusion. The ability to manipulate translocation of peptides and large proteins across the permeable cellular linings of the brain, as well as along intra-parenchymal routes, suggests numerous pharmacologic opportunities.

4. NEWLY EMERGING CONCEPTS OF VOLUME TRANSMISSION

A knowledge of the brain extracellular fluid characteristics is the key to understanding nonsynaptic (unconfined) communication among cells in the CNS.

There are two basic types of information transfer mechanisms for distributing signals throughout the brain: wiring transmission (WT) and volume transmission (VT). The classically appreciated WT involves nerve fibers and synapses, as well as diffusion of neurotransmitters across the microenvironment of synaptic clefts. In contrast, the more recently appreciated VT is concerned with the spreading of larger molecular signals (neuropeptides, proteins, hormones) by way of longer extracellular fluid pathways. Thus, VT phenomena are parasynaptic (endocrine-like) rather than synaptic. Due to the diffusive and bulk flow movements of peptides and proteins along macroscopic extracellular fluid conduits, a prerequisite for the development of pharmaceutical biotechnology for brain treatments is greater appreciation of the natural movement of macromolecules along circumscribed pathways in the CNS. Such knowledge will improve the effectiveness of eventual therapeutic regimens involving intracerebral and intraventricular injections of proteinaceous agents.

4.1. Classification According to Transmission Distance

Many publications over the past decade have dealt with the role of brain extracellular fluid as a pathway for electrical and chemical communication, i.e., VT. Classically, chemical communication between cells has been characterized on the basis of distance between sites of release of peptides and their target cells. Fuxe and Agnati (1991) view VT mainly as the conduction of messages along leaking (unconfined) pathways, some of which extend over considerable distances.

On the basis of the distance that the chemical signal travels, VT can be categorized as either: (1) *nonsynaptic* (a very-short to short-distance chemical phenomenon), (2) *paracrine* (a long-distance phenomenon), or (3) *neuroendocrine* (a very long distance between source and target cells). It is not the scope of this review to delve into the intricacies of each category. The reader is referred to Vizi (1984) and Christenson *et al.* (1990) for detailed discussions, respectively, on nonsynaptic and paracrine VT. The emphasis of this chapter is on the longer-span, neuroendocrine-like VT; the distances over which signal spreading occurs may be several millimeters or greater. The more lengthy transmission pathways have been implicated in long-term actions of peptides.

4.2. Chemophysical Characteristics of the Extracellular Space

As VT models have steadily gained acceptance, increasingly more attention has been directed toward the chemical and physical properties of the extracellular conduits and their interaction with the signals, i.e., peptides and proteins. The polyionic molecules of the interstitial matrix probably contribute to the effective tortuosity of the extracellular fluid pathways (Nicholson and Rice, 1986, 1991). Other chemical characteristics of the extracellular matrix include capabilities for buffering ions and altering the mobilities of macromolecules involved in VT (Fuxe and Agnati, 1991). Physical constraints on diffusing macromolecules are offered by glial cells and their processes. More information is needed to fill gaps in our knowledge about morphofunctional characteristics of extracellular fluid in the context of VT phenomena.

4.3. Other Factors Modifying Volume Transmission

Developmental, biochemical, and pharmacologic factors can substantially affect the nature of movement of neuropeptide and protein "messages" through the extracellular fluid pathways. Some of the more pronounced modulating effects are discussed below.

4.3.1. EARLY DEVELOPMENT AND AGING

Throughout perinatal development there are marked changes in the composition and dynamics of extracellular fluid in the CNS (Johanson, 1989a; Epstein and Johanson, 1987). The rat choroid plexus–CSF system undergoes rapid maturation during the 3 weeks after birth (Johanson et al., 1988; Parmelee and Johanson, 1989; Pershing and Johanson, 1982; Smith et al., 1982b). As fluid production by the choroid plexus is increased (Johanson and Woodbury, 1974), the capacity of the CSF drainage pathways is concomitantly enhanced (Jones et al., 1987). The progressively greater flow of CSF through the ventriculo-subarachnoid system has implications for VT phenomena in maturing organisms. Kozlowski (1982) has described neuropeptide distribution via large-cavity CSF in his "ventricular route" hypothesis. The more sluggish CSF flow and smaller CSF sink action (Parandoosh and Johanson, 1982) in neonates and infants of some mammalian species (e.g., mice, rats, and cats) would appear to significantly modify the distribution of peptides and proteins throughout extracellular fluid conduits in the immature CNS.

The volume of extracellular fluid in the brain, relative to volume of intracellular fluid, undergoes a continual decrease from birth through later stages of adulthood. During the first month post partum, the rat brain extracellular space volume decreases, on average, from about 25 to 15% (Johanson, 1989a). The loss of total water (as percent of brain wet weight), and the shrinkage in size of the extracellular space, likely have effects on movements of proteins and peptides throughout the interstitial space. Indeed, aged animals showed a marked decrease in the extent of chemical spread (Routtenberg, 1991), a finding that has been explained by an attenuated extracellular space.

Moreover, CSF formation rate slows down by nearly 20% in aged rats, i.e., 1 to 3 years old (Smith *et al.*, 1982a). This diminished secretory activity, coinciding temporally with a marked decrease in extracellular fluid volume (Bondareff and Narotzky, 1972), would be expected to impede removal of high-molecular-weight substances from the brain. Such factors deserve more attention due to the need for a better understanding of CNS fluid turnover in geriatric patients, some of whom with degenerative diseases may eventually be candidates for intracerebral administration of enzymes or other therapeutic agents.

4.3.2. DRUGS THAT ALTER FLUID DYNAMICS IN THE CNS

Agents and treatments that alter the physical configuration of extracellular conduits in CNS could interfere with VT. Neurosurgeons use osmotic agents (e.g., mannitol) to dehydrate edematous brain. The rationale is that hypertonic plasma pulls water from brain, more specifically from the extracellular compartment of nervous tissue (Cserr and Patlak, 1991). Routtenberg (1991) and colleagues have noted that agents used in neurosurgery to reduce swelling also decrease the spread of chemicals injected into the brain. Thus, one of the functions of the extracellular distribution systems, i.e., VT, is probably perturbed by treatments that effect redistribution of fluid among the CSF, ISF, and plasma compartments.

Protein movement into and out of CSF is intimately related to the movement of water (fluid) into and out of the CSF system. Acetazolamide, an inhibitor of CSF flow, significantly curtails the rate of transport of protein, in the direction from arterial blood to CSF (Hochwald and Wallenstein, 1967) as well as from CSF back to venous blood (Van Wart *et al.*, 1961). These basic studies with globulin and albumin demonstrated that protein concentrations in the extracellular fluid of the CNS are dependent on the rate of fluid transport across choroid plexus into CSF, as well as drainage mechanisms which clear protein from CSF to blood. Clearly, alterations in the hydrodynamics of brain fluids can markedly affect the concentrations of proteins, and probably peptides, in brain ISF and CSF. The relationship between CSF dynamics and CSF peptide concentrations is one that awaits elucidation in the context of volume transmission.

4.3.3. INACTIVATION OF PROTEIN AND PEPTIDE SIGNALS

The strength of signaling by peptides and proteins is controlled in part either by *physical* elimination of the macromolecular signal from the CNS, or by *chemical* activation/deactivation enzymatic reactions that alter the concentration of the particular ligand in extracellular fluid. Most peptides and proteins are large and hydrophilic, and therefore do not readily diffuse from blood into brain or CSF; consequently, a particular peptide "signal" is effectively abolished upon clearance of that molecule by its bulk flow with CSF into venous blood, from which it is unlikely to return to brain by diffusion back across the blood–brain barrier. Section 3.1 describes the bulk flow drainage routes.

CSF proteases and peptidases probably also have a significant role in regulating the concentration of proteins and peptides, some of which are very potent ligands. Neuropeptide transmitters, unlike their biogenic amine neurotransmitter counterparts, are not rendered inactive by cellular reuptake mechanisms in nerve endings. Thus, neuropeptides and their converting enzymes, upon release from neurons (and perhaps glia), are dependent on extracellular mechanisms for their clearance or inactivation.

The CSF proteolytic enzymes generally have smaller molecular masses than corresponding enzymes extracted from cerebral homogenates. An example is angiotensin-converting enzyme, which exists as a 140-kDa species but also is present as two smaller species of 30 and 60 kDa, both of which display full activity. The smaller species can likely diffuse more rapidly and globally throughout the CNS. By acting as scavengers or modifiers of peptide transmission, these proteolytic enzymes could play an important role in VT (Terenius and Nyberg, 1988).

In consideration of pharmaceutical development of inhibitors of proteolysis, the CSF is a conveniently sampled fluid as an enzyme source. Clinicians have long recognized that CSF levels of substances are useful as markers or reflectors of brain cell metabolism. Enzymes that are abundant in CSF lend themselves to isolation efforts and inhibitor development. Once the inhibitors are available, they can then be used as tools for mapping peptidergic pathways and possibly as therapeutic agents for CNS disorders caused by peptide imbalances.

The choroid plexus also has a rich content of proteolytic enzymes, e.g., aminopeptidases, angiotensin-converting enzyme, and carboxypeptidase M (Bourne *et al.*, 1989). The functions of the choroidal peptidases are not well understood; however, their role in the regulation of peptide concentrations in CSF must be considered.

5. CHOROID PLEXUS (CP): A "TARGET" AND A "SOURCE" IN VOLUME TRANSMISSION

Electron micrographs and molecular analyses of CP reveal an epithelium that is very actively involved in the metabolism and transport of numerous peptides and proteins. Because CP function ultimately serves the brain, many intriguing questions have been raised about the role of CP in transporting peptides between blood and CSF (Johanson, 1989b); in receiving peptide signals from various regions of brain; and in manufacturing and secreting proteins and growth factors into CSF for eventual distribution to target cells in brain parenchyma. The burgeoning interest for research in this area (Spector and Johanson, 1989) can be attributed to many attractive hypotheses involving CP and having a role for peptides and proteins as signals for the regulation of CSF production and intracranial pressure; and for the growth and repair of brain cells.

5.1. CP Receptors for Peptides

The activity of many polypeptides is substantially augmented in CP following either intravenous or intrathecal administration of the tracer ligand. Mainly by autoradiographic techniques, receptors have been identified in CP for insulin, enkephalins, thyroid hormone (T4), melatonin, vasopressin, oxytocin, prolactin, and vasoactive intestinal polypeptide (see review by Johanson, 1989b). Some of these receptors are on the choroidal vasculature, e.g., those for arginine vasopressin have been linked to modulation of CP blood flow (Faraci *et al.*, 1988). Several of the peptidergic receptors have been localized to the choroidal epithelium, where they presumably have roles in regulating CP cellular metabolism and transport; e.g., insulin stimulates sodium uptake by CP parenchymal cells *in vivo* (Johanson and Murphy, 1990), whereas vasopressin and atrial peptide have opposing effects on the extrusion of chlorine from CP tissue *in vitro* (Johanson *et al.*, 1990). Still other peptide receptors in CP have been characterized as having a clearance or removal function, e.g., atrial peptide (Brown and Czarnecki, 1990).

CP is a pivotal target organ for centrally manufactured and released peptides. Strategically located in all four cerebroventricles, the CPs have a great influence on the volume and composition of the ventricular CSF. Evidence is accumulating that peptides and some neurotransmitters (e.g., serotonin) are released from hypothalamic nuclei and subependymal regions into CSF, in which medium they are transported by VT to target sites on the apical (CSF-facing) membrane of CP (Nilsson *et al.*, 1991). It is not yet clear what the exact functional interactions between brain and CP are, but this fascinating area of research should yield important insights about the significance of peptide distribution in the interior of the brain.

5.2. CP as a Generator of Protein Signals

CP synthesizes and secretes plasma proteins and growth factors into CSF (Dickson *et al.*, 1985, 1986; Stauder *et al.*, 1986; Stylianopoulou *et al.*, 1988). These proteins are then distributed into brain by convection (bulk flow) of CSF by mechanisms described in Section 3. From the pattern and intensity of plasma protein synthesis in CP, Schreiber and Aldred (1990) have postulated that CP has a significant role in maintaining extracellular protein homeostasis in the brain. Proteins of choroidal (and extrachoroidal) origin likely have trophic effects on brain cells both in development and possibly in repair of injured brain (Section 6). Growth factor receptor-mediated events culminate in axonal sprouting. Neuronal and glial target cell specificity and cellular biochemical cascades await delineation. It is becoming increasingly clear, however, that the observation of availability of CP-manufactured proteins to distant regions of brain affirms the pervasiveness of the extracellular fluid convection systems.

6. BRAIN CELLS AS TARGETS FOR PROTEINACEOUS SUBSTANCES

6.1. Types of Growth Factors in Brain

Growth factor proteins are synthesized in many regions of the CNS. Among the most widely studied growth factors are basic fibroblast growth factor (bFGF), nerve growth factor (NGF), and insulinlike growth factor (IGF). These proteinaceous factors exert trophic effects on various populations of neuronal systems (Ferrari *et al.*, 1989; Knusel *et al.*, 1990). For example, both bFGF and NGF have prevented cell death of cholinergic systems in experimental studies (Anderson *et al.*, 1988; Stromberg *et al.*, 1990). These agents show promise for reducing morbidity of neurons in traumatic disorders and neurodegenerative diseases. NGF has been the most widely analyzed (Thoenen *et al.*, 1987), and so research progress with this factor is emphasized below.

6.2. Temporal Expression of Nerve Growth Factor

Neuronal populations show variation in their need for NGF, a trophic signal with local paracrine actions. Certain cholinergic populations require NGF throughout life (e.g., cholinergic projections from basal forebrain to hippocampus); other systems of cells express NGF only during ontogeny or after tissue damage in adulthood (e.g., striatal cholinergic interneurons) (Ernfors *et al.*, 1989; Eckenstein, 1988). Thus, when the brain is stressed with disease or trauma at certain stages of early development or in later life, the level of expressed NGF (and other trophic factors) may not be sufficient to maintain neuronal viability.

6.3. Boosting the Level of Growth Factors in the CNS

There are lower levels of NGF and its mRNA in the brain of aged rats compared to younger adults (Larkfors *et al.*, 1987). Diminished NGF production may also occur in the senile dementia of Alzheimer's disease due to the degeneration of cholinergic projections to cortex cerebri from the nucleus basalis. The addition of exogenous NGF to target areas would probably rescue some of the malfunctioning cells. This is suggested by the finding that infusion of NGF improves neuronal survival and the transforming ability of grafted cells.

6.4. How Cerebral Levels of NGF Can Be Elevated

Strategies for enhancing the central levels of NGF have been offered by Olson *et al.*, (1991). The most straightforward approach is to administer the purified growth factor. Injections directly into brain parenchyma or ventricular CSF have been attempted. Intraparenchymal infusion, by use of dialysis fibers connected to osmotic minipumps, has been efficacious for stimulating cholinergic neurons (Hefti *et al.*, 1989). Alternatively, the less-localized intraventricular infusion via cannula in the lateral ventricle has proven effective in reclaiming the cholinergic function of magnocellular neurons (nucleus basalis) lesioned by excitotoxin (Mandel *et al.*, 1989). Although the more-widely penetrating intraventricular infusate containing NGF definitely improves cholinergic systems in forebrain, some concern has been expressed for possible unwanted side effects brought about by simultaneous stimulation of NGF-sensitive, noncholinergic sites in circumventricular areas (Olson *et al.*, 1991).

Two other strategies for increasing central levels of NGF have been suggested, i.e., by increasing endogenous levels of NGF, and by implanting cell systems capable of producing NGF. Unfortunately, there is sparse knowledge about natural regulation of NGF synthesis. Inflammatory mediators used in hippocampal culture studies have induced NGF mRNA and protein (Friedman *et al.*, 1990); the usefulness of the pharmacologic approach, however, needs to be corroborated in intact animals.

In regard to implantation of cell systems, a potential advantage is the achievement of long-term augmentation of NGF levels. Transplants of submandibular glands from mice stimulate cholinergic growth (Springer *et al.*, 1988), but this approach with normal (nontransformed) cells has been beset by problems with immunosuppression and graft survival. Genetically modified cells from established lines are promising if the risks of immunological mismatching and tumor formation can be overcome.

6.5. Time Course of NGF Spreading after Injection into Brain

Accurate mapping of permeation routes taken by injected proteins is essential for proper interpretation of the functional consequences arising from the injection. Recently the distribution of NGF, acutely injected into striatum, has been traced immunohistochemically by Olson *et al.* (1991). Their study utilized affinity-purified antibodies directed against mouse salivary NGF. From 2 to 1100 ng was dissolved in saline (4 μl) for stereotactic injection into the center of the striatum. NGF could be visualized by fluorescence resulting from injected quantities as small as 20 ng. Rapid diffusion of NGF throughout the striatal neuropil was noted, so that by 15 min the fluorescent halo covered approximately two-thirds of a coronal section of striatum. Over the next few hours, NGF spread even farther so that the fluorescence rapidly attenuated at the injection site. The small faint halo observed at 7 hr was gone by 24 hr. This study by Olson, *et al.* (1991), has convincingly demonstrated that intrastriatally injected NGF spreads relatively rapidly throughout the entire targeted area. Such findings are consistent with the free-flowing nature of the extracellular distribution systems described in Section 3.

7. PROSPECTS FOR RESEARCH ON PEPTIDE AND PROTEIN DISTRIBUTION IN CNS

Endocrinology is predicated on the fact that "informational substances" can exert their action on targets distantly removed from the release site, i.e., via "bulk flow" carriage of hormones through the vascular circulatory system. Milhorat (1987) has encouraged the concept of the CSF system as the "third circulation," as originally proposed by Cushing (1926). The "third circulation" is less well known than the blood and lymph circulations. Historically, the concept of the CSF (together with the brain ISF) system as circulatory has been longer in coming because early investigators were not convinced that cerebral ISF drained into the ventricles. Recent advances in "volume transmission" should rekindle interest in the CSF and ISF as convective fluids in dynamic exchange with each other.

This review has focused on the conveyance of protein and peptide "signals" by way of extracellular fluid (CSF plus ISF). Many proteins and peptides in extracellular fluid are "informational molecules" to initiate trophic effects, neuromodulation, secretory activity, and so forth. Obviously, some proteinaceous materials in extracellular fluid are not "signals" *per se*, but rather are metabolic or breakdown products, some of which have immunogenic potential if the protein in CSF drains into cervical lymph nodes (Cserr and Patlak, 1991). Whatever the target or function (or lack) of the proteinaceous material, it will move along extracellular conduits until chemically broken down by enzymes or physically removed. Brain extracellular fluid protein homeostasis, a challenging new concept with many pathophysiological implications (Schreiber and Aldred, 1990), must be a complex interplay of factors that include elaboration of proteins by neuroepithelial and choroidal secretory cells.

The theme of this chapter, i.e., generalized interregional accessibilities made possible by freely communicating components of the extracellular system, belies the existence of locations that are relatively impermeable to macromolecules. Regions in addition to the dentate gyrus of the hippocampus (Ruth and Routtenberg, 1980) may have barriers to diffusion of large molecules. Some circumventricular organs are surrounded by cells joined with tight junctions that more or less functionally isolate these regions from adjacent areas of nervous tissue. In certain regions of third ventricle CSF, some cells in the ependymal lining have tight rather than gap junctions between them. Additional immunohistochemical mappings of brain regions with protein probes are needed to identify loci having true diffusion barriers.

Peptides and proteins released or injected at points in the nervous system having modest junctional specialization (i.e., lacking zonulae occludentes) can spread for great distances beyond the point of origin (Bondareff et al., 1970, 1971). Thus, the consequent behavioral phenomena that are manifested must be interpreted with caution if multiple sites of cells are eventually brought into contact with the convected ligand. Moreover, untoward complications may result from therapeutic efforts if the administered substance is conveyed well beyond the locale of the targeted cells. There is plainly the need for systematic comparison of distributive phenomena associated with access to target cells following intraventricular (CSF region) versus intra-parenchymal (brain region) injections of peptides and proteins.

ACKNOWLEDGMENTS.   Acknowledgment is due to H. Cserr for critical reading of the manuscript, to Carole Thompson for help in preparing the manuscript, and to the NIH for RO1 funding (NS 13988 and NS 27601) that has enabled the author and colleagues to investigate transport, permeability, and secretory phenomena in the developing choroid plexus–CSF–brain system.

REFERENCES

Anderson, K. J., Dam, D., Lee, S., and Cotman, C. W., 1988, Basic fibroblast growth factor prevents death of lesioned cholinergic neurons in vivo, Nature 332:360–361.

Bondareff, W., and Narotzky, R., 1972, Age changes in the neuronal microenvironment, Science 176:1135–1136.

Bondareff, W., Narotzky, R., and Routtenberg, A., 1971, Intrastriatal spread of catecholamines in senescent rats, J. Gerontol. 26:163–167.

Bondareff, W., Routtenberg, A., Narotzky, R., and McLone, D. G., 1970, Intrastriatal spreading of biogenic amines, Exp. Neurol. 28:213–229.

Bourne, A., Barnes, K., Taylor, B. A., Turner, A. J., and Kenny, A. J., 1989, Membrane peptidases in the pig choroid plexus and on other cell surfaces in contact with the cerebrospinal fluid, Biochem. J. 259:69–80.

Bradbury, M. W. B., and Westrop, R. J., 1983, Factors influencing exit of substances from cerebrospinal fluid into deep cervical lymph of the rabbit, J. Physiol. (London) 339: 519–534.

Brightman, M. W., and Reese, T. S., 1969, Junctions between intimately apposed cell membranes in the vertebrate brain, J. Cell Biol. 40:648–677.

Brown, J., and Czarnecki, A., 1990, Distribution of atrial natriuretic peptide receptor subtypes in brain, Am. J. Physiol. 258:R1078–R1083.

Butler, A. B., 1989, Alteration of CSF outflow in experimental acute subarachnoid hemorrhage, in: Outflow of Cerebrospinal Fluid (F. Gjerris, S. E. Borgesen, and P. S. Sorensen, eds.), Munksgaard, Copenhagen, pp. 69–75.

Christenson, J., Cullheim, S., Grillner, S., and Hokfelt, T., 1990, 5-HT varicosities in the lamprey spinal cord have not synaptic specializations, *Brain Res.* **512**:201–209.

Cserr, H. F., 1971, Physiology of the choroid plexus, *Physiol. Rev.* **51**:273–311.

Cserr, H. F., 1981, Convection of brain interstitial fluid, in: *Advances in Physiological Science*, Volume 7 (A. G. B. Kovach, J. Hamar, and L. Szabo, eds.), Pergamon Press, Elmsford, N.Y., pp. 1219–1226.

Cserr, H., and Patlak, C., 1991, Regulation of brain volume under isosmotic and anisosmotic conditions, in: *Advances in Comparative and Environmental Physiology*, Volume 9, Springer-Verlag, Berlin, pp. 61–80.

Cserr, H. F., and Patlak, C. S., 1992, Secretion and bulk flow of interstitial fluid, in: *Handbook of Experimental Pharmacology, Physiology and Pharmacology of the Blood–Brain Barrier* (M. W. B. Bradbury, ed.), pp. 245–261.

Cserr, H. F., DePasquale, M., Patlak, C. S., Pettigrew, K. D., and Rice, M. E., 1991, Extracellular volume decreases while cell volume is maintained by ion uptake in rat brain during acute hypernatraemia. *J. Physiol. (London)* **442**:277–295.

Cushing, H., 1926, The third circulation and its channels, in: *Studies in Intracranial Physiology and Surgery*, Oxford University Press, London, p. 1.

Dickson, P. W., Aldred, A. R., Marley, P. D., Guo-Fen, T., Howlett, G. J., and Schreiber, G., 1985, High prealbumin and transferrin mRNA levels in the choroid plexus of rat brain, *Biochem. Biophys. Res. Commun.* **127**:890–895.

Dickson, P. W., Aldred, A. R., Marley, P. D., Bannister, D., and Schreiber, G., 1986, Rat choroid plexus specializes in the synthesis and the secretion of transthyretin (prealbumin)—Regulation of transthyretin synthesis in choroid plexus is independent from that in liver, *J. Biol. Chem.* **261**:3475–3478.

Eckenstein, F., 1988, Transient expression of NGF-receptor-like immunoreactivity in postnatal rat brain and spinal cord, *Brain Res.* **446**:149–154.

Epstein, M. H., and Johanson, C. E., 1987, The Dandy–Walker syndrome, in: *Handbook of Clinical Neurology*, Volume 6 (N. Myrianthopoulos, ed.), Elsevier, Amsterdam, pp. 1–14.

Erlich, S. S., McComb, J. G., Hyman, S., and Weiss, M. H., 1986, Ultrastructural morphology of the olfactory pathway for cerebrospinal fluid drainage in the rabbit, *J. Neurosurg.* **64**: 466–473.

Ernfors, P., Henschen, A., Olson, L., and Persson, H., 1989, Expression of nerve growth factor receptor mRNA is developmentally regulated and increased after axotomy in rat spinal cord motoneurons, *Neuron* **2**:1605–1613.

Faraci, F. M., Mayhan, W. G., Farrell, W. J., and Heistad, D. D., 1988, Humoral regulation of blood flow to choroid plexus: Role of arginine vasopressin, *Circ. Res.* **63**:373–379.

Fenstermacher, J. D., and Patlak, C. S., 1976, The movements of water and solutes in the brains of mammals, in: *Dynamics of Brain Edema* (H. M. Pappius and W. Feindel, eds.), Springer-Verlag, Berlin, pp. 87–94.

Ferrari, G., Minozzi, M. C., Toffano, G., Leon, A., and Skaper, S. D., 1989, Basic fibroblast growth factor promotes the survival and development of mesencephalic neurons in culture, *Dev. Biol.* **133**:140–147.

Friedman, W. J., Larkfors, L., Ayer-LeLievre, C., Ebendal, T., Olson, L., and Persson, H., 1990, Regulation of beta-nerve growth factor expression by inflammatory mediators in hippocampal cultures, *J. Neurosci. Res.* **27**:374–382.

Fuxe, K., and Agnati, L. F., 1991, Two principal modes of electrochemical communication in the brain: Volume versus wiring communication, in: *Volume Transmission in the Brain: Novel Mechanisms for Neural Transmission* (K. Fuxe and L. F. Agnati, eds.), Raven Press, New York, pp. 1–10.

Harling-Berg, C., 1989, The humoral immune response to human serum albumin infused into the cerebrospinal fluid of the rat, Ph.D. thesis, Brown University, Providence.

Hefti, F., Hartikka, J., and Knusel, B., 1989, Function of neurotrophic factors in the adult and aging brain and their possible use in the treatment of neurodegenerative diseases, *Neurobiol. Aging* **10**:515–533.

Hochwald, G. M., and Wallenstein, M. C., 1967, Exchange of γ-globulin between blood, cerebrospinal fluid and brain in the cat, *Exp. Neurol.* **19**:115–126.

Huang, J. F., 1982, Accumulation of peptide Tyr-D-Ala-Gly by choroid plexus during ventriculocisternal perfusion of rat brain, *Neurochem. Res.* **7**:1541–1548.

Johanson, C., 1989a, Ontogeny and phylogeny of the blood–brain barrier, in: *Implications of the Blood–Brain Barrier and Its Manipulation*, Volume 1 (E. Neuwelt, ed.), Plenum Press, New York, pp. 157–198.

Johanson, C., 1989b, Potential for pharmacological manipulation of the blood–cerebrospinal fluid barrier, in: *Implications of the Blood–Brain Barrier and Its Manipulation*, Volume 1 (E. Neuwelt, ed.), Plenum Press, New York, pp. 223–260.

Johanson, C. E., and Murphy, V. A., 1990, Acetazolamide and insulin alter choroid plexus epithelial cell [Na$^+$], pH and volume, *Am. J. Physiol.* **258**:F1538–F1546.

Johanson, C. E., and Woodbury, D. M., 1974, Changes in CSF flow and extracellular space in the developing rat, in: *Drugs and the Developing Brain* (A. Vernadakis and N. Weiner, eds.), Plenum Press, New York, pp. 281–287.

Johanson, C. E., Allen, J., and Withrow, C. D., 1988, Regulation of pH and HCO$_3$ in brain and CSF of developing mammalian central nervous system, *Dev. Brain Res.* **38**:255–264.

Johanson, C. E., Nashold, J. R. B., Preston, J. E., Dyas, M., and Knuckey, N., 1990, Effects of vasopressin and atriopeptin on chloride transport in choroid plexus, *Soc. Neurosci. Abstr.* **16**:45.

Jones, H., Deane, R., and Bucknall, R. M., 1987, Developmental changes in cerebrospinal fluid pressure and resistance to absorption in rats, *Dev. Brain Res.* **33**:23–30.

Knusel, B., Michel, P. P., Schwaber, J. S., and Hefti, F., 1990, Selective and nonselective stimulation of central cholinergic and dopaminergic development *in vitro* by nerve growth factor, basic fibroblast growth factor, epidermal growth factor, insulin and the insulin-like growth factors I and II, *J. Neurosci.* **10(2)**:558–570.

Kozlowski, G. P., 1982, Ventricular route hypothesis and peptide-containing structures of the cerebroventricular system, in: *Frontiers of Hormone Research*, Volume 9 (E. M. Rodriguez and T. B. van Wimersma Greidanus, eds.), Karger, Basel, pp. 105–118.

Larkfors, L., Ebendal, T., Whittemore, S. R., Persson, H., Hoffer, B., and Olson, L., 1987, Decreased level of nerve growth factor (NGF) and its messenger RNA in the aged rat brain, *Mol. Brain Res.* **3**:55–60.

Levin, V. A., Fenstermacher, J. D., and Patlak, C. S., 1970, Sucrose and inulin space measurements of cerebral cortex in four mammalian species, *Am. J. Physiol.* **219**:1528–1533.

Mandel, R. J., Gage, F. H., and Thal, L. J., 1989, Spatial learning in rats: Correlation with cortical choline acetyltransferase and improvement with NGF following NBM damage, *Exp. Neurol.* **104**:208–217.

Milhorat, T. H., 1987, The three circulations, in: *Cerebrospinal Fluid and the Brain Edemas*, Neuroscience Society of New York, pp. 1–9.

Murphy, V. A., and Johanson, C. E., 1989, Acidosis, acetazolamide, and amiloride: Effects on ^{22}Na transfer across the blood–brain and blood–CSF barriers, *J. Neurochem.* **52:**1058–1063.

Nicholson, C., and Rice, M. E., 1986, The migration of substances in the neuronal microenvironment, *Ann. N. Y. Acad. Sci.* **481:**55–68.

Nicholson, C., and Rice, M. E., 1991, Diffusion of ions and transmitters in the brain cell microenvironment, in: *Volume Transmission in the Brain: Novel Mechanisms for Neural Transmission* (K. Fuxe and L. F. Agnati, eds.), Raven Press, New York, pp. 279–294.

Nilsson, C., Lindvall-Axelsson, M., and Owman, C., 1991, Role of cerebrospinal fluid in volume transmission involving the choroid plexus, in: *Volume Transmission in the Brain: Novel Mechanisms for Neural Transmission* (K. Fuxe and L. F. Agnati, eds.), Raven Press, New York, pp. 307–315.

Olson, L., Wetmore, C., Stromberg, I., and Ebendal, T., 1991, Endogenous and exogenous nerve growth factor in the central nervous system, in: *Volume Transmission in the Brain: Novel Mechanisms for Neural Transmission* (K. Fuxe and L. F. Agnati, eds.), Raven Press, New York, pp. 455–462.

Parandoosh, Z., and Johanson, C. E., 1982, Ontogeny of the blood–brain barrier to, and cerebrospinal fluid sink action on, C-14 urea, *Am. J. Physiol.* **243:**R400–R407.

Parmelee, J. T., and Johanson, C. E., 1989, Development of potassium transport capability by choroid plexus of infant rats, *Am. J. Physiol.* **256:**R786–R791.

Pershing, L. K., and Johanson, C. E., 1982, Acidosis-induced enhanced activity of the Na–K exchange pump in the in vivo choroid plexus: An ontogenetic analysis of possible role in cerebrospinal fluid pH homeostasis, *J. Neurochem.* **38:**322–332.

Pullen, R. G. L., and Cserr, H. F., 1984, Pressure dependent penetration of CSF into brain, *Fed. Proc. Abstr.* **43:**715.

Pullen, R. G. L., DePasquale, M., and Cserr, H. F., 1987, Bulk flow of cerebrospinal fluid into brain in response to acute hyperosmolality, *Am. J. Physiol.* **253:**F538–F545.

Rall, D. P., Oppelt, W. W., and Patlak, C. S., 1962, Extracellular space of brain as determined by diffusion of inulin from the ventricular system, *Life Sci.* **2:**43–48.

Rosenberg, G. A., Kyner, W. T., and Estrada, E., 1980, Bulk flow of brain interstitial fluid under normal and hyperosmolar conditions, *Am. J. Physiol.* **238:**F42–F49.

Routtenberg, A., 1991, Action at a distance: The extracellular spread of chemicals in the nervous system, in: *Volume Transmission in the Brain: Novel Mechanisms for Neural Transmission* (K. Fuxe and L. F. Agnati, eds.), Raven Press, New York, pp. 295–298.

Ruth, R. and Routtenberg, A., 1980, Possible diffusion barriers within rat dentate gyrus, *Fed. Proc.* **30:**597.

Schreiber, G., and Aldred, A. R., 1990, Pathophysiological aspects of plasma protein formation in the choroid plexus, in: *Pathophysiology of the Blood–Brain Barrier* (B. B. Johansson, C. Owman, and H. Widner, eds.), Elsevier, Amsterdam, pp. 89–103.

Smith, Q. R., and Rapoport, S. I., 1986, Cerebrovascular permeability coefficients to sodium, potassium and chloride, *J. Neurochem.* **46:**1732–1742.

Smith, Q. R., Takasato, Y., and Rapoport, S. I., 1982a, Age-associated decrease in the rate of cerebrospinal fluid uptake of Na in the Fischer-344 rat, *Soc. Neurosci. Abstr.* **8:**443.

Smith, Q. R., Woodbury, D. M., and Johanson, C. E., 1982b, Kinetic analysis of Cl-36, Na-22 and H-3 mannitol uptake into the in vivo choroid plexus–cerebrospinal fluid system: Ontogeny of the blood–brain and blood–CSF barriers, *Dev. Brain Res.* **3**:181–198.

Spector, R., 1989, Micronutrient homeostasis in mammalian brain and cerebrospinal fluid, *J. Neurochem.* **53**:1667–1674.

Spector, R., and Johanson, C. E., 1989, The mammalian choroid plexus, *Sci. Am.* **260 (11)**: 68–74.

Springer, J. E., Collier, T. J., Notter, M. F. D., Loy, R., and Sladek, J. R., Jr., 1988, Central nervous system grafts of nerve growth factor-rich tissue as an alternative source of trophic support for axotomized cholinergic neurons, *Prog. Brain Res.* **78**:401–407.

Stauder, A. J., Dickson, P. W., Aldred, A. R., Schreiber, G., Mendelsohn, F. A. O., and Hudson, P., 1986, Synthesis of transthyretin (prealbumin) mRNA in choroid plexus epithelial cells, localized by in situ hybridization in the rat brain, *J. Histochem. Cytochem.* **34**:949–952.

Stromberg, I., Wetmore, C. J., Ebendal, T., Ernfors, P., Persson, H., and Olson, L., 1990, Rescue of forebrain cholinergic neurons after implantation of genetically modified cells producing recombinant NGF, *J. Neurosci. Res.* **25**:405–411.

Stylianopoulou, F., Herbert, J., Soares, M. B., and Efstratiadis, A., 1988, Expression of the insulin-like growth factor II gene in the choroid plexus and the leptomeninges of the adult rat central nervous system, *Proc. Natl. Acad. Sci. USA* **85**:141–145.

Szentistvanyi, I., Patlak, C. S., Ellis, R. A., and Cserr, H. F., 1984, Drainage of interstitial fluid from different regions of rat brain, *Am. J. Physiol.* **246**:F835–F844.

Terenius, L., and Nyberg, F., 1988, Neuropeptide-processing, -converting, and -inactivating enzymes in human cerebrospinal fluid, *Int. Rev. Neurobiol.* **30**:101–121.

Thoenen, H., Bandtlow, C., and Heumann, R., 1987, The physiological function of nerve growth factor in the central nervous system: Comparison with the periphery, *Rev. Physiol. Biochem. Pharmacol.* **109**:145–178.

Van Deurs, B., 1976, Choroid plexus absorption of horseradish peroxidase from the cerebral ventricles, *J. Ultrastruct. Res.* **55**:400–415.

Van Deurs, B., 1978, Microperoxidase uptake into the rat choroid plexus epithelium, *J. Ultrastruct. Res.* **62**:168–180.

Van Wart, C. A., Dupont, J. R., and Kraintz, L., 1961, Effect of acetazolamide on passage of protein from cerebrospinal fluid to plasma, *Proc. Soc. Exp. Biol. Med.* **106**:113–114.

Vizi, E. S., 1984, Non-synaptic interactions between neurons: Modulation of neurochemical transmission, in: *Pharmacological and Clinical Aspects*, Wiley, New York.

Yamada, S., DePasquale, M., Patlak, C. S., and Cserr, H. F., 1991, Albumin outflow into deep cervical lymph from different regions of rabbit brain, *Am. J. Physiol.* **261**:H1197–1204.

Index